Cambridge
International AS and A Level

Geography

SECOND EDITION

Garrett Nagle
Paul Guinness

HODDER
EDUCATION
AN HACHETTE UK COMPANY

To Angela, Rosie, Patrick and Bethany for their support and good humour.

Orders: please contact Hachette UK Distribution, Hely Hutchinson Centre, Milton Road, Didcot, Oxfordshire, OX11 7HH. Telephone: +44 (0)1235 827827. Email education@hachette.co.uk Lines are open from 9 a.m. to 5 p.m., Monday to Friday. You can also order through our website: www.hoddereducation.com.

© Paul Guinness & Garrett Nagle 2016

First published in 2016 by

Hodder Education,

An Hachette UK Company

Carmelite House

50 Victoria Embankment

London EC4Y 0DZ

www.hoddereducation.com

Impression number   11

Year   2024

Cover photo © Paul Guinness. Photo taken in 2014 at the Border Ceremony on the Indian/Pakistan border to the west of the Indian city of Amritsar, on the Indian side of the border. The ceremony takes place late every afternoon before the border gates are closed for the day. Both sets of border guards are in full ceremonial uniform and carry out well-practised drills.

Illustrations by Aptara Inc.

Typeset in Caecilia LT Std 9.5/12pt by Aptara Inc.

Printed in India

A catalogue record for this title is available from the British Library.

ISBN: 978 1 4718 6856 6

Contents

Advanced Human Geography Options

Introduction

The aim of this book is to enable you to achieve your potential. In doing so, we hope to help you gain:

- an awareness of the usefulness of geographical analysis to understand and solve contemporary human and environmental problems
- a sense of relative location, including an appreciation of the complexity and variety of natural and human environments
- an understanding of the principal processes operating within physical and human geography
- an understanding of the causes and effects of change on the natural and human environments.

Any study of Geography involves a number of aspects, including:

- elements of physical and human geography and the inter-relationships between these components
- processes operating at different scales within physical and human geography
- a sense of relative location, including an appreciation of the complexity and variety of natural and human environments
- a demonstration and explanation of the causes and effects of change over space and time on the natural and human environments.

In addition, you will use many skills, including:

- exploring primary (fieldwork) sources and secondary sources (for example statistical data)
- interpreting a range of map and diagram techniques displaying geographical information
- assessing methods of enquiry and considering the limitations of evidence
- demonstrating skills of analysis and synthesis
- using geographical understanding to develop your own explanations and hypotheses.

The structure of the book

The Cambridge Assessment International Education AS and A Level Geography syllabus is presented in sections. The contents of this book follow the syllabus sequence, with each section the subject of a separate topic.

Topics 1 to 6 cover Sections 1 to 6 of the AS Level syllabus and are for all students. AS students are assessed only on these topics.

Topics 7 to 14 cover Sections 7 to 14, the additional sections of the syllabus for A Level students only.

A new feature of the syllabus is Key concepts. These are the essential ideas, theories, principles or mental tools that help learners to develop a deep understanding of their subject, and make links between different topics. An icon indicates where each Key concept is covered.

 Space

The spatial distributions and patterns of a range of physical and human geographical phenomena.

 Scale

The significance of spatial scale in interpreting environments, features and places from local to global, and time scale in interpreting change from the geological past to future scenarios.

 Place

The physical and human characteristics that create distinctive places with different opportunities and challenges.

 Environment

The interactions between people and their environment that create the need for environmental management and sustainability.

 Interdependence

The complex nature of interacting physical systems, human systems and processes that create links and interdependencies.

 Diversity

The similarities and differences between places, environments, cultures and people.

 Change

The dynamic nature of places, environments and systems.

Geographical Skills Workbook and Teacher's CD-ROM

In addition to this book there is a Geographical Skills Workbook, which enables you to revise and test your knowledge and to practise key skills. It includes key features of the geographical skills needed for your Geography course and further practice working with maps and satellite images. The Geographical Skills Workbook also contains a number of skills-based exercises to accompany the AS and A2 content covered in this book, relating to Papers 1 to 4.

Answers to the exercises in the Geographical Skills Workbook are provided on the Teacher's CD-ROM. Interactive multiple-choice tests, definitions of the key terms highlighted in blue in this book, chapter summaries, suggested useful websites and extension exercises are also available on the Teacher's CD-ROM. The interactive multiple-choice tests are also available on the student etextbook version of this book.

About the authors

Paul Guinness is a highly experienced author, teacher and examiner. Until recently, he was Head of Geography at King's College School, Wimbledon, London. He has written or co-authored over 35 textbooks and produced over 100 articles for A Level, IB and GCSE courses. In 2010, he was accredited with the title of Geography Consultant by the Geographical Association in the UK.

Garrett Nagle teaches at St Edward's School, Oxford, where he has been Head of Geography. He is also an experienced examiner, moderator and question setter. He has written and co-authored numerous books and articles for geography and environmental science.

Structure of the syllabus

Cambridge International AS and A Level Geography (syllabus code 9696 – for examination from 2018)

- Candidates for Advanced Subsidiary (AS) certification take Papers 1 and 2 only.
- Candidates who already have AS certification and wish to achieve the full Advanced Level qualification may carry their AS marks forward and take Papers 3 and 4 in the exam session in which they require certification.
- Candidates taking the complete Advanced Level qualification take all four papers.

☐ Paper 1: Core Physical Geography
1 hour 30 minutes

- Hydrology and fluvial geomorphology
- Atmosphere and weather
- Rocks and weathering

In Section A, candidates answer three data-response questions. In Section B, candidates answer one structured question from a total of three questions. All questions are based on the Core Physical Geography syllabus and are for a total of 60 marks.

50% of total marks at AS Level; 25% of total marks at A Level

☐ Paper 2: Core Human Geography
1 hour 30 minutes

- Population
- Migration
- Settlement dynamics

In Section A, candidates answer three data-response questions. In Section B, candidates answer one structured question from a total of three questions. All questions are based on the Core Human Geography syllabus and are for a total of 60 marks.

50% of total marks at AS Level; 25% of total marks at A Level

☐ Paper 3: Advanced Physical Geography Options
1 hour 30 minutes

- Tropical environments
- Coastal environments
- Hazardous environments
- Arid and semi-arid environments

Candidates answer questions on two topics. Each topic consists of one structured question (10 marks) and a choice of essay questions (20 marks), for a total of 60 marks.

25% of total marks at A Level

☐ Paper 4: Advanced Human Geography Options
1 hour 30 minutes

- Production, location and change
- Environmental management
- Global interdependence
- Economic transition

Candidates answer questions on two topics. Each topic consists of one structured question (10 marks) and a choice of essay questions (20 marks), for a total of 60 marks.

25% of total marks at A Level

1 Hydrology and fluvial geomorphology

1.1 The drainage basin system

The **hydrological cycle** refers to the cycle of water between **atmosphere**, lithosphere and biosphere (Figure 1.1). At a local scale – the drainage basin (Figure 1.2) – the cycle has a single input, **precipitation (PPT)**, and two major losses (outputs): **evapotranspiration (EVT)** and runoff. A third output, leakage, may also occur from the deeper subsurface to other basins. The drainage basin system is an **open system** as it allows the movement of energy and matter across its boundaries.

Water can be stored at a number of stages or levels within the cycle. These stores include vegetation, surface, soil moisture, **groundwater** and water **channels**.

Human modifications are made at every scale. Relevant examples include large-scale changes of channel flow and storage, irrigation and land drainage, and large-scale **abstraction** of groundwater and surface water for domestic and industrial use.

☐ Outputs

Evaporation

Evaporation is the process by which a liquid is changed into a gas. The process by which a solid is changed into a gas is sublimation. These terms refer to the conversion of solid and liquid precipitation (snow, ice and water) to **water vapour** in the atmosphere. Evaporation is most important from oceans and seas. It increases under warm, dry conditions and decreases under cold, calm conditions. Evaporation losses are greater in arid and semi-arid climates than in polar regions.

Factors affecting evaporation include meteorological factors such as temperature, **humidity** and wind speed. Of these, temperature is the most important factor. Other factors include the amount of water available, vegetation cover and colour of the surface (**albedo** or reflectivity of the surface).

Evapotranspiration

Transpiration is the process by which water vapour escapes from a living plant, principally the leaves, and enters the atmosphere. The combined effects of evaporation and transpiration are normally referred to as evapotranspiration (EVT). EVT represents the most important aspect of water loss, accounting for the loss of nearly 100 per cent of the annual precipitation in arid areas and 75 per cent in humid areas. Only over ice and snow fields, bare rock **slopes**, desert areas, water surfaces and bare soil will purely evaporative losses occur.

Potential evapotranspiration (P.EVT)

The distinction between actual EVT and P.EVT lies in the concept of **moisture availability**. Potential evapotranspiration is the water loss that would occur if there were an unlimited supply of water in the soil for use by the vegetation. For example, the actual evapotranspiration rate in Egypt is less than 250 mm,

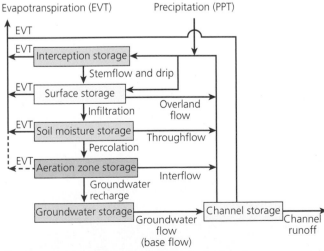

Source: *Advanced Geography: Concepts & Cases*
by P. Guinness & G. Nagle (Hodder Education, 1999), p.245

Figure 1.1 The global hydrological cycle

Figure 1.2 The drainage basin hydrological cycle

because there is less than 250 mm of rain annually. However, given the high temperatures experienced in Egypt, if the **rainfall** were as high as 2000 mm, there would be sufficient heat to evaporate that water. Hence the potential evapotranspiration rate there is 2000 mm. The factors affecting evapotranspiration include all those that affect evaporation. In addition, some plants, such as cacti, have adaptations to help them reduce moisture loss.

River discharge

River **discharge** refers to the movement of water in channels such as streams and rivers. The water may enter the river as direct channel precipitation (it falls on the channel) or it may reach the channel by surface runoff, groundwater flow (baseflow), or throughflow (water flowing through the soil).

☐ Stores

Interception

Interception refers to water that is caught and stored by vegetation. There are three main components:

- **interception loss** – water that is retained by plant surfaces and that is later evaporated away or absorbed by the plant
- **throughfall** – water that either falls through gaps in the vegetation or that drops from leaves or twigs
- **stemflow** – water that trickles along twigs and branches and finally down the main trunk.

Interception loss varies with different types of vegetation (Figure 1.3). Interception is less from grasses than from deciduous woodland owing to the smaller surface area of the grass shoots. From agricultural crops, and from cereals in particular, interception increases with crop density. Coniferous trees intercept more than deciduous trees in winter, but this is reversed in summer.

Soil water

Soil water (soil moisture) is the subsurface water in soil and subsurface layers above the water table. From here water may be:

- absorbed
- held
- transmitted downwards towards the water table, or
- transmitted upwards towards the soil surface and the atmosphere.

In coarse-textured soils much of the water is held in fairly large pores at fairly low suctions, while very little is held in small pores. In the finer-textured clay soils the range of pore sizes is much greater and, in particular, there is a higher proportion of small pores in which the water is held at very high suctions.

Field capacity refers to the amount of water held in the soil after excess water drains away; that is, saturation or near saturation. **Wilting point** refers to the range of moisture content in which permanent wilting of plants occurs.

There are a number of important seasonal variations in soil moisture budgets (Figure 1.4):

- **Soil moisture deficit** is the degree to which soil moisture falls below field capacity. In temperate areas, during late winter and early spring, soil moisture deficit is very low, due to high levels of precipitation and limited evapotranspiration.
- **Soil moisture recharge** occurs when precipitation exceeds potential evapotranspiration – there is some refilling of water in the dried-up pores of the soil.
- **Soil moisture surplus** is the period when soil is saturated and water cannot enter, and so flows over the surface.
- **Soil moisture utilisation** is the process by which water is drawn to the surface through capillary action.

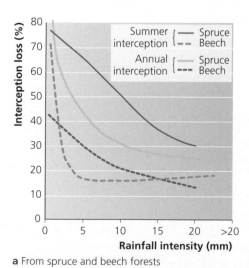

a From spruce and beech forests

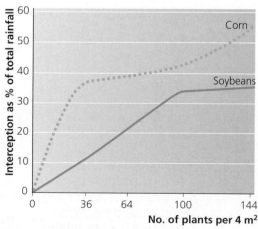

b By two agricultural crops

Source: *Advanced Geography: Concepts & Cases* by P. Guinness & G. Nagle (Hodder Education, 1999), p.245

Figure 1.3 Interception losses for different types of vegetation

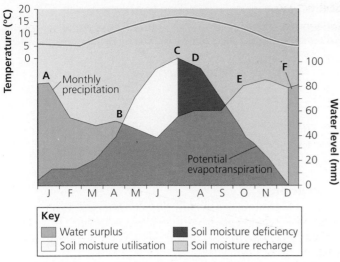

Key
- ■ Water surplus
- □ Soil moisture utilisation
- ■ Soil moisture deficiency
- ■ Soil moisture recharge

A Precipitation > potential evapotranspiration. Soil water store is full and there is a soil moisture surplus for plant use, runoff and groundwater recharge.

B Potential evapotranspiration > precipitation. Water store is being used up by plants or lost by evaporation (soil moisture utilisation).

C Soil moisture store is now used up. Any precipitation is likely to be absorbed by the soil rather than produce runoff. River levels will fall or dry up completely.

D There is a deficiency of soil water as the store is used up and potential evapotranspiration > precipitation. Plants must adapt to survive; crops must be irrigated.

E Precipitation > potential evapotranspiration. Soil water store will start to fill again (soil moisture recharge).

F Soil water store is full. Field capacity has been reached. Additional rainfall will percolate down to the water table and groundwater stores will be recharged.

Figure 1.4 Soil moisture status

Surface water

There are a number of types of surface water, some of which are temporary and some are permanent. Temporary sources include small puddles following a rainstorm and turloughs (seasonal lakes in limestone in the west of Ireland), while permanent stores include lakes, wetlands, swamps, peat bogs and marshes.

Groundwater

Groundwater refers to subsurface water that is stored under the surface in rocks. Groundwater accounts for 96.5 per cent of all freshwater on the Earth (Table 1.1). However, while some soil moisture may be recycled by evaporation into atmospheric moisture within a matter of days or weeks, groundwater may not be recycled for as long as 20 000 years. Recharge refers to the refilling of water in pores where the water has dried up or been extracted by human activity. Hence, in some places where recharge is not taking place, groundwater is considered a non-renewable resource.

Table 1.1 Global water reservoirs

Reservoir		Value (km^{-3} × 10^{-3})	% of total
Ocean		1 350 000.0	97.403
Atmosphere		13.0	0.000 94
Land		35 977.8	2.596
Of which	Rivers	1.7	0.000 12
	Freshwater lakes	100.0	0.007 2
	Inland seas	105.0	0.007 6
	Soil water	70.0	0.005 1
	Groundwater	8 200.0	0.592
	Ice caps/glaciers	27 500.0	1.984
	Biota	1.1	0.000 88

Channel storage

Channel storage refers to all water that is stored in rivers, streams and other drainage channels. Some rivers are seasonal, and some may disappear underground either naturally, such as in areas of **Carboniferous limestone**, or in urban areas, where they may be covered (culverted).

☐ Flows

Above ground

Throughfall refers to water that either falls through gaps in vegetation or that drops from leaves or twigs. Stemflow refers to water that trickles along twigs and branches and finally down the main trunk.

Overland flow (surface runoff) is water that flows over the land's surface. Surface runoff (or overland flow) occurs in two main ways:

- when precipitation exceeds the infiltration rate
- when the soil is saturated (all the pore spaces are filled with water).

In areas of high precipitation intensity and low infiltration capacity, overland runoff is common. This is clearly seen in semi-arid areas and in cultivated fields. By contrast, where precipitation intensity is low and infiltration is high, most overland flow occurs close to streams and river channels.

Channel flow or stream flow refers to the movement of water in channels such as streams and rivers. The water may have entered the stream as a result of direct precipitation, overland flow, groundwater flow (baseflow) or throughflow (water flowing through the soil).

Below ground

Porosity is the capacity of a rock to hold water, for example sandstone has a porosity (pore space) of 5–15 per cent, whereas clay may be up to 50 per cent. **Permeability** is the ability to transmit water through a rock via joints and fissures.

Infiltration

Infiltration is the process by which water soaks into or is absorbed by the soil. The **infiltration capacity** is the maximum rate at which rain can be absorbed by a soil in a given condition.

Infiltration capacity decreases with time through a period of rainfall until a more or less constant value is reached (Figure 1.5). Infiltration rates of 0–4mm/hour are common on clays, whereas 3–12mm/hour are common on sands. Vegetation also increases infiltration. This is because it intercepts some rainfall and slows down the speed at which it arrives at the surface. For example, on bare soils where rainsplash impact occurs, infiltration rates may reach 10mm/hour. On similar soils covered by vegetation, rates of between 50 and 100mm/hour have been recorded. Infiltrated water is chemically rich as it picks up minerals and organic acids from vegetation and soil.

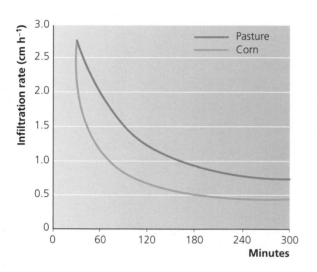

Source: *Advanced Geography: Concepts & Cases*
by P. Guinness & G. Nagle (Hodder Education, 1999), p.247

Figure 1.5 Infiltration rates under vegetation

Infiltration is inversely related to overland runoff and is influenced by a variety of factors, such as duration of rainfall, **antecedent soil moisture** (pre-existing levels of soil moisture), soil porosity, vegetation cover (Table 1.2), raindrop size and slope angle (Figure 1.6). In contrast, **overland flow** is water that flows over the land's surface.

Table 1.2 Influence of ground cover on infiltration rates

Ground cover	Infiltration rate (mm/hour)
Old permanent pasture	57
Permanent pasture: moderately grazed	19
Permanent pasture: heavily grazed	13
Strip-cropped	10
Weeds or grain	9
Clean tilled	7
Bare, crusted ground	6

Percolation
Water moves slowly downwards from the soil into the bedrock – this is known as **percolation**. Depending on the permeability of the rock, this may be very slow or in some rocks, such as Carboniferous limestone and chalk, it may be quite fast, locally.

Throughflow
Throughflow refers to water flowing through the soil in natural pipes and **percolines** (lines of concentrated water flow between soil horizons).

Groundwater and baseflow
Most groundwater is found within a few hundred metres of the surface but has been found at depths of up to 4kilometres beneath the surface. **Baseflow** refers to the part of a river's discharge that is provided by groundwater seeping into the bed of a river. It is a relatively constant flow although it increases slightly following a wet period.

☐ Underground water
The permanently saturated zone within solid rocks and sediments is known as the phreatic zone. The upper layer of this is known as the **water table**. The water table varies seasonally. In temperate zones it is higher in winter following increased levels of precipitation. The zone that is seasonally wetted and seasonally dries out is known as the aeration zone.

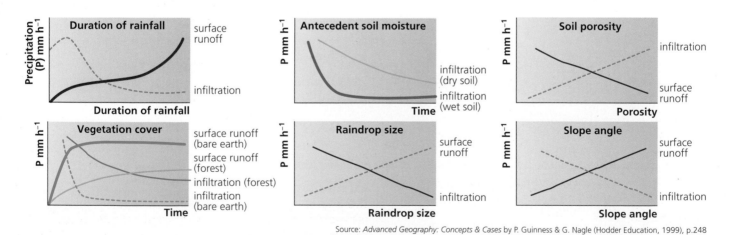

Source: *Advanced Geography: Concepts & Cases* by P. Guinness & G. Nagle (Hodder Education, 1999), p.248

Figure 1.6 Factors affecting infiltration and surface runoff

1 Hydrology and fluvial geomorphology

Aquifers (rocks that contain significant quantities of water) provide a great reservoir of water. Aquifers are permeable rocks such as sandstones and limestones. The water in aquifers moves very slowly and acts as a natural regulator in the hydrological cycle by absorbing rainfall that otherwise would reach streams rapidly. In addition, aquifers maintain stream flow during long dry periods. Where water flow reaches the surface (as shown by the discharge areas in Figure 1.7), springs may be found. These may be substantial enough to become the source of a stream or river.

a In humid regions

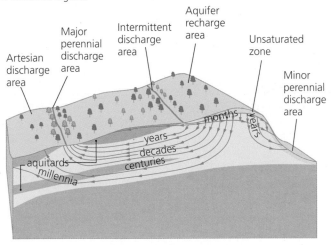

b In semi-arid regions

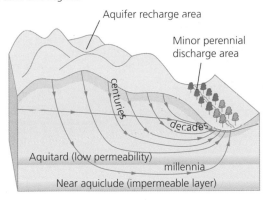

Source: *Advanced Geography: Concepts & Cases*
by P. Guinness & G. Nagle (Hodder Education, 1999), p.248

Figure 1.7 Groundwater and aquifer characteristics

Groundwater recharge occurs as a result of:
- **infiltration** of part of the total precipitation at the ground surface
- **seepage** through the banks and bed of surface water bodies such as ditches, rivers, lakes and oceans
- **groundwater leakage and inflow** from adjacent rocks and aquifers
- **artificial recharge** from irrigation, reservoirs, and so on.

Losses of groundwater result from:
- **evapotranspiration**, particularly in low-lying areas where the water table is close to the ground surface
- **natural discharge**, by means of spring flow and seepage into surface water bodies
- **groundwater leakage and outflow**, along aquicludes and into adjacent aquifers
- **artificial abstraction**, for example the water table near Lubbock on the High Plains of Texas (USA) has declined by 30–50 m in just 50 years, and in Saudi Arabia the groundwater reserve in 2010 was 42 per cent less than in 1985.

Section 1.1 Activities

1 Define the following hydrological characteristics:
   a interception   **b** evaporation   **c** infiltration.
2 Study Figure 1.2.
   a Define the terms *overland flow* and *throughflow*.
   b Compare the nature of water movement in these two flows.
   c Suggest reasons for the differences you have noted.
3 Figure 1.3 shows interception losses from spruce and beech forests and from three agricultural crops. Describe and comment on the relationship between the number of plants and interception, and the type of plants and interception.
4 Figure 1.6 shows the relationship between infiltration, overland flow (surface runoff) and six factors. Write a paragraph on each of the factors, describing and explaining the effect it has on infiltration and overland runoff.
5 Comment on the relationship between ground cover and infiltration, as shown in Table 1.2.
6 Define the terms *groundwater* and *baseflow*.
7 Outline the ways in which human activities have affected groundwater.

1.2 Discharge relationships within drainage basins

☐ Hydrographs

A **storm hydrograph** shows how the discharge of a river varies over a short time (Figure 1.8). Normally it refers to an individual storm or group of storms of not more than a few days in length. Before the storm starts, the main supply of water to the stream is through groundwater flow or baseflow. This is the main supplier of water to rivers. During the storm, some water infiltrates into the soil while some flows over the surface as overland flow or runoff. This reaches the river quickly as **quickflow**, which causes a rapid rise in the level of the river. The **rising limb** shows us how quickly the **flood** waters begin to rise, whereas the **recessional limb** is the speed with which the water level in the river declines after the peak. The **peak flow** is the maximum discharge of the river as a result of the storm, and the **time lag** is the time between the height of the storm (not the start or the end) and the maximum flow in the river.

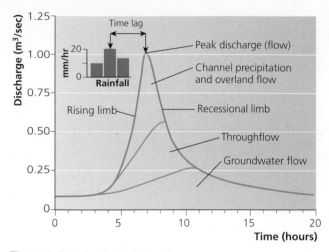

Figure 1.8 A simple hydrograph

In contrast, a **river regime** is the annual variation in the discharge of a river. Stream discharge occurs as a result of overland runoff and groundwater springs, and from lakes and meltwater in mountainous or sub-polar environments. The character or **regime** of the resulting stream or river is influenced by several variable factors:

- the amount and nature of precipitation
- the local rocks, especially porosity and permeability
- the shape or morphology of the drainage basin, its area and slope
- the amount and type of vegetation cover
- the amount and type of soil cover.

On an annual basis, the most important factor determining stream regime is climate. Figure 1.9 shows generalised regimes for Europe. Notice how the regime for the Shannon at Killaloe (Ireland) has a typical temperate regime, with a clear winter maximum. By contrast, Arctic areas such as the Gloma in Norway and the Kemi in Finland have a peak in spring associated with snowmelt. Others, such as the Po near Venice, have two main maxima – autumn and winter rains (Mediterranean climate) and spring snowmelt from Alpine tributaries.

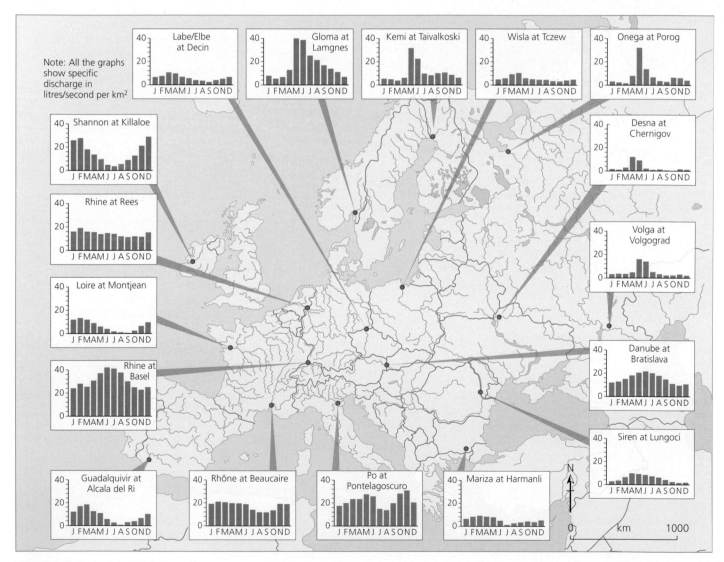

Figure 1.9 River regimes in Europe

1 Hydrology and fluvial geomorphology

a A simple river regime

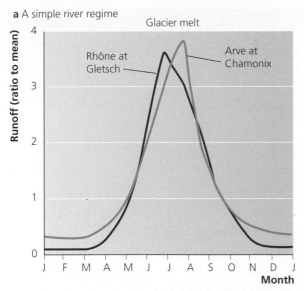

b Complex regime of the Rhine River

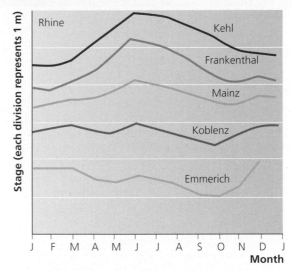

Source: *Advanced Geography: Concepts & Cases* by P. Guinness & G. Nagle (Hodder Education, 1999), p.262

Figure 1.10 Simple and complex river regimes

Figure 1.10a shows a simple regime, based upon a single river with one major peak flow. By contrast, Figure 1.10b shows a complex regime for the River Rhine. It has a number of large tributaries that flow in a variety of environments, including alpine, Mediterranean and temperate. By the time the Rhine has travelled downstream, it is influenced by many, at times contrasting, regimes.

☐ Influences on hydrographs

The effect of urban development on hydrographs is to increase peak flow and decrease time lag (Figure 1.11). This is due to an increase in the proportion of impermeable ground in a drainage basin, as well as an increase in the drainage density. Storm hydrographs also vary, with a number of other factors (Table 1.3) such as basin shape, drainage density and gradient.

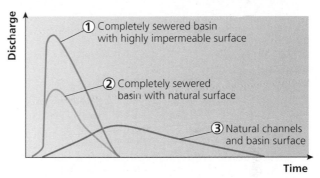

Source: *Advanced Geography: Concepts & Cases* by P. Guinness & G. Nagle (Hodder Education, 1999), p.255

Figure 1.11 The effects of urban development on storm hydrographs

Table 1.3 Factors affecting storm hydrographs

Factor	Influence on storm hydrograph
Climate	
Precipitation type and intensity	Highly intensive rainfall is likely to produce overland flow, a steep rising limb and high peak flow. Low-intensity rainfall is likely to infiltrate into the soil and percolate slowly into the rock, thereby increasing the time lag and reducing the peak flow. Precipitation that falls as snow sits on the ground until it melts. Sudden, rapid melting can cause flooding and lead to high rates of overland flow, and high peak flows.
Temperature, evaporation, transpiration and evapotranspiration	Not only does temperature affect the type of precipitation, it also affects the evaporation rate (higher temperatures lead to more evaporation and so less water getting into rivers). On the other hand, warm air can hold more water so the potential for high peak flows in hot areas is raised. Increased vegetation cover intercepts more rainfall and may return a proportion of it through transpiration, thereby reducing the amount of water reaching stream channels. The greater the return through evapotranspiration, the less water is able to reach stream channels, and therefore the peak of the hydrograph is reduced.
Antecedent moisture	If it has been raining previously and the ground is saturated or nearly saturated, rainfall will quickly produce overland flow, a high peak flow and short time lag.

Factor	Influence on storm hydrograph
Drainage basin characteristics	
Drainage basin size and shape	Smaller drainage basins respond more quickly to rainfall conditions. For example, the Boscastle (UK) floods of 2004 drained an area of less than 15 km². This meant that the peak of the flood occurred soon after the peak of the storm. In contrast, the Mississippi River is over 3700 km long – it takes much longer for the lower part of the river to respond to an event that might occur in the upper course of the river. Circular basins respond more quickly than linear basins, where the response is more drawn out.
Drainage density	Basins with a high drainage density, such as urban basins with a network of sewers and drains, respond very quickly. Networks with a low drainage density have a very long time lag.
Porosity and impermeability of rocks and soils	Impermeable surfaces cause more water to flow overland. This causes greater peak flows. Urban areas contain large areas of impermeable surfaces. In contrast, rocks such as chalk and gravel are permeable and allow water to infiltrate and percolate. This reduces the peak flow and increases the time lag. Sandy soils allow water to infiltrate, whereas clay is much more impermeable and causes water to pass overland.
Rock type	Impermeable rocks such as granite and clay produce greater peak flows with a more flashy response. In contrast, more permeable rocks such as chalk and limestone produce storm hydrographs with a much lower peak flow (if at all) and with a much delayed/ less flashy response (greater time lag).
Slopes	Steeper slopes create more overland flow, shorter time lags and higher peak flows.
Vegetation type	Forest vegetation intercepts more rainfall, especially in summer, and so reduces the amount of overland flow and peak flow and increases time lag. In winter, deciduous trees lose their leaves and so intercept less.
Land use	Land uses that create impermeable surfaces, or reduce vegetation cover, reduce interception and increase overland flow. If more drainage channels are built (sewers, ditches, drains), the water is carried to rivers very quickly. This means that peak flows are increased and time lags reduced.

Section 1.2 Activities

1 Compare the river regimes of the Gloma (Norway), Shannon (Ireland) and Rhine (Switzerland). Suggest reasons for their differences.
2 Table 1.4 shows precipitation and runoff data for a storm on the Delaware River, New York. Using this data, plot the storm hydrograph for this storm. Describe the main characteristics of the hydrograph you have drawn.
3 Define the terms *river regime* and *storm hydrograph*.
4 Study Figure 1.11, which shows the impact of urbanisation on storm hydrographs. Describe and explain the differences in the relationship between discharge and time.

Table 1.4 Precipitation and runoff data for a storm on the Delaware River, New York

Date	Time	Duration of rainfall	Total (cm)
29 September	6 a.m.	12 hours	0.1
29 September	6 p.m.	12 hours	0.9
30 September	6 p.m.	24 hours	3.7
30 September	12 p.m.	6 hours	0.1
		Total	**4.8**

Date	Stream runoff (m³/s)
28 September	28.3 (baseflow)
29 September	28.3 (baseflow)
30 September	339.2
1 October	2094.2
2 October	1330.1
3 October	594.3
4 October	367.9
5 October	254.2
6 October	198.1
7 October	176.0
8 October	170.0
9 October	165.2 (baseflow)

1.3 River channel processes and landforms

☐ Erosion

Abrasion (corrasion) is the wearing away of the bed and bank by the load carried by a river. It is the mechanical impact produced by the debris eroding the river's bed and banks. In most rivers it is the principal means of erosion. The effectiveness of abrasion depends on the concentration, hardness and energy of the impacting particles and the resistance of the bedrock. Abrasion increases as velocity increases (kinetic energy is proportional to the square of velocity).

Attrition is the wearing away of the load carried by a river. It creates smaller, rounder particles.

Hydraulic action is the force of air and water on the sides of rivers and in cracks. It includes the direct force of flowing water, and **cavitation** – the force of air exploding. As fluids accelerate, pressure drops and may cause air bubbles to form. Cavitation occurs as bubbles implode and evict tiny jets of water with velocities of up to 130 m/s. These can damage solid rock. Cavitation is an important process in rapids and waterfalls, and is generally accompanied by abrasion.

Corrosion or **solution** is the removal of chemical ions, especially calcium. The key factors controlling the rate of corrosion are bedrock, solute concentration of the stream water, discharge and velocity. Maximum rates of corrosion occur where fast-flowing, undersaturated streams pass over soluble rocks – humid zone streams flowing over mountain limestone.

There are a number of factors affecting rates of erosion. These include:

- **load** – the heavier and sharper the load the greater the potential for erosion
- **velocity** – the greater the velocity the greater the potential for erosion (see Figure 1.13)
- **gradient** – increased gradient increases the rate of erosion
- **geology** – soft, unconsolidated rocks such as sand and gravel are easily eroded
- **pH** – rates of solution are increased when the water is more acidic
- **human impact** – deforestation, dams and bridges interfere with the natural flow of a river and frequently end up increasing the rate of erosion.

Erosion by the river will provide loose material. This eroded material (plus other weathered material that has moved downslope from the upper valley sides) is carried by the river as its load.

Global sediment yield

It is possible to convert a value of mean annual sediment and solute load to an estimate of the rate of land surface lowering by fluvial denudation. This gives a combined sediment and solute load of 250 tonnes/km² per year – that is, an annual rate of lowering of the order of 0.1 mm per year. There is a great deal of variation in sediment yields. These range from 10 tonnes/km² per year in such areas as northern Europe and parts of Australia to in excess of 10 000 tonnes/km² per year in certain areas where conditions are especially conducive to high rates of erosion (Figure 1.12). These include Taiwan, New Zealand's South Island and the Middle Yellow River basin in China.

In the first two cases, steep slopes, high rainfall and tectonic instability are major influences, while in the last case the deep loess deposits and the almost complete lack of natural vegetation cover are important. Rates of land surface lowering vary from less than 0.004 mm per year to over 4 mm per year. The broad pattern of global suspended sediment is shown on the map and it reflects the influence of a wide range of factors, including climate, relief, geology, vegetation cover and land use.

Load transport

Load is transported downstream in a number of ways:

- The smallest particles (silts and clays) are carried in suspension as the **suspended load**.
- Larger particles (sands, gravels, very small stones) are transported in a series of 'hops' as the **saltated load**.
- Pebbles are shunted along the bed as the **bed** or **tracted load**.
- In areas of calcareous rock, material is carried in **solution** as the dissolved load.

The load of a river varies with discharge and velocity. The **capacity** of a stream refers to the largest amount of debris that a stream can carry, while the **competence** refers to the diameter of the largest particle that can be carried.

Deposition and sedimentation

There are a number of causes of deposition, such as:

- a shallowing of gradient, which decreases velocity and energy
- a decrease in the volume of water in the channel
- an increase in the friction between water and channel.

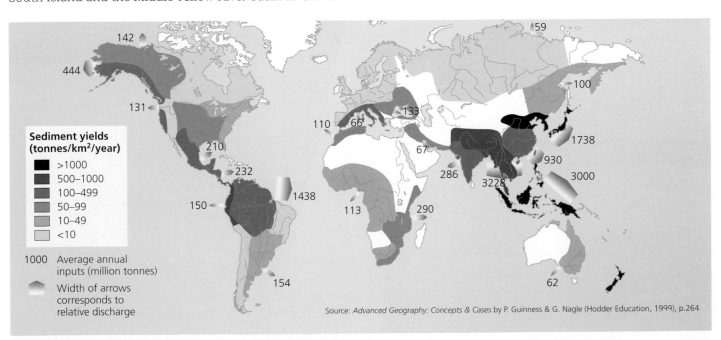

Source: *Advanced Geography: Concepts & Cases* by P. Guinness & G. Nagle (Hodder Education, 1999), p.264

Figure 1.12 Global sediment yield

The Hjülstrom curve

The **critical erosion velocity** is the lowest velocity at which grains of a given size can be moved. The relationship between these variables is shown by means of a **Hjülstrom curve** (Figure 1.13). For example, sand can be moved more easily than silt or clay, as fine-grained particles tend to be more cohesive. High velocities are required to move gravel and cobbles because of their large size. The critical velocities tend to be an area rather than a straight line on the graph.

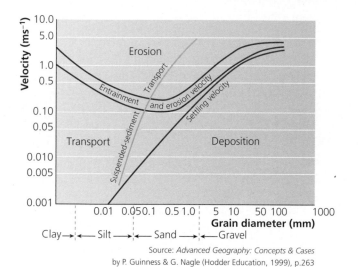

Figure 1.13 Hjülstrom curve

Source: *Advanced Geography: Concepts & Cases*
by P. Guinness & G. Nagle (Hodder Education, 1999), p.263

There are three important features on Hjülstrom curves:
- The smallest and largest particles require high velocities to lift them. For example, particles between 0.1 mm and 1 mm require velocities of around 100 mm/s to be entrained, compared with values of over 500 mm/s to lift clay (0.01 mm) and gravel (over 2 mm). Clay resists entrainment due to its cohesion; gravel due to its weight.
- Higher velocities are required for entrainment than for transport.
- When velocity falls below a certain level (settling or fall velocity), those particles are deposited.

Section 1.3 Activities

Study Figure 1.13.

1 Describe the work of the river when sediment size is 1 mm.
2 Comment on the relationship between velocity, sediment size and river process when the river is moving at 0.5 m/s⁻¹.

☐ River flow

Velocity and discharge

River flow and associated features of erosion are complex. The velocity and energy of a stream are controlled by:

- the gradient of the channel bed
- the volume of water within the channel, which is controlled largely by precipitation in the drainage basin (for example **bankfull** gives rapid flow, whereas low levels give lower flows)
- the shape of the channel
- channel roughness, including friction.

Manning's Equation

$Q = (AR^{2/3} S^{1/2})/n$

where Q = discharge, A = cross-sectional area, R = hydraulic radius, S = channel slope (as a fraction), n = coefficient of bed roughness (the rougher the bed the higher the value). As water flows over riffles, for example, there are changes in cross-sectional area, slope and hydraulic radius. Slope and velocity increase but depth decreases. Discharge remains the same.

Manning's 'n'

Mountain stream, rocky bed	0.04–0.05
Alluvial channel (large dunes)	0.02–0.035
Alluvial channel (small ripples)	0.014–0.024

Patterns of flow

There are three main types of flow: laminar, turbulent and helicoidal. For **laminar flow**, a smooth, straight channel with a low velocity is required. This allows water to flow in sheets, or laminae, parallel to the channel bed. It is rare in reality and most commonly occurs in the lower reaches. However, it is more common in groundwater, and in glaciers when one layer of ice moves over another.

Turbulent flow occurs where there are higher velocities and complex channel morphology such as a meandering channel with alternating pools and riffles. Bed roughness also increases turbulence, for example mountain streams with rocky beds create more turbulence than alluvial channels. Turbulence causes marked variations in pressure within the water. As the turbulent water swirls (eddies) against the bed or bank of the river, air is trapped in pores, cracks and crevices and put momentarily under great pressure. As the eddy swirls away, pressure is released; the air expands suddenly, creating a small explosion that weakens the bed or bank material. Thus turbulence is associated with hydraulic action.

Vertical turbulence creates hollows in the channel bed. Hollows may trap pebbles that are then swirled by eddying, grinding at the bed. This is a form of vertical corrasion or abrasion and given time may create potholes (Figure 1.14). Cavitation and vertical abrasion may help to deepen the channel, allowing the river to down-cut its valley. If the down-cutting is dominant over the other forms of erosion (vertical erosion exceeds lateral erosion), then a gulley or gorge will develop.

Figure 1.14 Potholes as seen by the areas occupied by water (dark patches)

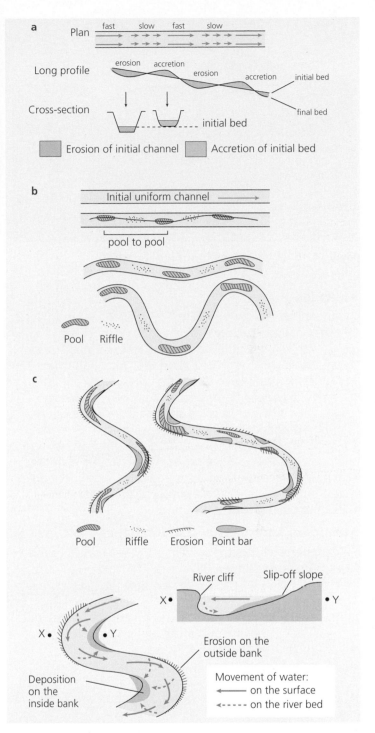

Figure 1.15 Meander formation

Horizontal turbulence often takes the form of **helicoidal flow**, a 'corkscrewing' motion. This is associated with the presence of alternating pools and riffles in the channel bed, and where the river is carrying large amounts of material. The erosion and deposition by helicoidal flow creates meanders (Figure 1.15). The thalweg is the line of maximum velocity and it travels from outside bank to outside bank of the meanders. The main current strikes the outer bank and creates a return flow to the inner bank, close to the channel bend. The movement transports sediment from the outer bank to the inner bank where it is deposited as a sand bar.

☐ Channel types

Sinuosity is the length of a stream channel expressed as a ratio of the valley length. A low sinuosity has a value of 1.0 (that is, it is straight) whereas a high sinuosity is above 4.4. The main groupings are **straight channels** (<1.5) and **meandering** (>1.5). Straight channels are rare. Even when they do occur the thalweg (line of maximum velocity) moves from side to side. These channels generally have a central ridge of deposited material, due to the water flow pattern.

Braiding occurs when the channel is divided by islands or bars (Figure 1.16). Islands are vegetated and long-lived, whereas bars are unvegetated, less stable and often short-term features. Braided channels are formed by various factors, for example:

- a steep channel gradient
- a large proportion of coarse material
- easily erodable bank material
- highly variable discharge.

Braiding tends to occur when a stream does not have the capacity to transport its load in a single channel, whether it is straight or meandering. It occurs when river discharge is very variable and banks are easily erodable. This gives abundant sediment. It is especially common in periglacial and semi-arid areas.

Braiding begins with a mid-channel bar that grows downstream. As the discharge decreases following a flood, the coarse bed load is first to be deposited. This forms the basis of bars that grow downstream and, as the flood is reduced, finer sediment is deposited. The upstream end becomes stabilised with vegetation. This island localises and narrows the channel in an attempt to increase the velocity to a point where it can transport its load. Frequently, subdivision sets in.

Figure 1.16 A braided river, Mýrdalsjökull, Iceland

Meanders

Meanders are complex (Figure 1.15). There are a number of relationships, although the reasons are not always very clear. However, they are *not* the result of obstructions in the floodplain. Meandering is the normal behaviour of fluids and gases in motion. Meanders can occur on a variety of materials, from ice to solid rock. Meander development occurs in conditions where channel slope, discharge and load combine to create a situation where meandering is the only way that the stream can use up the energy it possesses equally throughout the channel reach. The wavelength of the meander is dependent upon three main factors: channel width, discharge and the nature of the bed and banks.

Meanders and channel characteristics

- Meander wavelengths are generally 6–10 times channel width and discharge.
- Meander wavelengths are generally 5 times the radius of curvature.
- The meander belt (peak-to-peak amplitude) is generally 14–20 times the channel width.
- Riffles occur at about 6 times the channel width.
- Sinuosity increases as depth of channel increases in relation to width.
- Meandering is more pronounced when the bed load is varied.
- Meander wavelength increases in streams that carry coarse debris.
- Meandering is more likely on shallow slopes.
- Meandering best develops at or near bankfull state.

Natural meanders are rarely 'standard'. This is due, in part, to variations in bed load; where the bed load is coarse, meanders are often very irregular.

Causes of meanders

There is no simple explanation for the creation of meanders, and a number of factors are likely to be important.

- **Friction** with the channel bed and bank causes turbulence, which makes stream flow unstable. This produces bars along the channel, and a helicoidal flow (corkscrew motion), with water being raised on the outer surfaces of pools, and the return flow occurring at depth.
- **Sand bars** in the channel may cause meandering.
- **Sinuosity** is best developed on moderate angles. There is a critical minimum gradient below which straight channels occur. At very low energy (low gradient), helicoidal flow is insufficient to produce alternating **pools** and **riffles**. In addition, high-velocity flows in steep gradient channels are too strong to allow cross-channel meandering and the development of alternating pools and riffles. In such circumstances, braided channels are formed.
- **Helicoidal flow** (corkscrewing) causes the line of fastest flow to move from side to side within the channel. This increases the amplitude of the meander.

Change over time

There are a number of possibilities:

- Meanders may migrate downstream and erode **river cliffs**.
- They may migrate laterally (sideways) and erode the floodplain.
- They may become exaggerated and become cut-offs (ox-bow lakes).
- Under special conditions, they may become intrenched or ingrown.

Intrenched and ingrown meanders

The term **incised meanders** describes meanders that are especially well developed on horizontally bedded rocks, and form when a river cuts through alluvium and into underlying bedrock. Two main types occur – intrenched and ingrown meanders. Intrenched meanders are symmetrical, and occur when down-cutting is fast enough to offset the lateral migration of meanders. This frequently occurs when there is a significant fall in base level (generally sea level). The Goosenecks of the San Juan in the USA are classic examples of intrenched meanders. Ingrown meanders are the result of lateral meander migration. They are asymmetric in cross-section – examples can be seen in the lower Seine in France.

☐ Landforms

Meanders

Meanders have an asymmetric cross-section (Figure 1.15b). They are deeper on the outside bank and shallower on the inside bank. In between meanders they are more symmetrical. They begin with the development of pools and riffles in a straight channel and the thalweg begins to flow from side to side. Helicoidal flow occurs, whereby surface water flows towards the outer banks, while the bottom flow is towards the inner bank. This causes the variations in the cross-section and variations in erosion and deposition. These variations give rise to river cliffs on the outer bank and **point bars** on the inner bank.

Pools and riffles

Pools and riffles are formed by turbulence. Eddies cause the deposition of coarse sediment (riffles) at high velocity points and fine sediment (pools) at low velocity. Riffles have a steeper gradient than pools, which leads to variations in subcritical and supercritical flow, and therefore erosion and deposition.

Riffles are small ridges of material deposited where the river velocity is reduced midstream, in between pools (the deep parts of a meander).

Braided rivers

A braided river channel consists of a number of interconnected shallow channels separated by alluvial and shingle bars (islands). These may be exposed during low flow conditions. They are formed in rivers that are heavily laden with sediment and have a pronounced seasonal flow. There are excellent examples on the Eyjafjörður in northern Iceland.

Section 1.3 Activities

Study Figure 1.15.

1 Compare the main characteristics of river cliffs with those of point bars.
2 Briefly explain the meaning of the term *helicoidal flow*.
3 Describe and explain the role of pools and riffles in the development of meanders in a river channel.

Waterfalls and gorges

Waterfalls occur where the river spills over a sudden change in gradient, undercutting rocks by hydraulic impact and abrasion, thereby creating a waterfall (Figures 1.17 and 1.18). There are many reasons for this sudden change in gradient along the river:

(1) undercutting before collapse
(2) weight of water causes pressure on the unsupported Whin Sill
(3) pieces of Whin Sill – hard, igneous rock – are used to erode the limestone
(4) hydraulic action by force of falling water
(5) organic-rich waters help dissolve the limestone

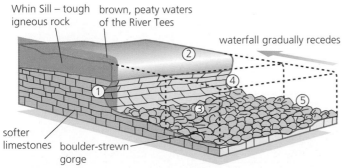

Whin Sill – tough igneous rock
brown, peaty waters of the River Tees
waterfall gradually recedes
softer limestones
boulder-strewn gorge

Source: Goudie, A. and Gardner, R., *Discovering Landscapes in England and Wales*, Unwin 1985

Figure 1.17 Waterfall formation

- a band of resistant strata such as the resistant limestones at Niagara Falls
- a plateau edge such as Victoria Falls on the Zimbabwe–Zambia border
- a fault scarp such as at Gordale, Yorkshire (UK)
- a hanging valley such as at Glencoyne, Cumbria in the UK
- coastal cliffs.

The undercutting at the base of a waterfall creates a precarious overhang, which will ultimately collapse. Thus a waterfall may appear to migrate upstream, leaving a gorge of recession downstream. The Niagara Gorge is 11 kilometres long due to the retreat of Niagara Falls.

Gorge development is common, for example where the local rocks are very resistant to **weathering** but susceptible to the more powerful river erosion. Similarly, in arid areas where the water necessary for weathering is scarce, gorges are formed by periods of river erosion. A rapid acceleration in down-cutting is also associated when a river is rejuvenated, again creating a gorge-like landscape. Gorges may also be formed as a result of:

- antecedent drainage (Rhine Gorge)
- glacial overflow channelling (Newtondale, UK)
- the collapse of underground caverns in Carboniferous limestone areas
- surface runoff over limestone during a periglacial period
- the retreat of waterfalls (Niagara Falls)
- superimposed drainage (Avon Gorge, UK).

Figure 1.18 Axara waterfall, Iceland

Case Study: Niagara Falls

Most of the world's great waterfalls are the result of the undercutting of resistant cap rocks, and the retreat or recession that follows. The Niagara River flows for about 50 kilometres between Lake Erie and Lake Ontario. In that distance it falls just 108 metres, giving an average gradient of 1:500. However, most of the descent occurs in the 1.5 kilometres above the Niagara Falls (13 metres) and at the Falls themselves (55 metres). The Niagara River flows in a 2 kilometre-wide channel just 1 kilometre above the Falls, and then into a narrow 400 metre-wide gorge, 75 metres deep and 11 kilometres long. Within the gorge the river falls a further 30 metres.

The course of the Niagara River was established about 12 000 years ago when water from Lake Erie began to spill northwards into Lake Ontario. In doing so, it passed over the highly resistant dolomitic (limestone) escarpment. Over the last 12 000 years the Falls have retreated 11 kilometres, giving an average rate of retreat of about 1 metre per year. Water velocity accelerates over the Falls, and decreases at the base of the Falls. Hydraulic action and abrasion have caused the development of a large plunge pool at the base, while the fine spray and eddies in the river help to remove some of the softer rock underneath the resistant dolomite. As the softer rocks are removed, the dolomite is left unsupported and the weight of the water causes the dolomite to collapse. Hence the waterfall retreats, forming a gorge of recession.

In the nineteenth century, rates of recession were recorded at 1.2 metres per year. However, now that the amount of water flowing over the Falls is controlled (due to the construction of hydro-electric power stations), rates of recession have been reduced. In addition, engineering works in the 1960s reinforced parts of the dolomite that were believed to be at risk of collapse. The Falls remains an important tourist attraction and local residents and business personnel did not want to lose their prized asset!

Section 1.3 Activities

Draw a labelled diagram to show the formation of a waterfall.

Levees, floodplains and bluffs

Levees and floodplain deposits are formed when a river bursts its banks over a long period of time. Water quickly loses velocity, leading to the rapid deposition of coarse material (heavy and difficult to move a great distance) near the channel edge. These coarse deposits build up to form embankments called **levees** (Figure 1.19). The finer material is carried further away to be dropped on the **floodplain** (Figure 1.20), sometimes creating **backswamps**. Repeated annual flooding slowly builds up the floodplain. Old floodplains may be eroded – the remnants are known as terraces. At the edge of the terrace is a line of relatively steep slopes known as **river bluffs**.

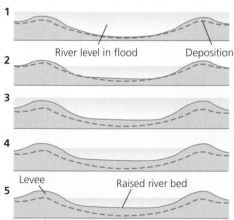

1 When the river floods, it bursts its banks. It deposits its coarsest load (gravel and sand) closest to the bank and the finer load (silt and clay) further away.
2, 3, 4 This continues over a long time, for centuries.
5 The river has built up raised banks, called levees, consisting of coarse material, and a floodplain of fine material.

Figure 1.19 The formation of levees

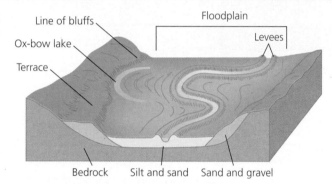

Figure 1.20 Floodplains, levees and bluffs

Ox-bow lakes

Ox-bow lakes are the result of both erosion and deposition. Lateral erosion, caused by helicoidal flow, is concentrated on the outer, deeper bank of a meander. During times of flooding, erosion increases. The river breaks through and creates a new steeper channel. In time, the old meander is closed off by deposition to form an ox-bow lake.

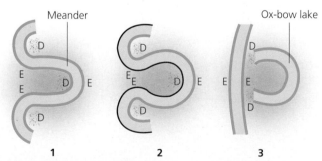

1 Erosion (E) and deposition (D) around a meander (a bend in a river).
2 Increased erosion during flood conditions. The meander becomes exaggerated.
3 The river breaks through during a flood. Further deposition causes the old meander to become an ox-bow lake.

Figure 1.21 Formation of an ox-bow lake

Deltas

Deltas are river sediments deposited when a river enters a standing body of water such as a lake, a lagoon, a sea or an ocean (Figure 1.22). They are the result of the interaction of fluvial and marine processes. For a delta to form there must be a heavily laden river, such as the Nile or the Mississippi, and a standing body of water with negligible currents, such as the Mediterranean or the Gulf of Mexico. Deposition is enhanced if the water is saline, because salty water causes small clay particles to flocculate or adhere together. Other factors include the type of sediment, local geology, sea-level changes, plant growth and human impact.

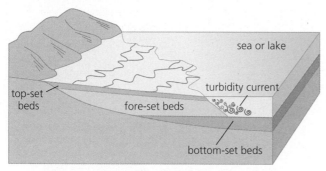

Source: *Advanced Geography: Concepts & Cases* by P. Guinness & G. Nagle (Hodder Education, 1999), p.268

Figure 1.22 Model of a simple delta

The material deposited as a delta can be divided into three types:

- **Bottomset beds** – the lower parts of the delta are built outwards along the sea floor by turbidity currents (currents of water loaded with material). These beds are composed of very fine material.
- **Foreset beds** – over the bottomset beds, inclined/sloping layers of coarse material are deposited. Each bed is deposited above and in front of the previous one, the material moving by rolling and saltation. Thus the delta is built seaward.
- **Topset beds** – composed of fine material, they are really part of the continuation of the river's floodplain. These topset beds are extended and built up by the work of numerous distributaries (where the main river has split into several smaller channels).

The character of any delta is influenced by the complex interaction of several variables (Figure 1.23):

- the rate of river deposition
- the rate of stabilisation by vegetation growth
- tidal currents
- the presence (or absence) of longshore drift
- human activity (deltas often form prime farmland when drained).

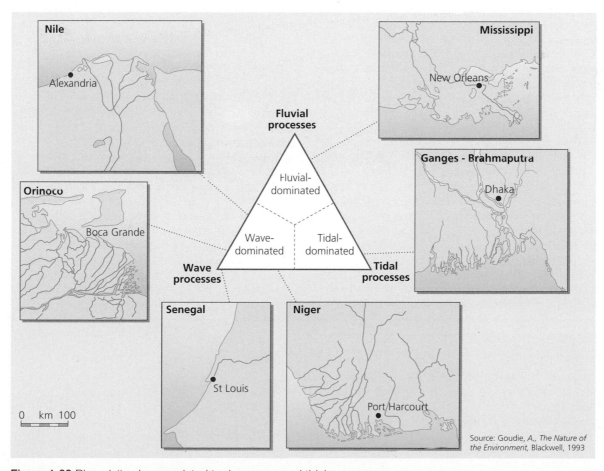

Source: Goudie, A., *The Nature of the Environment*, Blackwell, 1993

Figure 1.23 River delta shapes related to river, wave and tidal processes

There are many types of delta, but the three 'classic' ones are:

- **Arcuate delta**, or fan-shaped – these are found in areas where regular longshore drift or other currents keep the seaward edge of the delta trimmed and relatively smooth in shape, such as the Nile and Rhône deltas.
- **Cuspate delta** – pointed like a tooth or cusp, for example the Ebro and Tiber deltas, shaped by regular but opposing gentle water movement.
- **Bird's-foot delta** – where the river brings down enormous amounts of fine silt, deposition can occur

in a still sea area, along the edges of the distributaries for a very long distance offshore, such as the Mississippi delta.

Deltas can also be formed inland. When a river enters a lake it will deposit some or all of its load, so forming a **lacustrine delta**. As the delta builds up and out, it may ultimately fill the lake basin. The largest lacustrine deltas are those that are being built out into the Caspian Sea by the Volga, Ural, Kura and other rivers.

Case Study: The future of the Nile delta

The Nile delta is under threat from rising sea levels. Without the food it produces, Egypt faces much hardship. The delta is one of the most fertile tracts of land in the world. However, coastal erosion is steadily eroding it in some places at a rate of almost 100 metres a year. This is partly because the annual deposits from the Nile floods – which balanced coastal erosion – no longer reach the delta, instead being trapped behind the Aswan High Dam. However, erosion of the delta continues, and may be increasing, partly as a result of **global warming** and rising sea levels. The delta is home to about 50 million people, living at densities of up to 4000 people per km².

The Intergovernmental Panel on Climate Change has declared Egypt's Nile delta to be among the top three areas most vulnerable to a rise in sea level. Even a small temperature increase will displace millions of Egyptians from one of the most densely populated regions on Earth.

The delta stretches out from the northern reaches of Cairo into 25 000 km² of farmland fed by the Nile's branches. It is home to two-thirds of the country's rapidly growing population, and responsible for more than 60 per cent of its food supply. About 270 kilometres of the delta's coastline is at a dangerously low level and a 1 metre rise in the sea level would drown 20 per cent of the delta.

The delta is also suffering from a number of environmental crises, including flooding, coastal erosion, salinisation, industrial/agricultural **pollution** and urban encroachment. Egypt's population of 83 million is set to increase to more than 110 million in the next two decades. More people in the delta means more cars, more pollution and less land to feed them all on, just at a time when increased crop production is needed most.

Saltwater intrusion is destroying crops. Coastal farmland has always been threatened by salt water, but salinity has traditionally been kept at bay by plentiful supplies of fresh water flushing out the salt. It used to happen naturally with the Nile's seasonal floods; after the construction of Egypt's High Dam,

these seasonal floods came to an end, but a vast network of irrigation canals continued to bring enough fresh water to ensure salinity levels remained low.

Today, however, Nile water barely reaches the end of the delta. A growing population has extracted water supplies upstream, and what water does make it downriver is increasingly polluted with toxins and other impurities.

The impact of **climate change** is likely to be a 70 per cent drop in the amount of Nile water reaching the delta over the next 50 years, due to increased evaporation and heavier demands on water use upstream. The consequences for food production are ominous: wheat and maize yields could be down 40 per cent and 50 per cent respectively, and farmers could lose around $1000 per hectare for each degree rise in the average temperature.

While politicians, scientists and community workers are trying to educate Egyptians about the dangers of climate change, there is confusion over whether the focus should be on promoting ways to combat climate change, or on accepting climate change as inevitable and instead encouraging new forms of adaptation to the nation's uncertain future.

Egypt's contribution to global carbon emissions is just 0.5 per cent – nine times less per person than for the USA. However, the consequences of climate change are disproportionate and potentially disastrous.

The scale of the crisis – more people, less land, less water, less food – is overwhelming. As a result, many now believe that Egypt's future lies far away from the delta, in land newly reclaimed from the desert. Since the time of the pharaohs, when the delta was first farmed, Egypt's political leaders have tried to harness the Nile. The Egyptian government is creating an array of canals and pumping stations that draw water from the Nile into sandy valleys to the east and west, where the desert is slowly being turned green. The Nile delta may well become history – as a landform and for the people who live and work there.

Section 1.3 Activities

1 Outline the main conditions needed for delta formation.
2 Suggest reasons for the variety of deltas, as shown in Figure 1.23.
3 a Outline the natural and human processes that are operating on the Nile delta.
  b Comment on the advantages and disadvantages for people living in the delta.

1.4 The human impact

☐ Modifications to catchment stores and flows, and to channel flows

Evaporation and evapotranspiration

The human impact on evaporation and evapotranspiration is relatively small in relation to the rest of the hydrological cycle but is nevertheless important. There are a number of impacts:

- **Dams** – there has been an increase in evaporation due to the construction of large dams. For example, Lake Nasser behind the Aswan Dam loses up to a third of its water due to evaporation. Water loss can be reduced by using chemical sprays on the water, by building sand-fill dams and by covering the dams with some form of plastic.
- **Urbanisation** leads to a huge reduction in evapotranspiration due to the lack of vegetation. There may also be a slight increase in evaporation because of higher temperatures and increased surface storage (see Figure 1.24).

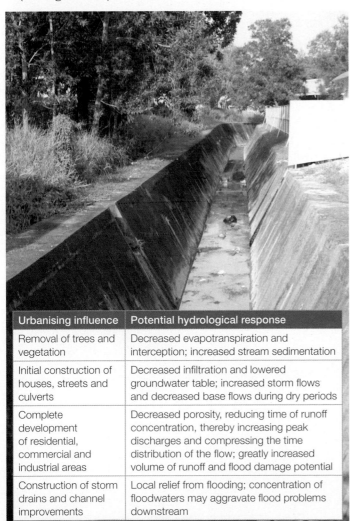

Urbanising influence	Potential hydrological response
Removal of trees and vegetation	Decreased evapotranspiration and interception; increased stream sedimentation
Initial construction of houses, streets and culverts	Decreased infiltration and lowered groundwater table; increased storm flows and decreased base flows during dry periods
Complete development of residential, commercial and industrial areas	Decreased porosity, reducing time of runoff concentration, thereby increasing peak discharges and compressing the time distribution of the flow; greatly increased volume of runoff and flood damage potential
Construction of storm drains and channel improvements	Local relief from flooding; concentration of floodwaters may aggravate flood problems downstream

Figure 1.24 Potential hydrological effects of urbanisation

Interception

Interception is determined by vegetation, density and type. Most vegetation is not natural but represents some disturbance by human activity. In farmland areas, for example, cereals intercept less than broad leaves. Row crops, such as wheat or corn, leave a lot of soil bare. For example, in the Mississippi basin, while sediment yields in woodland areas are just 1 unit, sediment from soil covered by pasture produces 30 units and areas under corn produce 350 units of sediment. **Deforestation** leads to:

- a reduction in evapotranspiration
- an increase in surface runoff
- a decline of surface storage
- a decline in time lag.

Afforestation is believed to have the opposite effect, although the evidence does not necessarily support it. For example, in parts of the Severn catchment, sediment loads increased four times after afforestation. Why was this? The result is explained by a combination of an increase in overland runoff, little ground vegetation, young trees, access routes for tractors, and fire- and wind-breaks. All of these allowed a lot of bare ground. However, after only five years the amount of erosion declined.

Infiltration and soil water

Human activity has a great impact on infiltration and soil water. Land-use changes are important. Urbanisation creates an impermeable surface with compacted soil. This reduces infiltration and increases overland runoff and flood peaks (Figure 1.25).

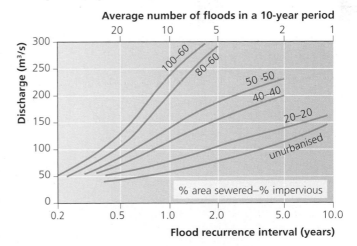

Figure 1.25 Flood frequency and urbanisation

Infiltration is up to five times greater under forest compared with grassland. This is because the forest trees channel water down their roots and stems. With deforestation there is reduced interception, increased soil compaction and more overland flow. Land-use practices are also important. Grazing leads to a decline in infiltration due to compaction and ponding of the soil. By contrast, ploughing increases infiltration because it loosens soils.

Waterlogging and salinisation are common if there is poor drainage. When the water table is close to the surface, evaporation of water leaves salts behind and may form an impermeable crust. Human activity also has an increasing impact on surface storage. There is increased surface storage due to the building of large-scale dams. These dams are being built in increasing numbers, and they are also larger in terms of general size and volume. This leads to:

- increased storage of water
- decreased flood peaks
- low flows in rivers
- decreased sediment yields (clear-water erosion)
- increased losses due to evaporation and seepage, leading to changes in temperature and salinity of the water
- decreased flooding of the land
- triggering of earthquakes
- salinisation, for example in the Indus Valley in Pakistan, 1.9 million hectares are severely saline and up to 0.4 million hectares are lost per annum to salinity
- large dams can cause local changes in climate.

In other areas there is a decline in the surface storage, for example in urban areas water is channelled away very rapidly over impermeable surfaces into drains and gutters.

Section 1.4 Activities

Study Figure 1.25. Describe and explain the changes in flood frequency and flood magnitude that occur as urbanisation increases.

Abstraction

Water availability problems occur when the demand for water exceeds the amount available during a certain period. This happens in areas with low rainfall and high population density, and in areas where there is intensive agricultural or industrial activity. Over-abstraction may lead to the drying up of rivers, falling water tables and saltwater intrusion in coastal areas.

In many parts of Europe, groundwater is the main source of fresh water. However, in many places water is being taken from the ground faster than it is being replenished.

Saline intrusion is widespread along the Mediterranean coastlines of Italy, Spain and Turkey (Figure 1.26), where the demands of tourist resorts are the major cause of over-abstraction. In Malta, most groundwater can no longer be used for domestic consumption or irrigation because it has been contaminated by saline intrusion. Consequently, Malta now has to use desalinated water. Intrusion of saline water due to excessive extraction of water is also a problem in northern countries, notably Denmark.

Irrigation is the main cause of groundwater overexploitation in agricultural areas. In Italy, overexploitation of the Po River in the region of the Milan aquifer has led to a 25 metre decrease in groundwater levels over the last 80 years.

Changing groundwater

Human activity has seriously reduced the long-term viability of irrigated agriculture in the High Plains of Texas. Before irrigation development started in the 1930s, the High Plains groundwater system was stable, in a state of dynamic equilibrium with long-term recharge equal to long-term discharge. However, groundwater is now being used at a rapid rate to supply **centre-pivot irrigation schemes**. In under 50 years, the water level has declined by 30–50 metres in a large area to the north of Lubbock, Texas. The aquifer has narrowed by more than 50 per cent

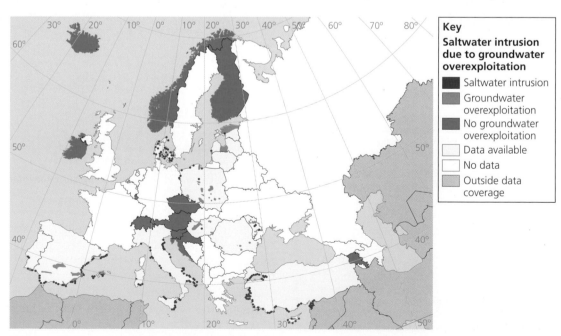

Figure 1.26 Groundwater abstraction and saline intrusion in Western Europe

in large parts of certain counties, and the area irrigated by each well is contracting as well as yields falling.

By contrast, in some industrial areas, recent reductions in industrial activity have led to less groundwater being taken out of the ground. As a result, groundwater levels in such areas have begun to rise, adding to the problem caused by leakage from ancient, deteriorating pipe and sewerage systems. Such a rise has numerous implications, including:

- increase in spring and river flows
- re-emergence of flow from 'dry springs'
- surface water flooding
- pollution of surface waters and the spread of underground pollution
- flooding of basements
- increased leakage into tunnels
- reduction in stability of slopes and retaining walls

- reduction in bearing capacity of foundations and piles
- swelling of clays as they absorb water
- chemical attack on building foundations.

There are various methods of recharging groundwater resources, provided that sufficient surface water is available. Where the materials containing the aquifer are permeable (as in some alluvial fans, coastal sand dunes or glacial deposits), water-spreading (a form of infiltration and seepage) is used. By contrast, in sediments with impermeable layers, such water-spreading techniques are not effective, and the appropriate method may then be to pump water into deep pits or into wells. This method is used extensively on the heavily settled coastal plain of Israel, both to replenish the groundwater reservoirs when surplus irrigation water is available, and in an attempt to diminish the problems associated with saltwater intrusions from the Mediterranean.

Case Study: Changing hydrology of the Aral Sea

The Aral Sea began shrinking in the 1960s when Soviet irrigation schemes took water from the Syr Darya and the Amu Darya rivers. This greatly reduced the amount of water reaching the Aral Sea. By 1994, the shorelines had fallen by 16 metres, the surface area had declined by 50 per cent and the volume had been reduced by 75 per cent (Figure 1.27). By contrast, salinity levels had increased by 300 per cent.

Increased salinity levels killed off the fishing industry. Moreover, ports such as Muynak are now tens of kilometres from the shore. Salt from the dry seabed has reduced soil fertility and frequent dust storms are ruining the region's cotton production. Drinking water has been polluted by pesticides and fertilisers and the air has been affected by dust and salt. There has been a noticeable rise in respiratory and stomach disorders and the region has one of the highest infant mortality rates in the former Soviet Union.

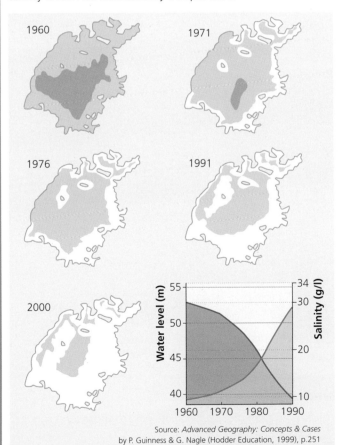

Source: *Advanced Geography: Concepts & Cases* by P. Guinness & G. Nagle (Hodder Education, 1999), p.251

Figure 1.27 The changing hydrology of the Aral Sea

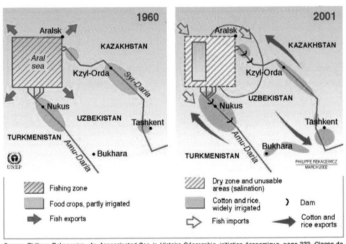

Source: Philippe Rekacewicz, *An Assassinated Sea*, in Histoire-Géographie, initiation économique, page 333, Classe de Troisième, Hatier, Paris, 1993 (data updated in 2002); L'état du Monde, 1992 and 2001 editions, La Découverte, Paris.

Source: quoted at www.columbia.edu/~tmt2120/impacts%20to%20life%20in%20the%20region.htm

Figure 1.28 The economic impacts of the shrinking sea

Study Figures 1.27 and 1.28.

1 Why do you think the Former Soviet Union (FSU) embarked on such a programme of large-scale irrigation? Use an atlas to produce detailed information.
2 Why have salinity levels increased so much?
3 What problems does the shrinking of the Aral Sea cause for towns such as Aralsk and Muynak?
4 What is the likely effect of the irrigation scheme on the two rivers in terms of velocity, erosion, sediment transport and deposition?

Water storage – dams

The number of large dams (more than 15 metres high) that are being built is increasing rapidly and is reaching a level of almost two completions every day (Figure 1.29).

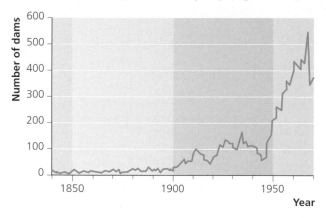

Figure 1.29 The trend in building large dams

The advantages of dams are numerous, as the following examples from the Aswan High Dam on the River Nile, Egypt, show:

■ **flood and drought control** – dams allow good crops in dry years as, for example, in Egypt in 1972 and 1973
■ **irrigation** – 60 per cent of water from the Aswan Dam is used for irrigation and up to 4000 km of the desert are irrigated
■ **hydro-electric power** – this accounts for 7000 million kW hours each year
■ **improved navigation**
■ **recreation and tourism**.

It is estimated that the value of the Aswan High Dam is about $500 million to the Egyptian economy each year.

On the other hand, there are numerous disadvantages. For example:

■ **water losses** – the dam provides less than half the amount of water expected
■ **salinisation** – crop yields have been reduced on up to one-third of the area irrigated by water from the Aswan Dam, due to salinisation

■ **groundwater changes** – seepage leads to increased groundwater levels and may cause secondary salinisation
■ **displacement of population** – up to 100 000 Nubian people have been removed from their ancestral homes
■ **drowning of archaeological sites** – the tombs of Ramases II and Nefertari at Abu Simbel had to be removed to safer locations – however, the increase in the humidity of the area has led to an increase in the weathering of ancient monuments
■ **seismic stress** – the earthquake of November 1981 is believed to have been caused by the Aswan Dam; as water levels in the Dam increase so too does seismic activity
■ **deposition within the lake** – infilling is taking place at about 100 million tonnes each year
■ **channel erosion (clear-water erosion) on the channel bed** – lowering the channel by 25 mm over 18 years, a modest amount
■ **erosion of the Nile delta** – this is taking place at a rate of about 2.5 cm each year
■ **loss of nutrients** – it is estimated that it costs $100 million to buy commercial fertilisers to make up for the lack of nutrients each year
■ **decreased fish catches** – sardine yields are down 95 per cent and 3000 jobs in Egyptian fisheries have been lost
■ **diseases have spread** – such as schistosomiasis (bilharzia).

Figure 1.30 Paphos dam, Cyprus

1 Study Figure 1.29. Describe the pattern shown and suggest reasons to explain the trend.
2 Evaluate the effectiveness of large dams.

☐ Flood risk

Floods are one of the most common of all environmental hazards. This is because so many people live in fertile river valleys and in low-lying coastal areas. For much of the time, rivers act as a resource. However, extremes of too much water – or too little – can be considered a hazard (Figure 1.31).

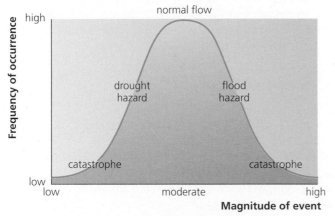

Figure 1.31 River discharge and frequency

In addition, extreme events occur infrequently. Many urban areas are designed to cope with floods that occur on a regular basis, perhaps annually or once in a decade. Most are ill-equipped to deal with the low-frequency/high-magnitude event that may occur once every 100 years or every 500 years (Figure 1.32). The **recurrence interval** refers to the regularity of a flood of a given size. Small floods may be expected to occur regularly. Larger floods occur less often. A 100-year flood is the flood that is expected to occur, *on average*, once every 100 years. Increasingly, larger floods are less common, but more damaging.

The nature and scale of flooding varies greatly. For example, less than 2 per cent of the population of England and Wales and in Australia live in areas exposed to flooding, compared with 10 per cent of the US population. The worst problems occur in Asia where floods damage about 4 million hectares of land each year and affect the lives of over 17 million people. Worst of all is China, where over 5 million people have been killed in floods since 1860.

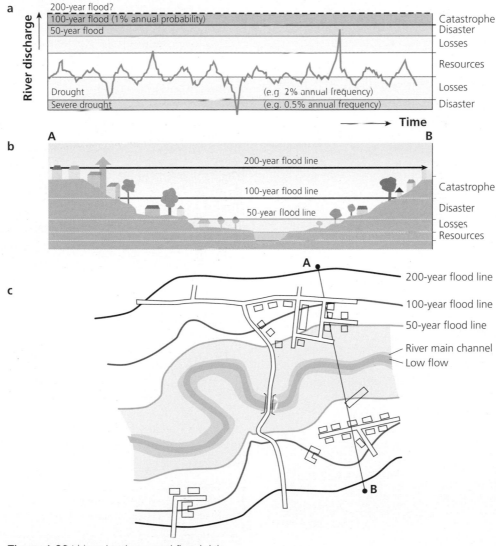

Figure 1.32 Urban land use and flood risk

Some environments are more at risk than others. The most vulnerable include the following:

- Low-lying parts of active floodplains and river estuaries. For example, in Bangladesh 110 million people living on the floodplain of the Ganges and Brahmaputra rivers are relatively unprotected. Floods caused by the monsoon regularly cover 20–30 per cent of the flat delta. In very high floods, up to half of the country may be flooded. In 1988, 46 per cent of the land was flooded and more than 1500 people were killed.
- Small basins subject to **flash floods**. These are especially common in arid and semi-arid areas. In tropical areas, some 90 per cent of lives lost through drowning are the result of intense rainfall on steep slopes.
- Areas below unsafe dams. In the USA, there are about 30 000 large dams and 2000 communities are at risk from dams. Following the 2008 Sichuan earthquake in China, some 35 quake dams were created by landslides blocking river routes. These were eventually made safe by engineers and the Chinese military.
- Low-lying inland shorelines such as along the Great Lakes and the Great Salt Lake in the USA.

In most high-income countries (HICs), the number of deaths from floods is declining, while in contrast the economic cost of flood damage has been increasing. In low-income countries (LICs), on the other hand, the death rate due to flooding is much greater, although the economic cost is not as great. It is likely that the hazard in LICs will increase over time as more people migrate and settle in low-lying areas and river basins. Often newer migrants are forced into the more hazardous zones.

Since the Second World War (1939–45), there has been a change in the understanding of the flood hazard, in the attitude towards floods and in the policy towards reducing the flood hazard. The response to hazards has moved away from physical control (engineering structures) towards reducing vulnerability through non-structural approaches.

☐ Causes of flooding

A flood is a high flow of water that overtops the bank of a river. The main causes of floods are climatic forces, whereas the flood-intensifying conditions tend to be drainage basin specific (Figure 1.33). Most floods in the UK, for example, are associated with deep **depressions** (low pressure systems) that are both long-lasting and cover a wide area. By contrast, in India up to 70 per cent of the annual rainfall occurs in three months during the summer monsoon. In Alpine and Arctic areas, melting snow is responsible for widespread flooding.

Flood-intensifying conditions cover a range of factors, which alter the drainage basin response to a given storm (Figure 1.34). The factors that influence the storm hydrograph determine the response of the basin to the storm. These factors include topography, vegetation, soil type, rock type and characteristics of the drainage basin.

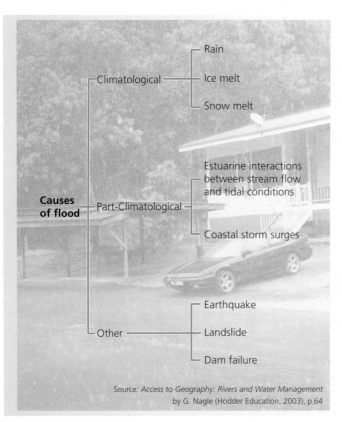

Source: *Access to Geography: Rivers and Water Management* by G. Nagle (Hodder Education, 2003), p.64

Figure 1.33 The causes of floods

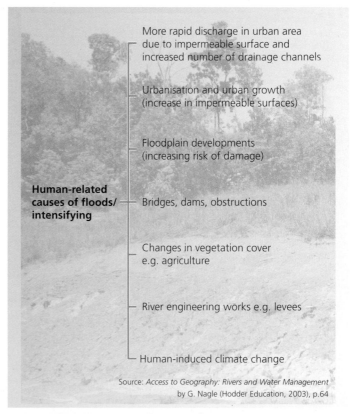

Source: *Access to Geography: Rivers and Water Management* by G. Nagle (Hodder Education, 2003), p.64

Figure 1.34 Flood-intensifying conditions

1 Hydrology and fluvial geomorphology

The potential for damage by floodwaters increases exponentially with velocity. The physical stresses on buildings are increased even more when rough, rapidly flowing water contains debris such as rocks, sediment and trees.

Other conditions that intensify floods include changes in land use. Urbanisation, for example, increases the magnitude and frequency of floods in at least three ways:

- creation of highly impermeable surfaces, such as roads, roofs, pavements
- smooth surfaces served with a dense network of drains, gutters and underground sewers increase drainage density
- natural river channels are often constricted by bridge supports or riverside facilities, reducing their carrying capacity.

Deforestation is also a cause of increased flood runoff and a decrease in channel capacity. This occurs due to an increase in deposition within the channel. However, the evidence is not always conclusive. In the Himalayas, for example, changes in flooding and increased deposition of silt in parts of the lower Ganges–Brahmaputra are due to the combination of high monsoon rains, steep slopes and the seismically unstable terrain. These ensure that runoff is rapid and sedimentation is high, irrespective of the vegetation cover.

☐ The prevention and amelioration of floods

Forecasting and warning

During the 1980s and 1990s, flood forecasting and warning had become more accurate and these are now among the most widely used measures to reduce the problems caused by flooding. Despite advances in **weather satellites** and the use of radar for forecasting, over 50 per cent of all unprotected dwellings in England and Wales have less than six hours of flood warning time. In most LICs there is much less effective flood forecasting. An exception is Bangladesh. Most floods in Bangladesh originate in the Himalayas, so authorities have about 72 hours' warning.

According to the United Nations Environment Programme's publication *Early Warning and Assessment*, there are a number of things that could be done to improve flood warnings. These include:

- improved rainfall and snow pack estimates, and better and longer forecasts of rainfall
- better gauging of rivers, collection of meteorological information and mapping of channels
- better and current information about human populations and infrastructure; elevation and stream channels need to be incorporated into flood-risk assessment models

- better sharing of information is needed between forecasters, national agencies, relief organisations and the general public
- more complete and timely sharing of information of meteorological and hydrological information is needed among countries within international drainage basins
- technology should be shared among all agencies involved in flood forecasting and risk assessment, both in the basins and throughout the world.

Loss sharing

Economic growth and population movements throughout the twentieth century have caused many floodplains to be built on. However, for people to live on floodplains there needs to be flood protection. This can take many forms, such as loss-sharing adjustments and event modifications.

Loss-sharing adjustments include disaster aid and insurance. **Disaster aid** refers to any aid, such as money, equipment, staff and technical assistance, that is given to a community following a disaster. In HICs, **insurance** is an important loss-sharing strategy. However, not all flood-prone households have insurance and many of those that are insured may be underinsured.

Hard engineering

Traditionally, floods have been managed by methods of 'hard engineering'. This largely means dams, levees, wing dykes and **straightened channels** that are wider and deeper than the ones they replace. In some cases, new diversion spillways (flood-relief channels and intercepting channels) may be built (Figure 1.35). Although hard engineering may reduce floods in some locations, it may cause unexpected effects elsewhere in the drainage basin, for example decreased water quality, increased sedimentation, bed and bank erosion and loss of habitats.

Levees are the most common form of river engineering. They can also be used to divert and restrict water to low-value land on the floodplain. Over 4500 kilometres of the Mississippi River have levees. Channel improvements such as channel enlargement will increase the carrying capacity of the river. **Reservoirs** store excess rainwater in the upper drainage basin. However, this may only be appropriate in small drainage networks. It has been estimated that some 66 billion m³ of storage is needed to make any significant impact on major floods in Bangladesh!

Hazard-resistant design

Flood-proofing includes any adjustments to buildings and their contents that help reduce losses. Some are temporary, such as:

- blocking up entrances
- sealing doors and windows
- removal of damageable goods to higher levels
- use of sandbags.

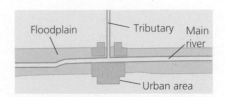

Floodplain | Tributary | Main river
Urban area

1 Flood embankments with sluice gates. The main problem with this is that it may raise flood levels up and down.

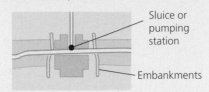

Sluice or pumping station
Embankments

2 Channel enlargement to accommodate larger discharges. One problem with such schemes is that as the enlarged channel is only rarely used it becomes clogged with weed.

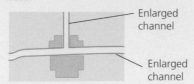

Enlarged channel
Enlarged channel

3 Flood relief channels. This is appropriate where it is impossible to modify the original channel due to cost, e.g. the flood relief channels around Oxford.

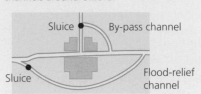

Sluice | By-pass channel
Sluice | Flood-relief channel

4 Intercepting channels. These are in use during times of flood, diverting part of the flow away, allowing flow for town and agricultural use, e.g. the Great Ouse Protection Scheme in the Fenlands

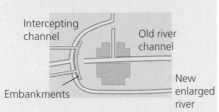

Intercepting channel
Old river channel
Embankments
New enlarged river

5 Flood storage reservoirs. This solution is widely used, especially as many reservoirs created for water-supply purposes may have a secondary flood control, e.g. the intercepting channels along the Loughton Brook.

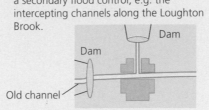

Dam
Dam
Old channel

6 The removal of settlements. This is rarely used because of cost, although many communities were forced to leave as a result of the 1993 Mississippi floods.

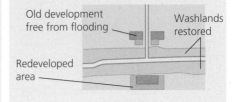

Old development free from flooding
Washlands restored
Redeveloped area

Figure 1.35 Channel diversions

Source: *Access to Geography: Rivers and Water Management* by G. Nagle (Hodder Education, 2003), p.65

Figure 1.36 Flood-relief channel, Zermatt, Switzerland

By contrast, long-term measures include moving the living spaces above the likely level of the floodplain. This normally means building above the flood level, but could also include building homes on stilts.

Land-use zoning

Most land-use zoning and land-use planning has been introduced since the Second World War. In the USA, land-use management has been effective in protecting new housing developments from 1 in 100-year floods (that is, the size of flood that we would expect to occur once every century).

One example where partial urban relocation has occurred is at Soldier's Grove on the Kickapoo River in south-western Wisconsin, USA. The town experienced a series of floods in the 1970s, and the Army Corps of Engineers proposed building two levees and moving part of the urban area. Following floods in 1978, they decided that relocation of the entire business district would be better than just flood-damage reduction. Although levees would have protected the village from most floods, they would not have provided other opportunities. Relocation allowed energy conservation and an increase in commercial activity in the area.

Soft engineering

Soft engineering generally refers to working with natural processes and features rather than attempts to control them. They include the management of whole catchments (catchment management plans), wetland conservation and river restoration.

Event modification adjustments include environmental control and hazard-resistant design. Physical control of floods depends on two measures: flood abatement and flood diversion. **Flood abatement** involves decreasing the amount of runoff, thereby reducing the flood peak in a **drainage basin**. There are a number of ways of reducing flood peaks. These include:

- reforestation
- reseeding of sparsely vegetated areas to increase evaporative losses
- treatment of slopes such as by contour ploughing or terracing to reduce runoff
- comprehensive protection of vegetation from wildfires, overgrazing and clear-cutting of forests

- clearance of sediment and other debris from headwater streams
- construction of small water- and sediment-holding areas
- preservation of natural water-storage zones, such as lakes.

Flood diversion refers to the practice of allowing certain areas, such as wetlands and floodplains, to be flooded to a greater extent. Natural flooding may be increased through the use of flood-relief channels (diversion spillways) to direct more water into these areas during times of flood.

River restoration

Case Study: Costs and benefits of the Kissimmee River restoration scheme

Figure 1.37 Part of the restored Kissimmee Restoration Scheme, Florida, USA

The 165 kilometre Kissimmee River once meandered through central Florida. Its floodplain, reaching up to 5 kilometres wide, was inundated for long periods by heavy seasonal rains. Wetland plants, wading birds and fish thrived there, but the frequent, prolonged flooding caused a severe impact on people.

Between 1962 and 1971, engineering changes were made to deepen, straighten and widen the river, which was transformed into a 90 kilometre, 10 metre-deep drainage canal. The river was **channelised** to provide an outlet canal for draining floodwaters from the developing upper Kissimmee lakes basin, and to provide flood protection for land adjacent to the river.

Impacts of channelisation

The channelisation of the Kissimee River had several unintended impacts:

- the loss of 2000 to 14 000 hectares of wetlands
- a 90 per cent reduction in wading bird and waterfowl usage
- a continuing long-term decline in game fish populations.

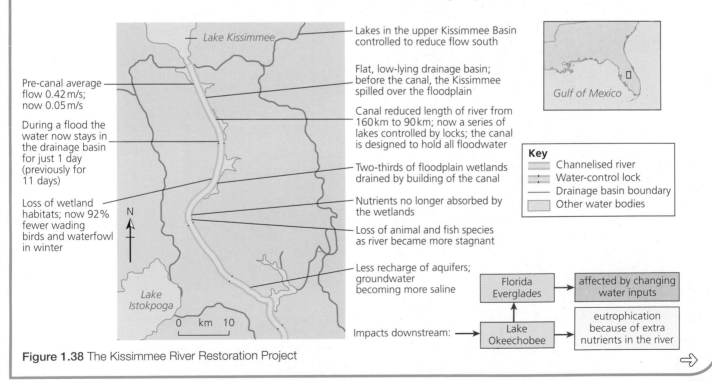

Figure 1.38 The Kissimmee River Restoration Project

Concerns about the **sustainability** of existing ecosystems led to a state and federally supported restoration study. The result was a massive restoration project, on a scale unmatched elsewhere.

The project restored over 100 km² of river and associated floodplain wetlands. It was started in 1999 and completed in 2015. It benefits over 320 fish and wildlife species, including the endangered bald eagle, wood stork and snail kite. It has created over 11 000 hectares of wetlands. Seasonal rains and flows now inundate the floodplain in the restored areas.

Restoration of the river and its associated natural resources required **dechannelisation**. This entailed backfilling approximately half of the flood-control channel and re-establishing the flow of water through the natural river channel. In residential areas, the flood-control channel will remain in place.

The costs of restoration

It is estimated that the project cost over $400 million (initial channelisation cost $20 million), a bill being shared by the state of Florida and the federal government.

Restoration of the river's floodplain could result in higher losses of water due to evapotranspiration during wet periods. In extremely dry spells, navigation may be impeded in some sections of the restored river. It is, however, expected that navigable depths will be maintained at least 90 per cent of the time.

Benefits of restoration

- Higher water levels should ultimately support a natural river ecosystem again.
- Re-establishment of floodplain wetlands and the associated nutrient filtration function is expected to result in decreased nutrient loads to Lake Okeechobee.
- Populations of key avian species, such as wading birds and waterfowl, have returned to the restored area, and in some cases numbers have more than tripled.
- Dissolved oxygen levels have doubled, which is critical for the survival of fish and other aquatic species.
- Potential revenue associated with increased recreational usage (such as hunting and fishing) and ecotourism on the restored river could significantly enhance local and regional economies.

Section 1.4 Activities

1 Outline the natural and human causes of floods.
2 Compare and contrast methods of flood management.
3 To what extent can flood frequency and magnitude be predicted?

4 Outline the disadvantages of channelisation as shown in Figure 1.36.
5 Outline the benefits of wetlands.
6 What is meant by *river restoration*? What are the benefits of river restoration?

Case Study: Flooding in Bangladesh

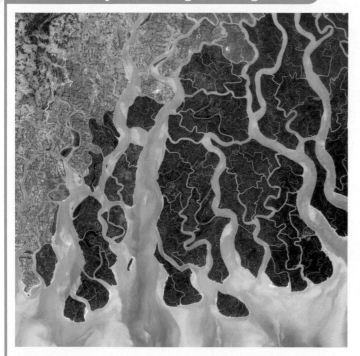

Figure 1.39 Satellite image of the 1998 floods

Bangladesh is a small, flat and low-lying country: 60 per cent is less than 6 metres above sea level. For this reason, tides affect one-third of land area. Banglasdesh is located where the Ganga, Brahmaputra (called *Jamuna* in Bangladesh) and Meghna rivers meet. The average gradient of the rivers is 6 cm/km. The country drains an area 12 times its own size. It has a high frequency of floods and cyclones.

Table 1.5 Bangladesh factfile

Area	143 998 km²
Population	166 million (2014)
Age structure	51 % under 25 years of age
Population density	1161/km²
Annual growth rate	0.6 %
Literacy	58.8 %
PPP	$3400
Life expectancy	70.65 years
Employment	Agriculture: 47 %
	Industry: 13 %
	Services: 40 %

Source: CIA World Factbook

Bangladesh has a high population density, low human development index (HDI) and a majority of the population is dependent on agriculture. An area of about 150 000 km² is shared by 123 million people.

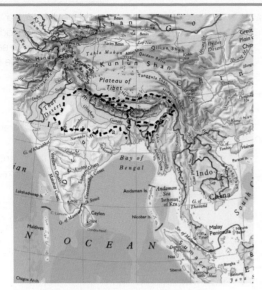

Source: Philip's Interactive Modern School Atlas by G. Nagle
(Hodder Education, 2006) © Philip's

Figure 1.40 The Ganges drainage basin

Several regions affect conditions within Bangladesh:

- **high plateau of Tibet** – the source of the Brahmaputra, where most of the river flow derives from snow melt and glacier melt
- **Himalayas** – source of the Ganga and many of the springs that feed into the Brahmaputra
- **Ganga Plain** – one of the largest lowland areas in the world, and a region of intense cultivation
- **Meghalaya Hills** – located between the floodplain of north east Bangladesh and the Indian lowlands of Assam; rise to a height of 2500 m and act as a barrier to the monsoon winds from the Indian Ocean; Cherrapungee has an annual rainfall of over 11 000 mm.

Table 1.6 Watershed characteristics of the Ganga and the Brahmaputra/Meghna (Br/M) rivers and a comparison with the Nile, the Amazon and the Mississippi

Characteristics	Ganga	Br/M	Nile	Amazon	Mississippi
Basin area (km²)	1016104	651334	3254555	6144727	3202230
Length (km)	2296	2772	5964	4406	4240
Average annual discharge (m³/s)	11365	19772	2760	176177	17600
Forest (%)	4	19	2	73	22
Cropland (%)	7	29	10	15	35
Cropland irrigated (%)	15	47	5	0	4
Grassland (%)	7	29	52	8	22
Large dams	6	0	7	2	2091

Source: Hofer, T. and Messerli, B, 2006, Floods in Bangladesh, *United Nations University Press, Table 2.2*

The Ganges and the Brahmaputra are two of the world's largest rivers by catchment size, length, amount of flow, sediment discharge (the Brahmaputra carries 540 million tons of sediment per year, the Ganges 520 million tons) and lateral shifting.

Causes of flooding

There are many causes of flooding in Bangladesh (Figure 1.41), which originate in three areas – the highlands (Himalayas); the Indian Plains of the Ganges and the Brahmaputra; and the floodplains of Bangladesh.

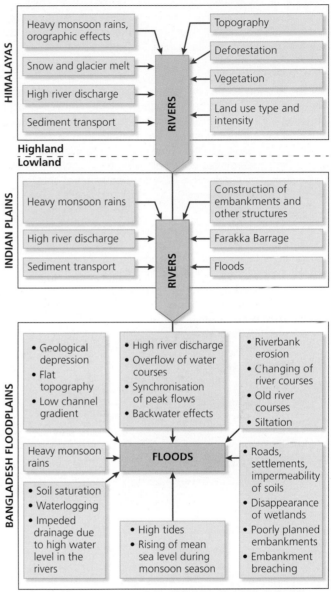

Source: Hofer, T and Messerli, B, 2006, *Floods in Bangladesh*, United Nations University Press

Figure 1.41 The causes of floods in the Ganges

Flooding in Bangladesh alternates between periods of high flood frequency and low flood frequency.

- Floods in the western part of Bangladesh were more intensive in the eighteenth and nineteenth centuries than in the twentieth or twenty-first centuries.
- Massive floods occurred regularly long before human impact on the watershed began; there is no evidence that flood frequency is increasing.
- The variation in the extent of flooding year by year has been increasing since the 1950s.

- There is increasing monsoon rain, particularly in the Brahmaputra–Meghna system.
- There is a worsening impact on the Bangladeshi people, but human influence in the Himalayas is not thought to be increasing flooding in Bangladesh.

Flooding is viewed very differently by rural people and by politicians and engineers. For many rural people, flooding is a short-term necessity for their crops; engineers and politicians see the damage it causes to infrastructure and the economy.

The 1998 floods

These were the longest lasting and most devastating floods in 100 years; 1998 was a La Niña year, in which normal circulatory patterns are intensified. The most-affected areas of Bangladesh included the capital Dhaka and other areas close to the main rivers; 53 of the 64 districts of the country – that is, about 50 per cent – were affected, by up to 3 metres of water for up to 67 days. The flooding on 7 September was probably the worst of the twentieth century.

The main causes were:

- the high peaks on all three main rivers occurring at the same time
- high tides causing the river floods to back up
- a strong monsoon that caused excessive flooding, and obstructions by man-made infrastructure.

Table 1.7 Major impacts of the 1998 floods

Number of people affected	c.30 million
Number of deaths	c.780–1500
Number facing malnutrition	25 million
Rice production loss	2.2 million tons
Damage to cultivated area	1.5 million ha
Loss of livestock sector	$500 million
Roads damaged	15 000 km
Embankments damaged	c.4500 km
Bridges/culverts damaged	>20 000
Villages damaged	30 000
Houses damaged	550 000–900 000

Source: Hofer, T. and Messerli, B, 2006, Floods in Bangladesh, *United Nations University Press, Table 2.2*

Coping with flooding in Bangladesh

- Many houses, and also many roads, are built on raised platforms, above the level of the average flood. People who live on islands mainly use bamboo and reeds for their houses, which can be dismantled in about an hour in an emergency.
- Rural people cultivate different varieties of rice, some of which can grow in floodwaters of 1 metre and grow up to

20 centimetres a day to keep up with the rising water level; jute and sugar cane can also withstand submergence.

- It takes up to three days for floodwaters to rise, giving people some time to prepare, such as raising platforms in their homes so that they can sleep on dry ground.
- Levees can prevent overflow but may cause deposition in the channel, which raises the river bed and reduces the capacity of the river.
- Levees can give a false sense of security – they protect against minor floods but not against major ones.
- The Flood Action Plan 1989–95 led to the development of the Bangladesh Water and Flood Management Strategy Report, which stated that there are three main water-resource development options:
  Minimum intervention – improve forecasting and improve existing flood schemes but do not create new ones
  Selective intervention – protect densely populated areas, key infrastructure and water supplies
  Major intervention – build large-scale engineering works on all main rivers.
- In terms of existing measures, there are currently over 10 000 kilometres of levees and a number of raised flood and cyclone shelters.
- Groynes in rivers protect important townships.
- Non-structural measures include flood forecasting, preparation and relief. The Flood Forecasting Warning Centre issues five-day forecasts during the monsoon season.
- Up to 20 per cent of the population is at risk from lateral erosion, which is more predictable on the Ganga than on the braided Brahmaputra. Many families may be forced to move 10 to 15 times during their lifetime.

Social problems

- **Loss of land**, leading to loss of social status and poverty, which can prevent a family's children from being able to marry.
- **Food shortages**, leading to reliance on relatives and neighbours.

Section 1.4 Activities

1 Study Table 1.6. Outline the main differences in watershed characteristics of the rivers. Suggest how these differences may affect the flood hazard.
2 Study Figure 1.41. Outline the main physical and human causes of flooding in Bangladesh.
3 Describe the main impacts of the 1998 floods in Bangladesh. Why were the impacts so great?
4 Evaluate the opportunities for flood control in Bangladesh.

2 Atmosphere and weather

2.1 Diurnal energy budgets

An **energy budget** refers to the amount of energy entering a system, the amount leaving the system and the transfer of energy within the system. Energy budgets are commonly considered at a global scale (macro-scale) and at a local scale (micro-scale). However, the term **microclimate** is sometimes used to describe regional climates, such as those associated with large urban areas, coastal areas and mountainous regions.

Figure 2.1 shows a classification of climate and weather phenomena at a variety of spatial and temporal scales. Phenomena vary from small-scale turbulence and eddying (such as dust devils) that cover a small area and last for a very short time, to large-scale **anticyclones** (high-pressure zones) and **jet streams** that affect a large area and may last for weeks. The jet stream that carried volcanic dust from underneath the Eyjafjallajökull glacier in Iceland to northern Europe in 2010 is a good example of jet-stream activity (Figure 2.2).

Flights over northern Europe were disrupted after a cloud of volcanic ash from Iceland's Eyjafjallsjökull volcano – drifting from 6000 to 11 000 metres high – closed airports and caused flights to be cancelled.

Figure 2.2 Jet-stream activity and the transfer of dust from Eyjafjallajökull, Iceland

These different scales should not be considered as separate scales but as a hierarchy of scales in which smaller phenomena may exist within larger ones. For example, the temperature surrounding a building will be affected by the nature of the building and processes that are taking place within the building. However, it will also be affected by the wider synoptic (weather) conditions, which are affected by latitude, **altitude**, **cloud** cover and season, for example.

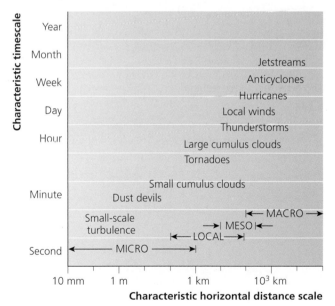

Figure 2.1 Classification of climate and weather phenomena at a variety of spatial and temporal scales

☐ Daytime and night-time energy budgets

There are six components to the daytime energy budget:

- incoming (shortwave) solar **radiation** (insolation)
- reflected solar radiation
- surface absorption
- **sensible heat transfer**
- **long-wave radiation** (Figure 2.3)
- latent heat (evaporation and condensation).

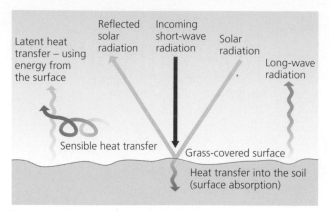

Figure 2.3 Local energy budget – daytime

These influence the gain or loss of energy for a point at the Earth's surface. The daytime energy budget assumes a horizontal surface with grass-covered soil and can be expressed by the formula:

energy available at the surface = incoming solar radiation – (reflected solar radiation + surface absorption + sensible heat transfer + long-wave radiation + latent heat transfers)

In contrast, the night-time energy budget consists of four components:

- long-wave Earth radiation
- **latent heat transfer** (condensation)
- absorbed energy returned to Earth (sub-surface supply)
- sensible heat transfer (Figure 2.4).

Incoming (shortwave) solar radiation

Incoming solar radiation (insolation) is the main energy input and is affected by latitude, season and cloud cover (Section 2.2). Figure 2.5 shows how the amount of insolation received varies with the angle of the Sun and with cloud type. For example, with strato-cumulus clouds (like those in Figure 2.6) when the Sun is low in the sky, about 23 per cent of the total radiation transmitted is received at the Earth's surface – about 250 watts per m². When the Sun is high in the sky, about 40 per cent is received – just over 450 watts per m². The less cloud cover there is, and/or the higher the cloud, the more radiation reaches the Earth's surface.

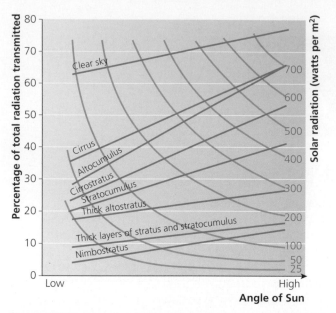

Figure 2.5 Energy, cloud cover/type and the angle of the Sun

Figure 2.6 Stratocumulus clouds

Reflected solar radiation

The proportion of energy that is reflected back to the atmosphere is known as the albedo. The albedo varies with colour – light materials are more reflective than dark materials (Table 2.1). Grass has an average albedo of 20–30 per cent, meaning that it reflects back about 20–30 per cent of the radiation it receives.

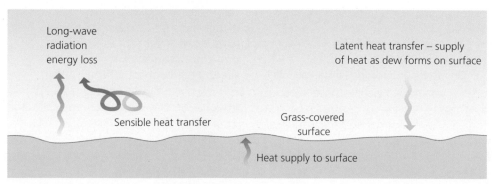

Figure 2.4 Night-time energy budget

Table 2.1 Selected albedo values

Surface	Albedo (%)
Water (Sun's angle over 40°)	2–4
Water (Sun's angle less than 40°)	6–80
Fresh snow	75–90
Old snow	40–70
Dry sand	35–45
Dark, wet soil	5–15
Dry concrete	17–27
Black road surface	5–10
Grass	20–30
Deciduous forest	10–20
Coniferous forest	5–15
Crops	15–25
Tundra	15–20

Section 2.1 Activities

1 The model for the daytime energy budget assumes a flat surface with grass-covered soil. Suggest reasons for this assumption.
2 Study Table 2.1.
  a What is meant by the term *albedo*?
  b Why is albedo important?

Surface and sub-surface absorption

Energy that reaches the Earth's surface has the potential to heat it. Much depends on the nature of the surface. For example, if the surface can conduct heat to lower layers, the surface will remain cool. If the energy is concentrated at the surface, the surface warms up.

The heat transferred to the soil and bedrock during the day may be released back to the surface at night. This can partly offset the night-time cooling at the surface.

Sensible heat transfer

Sensible heat transfer refers to the movement of parcels of air into and out of the area being studied. For example, air that is warmed by the surface may begin to rise (**convection**) and be replaced by cooler air. This is known as a convective transfer. It is very common in warm areas in the early afternoon. Sensible heat transfer is also part of the night-time energy budget: cold air moving into an area may reduce temperatures, whereas warm air may supply energy and raise temperatures.

Long-wave radiation

Long-wave radiation refers to the radiation of energy from the Earth (a cold body) into the atmosphere and, for some of it, eventually into space. There is, however, a downward movement of long-wave radiation from particles in the atmosphere. The difference between the two flows is known as the net long-wave radiation balance. During the day, the outgoing long-wave radiation transfer is greater than the incoming long-wave radiation transfer, so there is a net loss of energy from the surface.

During a cloudless night, there is a large loss of long-wave radiation from the Earth. There is very little return of long-wave radiation from the atmosphere, due to the lack of clouds. Hence there is a net loss of energy from the surface. In contrast, on a cloudy night the clouds return some long-wave radiation to the surface, hence the overall loss of energy is reduced. Thus in hot desert areas, where there is a lack of cloud cover, the loss of energy at night is maximised. In contrast, in cloudy areas the loss of energy (and change in daytime and night-time temperatures) is less noticeable.

Latent heat transfer (evaporation and condensation)

When liquid water is turned into water vapour, heat energy is used up. In contrast, when water vapour becomes a liquid, heat is released. Thus when water is present at a surface, a proportion of the energy available will be used to evaporate it, and less energy will be available to raise local energy levels and temperature.

During the night, water vapour in the air close to the surface can condense to form water, since the air has been cooled by the cold surface. When water condenses, latent heat is released. This affects the cooling process at the surface. In some cases, evaporation may occur at night, especially in areas where there are local sources of heat.

Dew

Dew refers to condensation on a surface. The air is saturated, generally because the temperature of the surface has dropped enough to cause condensation. Occasionally, condensation occurs because more moisture is introduced, for example by a sea breeze, while the temperature remains constant.

Absorbed energy returned to Earth

The insolation received by the Earth will be reradiated as long-wave radiation. Some of this will be absorbed by water vapour and other **greenhouse gases**, thereby raising the temperature.

☐ Temperature changes close to the surface

Ground-surface temperatures can vary considerably between day and night. During the day, the ground heats the air by radiation, **conduction** (contact) and convection. The ground radiates energy and as the air receives more radiation than it emits, the air is warmed. Air close to the ground is also warmed through conduction. Air movement at the surface is slower due to friction with the surface,

2.2 The global energy budget

☐ The latitudinal pattern of radiation: excesses and deficits

The atmosphere is an open energy system, receiving energy from both Sun and Earth. Although the latter is very small, it has an important local effect, as in the case of urban climates. **In**coming **sol**ar rad**iation** is referred to as **insolation**.

The atmosphere constantly receives solar energy, yet until recently the atmosphere was not getting any hotter. Therefore there has been a balance between inputs (insolation) and outputs (re-radiation) (Figure 2.9). Under 'natural' conditions the balance is achieved in three main ways:

- **radiation** – the emission of electromagnetic waves such as X-ray, short- and long-wave; as the Sun is a very hot body, radiating at a temperature of about 5700°C, most of its radiation is in the form of very short wavelengths such as ultraviolet and visible light
- **convection** – the transfer of heat by the movement of a gas or liquid
- **conduction** – the transfer of heat by contact.

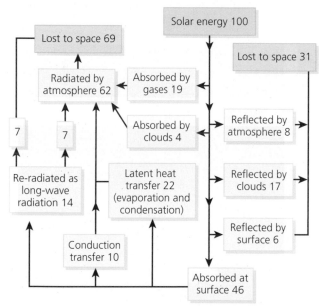

Source: *Access to Geography – Climate and Society* by G. Nagle (Hodder Education, 2002) p.15

Figure 2.9 The Earth's energy budget

Of incoming radiation, 19 per cent is absorbed by atmospheric gases, especially oxygen and ozone at high altitudes, and carbon dioxide and water vapour at low altitudes. Reflection by the atmosphere accounts for a net loss of 8 per cent, and clouds and water droplets reflect 23 per cent. Reflection from the Earth's surface (known as the **planetary albedo**) is generally about 6 per cent. About 36 per cent of insolation is reflected back to space and a further 19 per cent is absorbed by

atmospheric gases. So only about 46 per cent of the insolation at the top of the atmosphere actually gets through to the Earth's surface.

Energy received by the Earth is re-radiated at long wavelength. (Very hot bodies such as the Sun emit short-wave radiation, whereas cold bodies such as the Earth emit long-wave radiation.) Of this, 8 per cent is lost to space. Some energy is absorbed by clouds and re-radiated back to Earth. Evaporation and condensation account for a loss of heat of 22 per cent. There is also a small amount of condensation (carried up by turbulence). Thus heat gained by the atmosphere from the ground amounts to 32 per cent of incoming radiation.

The atmosphere is largely heated from below. Most of the incoming short-wave radiation is let through, but some outgoing long-wave radiation is trapped by greenhouse gases. This is known as the greenhouse principle or **greenhouse effect**.

There are important variations in the receipt of solar radiation with latitude and season (Figure 2.10). The result is an imbalance: an excess of radiation (positive budget) in the tropics; a **deficit** of radiation (negative balance) at higher latitudes (Figure 2.11). However, neither region is getting progressively hotter or colder. To achieve this balance, the horizontal transfer of energy from the equator to the poles takes place by winds and ocean currents. This gives rise to an important second energy budget in the atmosphere: the horizontal transfer between low latitudes and high latitudes to compensate for differences in global insolation.

Latitude
Areas that are close to the equator receive more heat than areas that are close to the poles. This is due to two reasons:

1. incoming solar radiation (insolation) is concentrated near the equator, but dispersed near the poles.

2. insolation near the poles has to pass through a greater amount of atmosphere and there is more chance of it being reflected back out to space.

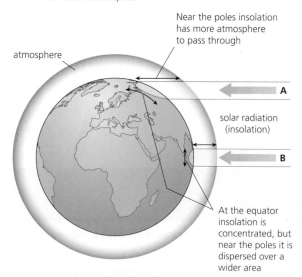

Source: Nagle, G., *Geography through diagrams*, OUP, 1998

Figure 2.10 Latitudinal contrasts in insolation

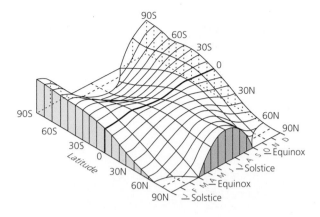

The variations of solar radiation with latitude and season for the whole globe, assuming no atmosphere. This assumption explains the abnormally high amounts of radiation received at the poles in summer, when daylight lasts for 24 hours each day.

Source: Barry, R. and Chorley, R., *Atmosphere, Weather and Climate*, Routledge, 1998

Figure 2.11 Contrasts in insolation by season and latitude

Section 2.2 Activities

1 Outline the main thermal differences between short-wave and long-wave radiation.
2 Study Figures 2.10 and 2.11. Comment on latitudinal differences in the receipt of solar radiation.

Annual temperature patterns

There are important large-scale north–south temperature zones (Figure 2.12). For example, in January highest temperatures over land (above 30 °C) are found in Australia and southern Africa. By contrast, the lowest temperatures (less than –40 °C) are found over parts of Siberia, Greenland and the Canadian Arctic. In general, there is a decline in temperatures northwards from the Tropic of Capricorn, although there are important anomalies, such as the effect of the Andes in South America, and the effect of the cold current off the coast of Namibia. In July, maximum temperatures are found over the Sahara, Near East, northern India and parts of southern USA and Mexico. By contrast, areas in the southern hemisphere are cooler than in January.

These patterns reflect the general decrease of insolation from the equator to the poles. There is little seasonal variation at the equator, but in mid or high latitudes large seasonal differences occur due to the decrease in insolation from the equator to the poles, and changes in the length of day. There is also a time lag between the overhead Sun and the period of maximum insolation – up to two months in some places – largely because the air is heated from below, not above. The coolest period is after the winter solstice (the shortest day), since the ground continues to lose heat even after insolation has resumed. Over oceans, the lag time is greater than over the land, due to differences in their specific heat capacities.

Section 2.2 Activities

Describe the differences in temperature as shown in Figure 2.12. Suggest reasons for these contrasts.

☐ Atmospheric transfers

There are two main influences on atmospheric transfer: pressure variations and ocean currents. Air blows from high pressure to low pressure, and is important in redistributing heat around the Earth. In addition, the atmosphere is influenced by ocean currents – warm currents raise the temperature of overlying air, while cold currents cool the air above them (see pages 39–40).

Pressure variations

Pressure is measured in millibars (mb) and is represented by isobars, which are lines of equal pressure. On maps, pressure is adjusted to mean sea level (MSL), therefore eliminating elevation as a factor. MSL pressure is 1013 mb, although the mean range is from 1060 mb in the Siberian winter high-pressure system to 940 mb (although some intense low pressure storms may be much lower). The trend of pressure change is more important than the actual reading itself. Decline in pressure indicates poorer weather, and rising pressure better weather.

Surface pressure belts

Sea-level pressure conditions show marked differences between the hemispheres. In the northern hemisphere there are greater seasonal contrasts, whereas in the southern hemisphere much simpler average conditions exist (see Figure 2.13). Over Antarctica there is generally high pressure over the 3–4 kilometre-high eastern Antarctic Plateau, but the high pressure is reduced by altitude. The differences are largely related to unequal distribution of land and sea, because ocean areas are much more equable in terms of temperature and pressure variations.

One of the more permanent features is the subtropical high-pressure (STHP) belts, especially over ocean areas. In the southern hemisphere these are almost continuous at about 30° latitude, although in summer over South Africa and Australia they tend to be broken. Generally pressure is about 1026 mb. In the northern hemisphere, by contrast, at 30° the belt is much more discontinuous because of the land. High pressure only occurs over the ocean as discrete cells such as the Azores and Pacific highs. Over continental areas such as south-west USA, southern Asia and the Sahara, major fluctuations occur: high pressure in winter, and summer lows because of overheating.

Over the equatorial trough, pressure is low: 1008–1010 mb. The trough coincides with the zone of maximum insolation. In the northern hemisphere (in July) it is well north of the equator (25 °C over India), whereas in the southern hemisphere (in January) it is just south of the equator because land masses in the southern hemisphere are not

Latitudinal variations in the ITCZ occur as a result of the movement of the overhead Sun. In June the ITCZ lies further north, whereas in December it lies in the southern hemisphere. The seasonal variation in the ITCZ is greatest over Asia, owing to its large land mass. By contrast, over the Atlantic and Pacific Oceans its movement is far less. Winds at the ITCZ are generally light (the doldrums), occasionally broken by strong westerlies, generally in the summer months.

Low-latitude winds between 10° and 30° are mostly easterlies; that is, they flow towards the west. These are the reliable trade winds; they blow over 30 per cent of the world's surface. The weather in this zone is fairly predictable: warm, dry mornings and showery afternoons, caused by the continuous evaporation from tropical seas. Showers are heavier and more frequent in the warmer summer season.

Occasionally there are disruptions to the pattern; easterly waves are small-scale systems in the easterly flow of air. The flow is greatest not at ground level but at the 700 mb level. Ahead of the easterly wave, air is subsiding; hence there is surface divergence. At the easterly wave, there is convergence of air, and ascent – as in a typical low pressure system. Easterly waves are important for the development of tropical cyclones (Section 9.3).

Westerly winds dominate between 35° and 60° of latitude, which accounts for about a quarter of the world's surface. However, unlike the steady trade winds, these contain rapidly evolving and decaying depressions.

The word 'monsoon' means 'reverse'; the monsoon is reversing wind systems. For example, the south-east trades from the southern hemisphere cross the equator in July. Owing to the Coriolis force, these south-east trades are deflected to the right in the northern hemisphere and become south-west winds. The monsoon is induced by Asia – the world's largest continent – which causes winds to blow outwards from high pressure in winter, but pulls the southern trades into low pressure in the summer.

The monsoon is therefore influenced by the reversal of land and sea temperatures between Asia and the Pacific during the summer and winter. In winter, surface temperatures in Asia may be as low as −20°C. By contrast, the surrounding oceans have temperatures of 20°C. During the summer, the land heats up quickly and may reach 40°C. By contrast, the sea remains cooler at about 27°C. This initiates a land–sea breeze blowing from the cooler sea (high pressure) in summer to the warmer land (low pressure), whereas in winter air flows out of the cold land mass (high pressure) to the warm water (low pressure). The presence of the Himalayan Plateau also disrupts the strong winds of the upper atmosphere, forcing winds either to the north or south and consequently deflecting surface winds.

The uneven pattern shown in Figure 2.14 is the result of seasonal variations in the overhead Sun. Summer in the southern hemisphere means that there is a cooling in the northern hemisphere, thereby increasing the differences between polar and equatorial air. Consequently, high-level westerlies are stronger in the northern hemisphere in winter.

> ## Section 2.2 Activities
>
> Describe the main global wind systems shown in Figure 2.14.

☐ Explaining variations in temperature, pressure and winds

Latitude

On a global scale, latitude is the most important factor determining temperature (Figure 2.10). Two factors affect the temperature: the angle of the overhead Sun and the thickness of the atmosphere. At the equator, the overhead Sun is high in the sky, so the insolation received is of a greater quality or intensity. At the poles, the overhead Sun is low in the sky, so the quality of energy received is poor. Secondly, the thickness of the atmosphere affects temperature. Energy has more atmosphere to pass through at point A on Figure 2.10, so more energy is lost, scattered or reflected by the atmosphere than at B – therefore temperatures are lower at A than at B. In addition, the albedo (reflectivity) is higher in polar regions. This is because snow and ice are very reflective, and low-angle sunlight is easily reflected from water surfaces. However, variations in length of day and season partly offset the lack of intensity in polar and arctic regions. The longer the Sun shines, the greater the amount of insolation received, which may overcome in part the lack of intensity of insolation in polar regions. (On the other hand, the long polar nights in winter lose vast amounts of energy.)

Land–sea distribution

There are important differences in the distribution of land and sea in the northern hemisphere and southern hemisphere. There is much more land in the northern hemisphere. Oceans cover about 50 per cent of the Earth's surface in the northern hemisphere but about 90 per cent of the southern hemisphere (Figure 2.15). This is not always clear when looking at conventional map projections such as the Mercator projection.

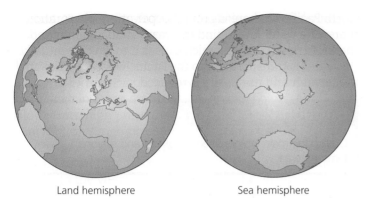

Land hemisphere   Sea hemisphere

Figure 2.15 Land and sea hemispheres

The distribution of land and sea is important because land and water have different thermal properties. The specific heat capacity is the amount of heat needed to raise the temperature of a body by 1°C. There are important differences between the heating and cooling of water. Land heats and cools more quickly than water. It takes five times as much heat to raise the temperature of water by 2°C as it does to raise land temperatures.

Water heats more slowly because:

- it is clear, so the Sun's rays penetrate to great depth, distributing energy over a wider area
- tides and currents cause the heat to be further distributed.

Therefore a larger volume of water is heated for every unit of energy than the volume of land, so water takes longer to heat up. Distance from the sea has an important influence on temperature. Water takes up heat and gives it back much more slowly than the land. In winter, in mid-latitudes sea air is much warmer than the land air, so onshore winds bring heat to the coastal lands. By contrast, during the summer coastal areas remain much cooler than inland sites. Areas with a coastal influence are termed **maritime** or **oceanic**, whereas inland areas are called **continental**.

Ocean currents

Surface ocean currents are caused by the influence of **prevailing winds** blowing steadily across the sea. The dominant pattern of surface ocean currents (known as **gyres**) is roughly a circular flow. The pattern of these currents is clockwise in the northern hemisphere and anti-clockwise in the southern hemisphere. The main exception is the circumpolar current that flows around Antarctica from west to east. There is no equivalent current in the northern hemisphere because of the distribution of land and sea there. Within the circulation of the gyres, water piles up into a dome. The effect of the rotation of the Earth is to cause water in the oceans to push westward; this piles up water on the western edge

of ocean basins – rather like water slopping in a bucket. The return flow is often narrow, fast-flowing currents such as the Gulf Stream. The Gulf Stream in particular transports heat northwards and then eastwards across the North Atlantic; the Gulf Stream is the main reason that the British Isles have mild winters and relatively cool summers (Figure 2.16).

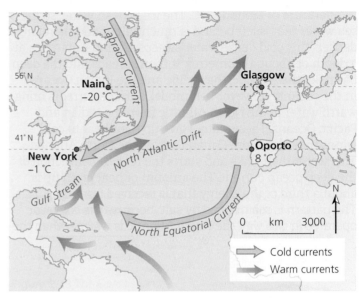

The effect of an ocean current depends upon whether it is a warm current or a cold current. Warm currents move away from the equator, whereas cold currents move towards it. The cold Labrador Current reduces the temperatures of the western side of the Atlantic, while the warm North Atlantic Drift raises temperatures on the eastern side.

Source: Nagle, G., *Geography through diagrams*, OUP, 1998

Figure 2.16 The effects of the North Atlantic Drift/Gulf Stream

The effect of ocean currents on temperatures depends upon whether the current is cold or warm. Warm currents from equatorial regions raise the temperature of polar areas (with the aid of prevailing westerly winds). However, the effect is only noticeable in winter. For example, the North Atlantic Drift raises the winter temperatures of north-west Europe. By contrast, other areas are made colder by ocean currents. Cold currents such as the Labrador Current off the north-east coast of North America may reduce summer temperatures, but only if the wind blows from the sea to the land.

In the Pacific Ocean, there are two main atmospheric states. The first is warm surface water in the west with cold surface water in the east; the other is warm surface water in the east with cold in the west. In both cases, the warm surface causes low pressure. As air blows from high pressure to low pressure, there is a movement of water from the colder area to the warmer area. These winds push warm surface water into the warm region, exposing colder deep water behind them and maintaining the pattern.

General circulation model

In general:

- warm air is transferred polewards and is replaced by cold air moving towards the equator
- air that rises is associated with low pressure, whereas air that sinks is associated with high pressure
- low pressure produces rain; high pressure produces dry conditions.

Any circulation model must take into account the meridional (north/south) transfer of heat, and latitudinal variations in rainfall and winds. (Any model is descriptive and static – unlike the atmosphere.) In 1735, George Hadley described the operation of the Hadley cell, produced by the direct heating over the equator. The air here is forced to rise by convection, travels polewards and then sinks at the subtropical anticyclone (high-pressure belt). Hadley suggested that similar cells might exist in mid-latitudes and high latitudes. William Ferrel suggested that Hadley cells interlink with a mid-latitude cell, rotating it in the reverse direction, and these cells in turn rotate the polar cell.

There are very strong differences between surface and upper winds in tropical latitudes. Easterly winds at the surface are replaced by westerly winds above, especially in winter. At the ITCZ, convectional storms lift air into the atmosphere, which increases air pressure near the tropopause, causing winds to diverge at high altitude. They move out of the equatorial regions towards the poles, gradually losing heat by radiation. As they contract, more air moves in and the weight of the air increases the air pressure at the subtropical high-pressure zone (Figure 2.20). The denser air sinks, causing subsidence (**stability**). The north/south component of the Hadley cell is known as a meridional flow. The Ferrel Cell was originally considered to be a thermally indirect cell (driven by the Hadley cell and polar cell). Now it is known to be more complex, and there is some equator-ward movement of air related to temperate high- and low-pressure systems. These are related to Rossby waves and jet streams (Figure 2.20c).

The zonal flow (east–west) over the Pacific was discovered by Gilbert Walker in the 1920s. The Southern Oscillation Index (SOI) is a measure of how far temperatures vary from the 'average'. A high SOI is associated with strong westward trades (because winds near the equator blow from high pressure to low pressure and are unaffected by the Coriolis force). Tropical cyclones are more common in the South Pacific when there is an **El Niño** Southern Oscillation warm episode.

The polar cell is found in high latitudes. Winds at the highest latitudes are generally easterly. Air over the North Pole continually cools; and being cold, it is dense and therefore it subsides, creating high pressure. Air above the polar front flows back to the North Pole, creating a polar cell. In between the Hadley cell and the polar cell is an indirect cell, the Ferrel cell, driven by the movement of the other two cells, rather like a cog in a chain.

In the early twentieth century, researchers investigated patterns and mechanisms of upper winds and clouds at an altitude of between 3 and 12 kilometres. They identified large-scale fast-moving belts of westerly winds, which follow a ridge and trough wave-like pattern known as Rossby waves or planetary waves (Figure 2.21). The presence of these winds led to Rossby's 1941 model of the atmosphere. This suggested a three-cell north/south (meridional) circulation, with two thermally direct cells and one thermally indirect cell. The thermally direct cell is driven by the heating at the equator (the Hadley cell) and by the sinking of cold air at the poles (the polar cell). Between them lies the thermally indirect cell whose energy is obtained from the cells to either side by the mixing of the atmosphere at upper levels. The jet streams are therefore key locations in the transfer of energy through the atmosphere. Further modifications of Rossby's models were made by Palmen in 1951.

New models change the relative importance of the three convection cells in each hemisphere. These changes are influenced by jet streams and Rossby waves:

- Jet streams are strong, regular winds that blow in the upper atmosphere about 10 km above the surface; they blow between the poles and tropics (100–300 km/h).
- There are two jet streams in each hemisphere – one between 30° and 50°; the other between 20° and 30°. In the northern hemisphere, the polar jet and the subtropical jet flow eastwards.
- Rossby waves are 'meandering rivers of air' formed by westerly winds. There are three to six waves in each hemisphere. They are formed by major relief barriers such as the Rockies and the Andes, by thermal differences and uneven land–sea interfaces.

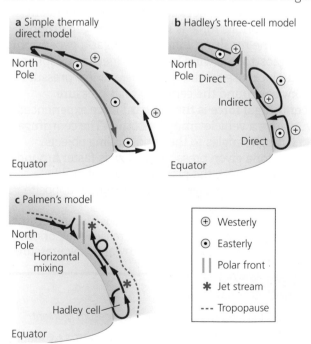

Figure 2.20 General circulation model

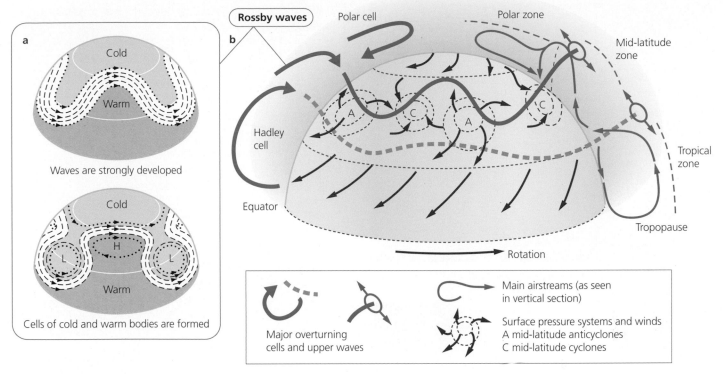

Figure 2.21 Rossby waves

- The jet streams result from differences in equatorial and sub-tropical air, and between polar and sub-tropical air. The greater the temperature difference, the stronger the jet stream.

Rossby waves are affected by major topographic barriers such as the Rockies and the Andes. Mountains create a wave-like pattern, which typically lasts six weeks. As the pattern becomes more exaggerated (Figure 2.21b), it leads to blocking anticyclones (blocking highs) – prolonged periods of unusually warm weather.

Jet streams and Rossby waves are an important means of mixing warm and cold air.

Section 2.2 Activities

1 Describe and explain how the Hadley cell operates.
2 Define the term *Rossby wave*. Suggest how an understanding of Rossby waves may help in our understanding of the general circulation.

2.3 Weather processes and phenomena

☐ Atmospheric moisture processes

Atmospheric moisture exists in all three states – vapour, liquid and solid (Figures 2.22–2.24). Energy is used in the change from one phase to another, for example between a liquid and a gas. In evaporation, heat is absorbed. It takes 600 calories of heat to change 1 gram of water from a liquid to a vapour. Heat loss during evaporation passes into the water as latent heat (of vaporisation). This would cool 1 kilogram of air by 2.5 °C. By contrast, when condensation occurs, latent heat locked in the water vapour is released, causing a rise in temperature. In the changes between vapour and ice, heat is released when vapour is converted to ice (solid), for example rime at high altitudes and high latitudes. In contrast, heat is absorbed in the process of sublimation, for example when snow patches disappear without melting. When liquid water turns to ice, heat is released and temperatures drop. In contrast, in melting ice heat is absorbed and temperatures rise.

Figure 2.22 Atmospheric moisture – condensation

Figure 2.23 Radiation fog in the lower part of alpine valleys

Figure 2.24 Moisture in its liquid state – Augher Lake, Gap of Dunloe, Killarney, Ireland

Factors affecting evaporation

Evaporation occurs when vapour pressure of a water surface exceeds that in the atmosphere. Vapour pressure is the pressure exerted by the water vapour in the atmosphere. The maximum vapour pressure at any temperature occurs when the air is saturated (Figure 2.25). Evaporation aims to equalise the pressures. It depends on three main factors:

- **initial humidity of the air** – if air is very dry then strong evaporation occurs; if it is saturated then very little occurs
- **supply of heat** – the hotter the air, the more evaporation that takes place
- **wind strength** – under calm conditions the air becomes saturated rapidly.

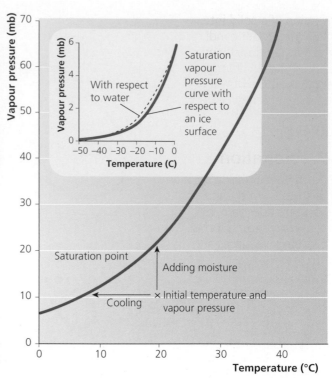

The curves demonstrate how much moisture the air can hold for any temperature. Below 0 °C the curve is slightly different for an ice surface than for a supercooled water droplet.

Source: Briggs et al., *Fundamentals of the physical environment*, Routledge, 1997

Figure 2.25 Maximum vapour pressure

Factors affecting condensation

Condensation occurs when either **a** enough water vapour is evaporated into an **air mass** for it to become saturated or **b** when the temperature drops so that dew point (the temperature at which air is saturated) is reached. The first is relatively rare; the second common. Such cooling occurs in three main ways:

- radiation cooling of the air
- contact cooling of the air when it rests over a cold surface
- adiabatic (expansive) cooling of air when it rises.

Condensation is very difficult to achieve in pure air. It requires some tiny particle or nucleus onto which the vapour can condense. In the lower atmosphere these are quite common, such as sea salt, dust and pollution particles. Some of these particles are hygroscopic – that is, water-seeking – and condensation may occur when the relative humidity is as low as 80 per cent.

Other processes

Freezing refers to the change of liquid water into a solid, namely ice, once the temperature falls below 0 °C. **Melting** is the change from a solid to a liquid when the air temperature rises above 0 °C. **Sublimation** is the conversion of a solid into a vapour with no intermediate liquid state. Under conditions of low humidity, snow can be evaporated directly into water vapour without entering the liquid state. Sublimation is also used to describe the direct **deposition** of water vapour onto ice. In some cases, water droplets may be deposited directly onto natural features (such as plants and animals) as well as built structures (for example buildings and vehicles).

☐ Precipitation

The term 'precipitation' refers to all forms of deposition of moisture from the atmosphere in either solid or liquid states. It includes rain, hail, snow and dew. Because rain is the most common form of precipitation in many areas, the term is sometimes applied to rainfall alone. For any type of precipitation, except dew, to form, clouds must first be produced.

When minute droplets of water are condensed from water vapour, they float in the atmosphere as clouds. If droplets coalesce, they form large droplets that, when heavy enough to overcome by gravity an ascending current, they fall as rain. Therefore cloud droplets must get much larger to form rain. There are a number of theories to suggest how raindrops are formed.

The Bergeron theory suggests that for rain to form, water and ice must exist in clouds at temperatures below 0°C. Indeed, the temperature in clouds may be as low as −40°C. At such temperatures, water droplets and ice droplets form. Ice crystals grow by condensation and become big enough to overcome turbulence and cloud updrafts, so they fall. As they fall, crystals coalesce to form larger snowflakes. These generally melt and become rain as they pass into the warm air layers near the ground. Thus, according to Bergeron, rain comes from clouds that are well below freezing at high altitudes, where the coexistence of water and ice is possible. The snow/ice melts as it passes into clouds at low altitude where the temperatures are above freezing level.

Other mechanisms must also exist as rain also comes from clouds that are not so cold. Mechanisms include:

- condensation on extra-large hygroscopic nuclei
- coalescence by sweeping, whereby a falling droplet sweeps up others in its path
- the growth of droplets by electrical attraction.

Causes of precipitation

The Bergeron theory relates mostly to snow-making. **Snow** is a single flake of frozen water. Rain and drizzle are found when the temperature is above 0°C (drizzle has a diameter of < 0.5 mm). **Sleet** is partially melted snow.

There are three main types of rainfall: **convectional**, **frontal (depressional)** and **orographic (relief)** (Figure 2.26).

Convectional rainfall

When the land becomes very hot, it heats the air above it. This air expands and rises. As it rises, cooling and condensation take place. If it continues to rise, rain will fall. It is very common in tropical areas (Figure 2.27) and is associated with the permanence of the ITCZ. In temperate areas, convectional rain is more common in summer.

Frontal or cyclonic rainfall

Frontal rain occurs when warm air meets cold air. The warm air, being lighter and less dense, is forced to rise over the cold, denser air. As it rises, it cools, condenses and forms rain. It is most common in middle and high latitudes where warm tropical air and cold polar air converge.

Orographic (or relief) rainfall

Air may be forced to rise over a barrier such as a mountain. As it rises, it cools, condenses and forms rain. There is often a **rainshadow** effect, whereby the leeward slope receives a relatively small amount of rain. Altitude is important, especially on a local scale. In general, there are increases of precipitation up to about 2 kilometres. Above this level, rainfall decreases because the air temperature is so low.

Thunderstorms (intense convectional rainfall)

Thunderstorms are special cases of rapid cloud formation and heavy precipitation in unstable air conditions. Absolute or **conditional instability** exists to great heights, causing strong updraughts to develop within cumulonimbus clouds. Air continues to rise as long as it is saturated (relative humidity is 100 per cent; that is, it has reached its dew point). Thunderstorms are especially common in tropical and warm areas where air can hold large amounts of water. They are rare in polar areas.

Several stages can be identified (Figure 2.28):

1 **Developing stage:** updraught caused by uplift; energy (latent heat) is released as condensation occurs; air becomes very unstable; rainfall occurs as cloud temperature is greater than 0°C; the great strength of uplift prevents snow and ice from falling.
2 **Mature stage:** sudden onset of heavy rain and maybe thunder and lightning; rainfall drags cold air down with it; upper parts of the cloud may reach the tropopause; the cloud spreads, giving the characteristic anvil shape.
3 **Dissipating stage:** downdraughts prevent any further convective instability; the new cells may be initiated by the meeting of cold downdraughts from cells some distance apart, triggering the rise of warm air in between.

Lightning occurs to relieve the tension between different charged areas, for example between cloud and ground or within the cloud itself. The upper parts of the cloud are positive, whereas the lower parts are negative. The very base of the cloud is positively charged. The origin of the charges is not very clear, although they are thought to be

a Frontal or cyclonic

Warm air rises over cold air; it expands, cools and condensation takes place, clouds and rain form

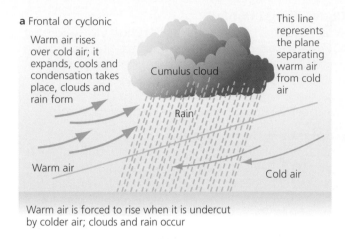

Cumulus cloud

This line represents the plane separating warm air from cold air

Rain

Warm air

Cold air

Warm air is forced to rise when it is undercut by colder air; clouds and rain occur

Figure 2.27 Convectional rain in Brunei

b Relief or orographic

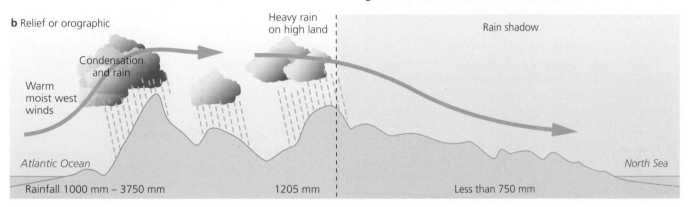

Heavy rain on high land

Rain shadow

Condensation and rain

Warm moist west winds

Atlantic Ocean

North Sea

Rainfall 1000 mm – 3750 mm

1205 mm

Less than 750 mm

c Convectional

When the land becomes hot it heats the air above it. This air expands and rises. As it rises, cooling and condensation takes place. If it continues to rise rain will fall. It is common in tropical areas. In the UK it is quite common in the summer, especially in the South East.

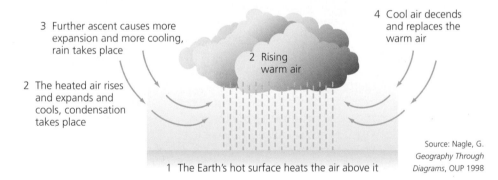

3 Further ascent causes more expansion and more cooling, rain takes place

4 Cool air decends and replaces the warm air

2 Rising warm air

2 The heated air rises and expands and cools, condensation takes place

1 The Earth's hot surface heats the air above it

Source: Nagle, G. *Geography Through Diagrams*, OUP 1998

Figure 2.26 Types of precipitation

due to condensation and evaporation. Lightning heats the air to very high temperatures. Rapid expansion and vibration of the column of air produces thunder.

Section 2.3 Activities

Using diagrams, explain the meaning of the terms
a *convectional rainfall*, **b** *orographic rainfall*, **c** *frontal rainfall*.

Clouds

Clouds are formed of millions of tiny water droplets held in suspension. They are classified in a number of ways, the most important being:

- **form or shape**, such as stratiform (layers) and cumuliform (heaped type)
- **height**, such as low (<2000 m), medium or alto (2000–7000 m) and high (7000–13 000 m).

There are a number of different types of clouds (Figure 2.29). High clouds consist mostly of ice crystals. Cirrus are wispy clouds, and include cirrocumulus (mackerel sky) and cirrostratus (halo effect around the Sun or Moon). Alto or middle-height clouds generally consist of water drops. They exist at temperatures lower than 0 °C. Low clouds indicate poor weather. Stratus clouds are dense, grey and low lying (Figure 2.30). Nimbostratus are those that produce rain ('nimbus' means 'storm'). Stratocumulus are long cloud rolls, and a mixture of stratus and cumulus (see Figure 2.6 in Section 2.1).

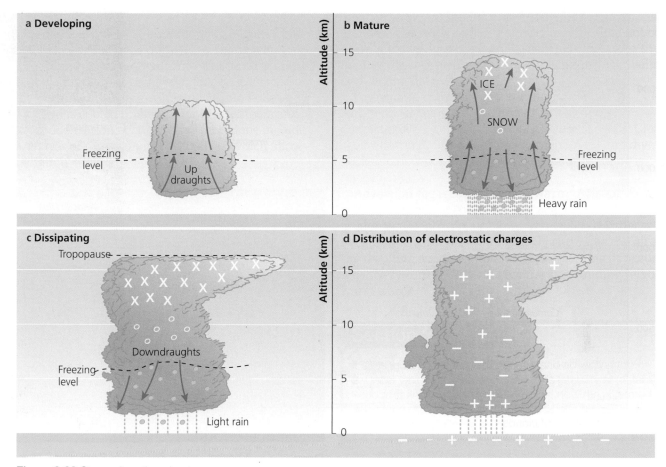

Figure 2.28 Stages in a thunderstorm

Vertical development suggests upward movement. Cumulus clouds are flat-bottomed and heaped. They indicate bright brisk weather. Cumulonimbus clouds produce heavy rainfall and often thunderstorms.

The important facts to keep in mind:

- In unstable conditions, the dominant form of uplift is convection and this may cause cumulus clouds.
- With stable conditions, stratiform clouds generally occur.
- Where fronts are involved, a variety of clouds exist.
- Relief or topography causes stratiform or cumuliform clouds, depending on the stability of the air.

Banner clouds

These are formed by orographic uplift (that is, air forced to rise, over a mountain for example) under stable air conditions. Uplifted moist air streams reach condensation only at the very summit, and form a small cloud. Further downwind the air sinks, and the cloud disappears. Wave clouds reflect the influence of the topography on the flow of air.

Types of precipitation

Rain

Rain refers to liquid drops of water with a diameter of between 0.5 millimetres and 5 millimetres. It is heavy enough to fall to the ground. Drizzle refers to rainfall with a diameter of less than 0.5 millimetres. Rainfall varies in terms of total amount, seasonality, intensity, duration and effectiveness; that is, whether there is more rainfall than potential evapotranspiration. (Refer back to page 45 for more information on the three main types of rainfall.)

Hail

Hail is alternate concentric rings of clear and opaque ice, formed by raindrops being carried up and down in vertical air currents in large cumulonimbus clouds. Freezing and partial melting may occur several times before the pellet is large enough to escape from the cloud. As the raindrops are carried high up in the cumulonimbus cloud they freeze. The hailstones may collide with droplets of supercooled water, which freeze on impact with and form a layer of opaque ice around the hailstone. As the hailstone falls, the outer layer may be melted but may freeze again with further uplift. The process can occur many times before the hail finally falls to ground, when its weight is great enough to overcome the strong updraughts of air.

Snow

Snow is frozen precipitation (Figure 2.31). Snow crystals form when the temperature is below freezing and water vapour is converted into a solid. However, very cold air contains a limited amount of moisture, so the heaviest snowfalls tend to occur when warm moist air is forced over very high mountains or when warm moist air comes into contact with very cold air at a front.

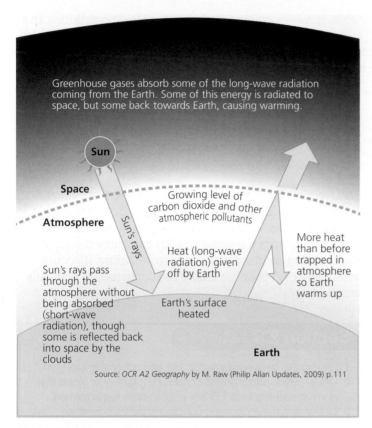

Greenhouse gases absorb some of the long-wave radiation coming from the Earth. Some of this energy is radiated to space, but some back towards Earth, causing warming.

Sun

Space

Atmosphere

Sun's rays

Growing level of carbon dioxide and other atmospheric pollutants

Heat (long-wave radiation) given off by Earth

More heat than before trapped in atmosphere so Earth warms up

Sun's rays pass through the atmosphere without being absorbed (short-wave radiation), though some is reflected back into space by the clouds

Earth's surface heated

Earth

Source: OCR A2 Geography by M. Raw (Philip Allan Updates, 2009) p.111

Figure 2.34 The greenhouse effect

The enhanced greenhouse effect is built up of certain greenhouse gases as a result of human activity (Table 2.2). **Carbon dioxide** (CO_2) levels have risen from about 315ppm (parts per million) in 1950 to over 400ppm and are expected to reach 600ppm by 2050. The increase is due to human activities: burning fossil fuels (coal, oil and natural gas) and deforestation. Deforestation of the tropical rainforest is a double blow – not only does it increase atmospheric CO_2 levels, it also removes the trees that convert CO_2 into oxygen.

Methane is the second largest contributor to global warming, and is increasing at a rate of between 0.5 and 2 per cent per annum. Cattle alone give off between 65 and 85 million tonnes of methane per year. Natural wetland and paddy fields are another important source – paddy fields emit up to 150 million tonnes of methane annually. As global warming increases, bogs trapped in permafrost will melt and release vast quantities of methane. **Chlorofluorocarbons** (CFCs) are synthetic chemicals that destroy ozone, as well as absorb long-wave radiation. CFCs are increasing at a rate of 6 per cent per annum, and are up to 10 000 times more efficient at trapping heat than CO_2.

As long as the amount of water vapour and carbon dioxide stay the same and the amount of solar energy remains the same, the temperature of the Earth should remain in equilibrium. However, human activities are upsetting the natural balance by increasing the amount of CO_2 in the atmosphere, as well as the other greenhouse gases.

How human activities add to greenhouse gases

Much of the evidence for the greenhouse effect has been taken from ice cores dating back 160 000 years. These show that the Earth's temperature closely paralleled the levels of CO_2 and methane in the atmosphere. Calculations indicate that changes in these greenhouse gases were part, but not all, of the reason for the large (5°–7°) global temperature swings between ice ages and interglacial periods.

Accurate measurements of the levels of CO_2 in the atmosphere began in 1957 in Hawaii. The site chosen was far away from major sources of industrial pollution and shows a good representation of unpolluted atmosphere. The trend in CO_2 levels shows a clear annual pattern, associated with seasonal changes in vegetation, especially those over the northern hemisphere. By the 1970s there was a second trend, one of a long-term increase in CO_2 levels, superimposed upon the annual trends.

Studies of cores taken from ice packs in Antarctica and Greenland show that the level of CO_2 between 10 000 years ago and the mid-nineteenth century was stable, at about 270ppm. By 1957, the concentration of CO_2 in the atmosphere was 315ppm, and it has since risen to about 360ppm. Most of the extra CO_2 has come

Table 2.2 Properties of key greenhouse gases

	Average atmospheric concentration (ppmv)	Rate of change (% per annum)	Direct global warming potential (GWP)	Lifetime (years)	Type of indirect effect
Carbon dioxide	400	0.5	1	120	None
Methane	1.72	0.6–0.75	11	10.5	Positive
Nitrous oxide	0.31	0.2–0.3	270	132	Uncertain
CFC-11	0.000255	4	3400	55	Negative
CFC-12	0.000453	4	7100	116	Negative
CO				Months	Positive
NOx					Uncertain

from the burning of fossil fuels, especially coal, although some of the increase may be due to the disruption of the rainforests. For every tonne of carbon burned, 4 tonnes of CO_2 are released.

By the early 1980s, 5 gigatonnes (5000 million tonnes, or 5 Gt) of fuel were burned every year. Roughly half the CO_2 produced is absorbed by natural sinks, such as vegetation and plankton.

Other factors have the potential to affect climate too. For example, a change in the albedo (reflectivity of the land brought about by desertification or deforestation) affects the amount of solar energy absorbed at the Earth's surface. Aerosols made from sulphur, emitted largely in fossil-fuel combustion, can modify clouds and may act to lower temperatures. Changes in ozone in the stratosphere due to CFCs may also influence climate.

Since the Industrial Revolution, the combustion of fossil fuels and deforestation have led to an increase of 26 per cent of CO_2 concentration in the atmosphere (Figure 2.35). Emissions of CFCs used as aerosol propellants, solvents, refrigerants and foam-blowing agents are also well known. They were not present in the atmosphere before their invention in the 1930s. The sources of methane and nitrous oxides are less well known. Methane concentrations have more than doubled because of rice production, cattle rearing, biomass burning, coal mining and ventilation of natural gas. Fossil fuel combustion may have also contributed through chemical reactions in the atmosphere, which reduce the rate of removal of methane. Nitrous oxide has increased by about 8 per cent since pre-industrial times, presumably due to human activities. The effect of ozone on climate is strongest in the upper troposphere and lower stratosphere.

- The increasing carbon dioxide in the atmosphere since the pre-industrial era, from about 280 to 382 ppmv (parts per million by volume), makes the largest individual contribution to greenhouse gas radiative forcing: $1.56 W/m^2$ (watts per square metre).
- The increase of methane (CH_4) since pre-industrial times (from 0.7 to 1.7 ppmv) contributes about $0.5 W/m^2$.
- The increase in nitrous oxide (NOx) since pre-industrial times, from about 275 to 310 ppbv3, contributes about $0.1 W/m^2$.
- The observed concentrations of halocarbons, including CFCs, have resulted in direct radiative forcing of about $0.3 W/m^2$.

Figure 2.35 Changes in greenhouse gases since pre-industrial times

Arguments surrounding global warming

There are many causes of global warming and climate change. Natural causes include:

- variations in the Earth's orbit around the Sun
- variations in the tilt of the Earth's axis
- changes in the aspect of the poles from towards the Sun to away from it
- variations in solar output (sunspot activity)
- changes in the amount of dust in the atmosphere (partly due to volcanic activity)
- changes in the Earth's ocean currents as a result of **continental drift**.

All of these have helped cause climate change, and may still be doing so, despite anthropogenic forces.

Complexity of the problem

Climate change is a very complex issue for a number of reasons:

- Scale – it includes the atmosphere, oceans and land masses across the world.
- Interactions between these three areas are complex.
- It includes natural as well as anthropogenic forces.
- There are feedback mechanisms involved, not all of which are fully understood.
- Many of the processes are long term and so the impact of changes may not yet have occurred.

The effects of increased global temperature change

The effects of global warming are varied (see Table 2.3). Much depends on the scale of the changes. For example, some impacts could include:

- a rise in sea levels, causing flooding in low-lying areas such as the Netherlands, Egypt and Bangladesh – up to 200 million people could be displaced
- 200 million people at risk of being driven from their homes by flood or drought by 2050
- 4 million km^2 of land, home to one-twentieth of the world's population, threatened by floods from melting glaciers
- an increase in storm activity, such as more frequent and intense hurricanes (owing to more atmospheric energy)
- changes in agricultural patterns, for example a decline in the USA's grain belt, but an increase in Canada's growing season
- reduced rainfall over the USA, southern Europe and the Commonwealth of Independent States (CIS), leading to widespread drought (Figure 2.36)
- 4 billion people could suffer from water shortages if temperatures rise by 2°C
- a 35 per cent drop in crop yields across Africa and the Middle East expected if temperatures rise by 3°C
- 200 million more people could be exposed to hunger if world temperatures rise by 2°C; 550 million if temperatures rise by 3°C
- 60 million more Africans could be exposed to malaria if world temperatures rise by 2°C
- extinction of up to 40 per cent of species of wildlife if temperatures rise by 2°C.

Urban climates

Urban climates occur as a result of extra sources of heat released from industry; commercial and residential buildings; as well as from vehicles, concrete, glass, bricks, tarmac – all of these act very differently from soil and vegetation. For example, the albedo (reflectivity) of tarmac is about 5–10 per cent, while that of concrete is 17–27 per cent. In contrast, that of grass is 20–30 per cent.

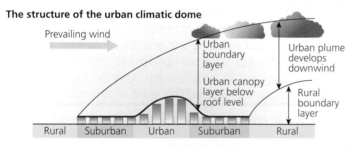

a Isolated buildings

Isolated building

Heat stored and re-radiated

Sunny side heated by insolation, reflected side insolation, radiation, and conduction

Shaded

c High buildings

Very little radiation reaches street level. Radiation reflected off lower walls after reflection from near tops of buildings

b Low buildings

Street collects reflected radiation

The structure of the urban climatic dome

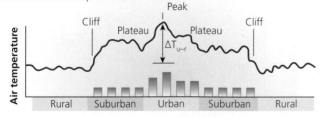

Prevailing wind

Urban boundary layer

Urban plume develops downwind

Urban canopy layer below roof level

Rural boundary layer

Rural | Suburban | Urban | Suburban | Rural

The morphology of the urban heat island

ΔT_{u-r} is the urban heat island intensity, i.e. the temperature difference between the peak and the rural air

Peak

Cliff

Plateau

Plateau

Cliff

ΔT_{u-r}

Air temperature

Rural | Suburban | Urban | Suburban | Rural

Airflow modified by a single building

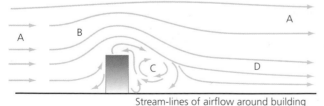

A

B

A

A

C

D

Stream-lines of airflow around building

Figure 2.39 Processes in the urban heat island

Some of these – notably dark bricks – absorb large quantities of heat and release them slowly by night (Figure 2.39). In addition, the release of **pollutants** helps trap radiation in urban areas. Consequently, urban microclimates can be very different from rural ones. Greater amounts of dust mean an increasing concentration of hygroscopic particles. There is less water vapour, but more carbon dioxide and higher proportions of noxious fumes owing to combustion of imported fuels. Discharge of waste gases by industry is also increased.

Urban heat budgets differ from rural ones. By day, the major source of heat is solar energy; and in urban areas brick, concrete and stone have high heat capacities. A kilometre of an urban area contains a greater surface area than a kilometre of countryside, and the greater number of surfaces in urban areas allow a greater area to be heated. There are more heat-retaining materials with lower albedo and better radiation-absorbing properties in urban areas than in rural ones.

Moisture and humidity

In urban areas, there is relative lack of moisture. This is due to:

■ a lack of vegetation
■ a high drainage density (sewers and drains), which removes water.

Thus there are decreases in relative humidity in inner cities due to the lack of available moisture and higher temperatures there. However, this is partly countered in very cold, stable conditions by early onset of condensation in low-lying districts and industrial zones.

Nevertheless, there are more intense storms, particularly during hot summer evenings and nights, owing to greater **instability** and stronger convection above built-up areas. There is a higher incidence of thunder (due to more heating and instability) but less snowfall (due to higher temperatures), and any snow that does fall tends to melt rapidly.

Hence little energy is used for evapotranspiration, so more is available to heat the atmosphere. This is in addition to the sources of heating produced by people, such as in industry and by cars.

At night, the ground radiates heat and cools. In urban areas, the release of heat by buildings offsets the cooling process, and some industries, commercial activities and transport networks continue to release heat throughout the night.

There is greater scattering of shorter-wave radiation by dust, but much higher absorption of longer waves owing to the surfaces and to carbon dioxide. Hence there is more diffuse radiation, with considerable local contrasts owing to variable screening by tall buildings in shaded narrow streets. There is reduced visibility arising from industrial haze.

There is a higher incidence of thicker cloud cover in summer because of increased convection, and radiation fogs or smogs in winter because of air pollution. The concentration of hygroscopic particles accelerates the onset of condensation. Daytime temperatures are, on average, 0.6 °C higher.

Urban heat island effect

The contrast between urban and rural areas is greatest under calm high-pressure conditions. The typical heat profile of an urban **heat island** shows a maximum at the city centre, a plateau across the suburbs and a temperature cliff between the suburban and rural areas (Figure 2.40). Small-scale variations within the urban heat island occur with the distribution of industries, open spaces, rivers, canals, and so on.

The heat island is a feature that is delimited by isotherms (lines of equal temperature), normally in an urban area. This shows that the urban area is warmer than the surrounding rural area, especially by dawn during anticyclonic conditions (Figure 2.41). The heat-island effect is caused by a number of factors:

- heat produced by human activity – a low level of radiant heat can be up to 50 per cent of incoming energy in winter
- changes of energy balance – buildings have a high thermal capacity in comparison to rural areas; up to six times greater than agricultural land
- the effect on airflow – turbulence of air may be reduced overall, although buildings may cause funnelling effects
- there are fewer bodies of open water, so less evaporation and fewer plants, therefore less transpiration
- the composition of the atmosphere – the blanketing effect of smog, smoke or haze
- reduction in thermal energy required for evaporation and evapotranspiration due to the surface character, rapid drainage and generally lower wind speeds
- reduction of heat diffusion due to changes in airflow patterns as a result of urban surface roughness.

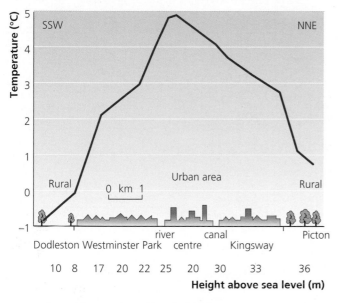

Source: Briggs, D. et al., *Fundamentals of the Physical Environment*, Routledge, 1997

Figure 2.40 The urban heat island (Chester, UK)

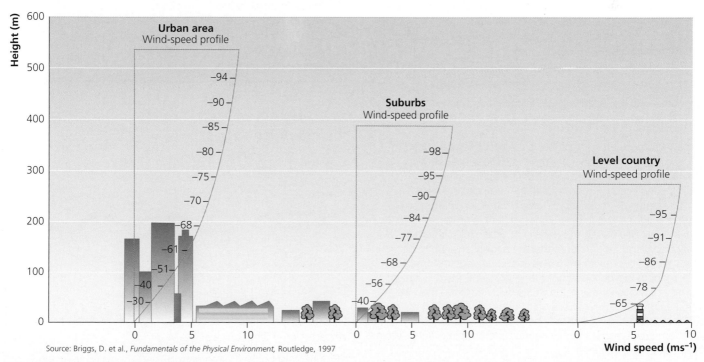

Source: Briggs, D. et al., *Fundamentals of the Physical Environment*, Routledge, 1997

Figure 2.41 The effect of terrain roughness on wind speed – with decreasing roughness, the depth of the affected layer becomes shallower and the profile steeper (numbers refer to wind strength as a percentage of maximum air speed)

Air flow

Urban areas may also develop a pollution dome. Highest temperatures are generally found over the city centre – or downwind of the city centre if there is a breeze present. Pollutants may be trapped under the dome. Cooler air above the dome prevents the pollutants from dispersing. These pollutants may prevent some incoming radiation from passing through, thereby reducing the impact of the heat island. By night, the pollutants may trap some long-wave radiation from escaping, thereby keeping urban areas warmer than surrounding rural areas.

Airflow over an urban area is disrupted; winds are slow and deflected over buildings (Figure 2.41). Large buildings can produce eddying. Severe gusting and turbulence around tall buildings causes strong local pressure gradients from windward to leeward walls. Deep narrow streets are much calmer unless they are aligned with prevailing winds to funnel flows along them – the 'canyon effect'.

The nature of urban climates is changing (Table 2.4). With the decline in coal as a source of energy, there is less sulphur dioxide pollution and so fewer hygroscopic nuclei; there is therefore less fog. However, the increase in cloud cover has occurred for a number of reasons:

- greater heating of the air (rising air, hence condensation)
- increase in pollutants
- frictional and turbulent effects on airflow
- changes in moisture.

Table 2.4 Average changes in climate caused by urbanisation

Factor		Comparison with rural environments
Radiation	Global	2–10% less
	Ultraviolet, winter	30% less
	Ultraviolet, summer	5% less
	Sunshine duration	5–15% less
Temperature	Annual mean	1°C more
	Sunshine days	2–6°C more
	Greatest difference at night	11°C more
	Winter maximum	1.5°C more
	Frost-free season	2–3 weeks more
Wind speed	Annual mean	10–20% less
	Gusts	10–20% less
	Calms	5–20% more
Relative humidity	Winter	2% less
	Summer	8–10% less
Precipitation	Total	5–30% more
	Number of rain days	10% more
	Snow days	14% less
Cloudiness	Cover	5–10% more
	Fog, winter	100% more
	Fog, summer	30% more
	Condensation nuclei	10 times more
	Gases	5–25 times more

Source: J. Tivy, Agricultural Ecology, *Longman 1990 p.372*

Section 2.4 Activities

1. Describe and account for the main differences in the climates of urban areas and their surrounding rural areas.
2. What is meant by the *urban heat island*?
3. Describe **one** effect that atmospheric pollution may have on urban climates.
4. Explain how buildings, tarmac and concrete can affect the climate in urban areas.
5. Why are microclimates, such as urban heat islands, best observed during high-pressure (anticyclonic) weather conditions?

Case Study: Urban microclimate – London

The heat island effect

Urban microclimates are perhaps the most complex of all microclimates. The general pattern in Figure 2.42 shows the highest temperatures in the city centre, reaching 10–11°C, compared with the rural fringe temperature of 5°C. Temperature falls more rapidly along the River Thames to the east of the City. The temperature gradient is more gentle in the west of the city, due to the density of urban infrastructure. Over steep temperature gradients there is a low density of urban infrastructure, for example the river and its vegetated banks. Where there is a gentle temperature gradient, there is a high density of urban infrastructure. Effectively from the map we can see that the east of London is less built up than the west. Temperature remains relatively constant for approximately 15 kilometres west of the city centre before rapidly falling within a 5–6 kilometre distance.

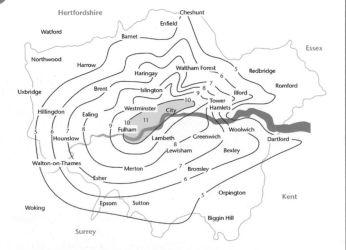

Figure 2.42 London's heat-island effect, showing minimum temperatures (°C) in mid-May

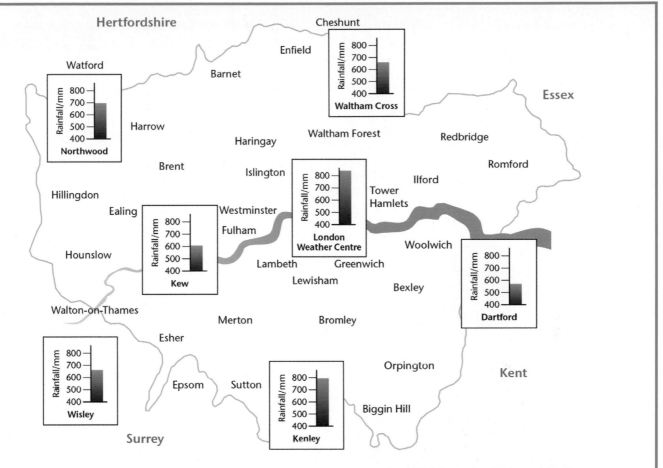

Source: National Meteorological Library and Archive Fact sheet 14 — Microclimates; Figure 16. Mean annual rainfall totals for a number of stations around London.
www.metoffice.gov.uk/media/pdf/n/9/Fact_sheet_No._14.pdf

Figure 2.43 Variations in mean annual rainfall around London

Recent research on London's heat island has shown that the pollution domes can also filter incoming solar radiation, thereby reducing the build-up of heat during the day. At night, the dome may trap some of the heat accrued during the day, so these domes might be reducing the sharp differences between urban and rural areas.

There is an absence of strong winds both to disperse the heat and to bring in cooler air from rural and suburban areas. Indeed, urban heat islands are often most clearly defined on calm summer evenings, often under blocking anticyclones.

The distribution of rainfall is very much influenced by topography, with the largest values occurring over the more hilly regions, and lowest values in more low-lying areas. Figure 2.43 illustrates this point quite clearly. Kenley on the North Downs, at an altitude of 170 metres above mean sea-level, has an average annual rainfall of nearly 800 millimetres, whereas London Weather Centre, at 43 metres above mean sea-level, has an average annual rainfall of less than 550 millimetres. Overall, humidity is lower in London than surrounding areas, partly due to higher temperatures (warm air can hold more moisture, hence relative humidity may be lower), but water is removed from large urban areas due to the combination of drains and sewers, the large amount of impermeable surfaces and the reduced vegetation cover.

The urban heat island creates the urban boundary layer, which is a dome of rising warm air and low pressure. As ground surfaces are heated, rapid evapotranspiration takes place. This evapotranspiration, although lower compared to rural areas, occurs more rapidly and can result in cumulus cloud and convectional weather patterns. Due to the low pressure caused by rising air, surface winds are drawn in from the surrounding rural fringe. This air then converges as it is forced to rise over the high urban canopy. The urban boundary essentially creates an orographic process similar to a mountain barrier. The movement of winds contributes to increased rainfall patterns over the city that are most pronounced to the leeward side of the city core. However, as air passes over the urban boundary layer it begins to sink, leading to lower precipitation at the leeward rural area. These differences are also more pronounced in the summer compared to the winter.

Some studies have demonstrated a pattern of increased rain through the week and have shown Saturday rain to be a result of a build-up of pollutants due to five consecutive commutes. By Monday, pollutants have fallen and rainfall is less likely to form.

The impact of river restoration on urban microclimates

In Seoul, capital of South Korea, there has been a very marked change in the urban microclimate following the removal of a large, downtown elevated motorway, and the restoration of a river and floodplain that had been built over. Since the restoration of the stream, air temperature has decreased by up to 10–13 per cent; that is, by 3-4 °C during the hottest days. Before the restoration, the area was showing a temperature about 5 °C higher than the average temperature of the city. The decrease in the number of vehicles passing by also contributed to the drop in the temperature. The heat island phenomenon used to be observed in the Cheong Gye Cheon Stream area under the impact of the heavy traffic, concentration of commercial facilities and the impermeable surface.

Following the completion of the restoration, the wind speed has become faster (by 2.2–7.1 per cent). The average wind speed measured at Cheong Gye Cheon is up to 7.8 per cent faster than before, apparently under the influence of the cool air forming along the stream.

Figure 2.44 Cheong Gye Cheon – **a** when the area was developed with an elevated highway and **b** after restoration

With increasing distance from the city centre, the amount of tree cover in a suburb decreases, while the amount of green space, such as lawns and parks, increases. In Melbourne, for every 10 kilometres from the city centre, the tree cover drops by more than 2 per cent. That means Melbourne's inner suburbs might have more than 15 per cent cover, but an outer suburb could have less than 10 per cent. A 5 per cent fall in urban tree cover can lead to a 1–2 °C rise in air temperature. This matters for community health and well-being, especially for the vulnerable – the elderly, young children and those with existing health issues.

Trees are missing from back gardens – partly because modern houses in the outer suburbs take up more space, leaving less room for trees – and they are missing from the streets. The property boom led to a gradual thinning out of tree cover in established suburbs, as residential plots were sub-divided.

Melbourne aims to increase tree cover by 75 per cent before 2040, Sydney by 50 per cent before 2030 and Brisbane is targeting tree cover for cycleways and footpaths.

☐ Microclimate mitigation

Increasingly, there are attempts to reverse urban microclimates. Heat-island mitigation strategies include urban forestry, living/green roofs and light surfaces.

In general, substantial reductions in surface and near-surface air temperature can be achieved by implementing heat-island mitigation strategies. Vegetation cools surfaces more effectively than increases in albedo, and curbside planting is the most effective mitigation strategy per unit area redeveloped. However, the greatest absolute temperature reductions are possible with light surfaces.

Table 2.5 Characteristics of the London Plane tree

Characteristic	The London Plane tree
Aesthetic value	A tall elegant tree providing pleasant shade in summer and a pleasing winter silhouette. Flaking bark creates attractive colours on trunk.
Does it make a mess?	Leaves, fruit and bark need clearing from streets and pavements.
Pollution tolerance	Very tolerant of air pollution. Hairs on young shoots and leaves help to trap particulate pollution.
Pests and diseases	Rarely affected by disease and pests (although some shoots are killed each year by fungal infection).
Soil conditions	Very tolerant of poor soil conditions, including compacted soil (although some stunting of growth is caused by road salt).
Space	Grows vigorously and is very tolerant of pruning.
Safety hazards	Trees rarely blow over or shed branches. Fine hairs on young shoots, leaves and fruit may cause irritation and even allergies in some people.
Microclimate	Open canopy produces light shade. Will intercept some rain, especially when in leaf.
Biodiversity	Provides valuable nesting sites for birds. Sufficient light below canopy to allow significant plant growth.

Source: Adapted from the Field Studies Council's Urban Ecosystems website www.field-studies-council.org/urbaneco

The London Plane tree – urban saviour

With an extensive and healthy urban forest, air quality can be drastically improved. Trees help to lower air temperatures through increasing evapotranspiration. This reduction of temperature not only lowers energy use, it also improves air quality, as the formation of ozone is dependent on temperature. Large shade trees can reduce local ambient temperatures by 3 to 5 °C. Maximum midday temperature reductions due to trees range from 0.04 °C to 0.2 °C per 1 per cent canopy cover increase.

Living roofs offer greater cooling per unit area than light surfaces, but less cooling per unit area than curbside planting.

Although street trees provide the greatest cooling potential per unit area, light surfaces provide the greatest overall cooling potential when available area is taken into account because there is more available area in which to implement this strategy compared to the other strategies.

3 Rocks and weathering

3.1 Plate tectonics

☐ The Earth's interior

The theory of plate tectonics states that the Earth is made up of a number of layers (Figure 3.1). On the outside, there is a very thin crust, and underneath is a mantle that makes up 82 per cent of the volume of the Earth. Deeper still is a very dense and very hot **core**. In general, these concentric layers become increasingly more dense towards the centre. Their density is controlled by temperature and pressure. Temperature softens or melts rocks.

Close to the surface, rocks are mainly solid and brittle. This upper surface layer is known as the **lithosphere**, which includes the **crust** and the upper **mantle**, and is about 70 kilometres deep. The Earth's crust is commonly divided up into two main types: **continental crust** and **oceanic crust** (Table 3.1). In continental areas, silica and aluminium are very common. When combined with oxygen they make up the most common type of rock: granitic. By contrast, below the oceans the crust consists mainly of basaltic rock in which silica, iron and magnesium are most common.

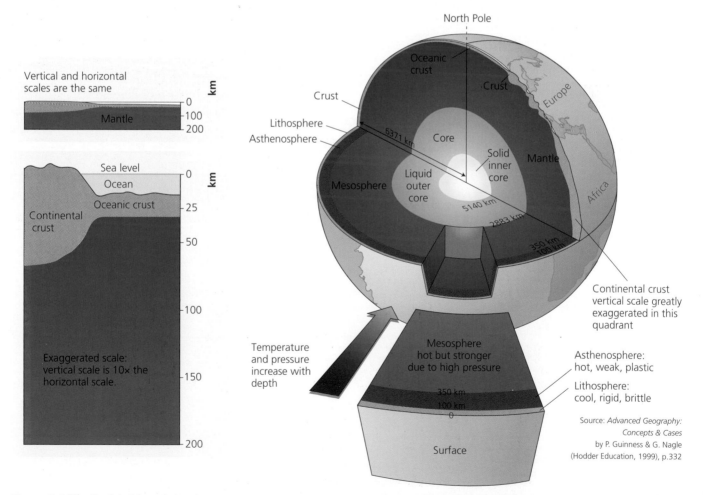

Source: *Advanced Geography: Concepts & Cases* by P. Guinness & G. Nagle (Hodder Education, 1999), p.332

Figure 3.1 The Earth's internal structure

Table 3.1 A comparison of oceanic crust and continental crust

Examples	Continental crust	Oceanic crust
Thickness	35–70 km on average	6–10 km on average
Age of rocks	Very old; mainly over 1500 million years	Very young; mainly under 200 million years
Colour and density of rocks	Lighter, with an average density of 2.6; light in colour	Heavier, with an average density of 3.0; dark in colour
Nature of rocks	Numerous types, many contain silica and oxygen; granitic is the most common	Few types, mainly basaltic

☐ The evidence for plate tectonics and their global patterns

In 1912, Alfred Wegener proposed the idea of continental drift. Others, such as Francis Bacon in 1620, had commented on how the shape of the coast of Africa was similar to that of South America. Wegener proposed that the continents were slowly drifting about the Earth. He suggested that, starting in the Carboniferous period some 250 million years ago, a large single continent, Pangaea, broke up and began to drift apart, forming the continents we know today. Wegener's theory provoked widespread debate initially, but with the lack of a mechanism to cause continental drift, his theory failed to receive widespread support.

In the mid-twentieth century, American Harry Hess suggested that convection currents would force molten rock (**magma**) to well up in the interior and to crack the crust above and force it apart. In the 1960s, research on rock magnetism supported Hess. The rocks of the Mid-Atlantic Ridge were magnetised in alternate directions in a series of identical bands on both sides of the ridge. This suggested that fresh magma had come up through the centre and forced the rocks apart. In addition, with increasing distance from the ridge the rocks were older. This supported the idea that new rocks were being created at the centre of the ridge and the older rocks were being pushed apart.

In 1965, Canadian geologist J. Wilson linked together the ideas of continental drift and **sea-floor spreading** into a concept of mobile belts and rigid plates, which formed the basis of plate tectonics.

The evidence of plate tectonics includes:

- the past and present distribution of earthquakes
- changes in the Earth's magnetic field
- the 'fit' of the continents: in 1620 Francis Bacon noted how the continents on either side of the Atlantic could be fitted together like a jigsaw (Figure 3.2)
- glacial deposits in Brazil match those in West Africa
- the fossil remains in India match those of Australia
- the geological sequence of sedimentary and igneous rocks in parts of Scotland match those found in Newfoundland

- ancient mountains can be traced from east Brazil to west Africa, and from Scandinavia through Scotland to Newfoundland and the Appalachians (eastern USA)
- fossil remains of a small aquatic reptile, Mesosaurus, which lived about 270 million years ago, are found only in a restricted part of Brazil and in south-west Africa – it is believed to be a poor swimmer!

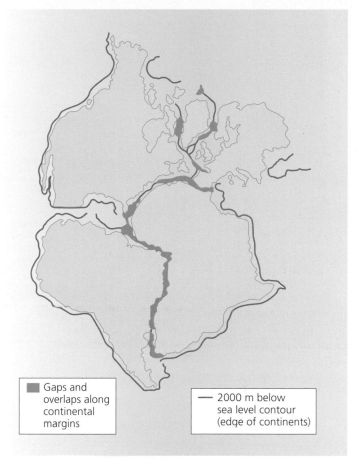

■ Gaps and overlaps along continental margins	— 2000 m below sea level contour (edge of continents)

Figure 3.2 Evidence for plate tectonics

☐ Plate boundaries

The zone of earthquakes around the world has helped to define six major plates and a number of minor plates (Figures 3.3 and 3.4). The boundaries between plates can be divided into three main types: **divergent** (constructive) boundaries, **convergent** plate boundaries (including destructive and collision boundaries) and **conservative** plate boundaries. Divergent (constructive) plate boundaries, where new crust is formed, are mostly in the middle of oceans (Figure 3.5a). These ridges are zones of shallow earthquakes (less than 50 kilometres below the surface). Where two plates converge, a deep-sea trench may be formed when one of the plates is **subducted** (forced downwards) into the mantle (Figure 3.5b). Deep earthquakes, up to 700 kilometres below the surface, are common. Good examples include the trenches off the Andes, and the Aleutian Islands that stretch out from Alaska.

Figure 3.3 Thingvellir, Iceland – a constructive plate boundary; here, the North American plate (left) is pulling away from the Eurasian plate (right of picture)

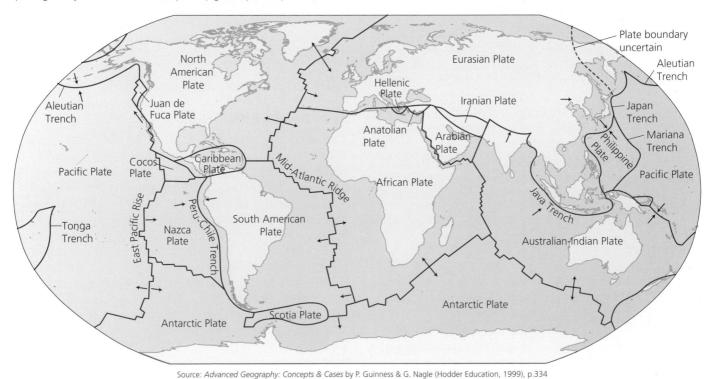

Source: *Advanced Geography: Concepts & Cases* by P. Guinness & G. Nagle (Hodder Education, 1999), p.334

Figure 3.4 Plate boundaries

a Divergent (constructive) boundary

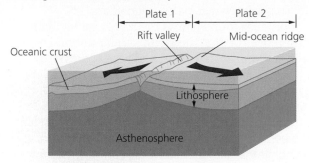

b Convergent (destructive) boundary

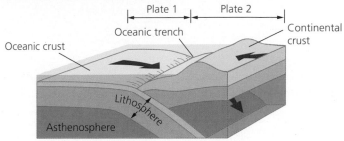

c Conservative boundary

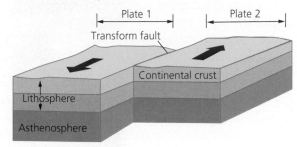

d Convergent (collision) boundary

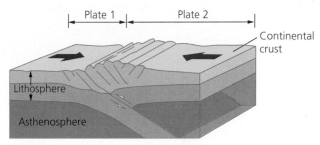

Source: *Advanced Geography: Concepts & Cases* by P. Guinness & G. Nagle (Hodder Education, 1999), p.334

Figure 3.5 Types of plate boundary

Along some plate boundaries, plates slide past one another to create a transform fault (fault zone) without converging or diverging (Figure 3.5c). Again, these are associated with shallow earthquakes, such as the San Andreas Fault in California. Where continents embedded in the plates collide with each other, there is no subduction but collision leading to crushing, and folding may create young fold mountains such as the Himalayas and the Andes (Figure 3.5d).

☐ The movement of plates

There are three main theories about movement:

1 **The convection current theory** – This states that huge convection currents occur in the Earth's interior. Hot magma rises through the core to the surface and then spreads out at **mid-ocean ridges**. The cold solidified crust sinks back into the Earth's interior, because it is heavier and denser than the surrounding material. The cause of the movement is radioactive decay in the core.

2 **The dragging theory** – Plates are dragged or subducted by their oldest edges, which have become cold and heavy. Plates are hot at the mid-ocean ridge but cool as they move away. Complete cooling takes about a million years. As cold plates descend at the trenches, pressure causes the rock to change and become heavier.

3 A **hotspot** is a plume of **lava** that rises vertically through the mantle. Most are found near plate margins and they may be responsible for the original rifting of the crust. However, the world's most abundant source of lava, the Hawaiian Hotspot, is not on a plate margin. Hotspots

can cause movement – the outward flow of viscous rock from the centre may create a drag force on the plates and cause them to move.

Section 3.1 Activities

1 Briefly outline the evidence for plate tectonics.
2 Describe how a convection current works. How does it help explain the theory of plate tectonics?
3 What happens at **a** a mid-ocean ridge and **b** a subduction zone?

☐ Processes and associated landforms

Sea-floor spreading

It was not until the early 1960s that R.S. Dietz and H.H. Hess proposed the mechanism of sea-floor spreading to explain continental drift. They suggested that continents moved in response to the growth of oceanic crust between them. Oceanic crust is thus created from the mantle at the crest of the mid-ocean ridge system.

Confirmation of the hypothesis of sea-floor spreading came with the discovery by F.J. Vine and D.H. Matthews that magnetic anomalies across the Mid-Atlantic Ridge were symmetrical on either side of the ridge axis (Figure 3.6). The only acceptable explanation for these magnetic anomalies was in terms of sea-floor spreading and the creation of new oceanic crust. When lava cools on the sea floor, magnetic grains in the rock acquire

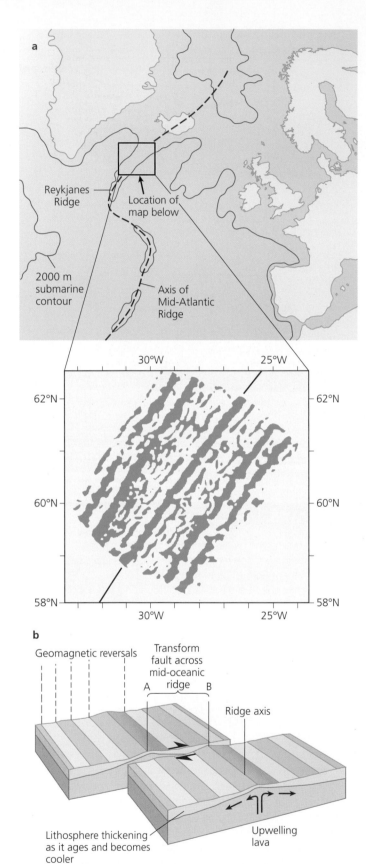

Figure 3.6 Sea-floor spreading and paleomagnetism

the direction of the Earth's magnetic field at the time of cooling. This is known as paleomagnetism.

The anomalies found across the Mid-Atlantic Ridge could, moreover, be matched with similar anomalies that had been discovered in Iceland and other parts of the world where young volcanic rocks could be dated.

The reason that the ridges are elevated above the ocean floor is that they consist of rock that is hotter and less dense than the older, colder plate. Hot mantle material wells up beneath the ridges to fill the gap created by the separating plates; as this material rises, it is decompressed and undergoes partial melting.

Spreading rates are not the same throughout the mid-ocean ridge system, but vary considerably from a few millimetres per year in the Gulf of Aden to 1 centimetre per year in the North Atlantic near Iceland and 6 centimetres per year for the East Pacific Rise. This variation in spreading rates appears to influence the ridge topography. Slow-spreading ridges, such as the Mid-Atlantic Ridge, have a pronounced rift down the centre. Fast-spreading ridges, such as the East Pacific Rise, lack the central rift and have a smooth topography. In addition, spreading rates have not remained constant through time.

The main reason for the differences in spreading rates is that the slow-spreading ridges are fed by small and discontinuous magma chambers, thereby allowing for the eruption of a comparatively wide range of basalt types. Fast-spreading ridges have large, continuous magma chambers that generate comparatively similar magmas. Because of the higher rates of magma discharge, sheet lavas are more common.

Although mid-ocean ridges appear at first sight to be continuous features within the oceans, they are all broken into segments by transverse fractures (faults) that displace the ridges by tens or even hundreds of kilometres.

Section 3.1 Activities

Briefly explain what is meant by **a** *paleomagnetism* and **b** *sea-floor spreading*.

Fractures are narrow, linear features that are marked by near-vertical fault planes.

Subduction zones and ocean trenches

Subduction zones form where an oceanic lithospheric plate collides with another plate, whether continental or oceanic (Figure 3.7). The density of the oceanic plate is similar to that of the **asthenosphere**, so it can be easily pushed down into the upper mantle. Subducted (lithospheric) oceanic crust remains cooler, and therefore denser than the surrounding mantle, for millions of years; so once initiated, subduction carries on, driven, in part, by the weight of the subducting crust. As the Earth has not grown significantly in size – not enough to accommodate

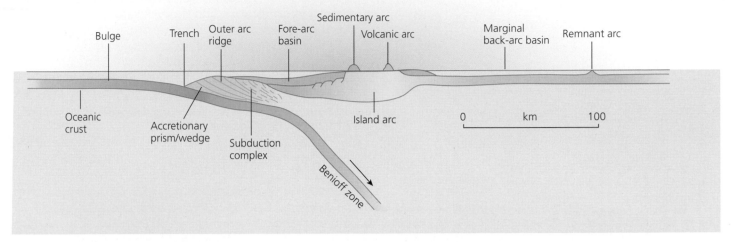

Figure 3.7 Ocean–ocean subduction zone

the new crustal material created at mid-ocean ridges – the amount of subduction roughly balances the amount of production at the constructive plate margins.

Subduction zones dip mostly at angles between 30° and 70°, but individual subduction zones dip more steeply with depth. The dip of the slab is related inversely to the velocity of convergence at the trench, and is a function of the time since the initiation of subduction. The older the crust, the steeper it dips. Because the downgoing slab of lithosphere is heavier than the plastic asthenosphere below, it tends to sink passively; and the older the lithosphere, the steeper the dip.

The evidence for subduction is varied:

- the existence of certain landforms such as deep-sea trenches and folded sediments – normally arc-shaped and containing volcanoes
- the **Benioff zone** – a narrow zone of earthquakes dipping away from the deep-sea trench
- the distribution of temperature at depth – the oceanic slab is surrounded by higher temperatures.

At the subduction zone, deep-sea **ocean trenches** are found. Deep-sea trenches are long, narrow depressions in the ocean floor with depths from 6000 metres to 11 000 metres. Trenches are found adjacent to land areas and associated with island arcs worldwide. They are more numerous in the Pacific Ocean. The trench is usually asymmetric, with the steep side towards the land mass. Where a trench occurs off a continental margin, the turbidites (sediments) from the slope are trapped, forming a hadal plain on the floor of the trench.

Benioff zone

A large number of events take place on a plane that dips on average at an angle of about 45° away from the underthrusting oceanic plate. The plane is known as the

Benioff (or Benioff-Wadati) zone, after its discoverer(s), and earthquakes on it extend from the surface, at the trench, down to a maximum depth of about 680 kilometres. For example, shallow, intermediate and deep-focus earthquakes in the south-western Pacific occur at progressively greater distances away from the site of underthrusting at the Tonga Trench.

Section 3.1 Activities

Describe the main characteristics of **a** mid-ocean ridges and **b** subduction zones.

Fold-mountain building

Plate tectonics is associated with mountain building. Linear or arcuate chains – sometimes called 'orogenic mountain belts' – are associated with convergent plate boundaries, and formed on land. Where an ocean plate meets a continental plate, the lighter, less dense continental plate may be folded and buckled into fold mountains, such as the Andes. Where two continental plates meet, both may be folded and buckled, as in the case of the Himalayas, formed by the collision of the Eurasian and Indian plates. Mountain building is often associated with crustal thickening, deformation and volcanic activity, although in the case of the Himalayas, volcanic activity is relatively unimportant.

The Indian subcontinent moved rapidly north during the last 70 million years, eventually colliding with the main body of Asia. A huge ocean (Tethys) has been entirely lost between these continental masses. Figure 3.8a shows the situation just prior to the elimination of the Tethys Ocean by subduction beneath Asia. Note the volcanic arc on the Asian continent (rather like the Andes today).

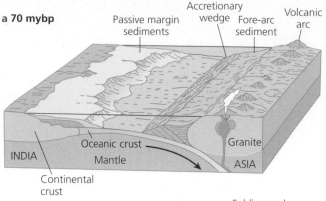

a 70 mybp

Passive margin sediments · Accretionary wedge · Fore-arc sediment · Volcanic arc

Oceanic crust · Granite · Mantle · INDIA · Continental crust · ASIA

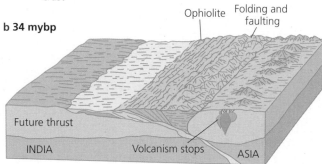

b 34 mybp

Ophiolite · Folding and faulting

Future thrust · INDIA · Volcanism stops · ASIA

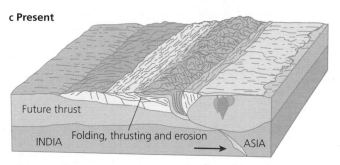

c Present

Future thrust · INDIA · Folding, thrusting and erosion · ASIA

Figure 3.8 Formation of the Himalayas

In Figure 3.8b, the Tethys Ocean has *just* closed. The leading edge of the Indian subcontinent and the sedimentary rocks of its continental shelf have been thrust beneath the edge of the Asian continent. (Ophiolite refers to pieces of ocean crust thrusted onto the edge of the continental crust.)

Finally, in Figure 3.8c the Indian subcontinent continues to move north-eastward relative to the rest of Asia. In the collision zone, the continental crust is thickened because Asia overrides India, and it is this crustal thickening that results in the uplift of the Himalayan mountain range. The red lines show the many locations in the collision zone where thrust faults are active to accommodate the deformation and crustal thickening.

In contrast, the Andes were formed as a result of the subduction of oceanic crust under continental crust. The Andes are the highest mountain range in the Americas, with 49 peaks over 6000 metres high. Unlike the Himalayas, the Andes contain many active volcanoes.

Before about 250 million years ago, the western margin of South America was a passive continental margin.

Sediments accumulated on the continental shelf and slope. With the break-up of Pangaea, the South American plate moved westward, and the eastward-moving oceanic lithosphere began subducting beneath the continent.

As subduction continued, rocks of the continental margin and trench were folded and faulted and became part of an accretionary wedge along the west coast of South America. Subduction also resulted in partial melting of the descending plate, producing andesitic volcanoes at the edge of the continent.

The Andes mountains comprise a central core of granitic rocks capped by andesitic volcanoes. To the west of this central core, along the coast, are the deformed rocks of the accretionary wedge; and to the east of the central core are sedimentary rocks that have been intensely folded. Present-day subduction, volcanism and seismicity indicate that the Andes mountains are still actively forming.

Figure 3.9 Fold mountains

Section 3.1 Activities

Compare and contrast the formation of the Andes and the formation of the Himalayas.

Ocean ridges

The longest linear, uplifted features of the Earth's surface are to be found in the oceans. They are giant submarine mountain chains with a total length of more than 60 000 kilometres, between 1000 and 4000 kilometres wide and have crests that rise 2 to 3 kilometres above the surrounding ocean basins, which are 5 kilometres deep. The average depth of water over their crests is thus about 2500 metres. These features are the mid-ocean ridges, famous now not only for their spectacular topography, but because it was with them, in the early 1960s, that the theory of ocean-floor spreading, the precursor of plate tectonic theory, began. We now know that it is at these mid-ocean ridges that new lithosphere is created.

Similar ridges occur at the margins of oceans; the East Pacific Rise is an example. There are other spreading ridges behind the volcanic arcs of subduction zones. These are usually termed 'back-arc spreading centres'. The first ridge to be discovered, the Mid-Atlantic Ridge, was found during attempts to lay a submarine cable across the Atlantic in the mid-nineteenth century.

Volcanic island arcs

Island arc systems are formed when oceanic lithosphere is subducted beneath oceanic lithosphere. They are consequently typical of the margins of shrinking oceans such as the Pacific, where the majority of island arcs are located. They also occur in the western Atlantic, where the Lesser Antilles (Caribbean) and Scotia arcs are formed at the eastern margins of small oceanic plates. The Lesser Antilles (Eastern Caribbean) Arc shows all the features of a typical island arc. Ocean–ocean subduction zones tend to be simpler than ocean–continental subduction zones. In a typical ocean–ocean subduction zone, there are a number of characteristic features (Figure 3.10):

- Ahead of the subduction zone, there is a low bulge on the sea floor (known as the **trench outer rise**) caused by the bending of the plate as it subducts. One of the best-known features is the trench that marks the boundary between the two plates. In the Eastern Caribbean, the trench associated with the subduction zone is largely filled with sediment from the Orinoco River. These sediments, more than 20 kilometres thick, have been deformed and folded into the Barbados Ridge, which emerges above the sea at Barbados.

- The **outer slope** of the trench is generally gentle, but broken by faults as the plate bends. The floor of the trench is often flat and covered by sediment (turbidites) and ash. The trench **inner slope** is steeper and contains fragments of the subducting plate, scraped off like shavings from a carpenter's plane. The **subduction complex** (also known as **accretionary prisms**) is the slice of the descending slab and may form significant landforms – for example in the Lesser Antilles, the islands of Trinidad, Tobago and Barbados are actually the top of the subduction complex.

- Most subduction zones contain an **island arc**, located parallel to a trench on the overriding plate. Typically they are found some 150–200 kilometres from the trench. Volcanic island arcs such as those in the Caribbean, including the islands from Grenada to St Kitts, are island arcs above sea level.

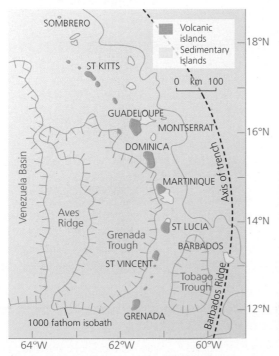

Figure 3.10 Island arcs in the Caribbean

3.2 Weathering and rocks

Weathering is the **decomposition** and **disintegration** of rocks *in situ*. Decomposition refers to **chemical weathering** and creates altered rock substances, such as kaolinite (china clay) from **granite**. By contrast, disintegration or **mechanical weathering** produces smaller, angular fragments of the same rock, such as **scree**. A third type, **biological weathering**, has been identified, whereby plants and animals chemically alter rocks and physically break rocks through their growth and movement. Biological weathering is not a separate type of weathering, but a form of disintegration and decomposition. It is important to note that these processes are **interrelated** rather than operating in isolation.

Weathering is central to landscape evolution, as it breaks down rock and enables erosion and transport. A number of key features can be recognised:

- Many minerals are formed under high pressure and high temperatures in the Earth's core. As they cool, they become more stable.
- Weathering produces irreversible changes in a rock. Some rocks change from a solid state to a fragmented or **clastic** state, such as scree. Others are changed to a pliable or **plastic** state, such as clay.
- Weathering causes changes in volume, density, grain size, surface area, permeability, consolidation and strength.
- Weathering forms new minerals and solutions.
- Some minerals, such as quartz, may resist weathering.
- Minerals and salts may be removed, transported, concentrated or consolidated.
- Weathering prepares rocks for subsequent erosion and transport.
- New landforms and features are produced.

☐ Physical/mechanical weathering

There are four main types of mechanical weathering: freeze–thaw (ice crystal growth), salt crystal growth, disintegration and pressure release. Mechanical weathering operates at or near the Earth's surface, where temperature changes are most frequent.

Freeze–thaw (also called 'ice crystal growth' or 'frost shattering') occurs when water in joints and cracks freezes at $0°C$. It expands by about 10 per cent and exerts pressure up to a maximum of $2100 \, kg/cm^2$ at $-22°C$. These pressures greatly exceed most rocks' resistance (Table 3.2). However, the average pressure reached in freeze–thaw is only $14 \, kg/cm^2$.

Table 3.2 Resistance to weathering

Rock	Resistance (kg/cm^2)
Marble	100
Granite	70
Limestone	35
Sandstone	7–14

Freeze–thaw is most effective in environments where moisture is plentiful and there are frequent fluctuations above and below freezing point. Hence it is most effective in periglacial and alpine regions. Freeze–thaw is most rapid when it operates in connection with other processes, notably pressure release and salt crystallisation.

Salt crystallisation causes the decomposition of rock by solutions of salt. There are two main types of **salt crystal growth**. First, in areas where temperatures fluctuate around $26–28°C$, sodium sulphate (Na_2SO_4) and sodium carbonate (Na_2CO_3) expand by about 300 per cent. This creates pressure on joints, forcing them to crack. Second, when water evaporates, salt crystals may be left behind. As the temperature rises, the salts expand and exert pressure on rock. Both mechanisms are frequent in hot desert regions where low rainfall and high temperatures cause salts to accumulate just below the surface. It may also occur in polar areas when salts are deposited from snowflakes.

Experiments investigating the effectiveness of saturated salt solutions have shown a number of results.

- The most effective salts are sodium sulphate, magnesium sulphate and calcium chloride.
- Chalk decomposes fastest, followed by limestone, sandstone and shale.
- The rate of disintegration of rocks is closely related to porosity and permeability.
- Surface texture and grain size control the rate of rock breakdown. This diminishes with time for fine materials and increases over time for coarse materials.
- Salt crystallisation is more effective than insolation weathering, hydration or freeze–thaw. However, a combination of freeze–thaw and salt crystallisation produces the highest rates of breakdown.

Heating and cooling may cause disintegration in hot desert areas where there is a large diurnal temperature range. In many desert areas, daytime temperatures exceed $40°C$, whereas night-time ones are little above freezing. Rocks heat up by day and contract by night. As rock is a poor conductor of heat, stresses occur only in the outer layers. This causes peeling or **exfoliation** to occur. Griggs (1936) showed that moisture is essential for this to happen. In the absence of moisture, temperature change alone did not cause the rocks to break down. The role of salt in insolation weathering has also been studied.

The expansion of many salts such as sodium, calcium, potassium and magnesium has been linked with exfoliation. However, some geographers find little evidence to support this view.

Pressure release (dilatation) is the process whereby overlying rocks are removed by erosion. This causes underlying rocks to expand and fracture parallel to the surface. The removal of a great weight, such as a glacier, has the same effect. Rocks are formed at very high pressure in confined spaces in the Earth's interior. The **unloading** of pressure by the removal of overlying rocks causes cracks or joints to form at right-angles to the unloading surface. These cracks are lines of weakness within the rock. For example, if overlying pressure is released, horizontal **pseudo-bedding planes** will be formed. By contrast, if horizontal pressure is released, as on a cliff face, vertical joints will develop. The size and spacing of cracks varies with distance from the surface: with increasing depth, the cracks become smaller and further apart. Hence the part of the rock that is broken the most is the part that is most subjected to denudation processes, namely at the surface.

Vegetation roots may also physically break down rocks. Figure 3.11 shows the impact of plants roots helping to break up rock.

Figure 3.11 Biological weathering – the physical impact of plant roots

Section 3.2 Activities

1 Define mechanical weathering.
2 Explain how freeze–thaw weathering operates.
3 Comment on the resistance to weathering (Table 3.2) compared with the pressure exerted by ice when it expands.
4 Describe the process of heating/cooling. Explain why it is common in hot, arid environments.

☐ Chemical weathering

Water is the key medium for chemical weathering. Unlike mechanical weathering, chemical weathering is most effective sub-surface since percolating water has gained organic acids from the soil and vegetation. Acidic water helps to break down rocks such as chalk, limestone and granite. The amount of water is important as it removes weathered products by solution. Most weathering therefore takes place above the water table, since weathered material accumulates in the water and saturates it. There are three main types of chemical weathering: carbonation-solution, hydrolysis and hydration.

Carbonation-solution occurs on rocks with calcium carbonate, such as chalk and limestone. Rainfall combines with dissolved carbon dioxide or organic acid to form a weak carbonic acid.

$CO_2 + H_2O \leftrightarrow H_2CO_3$ (carbonic acid)

Calcium carbonate (calcite) reacts with an acid water and forms **calcium bicarbonate** (also termed 'calcium hydrogen carbonate'), which is soluble and removed by percolating water:

$CaCO_3 + H_2CO_3 \rightarrow Ca(HCO_3)_2$

calcite + carbonic → calcium bicarbonate
acid

The effectiveness of solution is related to the pH of the water. For example, iron is highly soluble when the pH is 4.5 or less, and alumina (Al_2O_3) is highly soluble below 4.0 or above 9.0 but not in between.

Hydrolysis occurs on rocks with orthoclase feldspar, notably granite. Feldspar reacts with acid water and forms **kaolin** (also termed 'kaolinite' or 'china clay'), silicic acid and potassium hydroxyl:

$2KAlSi_3O_8 + 2 H_2O \rightarrow Al_2Si_2O_5 (OH)_4 + K_2O + 4 SiO_2$

orthoclase + water → kaolinite + potassium + silicic
feldspar   hydroxyl   acid

The acid and hydroxyl are removed in the solution, leaving kaolin behind as the end product. Other minerals in the granite, such as quartz and mica, remain in the kaolin. Hydrolysis also involves solution as the potassium hydroxyl is carbonated and removed in solution.

Hydration is the process whereby certain minerals absorb water, expand and change. For example, anhydrite is changed to gypsum. Although it is often classified as a type of chemical weathering, mechanical stresses occur as well. When anhydrite ($CaSO_4$) absorbs water to become gypsum ($CaSO_4.2H_2O$) it expands by about 0.5 per cent. More extreme is the increase in volume of up to 1600 per cent by shales and mudstones when clay minerals absorb water.

Section 3.2 Activities

1 Compare the character of rocks affected by mechanical weathering with those affected by chemical weathering.
2 Briefly explain the processes of carbonation-solution and hydrolysis.

☐ Controls of weathering

The following factors affect the type and rate of weathering that takes place.

Climate

In the simplest terms, the type and rate of weathering vary with climate (Figure 3.12), but it is very difficult to isolate the exact relationship, at any scale, between climate type and rate of process. Peltier's diagrams (1950) show how weathering is related to moisture availability and average annual temperature (Figure 3.13; see also Table 3.3). In general, frost-shattering increases as the number of freeze–thaw cycles increases. By contrast,

chemical weathering increases with moisture and heat. According to **Van't Hoff's Law**, the rate of chemical weathering increases 2–3 times for every increase in temperature of 10°C (up to a maximum temperature of 60°C). The efficiency of freeze–thaw, salt crystallisation and insolation weathering is influenced by:

- critical temperature changes
- frequency of cycles
- diurnal and seasonal variations in temperature.

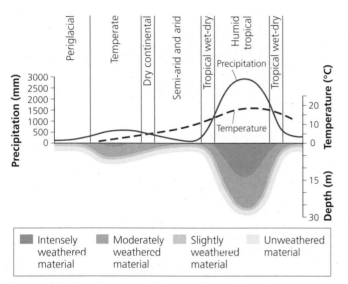

Figure 3.12 Depth of weathering profile and climate

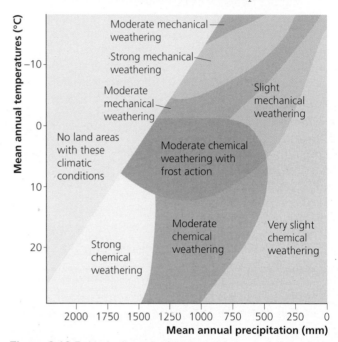

Figure 3.13 Peltier's diagram showing variations of chemical and mechanical weathering with climate

Table 3.3 Generalised weathering characteristics in four climatic regions

Climatic region	Characteristics	Examples – rates of weathering (mm yr⁻¹)
Glacial/ Periglacial	Frost very important. Susceptibility to frost increases with increasing grain size. **Taiga:** fairly high soil leaching, low rates organic matter decomposition. **Tundra:** low precipitation, low temperatures, permafrost – moist conditions, slow organic production and breakdown. May have slower chemical weathering. Algal, fungal, bacterial weathering may occur. Granular disintegration occurs. Hydrolytic action reduced on sandstone, quartzite, clay, calcareous shales, phyllites, dolerites. Hydration weathering common due to high moisture. CO_2 is more soluble at low temperatures.	Narvik 0.001 Spitzbergen 0.02–0.2 Alaska 0.04
Temperate	Precipitation and evaporation generally fluctuate. Both mechanical weathering and chemical weathering occur. Iron oxides leached and redeposited. Carbonates deposited in drier areas, leached in wetter areas. Increased precipitation, lower temperatures, reduced evaporation. Organic content moderate to high, breakdown moderate. Silicate clays formed and altered. **Deciduous forest areas:** abundant bases, high nutrient status, biological activity moderate to high. **Coniferous areas:** acidic, low biological activity, leaching common.	Askrigg 0.5–1.6 Austria 0.015–0.04
Arid/semi-arid	Evaporation exceeds precipitation. Rainfall low. Temperatures high, seasonal. Organic content low. Mechanical weathering, salt weathering, granular disintegration, dominant in driest areas. Thermal effects possible. Low organic input relative to decomposition. Slight leaching produces $CaCO_3$ in soil. Sulphates and chlorides may accumulate in driest areas. Increased precipitation and decreased evaporation toward semi-arid areas and steppes yield thick organic layers, moderate leaching and $CaCO_3$ accumulation.	Egypt 0.0001–2.0 Australia 0.6–1.0
Humid tropical	High rainfall often seasonal. Long periods of high temperatures. Moisture availability high. Weathering products **a** removed or **b** accumulate to yield red and black clay soils, ferruginous and aluminous soils (lateritic), calcium-rich soils. Calcareous rocks generally heavily leached where silica content is high, soluble weathering products removed and parent silica in stable products are sandy. Where products remain, iron and aluminium are common. Usually intense deep weathering, iron and alumina oxides and hydroxides predominate. Organic content high but decomposition high.	Florida 0.005

Rock type

Rock type influences the rate and type of weathering in many ways due to:

- chemical composition
- the nature of cements in sedimentary rock
- joints and bedding planes.

For example, limestone consists of calcium carbonate and is therefore susceptible to carbonation-solution. By contrast, granite is prone to hydrolysis because of the presence of feldspar. In sedimentary rocks, the nature of the cement is crucial. Iron-oxide based cements are prone to oxidation, whereas quartz cements are very resistant.

Rock structure

The effect of rock structure varies from large-scale folding and faulting to localised patterns of joints and bedding planes. Joint patterns exert a strong control on water movement. These act as lines of weakness, thereby creating **differential resistance** within the same rock type. Similarly, grain size influences the speed with which rocks weather. Coarse-grained rocks weather quickly owing to a large void space and high permeability (Table 3.4). On the other hand, fine-grained rocks offer a greater surface area for weathering and may be highly susceptible to weathering. The importance of individual minerals was stressed by Goldich in 1938. Rocks formed of resistant minerals, such as quartz, muscovite and feldspar in granite, will resist weathering (Figure 3.14). By contrast, rocks formed of weaker minerals will weather rapidly. The interrelationship of geology and climate on the development of landforms is well illustrated by limestone and granite.

Table 3.4 Average porosity and permeability for common rock types

Rock type	Porosity (%)	Relative permeability
Granite	1	1
Basalt	1	1
Shale	18	5
Sandstone	18	500
Limestone	10	30
Clay	45	10
Silt	40	–
Sand	35	1 100
Gravel	25	10 000

Source: D. Brunsden, 'Weathering processes' in C. Embleton and J. Thornes (eds), Processes in Geomorphology, Edward Arnold 1979

Stability	Dark-coloured minerals	Light-coloured minerals	Susceptibility
Least stable ↑	Olivine		Most susceptible ↑
		Lime plagioclase	
	Augite		
		Lime soda plagioclase	
	Hornblende	Soda lime plagioclase	
		Soda plagioclase	
	Biotite		
Most stable ↓	Orthoclase	Muscovite Quartz	Least susceptible ↓

Figure 3.14 Goldich's weathering system

Vegetation

The influence of vegetation is linked with the type of climate and the nature of the soil. Moisture content, root depth and acidity of humus will influence the nature and rate of weathering. Vegetation weathers rocks in two main ways: through the secretion of organic acids, it helps to chemically weather the soil; and through the growth of roots, it physically weathers the soil.

Depth of soil may have an effect on the amount of weathering that occurs. Soils may protect rocks from further breakdown – or they may increase the rate of breakdown due to the vegetation it supports.

Relief

For weathering to continue, weathered material needs to be removed. If the slope is too shallow, removal might not occur. If the slope is too steep, water may flow over the surface. Hence, intermediate slope angles may produce most weathering.

Aspect is also important, as there may be important temperature differences between south- and north-facing slopes. However, this is important only if the temperature differences are around a critical temperature, for example 0 °C for freeze–thaw weathering.

Section 3.2 Activities

1 **a** Define the terms *porosity* and *permeability*.
   b Choose a suitable method to show the relationship between porosity and permeability.
   c Describe the relationship between porosity and permeability.
   d What are the exceptions, if any, to this relationship?
2 Describe and explain how the type and intensity of mechanical weathering varies with climate.
3 Describe and explain how the type and intensity of chemical weathering varies with climate.
4 How useful are mean annual temperature and mean annual rainfall as a means of explaining variations in the type and intensity of weathering processes?
5 Describe two ways in which vegetation affects the type and rate of weathering.

3.3 Slope processes

☐ Introduction and definitions

The term 'slope' refers to:

- an inclined surface or **hillslope**
- an angle of inclination or **slope angle**.

Slopes therefore include any part of the solid land surface, including level surfaces of 0° (Figure 3.15). These can be **sub-aerial** (exposed) or **sub-marine** (underwater), **aggradational** (depositional), **degradational** (erosional), **transportational** or any mixture of these. Given the large scope of this definition, geographers generally study the hillslope. This is the area between the **watershed** (or drainage-basin divide) and the base. It may or may not contain a river or stream.

☐ Slope processes

Many slopes vary with **climate**. In humid areas, slopes are frequently rounder, due to chemical weathering, soil creep and fluvial transport. By contrast, in arid regions slopes are jagged or straight owing to mechanical weathering and sheetwash (Figure 3.16). **Climatic geomorphology** is a branch of geography that studies how different processes operate in different climatic zones, and produce different **slope forms** or shapes (see Table 3.3 in Section 3.2).

Geological structure is another important control on slope development. This includes faults, angle of dip and vulcanicity. These factors influence the strength of a rock and create lines of potential weakness within it. In addition, rock type and character affect vulnerability to weathering and the degree of resistance to downslope movement.

Geological structure can also influence the occurrence of landslips. Slopes composed of many different types of rock are often more vulnerable to landslides due to differential erosion; that is, less resistant rocks are worn away and can lead to the undermining of more resistant rocks.

Figure 3.15 Rounded slopes at Wytham, Oxfordshire, UK – a temperate region

Figure 3.16 Silent Valley, Dolomites, Italy

Soil can be considered as part of the **regolith**. Its structure and texture will largely determine how much water it can hold. Clay soils can hold more water than sandy soils. A deep clay on a slope where vegetation has been removed will offer very little resistance to **mass movement**.

Aspect refers to the direction in which a slope faces. In some areas, past climatic conditions varied depending on the direction a slope faced. During the cold periglacial period in the northern hemisphere, in an east–west valley, the southern slope which faced north, remained in the shade. Temperatures rarely rose above freezing. By contrast, the northern slope, facing south, was subjected to many more cycles of freeze–thaw. Solifluction and overland runoff lowered the level of the slope, and streams removed the debris from the valley. The result was an asymmetric valley.

Vegetation can decrease overland runoff through the interception and storage of moisture. Deforested slopes are frequently exposed to intense erosion and gullying. However, vegetation can also increase the chance of major landslips. Dense forests reduce surface wash, causing a build-up of soil between the trees, thus deepening the regolith and increasing the potential for failure.

Section 3.3 Activities

1. Briefly describe **two** ways in which climate affects slope development. What does the term *climatic geomorphology* mean?
2. Briefly describe **two** ways in which geology affects slope development.

☐ Mass movements

Mass movements include any large-scale movement of the Earth's surface that are not accompanied by a moving agent such as a river, glacier or ocean wave. They include:

- **very slow movements**, such as soil creep
- **fast movement**, such as **avalanches**
- **dry movement**, such as rockfalls
- **very fluid movements**, such as mudflows (Figure 3.17).

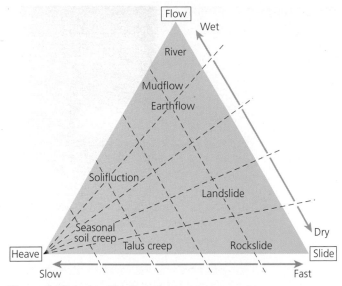

Figure 3.17 A classification of mass movements

A range of **slope processes** occur that vary in terms of magnitude, frequency and scale. Some are large and occur infrequently, notably rockfalls, whereas others are smaller and more continuous, such as soil creep.

The **types of processes** can be classified in a number of different ways:

- speed of movement (Figure 3.18)
- water content
- type of movement: **flows**, **slides**, **slumps**
- material.

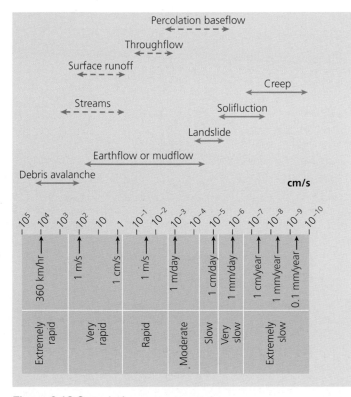

Figure 3.18 Speed of mass movements

Causes of mass movements

The likelihood of a slope failing can be expressed by its safety factor. This is the relative strength or resistance of the slope, compared with the force that is trying to move it. The most important factors that determine movement are gravity, slope angle and pore pressure.

Gravity has two effects. First, it acts to move the material downslope (a slide component). Second, it acts to stick the particle to the slope (a stick component). The downslope movement is proportional to the weight of the particle and slope angle. Water lubricates particles and in some cases fills the spaces between the particles. This forces them apart under pressure. Pore pressure will greatly increase the ability of the material to move. This factor is of particular importance in movements of wet material on low-angle slopes.

Shear strength and shear resistance

Slope failure is caused by two factors:

1 a reduction in the internal resistance, or **shear strength**, of the slope, or
2 an increase in **shear stress**; that is, the forces attempting to pull a mass downslope.

Both can occur at the same time.

Increases in shear stress can be caused by a multitude of factors (Table 3.5). These include material

Table 3.5 Increasing stress and decreasing resistance

Factor	Example
Factors that contribute to increased shear stress	
Removal of lateral support through undercutting or slope steepening	Erosion by rivers and glaciers, wave action, faulting, previous rockfalls or slides
Removal of underlying support	Undercutting by rivers and waves, subsurface solution, loss of strength by extrusion of underlying sediments
Loading of slope	Weight of water, vegetation, accumulation of debris
Lateral pressure	Water in cracks, freezing in cracks, swelling (especially through hydration of clays), pressure release
Transient stresses	Earthquakes, movement of trees in wind
Factors that contribute to reduced shear strength	
Weathering effects	Disintegration of granular rocks, hydration of clay minerals, dissolution of cementing minerals in rock or soil
Changes in pore-water pressure	Saturation, softening of material
Changes of structure	Creation of fissures in shales and clays, remoulding of sand and sensitive clays
Organic effects	Burrowing of animals, decay of tree roots

characteristics, weathering processes and changes in water availability. Weaknesses in rocks include joints, bedding planes and faults. Stress may be increased by:

- steepening or undercutting of a slope
- addition of a mass of regolith
- dumping of mining **waste**
- sliding from higher up the slope
- vibrational shock
- earthquakes.

Weathering may reduce cohesion and resistance. Consequently, material may be more susceptible to movement on slopes, even though the original material was stable.

Water can weaken a slope by increasing shear stress and decreasing shear resistance. The weight of a potentially mobile mass is increased by:

- an increase in the volume of water
- heavy or prolonged rain
- a rising water table
- saturated surface layers.

Moreover, water reduces the cohesion of particles by saturation. Water pressure in saturated soils (pore-water pressure) decreases the frictional strength of the solid material. This weakens the slope. Over time the safety factor for a particular slope will change. These changes may be gradual, for example percolation carrying away finer material. By contrast, some changes are rapid.

There are a number of ways that downslope movement can be opposed:

- **Friction** will vary with the weight of the particle and slope angle. Friction can be overcome on gentle slope angles if water is present. For example, solifluction can occur on slopes as gentle as 3°.
- **Cohesive forces** act to bind the particles on the slope. Clay may have high cohesion, but this may be reduced if the water content becomes so high that the clay liquefies, when it loses its cohesive strength.
- **Pivoting** occurs in the debris layers which contain material embedded in the slope.
- **Vegetation** binds the soil and thereby stabilises slopes. However, vegetation may allow soil moisture to build up and make landslides more likely (see pages 75–77).

Section 3.3 Activities

1 **a** Define the term *mass movement*.
   b Suggest how mass movements can be classified.
2 Define the terms *strength* and *shear stress*.
3 With the use of examples, explain why mass movements occur.

Types of mass movement

Heave or **creep** is a slow, small-scale process that occurs mostly in winter. It is one of the most important slope processes in environments where flows and slides are not common. **Talus creep** is the slow movement of fragments on a scree slope.

Individual soil particles are pushed or heaved to the surface by **a** wetting, **b** heating or **c** freezing of water (Figure 3.19). About 75 per cent of the soil-creep movement is induced by moisture changes and associated volume change. Nevertheless, freeze–thaw and normal temperature-controlled expansion and contraction are important in periglacial and tropical climates.

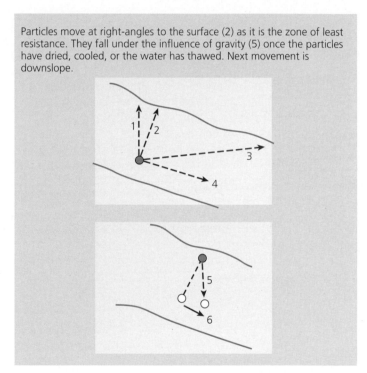

Particles move at right-angles to the surface (2) as it is the zone of least resistance. They fall under the influence of gravity (5) once the particles have dried, cooled, or the water has thawed. Next movement is downslope.

Figure 3.19 Soil creep

Rates of soil creep are slow, at 1–3 millimetres per year in temperate areas and up to 10 millimetres per year in tropical rainforest. They form terracettes. In well-vegetated humid temperate areas, soil creep can be ten times more important than slope wash. In periglacial areas, it can be as much as 300 millimetres per year. By contrast, in arid environments slope wash is more important. Small-scale variations in slope, compaction, cohesion and vegetation will have a significant effect on the rate of creep.

Observation of soil creep is difficult. Traditional qualitative evidence such as bent trees (Figure 3.20) is misleading and now largely discredited. The slow rate of movement may mean that measurement errors are serious.

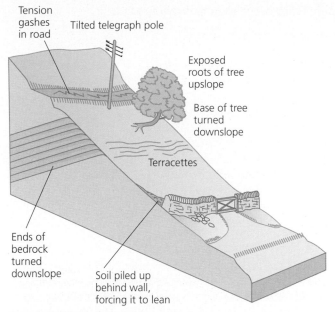

Figure 3.20 The evidence for soil creep

Slumps and flows

Slumps occur on weaker rocks, especially clay, and have a rotational movement along a curved slip plane (Figure 3.21). Clay absorbs water, becomes saturated and exceeds its liquid limit. It then flows along a slip plane. Frequently the base of a cliff has been undercut and weakened by erosion, thereby reducing its strength. By contrast, flows are more continuous, less jerky, and are more likely to contort the mass into a new form (Figure 3.22). Material is predominantly of a small size, such as deeply weathered clays. Particle size involved in flows is generally small, for example sand-sized and smaller.

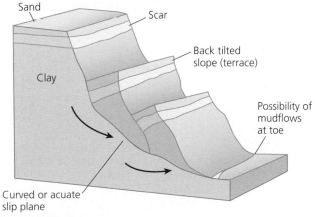

Figure 3.21 Slumps

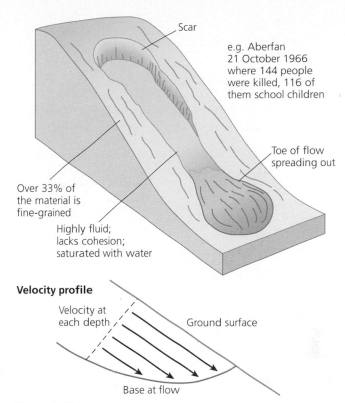

Figure 3.22 Flows

The speed of a flow varies: mudflows are faster and more fluid than earthflows, which tend to be thicker and deeper. A higher water content will enable material to flow across gentle angles.

Earthflows and mudflows can occur on the saturated toe (end) of a landslide, or may form a distinctive type of mass movement in their own right. Small flows may develop locally, whereas others may be larger and more rapid. In theory, mudflows give way to sediment-laden rivers – but the distinction is very blurred.

Case Study: Sidoarjo mudflow

Since May 2006, more than 50 000 people in Porong District, Indonesia, have been displaced by hot mud flowing from a natural well. Gas and hot mud began spewing out when a drill penetrated a layer of liquid sediment. The amount of material spilling out peaked at 135 000 m³/day in September 2006. By 2010, the main thoroughfare in Porong was raised 80 cm to avoid further mudflows. The Sidoarjo mudflow is an ongoing eruption of gas and mud.

Slides

Slides occur when an entire mass of material moves along a slip plane. These include:

- **rockslides and landslides** of any material, rock or regolith
- **rotational slides**, which produce a series of massive steps or terraces.

Slides commonly occur where there is a combination of weak rocks, steep slopes and active undercutting. Slides are often caused by a change in the water content of a slope or by very cold conditions. As the mass moves along the slip plane, it tends to retain its shape and structure until it hits the bottom of a slope (Figure 3.23). Slides range from small-scale slides close to roads, to large-scale movements that kill thousands of people.

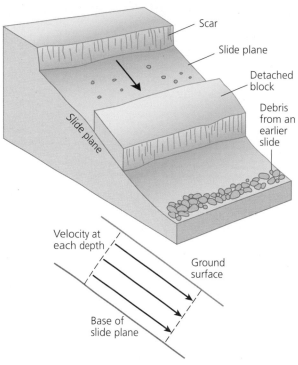

Figure 3.23 Slides

Slip planes occur for a variety of reasons:

- at the junction of two layers
- at a fault line
- where there is a joint
- along a bedding plane
- at the point beneath the surface where the shear stress becomes greater than the shear strength.

Weak rocks such as clay have little shear strength to start with, and are particularly vulnerable to the development of slip planes. The slip plane is typically a concave curve and as the slide occurs the mass will be rotated backwards.

Rockslides

In 1959, the sixth strongest earthquake ever to affect the USA occurred in Montana. Close to the epicentre of the earthquake, in the Madison River valley, a slope of schists and gneiss with slippery mica and clay was supported by a base of dolomite. The earthquake cleanly broke the dolomite. A huge volume of rock, 400 metres high and 1000 metres long, slid into the valley; 80 million tonnes of material moved in less than a minute! The Madison River was dammed and a lake 60 metres deep and 8 kilometres long was created.

Landslides

Loose rock, stones and soil all have a tendency to move downslope. They will do so whenever the downward force exceeds the resistance produced by friction and cohesion. When the material moves downslope as a result of shear failure at the boundary of the moving mass, the term 'landslide' is applied. This may include a flowing movement as well as straightforward sliding. Landslides are very sensitive to water content, which reduces the strength of the material by increasing the water pressure. This effectively pushes particles apart, thereby weakening the links between them. Moreover, water adds weight to the mass, increasing the downslope force.

Case Study: The Abbotsford landslide, Dunedin, New Zealand

The landslide that took place in East Abbotsford, South Island, New Zealand is a very good example of how human and physical factors can interact to produce a hazardous event. It also shows clearly how such hazards can be managed.

From 1978, several families in Abbotsford noticed hairline cracks appearing in their homes – in the brickwork, concrete floors and driveways. During 1979, workmen discovered that a leaking water main had been pulled apart. Geologists discovered that water had made layers of clay on the hill soft, and the sandstone above it was sliding on this slippery surface.

As a result, an early warning system was put in place. A civil defence emergency was declared on 6 August, although the

situation wasn't thought to be urgent as geologists believed that landslip would continue to move only slowly. However, on 8 August a 7 hectare section of Abbotsford started down the hill at a rate of over 3 metres a minute (Figure 3.24), with houses and 17 people on board. No-one was killed, although 69 homes were destroyed or damaged and over 200 people were displaced. The total cost from the destruction of the homes, infrastructure and relief operation amounted to over £7 million. An insurance scheme designed to cope with such disasters, and government and voluntary relief measures, meant that many of the residents were compensated for their loss. However, other costs, such as depressed house prices in the surrounding area, psychological trauma and the expense of a prolonged public enquiry, were not immediately appreciated.

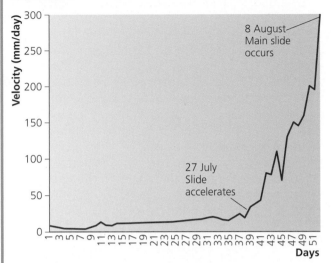

Figure 3.24 Abbotsford landslide, New Zealand

The landslide was essentially a block slide of sandstone resting on a bed of weaker clay. Displacement of 50 metres took place in about 30 minutes, leaving a small rift 30 metres deep at the head of the slope. Such geological conditions – in which a permeable hard rock rests on an impermeable soft rock – are commonly associated with landslides. In addition, the slope was dipping at an angle of 7°. Water collected in the impermeable clay, reduced its strength and cohesion, and caused the sandstone to slip along the boundary of the two rocks.

The landslide involved 5.4 million m³ of material. At first, the land moved as a slow creep, followed by a rapid movement with speeds of 1.7 metres per minute. Rapid sliding lasted for about 30 minutes. An area of about 18 hectares was affected.

However, other factors are also believed to have made a contribution. Deforestation in the area, even over a century before, had reduced evapotranspiration in the area and there was less binding of the soil by plant roots. Urbanisation in the previous 40 years had modified the slopes by cutting and infilling, and had altered surface drainage (speeding up the removal of surface water). Quarrying of material at the toe of the slope in the 1960s and 1970s had removed support from the base of the slope. The trigger of the landslide is believed to have been a combination of leaking water pipes and heavy rainfall.

A number of lessons can be learnt from the Abbotsford landslide:

- Dangerous landslides can occur on relatively gentle slopes if the right conditions exist.
- Attention to early warning can help preparedness and reduce the loss of life.
- Human activity can destabilise slopes.
- Low-frequency, high-magnitude events may be hard to predict, but mapping and dating of old hazards may indicate areas of potential risk – a regional landslide **hazard assessment** should be made where there is evidence of previous landslide activity.
- A landslide insurance scheme eased the cost of the event – however, money was available only after the event rather than beforehand, and the insurance only covered houses, not land damage.

Section 3.3 Activities

1 What were the causes of the Abbotsford landslide?
2 Describe the impacts of the Abbotsford landslide.
3 What lessons can be learnt from the Abbotsford landslide?

Case Study: Mexican landslides, 2010

In October 2010, mud buried part of a remote town in the southern Mexican state of Oaxaca when a large chunk of a nearby mountain collapsed after three days of relentless rain. Initially, it was thought that the landslide had caused a massive tragedy with up to 1000 people killed. However, the number of deaths was believed to be less than ten. The landslide happened at about four o'clock in the morning. The authorities were unsure how many houses had been buried because it was dark, so they estimated.

The rescue progress along the unpaved mountain road was hampered by smaller landslides and a collapsed bridge. Heavy cloud cover prohibited helicopters from getting a clear view of

the situation on the ground. When the first rescue workers and soldiers eventually reached the town, they found considerable destruction in one relatively small part of the town. Two houses were completely interred, two partially buried and thirty more in serious danger because they lay within the path of the still-unstable mudflow.

In 2010, Mexico experienced one of the most intense rainy seasons on record, with large areas under water in lowland regions of Oaxaca as well as in other southern states. Landslides are a major danger in mountainous parts of the country – particularly those, such as Oaxaca, that have long suffered from severe deforestation.

Falls

Falls occur on steep slopes (greater than 40°), especially on bare rock faces where joints are exposed. The initial cause of the fall may be weathering, such as freeze–thaw or disintegration, or erosion prising open lines of weakness. Once the rocks are detached, they fall under

the influence of gravity (Figure 3.25). If the fall is short, it produces a relatively straight scree. If it is long, it forms a concave scree. Falls are significant in producing the retreat of steep rock faces and in providing debris for scree slopes and talus slopes.

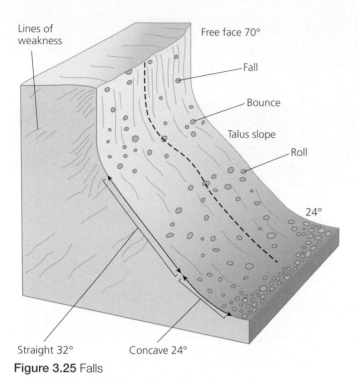

Figure 3.25 Falls

Section 3.3 Activities

1 Explain the terms *mass movement*, *soil creep* and *rotational slide*.
2 Outline the main characteristics of slumps and flows.

☐ Water and sediment movement on hillslopes

Surface wash occurs when the soil's infiltration capacity is exceeded. In the UK, this commonly occurs in winter as water drains across saturated or frozen ground, following prolonged or heavy downpours or the melting of snow. It is also common in arid and semi-arid regions where particle size limits percolation.

Sheetwash is the unchannelled flow of water over a soil surface. On most slopes, sheetwash breaks into areas of high velocity separated by areas of lower velocity. It is capable of transporting material dislodged by rainsplash (see the following section). Sheetwash occurs in the UK on footpaths and moorlands. For example, during the Lynmouth floods of 1952, sheetwash from the shallow moorland peat caused gullies 6 metres deep to form. In the semi-arid areas of south-west USA, it lowers surfaces by 2–5 millimetres per year compared with 0.01 millimetres per year on vegetated slopes in a temperate climate.

Sheetwash erosion of soil occurs through raindrop impact and subsequent transport by water flowing overland rather than in channels. The result is a relatively uniform layer of soil being eroded. A **rill** is a relatively shallow channel, generally less than tens of centimetres deep and carrying water and sediment for only a short period. Rills are common in agricultural areas, following the removal of vegetation during the harvest season, and the ground subsequently being left bare. They are also common in areas following deforestation or land-use changes. Ground compaction by machinery may also lead to the generation of rills during rainfall events.

Throughflow refers to water moving down through the soil. It is channelled into natural pipes in the soil. This gives it sufficient energy to transport material, and added to its solute load, may amount to a considerable volume.

Rainsplash erosion

Raindrops can have an erosive effect on hillslopes (Figure 3.26). On a 5° slope, about 60 per cent of the movement is downslope. This figure increases to 95 per cent on a 25° slope. The amount of erosion depends on the rainfall intensity, velocity and raindrop distribution. It is most effective on slopes of between 33° and 45° and at the start of a rainfall event when the soil is still loose.

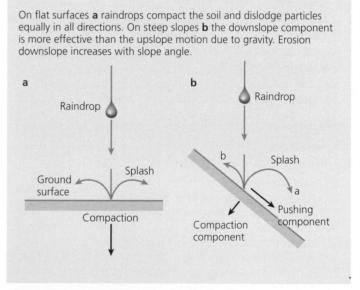

Figure 3.26 Rainsplash erosion

Section 3.3 Activities

1 Briefly explain how rainsplash erosion occurs.
2 Define the term *sheetwash*.
3 Under what conditions do rills occur?

3.4 The human impact

☐ Stability of slopes

Rates of mass movement can be altered by human activities, such as building or excavation, drainage or agriculture. Mass movements can be accelerated by destabilising slopes. Local erosion can be intensified by footpath trampling in recreational areas. Some mass movements are created by humans piling up waste soil and rock into unstable accumulations that move without warning. Landslides can be created by undercutting or overloading. Most changes to slopes caused by human activities have been very minor in relation to the scale of the natural land surface. Human interference with slopes tends to have been most effective in speeding up naturally occurring processes rather than creating new features.

In urban areas, the intensity of slope modification is often very high, given the need for buildings and roads to be constructed safely, using sound engineering principles. Almost all buildings with foundations cause some modification to the natural slope of the land, and even on flat sites, large modern buildings generally involve the removal of material to allow for proper foundations. Slope modification tends to increase as a construction moves on to steeper slopes. In these conditions, in order to provide a horizontal base plus reasonable access, a cut-and-fill technique is often used (Figure 3.27), thereby creating a small level terrace with an over-steepened slope at both ends. The steep slopes, devoid of soil and vegetation, are potentially much less stable than the former natural slope and are, in times of intense rainfall, susceptible to small but quite damaging landslips.

☐ Strategies to reduce mass movement

As well as causing mass movements, human activities can reduce them (Table 3.6).

Table 3.6 Examples of methods of controlling mass movement

Type of movement	Method of control
Falls	Flattening the slope Benching the slope Drainage Reinforcement of rock walls by grouting with cement, anchor bolts Covering of wall with steel mesh
Slides and flows	Grading or benching to flatten the slope Drainage of surface water with ditches Sealing surface cracks to prevent infiltration Subsurface drainage Rock or earth buttresses at foot Retaining walls at foot Pilings through the potential slide mass

Source: Goudie, 1993

Pinning is used to attach wire nets (or sometimes concrete blocks) to a rock face or slope so that the risk of rock falls is reduced or the risk of erosion is reduced. **Netting** may help collect fragments of scree, which can be safely removed at a later date. This is often used in areas where tourism is important, and where the risk of rock fall is high.

Grading refers to the re-profiling of slopes (see Figure 3.27) so that they become more stable. Afforestation is the planting of new forest in upper parts of a catchment to increase interception and reduce overland flow. They may take many years to be effective as the young, immature trees intercept relatively small amounts of water.

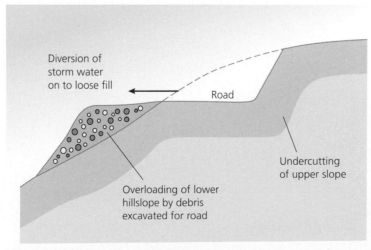

Figure 3.27 Slope instability caused by road building

Hong Kong has a long history of landslides – largely due to a combination of high rainfall (the wet season is from May to September), steep slopes and dense human developments on the islands (Figure 3.28). Between 1947 and 1997, more than 470 people died as a result of landslides.

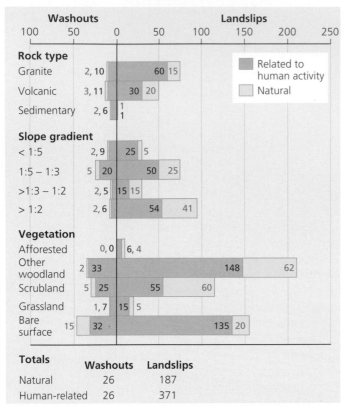

Number of mass movements per 100 km² chart:

	Washouts (Related to human activity, Natural)	Landslips (Related to human activity, Natural)
Rock type		
Granite	2, 10	60, 15
Volcanic	3, 11	30, 20
Sedimentary	2, 6	1, 1
Slope gradient		
< 1:5	2, 9	25, 5
1:5 – 1:3	5, 20	50, 25
>1:3 – 1:2	2, 5	15, 15
> 1:2	2, 6	54, 41
Vegetation		
Afforested	0, 0	6, 4
Other woodland	2, 33	148, 62
Scrubland	5, 25	55, 60
Grassland	1, 7	15, 5
Bare surface	15, 32	135, 20

Totals	Washouts	Landslips
Natural	26	187
Human-related	26	371

Figure 3.28 Number of mass movements per 100 km² in Hong Kong in the 1960s and 1970s

In June 1966, rainstorms triggered massive landslides that killed 64 people. Over 2500 people were made homeless and a further 8000 were evacuated. Rainfall had been high for the first ten days in June. Over 300 millimetres had fallen, compared with 130 millimetres in a normal year. On 11 and 12 June, over 400 millimetres fell – nearly a third of this occurred in just one hour! By 15 June, the area had received over 1650 millimetres of rain. Over 700 landslides were recorded in Hong Kong that month.

Some geographers believe that vegetation increased the problem. The trees held back many of the smaller landslides and allowed the larger ones, **washout**, to occur. Other forms of landslides included debris avalanches and rockslides.

At 1075 km², Hong Kong is one of the most densely populated urban areas worldwide, with a population of over 7 million (2015). It consists of the main island of Hong Kong, the peninsula of Kowloon, the New Territories and more than 230 islands with natural steep terrain and hills. The upper slopes are steeper than 30°. Most of the population is concentrated along the less steep urban areas on both sides of Victoria Harbour (Figure 3.30). With urban development, landslides are triggered by excavation and building works (Figures 3.31 and 3.32).

Figure 3.29 Hong Kong landscape

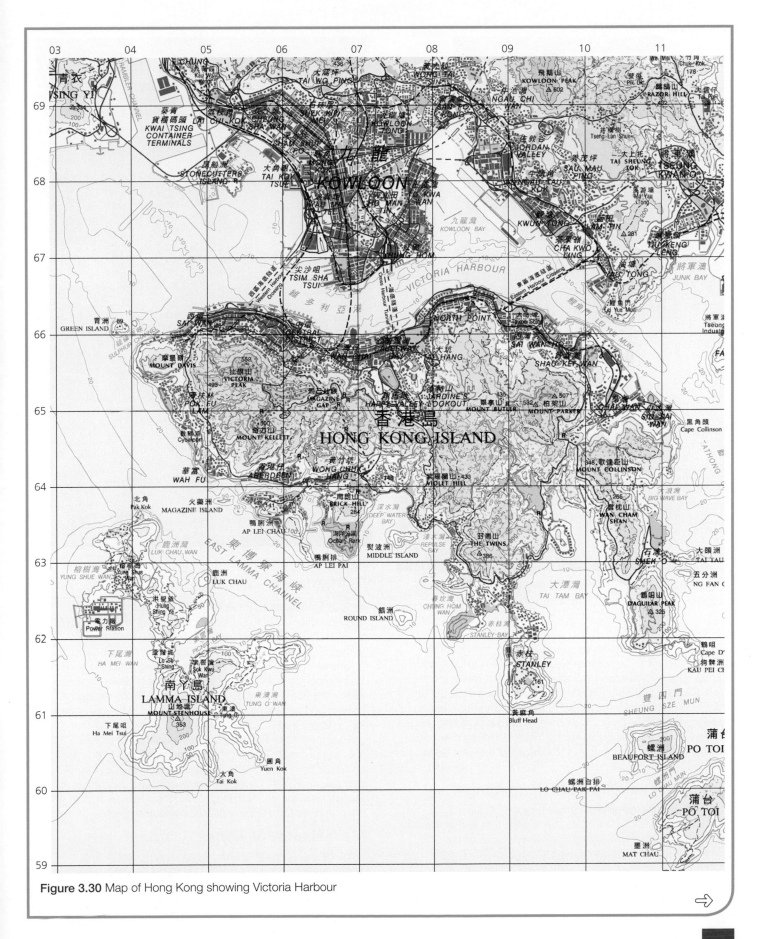

Figure 3.30 Map of Hong Kong showing Victoria Harbour

Figure 3.31 Landslide warning sign

Figure 3.32 Steep slopes and dense urban development combine to create a landslide risk

Figure 3.33 Drainage channel

Geology

The geology of Hong Kong is constructed mainly from three rock types: sedimentary rocks, granites and volcanic rocks. The sedimentary rocks generally form the lowlands. The granites and volcanic rocks, however, are situated on higher ground and are prone to failure. Both are seriously weathered, although granite rocks tend to be weathered more deeply than volcanic rocks. Volcanic rocks are more resilient and less prone to weathering and therefore less prone to slope failure.

Managing landslides in Hong Kong

The Hong Kong government has a responsibility to manage landslides. The Slope Safety System is managed by the Geotechnical Engineering Office (GEO) of the Civil Engineering Development Department (CEDD). The GEO has a staff establishment of over 700 for its wide range of activities. The GEO maintains its slope safety through investigating and researching the causes of significant and serious landslides to improve the Slope Safety Systems. The GEO is continuously updating, maintaining and disclosing the Catalogue of Slopes, which contains information of some 57 000 sizeable man-made slopes in Hong Kong.

One of Hong Kong's government interventions is to ensure that the private owners of slopes take responsibility for slope safety. If a slope owner does not comply with the regulation, prosecution will lead to a HK$50 000 fine, and to imprisonment for up to one year.

The government intervention in Hong Kong has had successful results. The risk from landslides has been reduced by 50 per cent since 1977. However, as a result of continued population growth, developers increasingly build further up the slopes. The risk from landslides, therefore, increases and the damages from a potential slide become greater.

Maintenance of slopes

Since heavy rainfall and surface runoffs are contributing to slope failure in Hong Kong, it is vital to remove excess water from slopes. Surface draining systems and protective covers are two methods used to protect slopes.

Surface drains are very vulnerable to blockage. Without proper drain maintenance, landslides are more common than on slopes without drains. Unfortunately, due to confusion over the responsibility, many drains are not properly maintained.

Man-made slopes are one of the main methods of slope stabilisation used in Hong Kong. These contain drains to intercept and direct water away from the slopes. The slope is usually protected from infiltration and the erosive effects of water by impermeable hard covers.

Greening techniques refer to the use of natural vegetation to reduce the risk of mass movements. There are three main types of greening techniques that are used in Hong Kong:

- The mulching system provides a protective cover that makes it possible for natural vegetation to grow on the slope; a natural vegetative cover is able to grow through the mat, securing it in place.
- The use of **long-rooting grass** is a fast and cost-effective system to cover man-made slopes. This system is applied by drilling planter holes into a hard cover. The drilled hole is then filled with soil mix and fertilisers, and finally the long-rooting grass is planted within.
- The **fibre reinforced soil system** is constructed by mixing polyester fibre into sandy soils. This mixture is capable of resisting tension.

Some of the advantages of greening techniques are outlined in Table 3.7.

Table 3.7 The advantages of greening techniques

Greening techniques	Advantages
Mulching system	Higher adhesive capacity on steep slopes High resistance to rain erosion High water-retaining capacity Long-lasting fertilisers Adaptable to rough surfaces
Planting long-rooting grass	Natural and environmentally friendly Cost-effective Fast and easy installation Can be applied on steep slopes Low maintenance
Fibre reinforced soil system	Self-sustained vegetation system with low maintenance Fibre strengthens soil particles to prevent erosion Visual improvement of the slope with various plant species Restoration of natural habitats on the slope

Section 3.4 Activities

1 Using the data in Table 3.8, draw a climate graph for Hong Kong. Describe the main characteristics of Hong Kong's climate.

Table 3.8 Climate data for Hong Kong

Month	Average temperature (°C)	Rainfall (mm)
January	16	30
February	15	60
March	18	70
April	22	133
May	25	332
June	28	479
July	28	286
August	28	415
September	27	364
October	25	33
November	21	46
December	17	17
Average/total	**23**	**2265**

2 Study Figure 3.28, which provides details of landslides in Hong Kong.
   a Using the data, describe and explain the relationship between mass movements and **i** rock type, **ii** gradient and **iii** vegetation.
   b What type of mass movement was most common in Hong Kong?
   c What do you think is the difference between a washout and a landslip? Give reasons for your answer.
   d Which type of rock was most affected by **a** washouts and **b** landslips?
   e What type of mass movement most affected **a** granite and **b** volcanic rocks? How do you explain these differences?
   f What is the relationship between gradient and mass movement? Give reasons for your answer.
   g What impact does vegetation have on the type and number of mass movements? Briefly explain your answer.
   h Briefly discuss the impact of human activity on mass movements. Use the evidence in Figure 3.28 to support your answer.
3 Study Figure 3.30, a map of Hong Kong. Using map evidence, suggest why landslides are a hazard in Hong Kong.
4 Suggest how population growth in Hong Kong contributes to the landslide hazard.
5 Describe the methods of landslide management that are used in Hong Kong.

4 Population

4.1 Natural increase as a component of population change

☐ Early humankind

The first hominids appeared in Africa around 5 million years ago, on a planet that is generally accepted to be 4600 million years old. They differed from their predecessors, the apes, in that they walked on two legs and did not use their hands for weight-bearing. During most of the period of early humankind, the global population was very small, reaching perhaps some 125 000 people a million years ago. It has been estimated that 10 000 years ago, when people first began to domesticate animals and cultivate crops, world population was no more than 5 million. Known as the Neolithic Revolution, this period of economic change significantly altered the relationship between people and their environments; but even then, the average annual growth rate was less than 0.1 per cent per year – extremely low compared with contemporary trends. This figure represents a rate of **natural increase** of one per thousand (1/1000). The rate of natural increase (or decrease) is the difference between the birth rate and the death rate. Most countries experience **natural increase** because of the excess of births over deaths.

However, as a result of technological advances, the **carrying capacity** of the land improved and population increased. By 3500 BCE, global population had reached 30 million, and by 0 CE this had risen to about 250 million (Figure 4.1).

Figure 4.1 The Great Wall of China – the history of the Great Wall goes back more than 2000 years when world population was only about 250 million

Demographers estimate that world population reached 500 million by about 1650. From this time, population grew at an increasing rate. By 1800, global population had doubled to reach 1 billion (Figure 4.2). Table 4.1 shows the time taken for each subsequent billion to be reached, with the global total reaching 6 billion in 1999. It had taken only 12 years for world population to increase from 5 billion to 6 billion – a year less than the time span required for the previous billion to be added. Global population reached 7 billion in October 2011, with another 12-year gap from the previous billion. The population in 2011 was double that in 1967. Alongside such rapid population growth has been much greater movement of population between countries, both on a short-term and long-term basis (Figure 4.3).

Figure 4.3 People from many countries at the Olympic Games in London, 2012

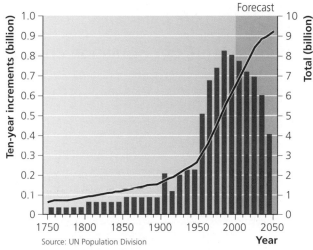
Figure 4.2 World population growth, 1750–2050

Table 4.1 World population growth by each billion

Each billion	Year	Number of years to add each billion
1st	1800	All of human history
2nd	1930	130
3rd	1960	30
4th	1974	14
5th	1987	13
6th	1999	12
7th	2011	12
8th	2024	13

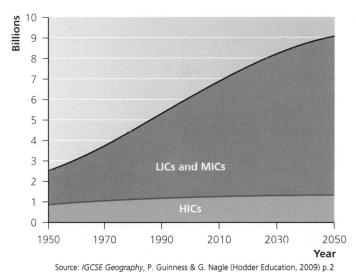
Source: *IGCSE Geography*, P. Guinness & G. Nagle (Hodder Education, 2009) p.2
Figure 4.4 Population growth in LICs and MICs and HICs, 1950–2050

✷ □ Recent demographic change

Figure 4.4 shows that both total population and the rate of population growth are much higher in the **low-income countries (LICs)** than in the **high-income countries (HICs)**. However, only since the Second World War has population growth in the LICs overtaken that in the HICs. The HICs had their period of high population growth in the nineteenth and early twentieth centuries, while for the LICs and MICs high population growth has occurred since 1950.

The highest ever global population growth rate was reached in the early to mid-1960s when population growth in the LICs and MICs peaked at 2.4 per cent a year. At this time, the term 'population explosion' was widely used to describe this rapid population growth, but by the late 1990s the rate of population growth was down to 1.8 per cent. However, even though the rate of growth has been falling for three decades, **population momentum** meant that the numbers being added each year did not peak until the late 1980s (Figure 4.5).

The demographic transformation, which took a century to complete in HICs, has occurred in a generation in some LICs and MICs. Fertility has dropped further and faster than most demographers foresaw 20 or 30 years ago. The exception is in Africa, where in over 20 countries families of at least five children are the norm and population growth is still around 2.5 per cent – this is a very high rate of natural increase.

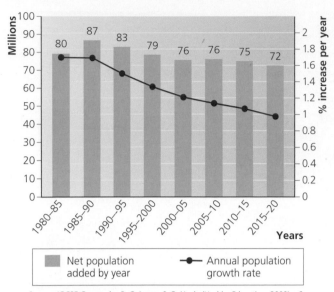

Figure 4.5 Population increase and growth rate in five-year periods, 1980–2020

Source: *IGCSE Geography*, P. Guinness & G. Nagle (Hodder Education, 2009) p.2

Table 4.2 shows the global population change in 2014. With 143.3 million births and 56.8 million deaths, global population increased by 86.6 million in 2014. Table 4.3 shows the ten most populous countries in the world in 2014 and the forecast for the top ten in 2050. Between them, China and India accounted for an astonishing 36.8 per cent of the world's population in 2014.

Table 4.2 World population clock, 2014

		World	HICs	LICs
Population		7 238 184 000	1 248 958 000	5 989 225 000
Births per	Year	143 341 000	13 794 000	129 547 000
	Day	392 714	37 792	354 923
	Minute	273	26	246
Deaths per	Year	56 759 000	12 328 000	44 432 000
	Day	155 505	33 775	121 730
	Minute	108	23	85
Natural increase (births – deaths/10) per	Year	86 581 000	1 466 000	85 115 000
	Day	237 209	4 017	233 193
	Minute	165	3	162
Infant deaths per	Year	5 507 000	72 000	5 435 000
	Day	15 087	197	14 890
	Minute	10	0.1	10

Source: 2014 World Population Data Sheet Population Reference Bureau

Table 4.3 Ten most populous countries in the world, 2014 and 2050

2014		2050 (projected)	
Country	Population (millions)	Country	Population (millions)
China	1364	India	1657
India	1296	China	1312
United States	318	Nigeria	396
Indonesia	251	United States	395
Brazil	203	Indonesia	365
Pakistan	194	Pakistan	348
Nigeria	177	Brazil	226
Bangladesh	158	Bangladesh	202
Russia	144	Democratic Republic of the Congo	194
Japan	127	Ethiopia	165

☐ The components of population change

Figure 4.6 illustrates the components of population change for world regions and smaller areas. In terms of the planet as a whole, natural change accounts for all population increase. Natural change is the balance between births and deaths, while **net migration** is the difference between immigration and emigration. The corrugated divide on Figure 4.6 indicates that the relative contributions of natural change and net migration can vary over time within a particular country, as well as between countries at any one point in time. The model is a simple graphical alternative to the population equation:

$$P = (B - D) \pm M$$

where P = population, B = births, D = deaths and M = migration.

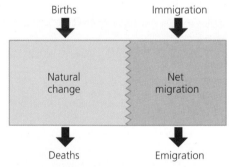

Source: *Advanced Geography: Concepts & Cases*, P. Guinness & G. Nagle (Hodder Education, 1999) p.17

Figure 4.6 Input–output model of population change

Natural change can be stated in absolute or relative terms. Absolute natural change gives the actual change in population as a result of the difference between the number of births and deaths, for example 200 000. Relative natural change is expressed as a rate per thousand, for example 3/1000. Table 4.4 shows natural change by world region for 2014.

Table 4.4 Birth rate, death rate and rate of natural change by world region, 2014

Region	Birth rate (per 1000)	Death rate (per 1000)	Rate of natural increase (%)
World	**20**	**8**	**1.2**
HICs	11	10	0.1
LICs, MICs	22	7	1.5
Africa	36	10	2.6
Asia	18	7	1.1
Latin America/ Caribbean	18	6	1.2
North America	12	8	0.4
Oceania	18	7	1.1
Europe	11	11	0

Source: Population Reference Bureau, 2014 World Population Data Sheet

Table 4.5 Variations in total fertility rate and the percentage of women using contraception by world region, 2014

Region	Total fertility rate	% of women using contraception (all methods)
World	**2.5**	**63**
HICs	1.6	70
MICs, LICs	2.6	61
Africa	4.7	34
Asia	2.2	66
Latin America/ Caribbean	2.2	73
North America	1.8	77
Oceania	2.4	62
Europe	1.6	70

Source: Population Reference Bureau, 2014 World Population Data Sheet

Section 4.1 Activities

1 Define the term *natural increase*.
2 Describe the change in world population since 1750 illustrated by Figure 4.2.
3 Describe and explain the trends shown in Figure 4.5.
4 Discuss the global variations in birth rate, death rate and rate of natural change shown in Table 4.4.

☐ The factors affecting levels of fertility

Fertility varies widely around the world. According to the 2014 World Population Data Sheet, the **crude birth rate**, the most basic measure of fertility, varied from a high of 50/1000 in Niger to a low of 6/1000 in Monaco. The word 'crude' means that the birth rate applies to the total population, taking no account of gender and age. The crude birth rate is heavily influenced by the age structure of a population. The crucial factor is the percentage of young women of reproductive age, as these women produce most children.

For more accurate measures of fertility, the 'fertility rate' and the 'total fertility rate' are used. The **fertility rate** is the number of live births per 1000 women aged 15–49 years in a given year. The **total fertility rate** is the average number of children that would be born alive to a woman (or group of women) during her lifetime, if she were to pass through her childbearing years conforming to the age-specific fertility rates of a given year. The total fertility rate varies from a high of 7.6 in Niger to a low of 1.1 in Taiwan. Table 4.5 shows the variations in total fertility rate by world region, alongside data for the percentage of women using contraception for each region. The latter is a major factor influencing fertility. Figure 4.7 shows in detail how the fertility rate varies by country around the world.

The factors affecting fertility can be grouped into four main categories:

- **Demographic** – Other population factors, particularly mortality rates, influence fertility. Where infant mortality is high, it is usual for many children to die before reaching adult life. In such societies, parents often have many children to compensate for these expected deaths.
- **Social/Cultural** – In some societies, particularly in Africa, tradition demands high rates of reproduction. Here, the opinion of women in the reproductive years may have little influence weighed against intense cultural expectations. Education, especially female literacy, is the key to lower fertility (Figure 4.8). With education comes a knowledge of birth control, greater social awareness, more opportunity for employment and a wider choice of action generally. In some countries, religion is an important factor. For example, the Muslim and Roman Catholic religions oppose artificial birth control. Most countries that have population policies have been trying to reduce their fertility by investing in birth-control programmes. Within LICs, it is usually the poorest neighbourhoods that have the highest fertility, due mainly to a combination of high infant mortality and low educational opportunities for women.
- **Economic** – In many LICs, children are seen as an economic asset because of the work they do and also because of the support they are expected to give their parents in old age. In HICs, the general perception is reversed and the cost of the child-dependency years is a major factor in the decision to begin or extend a family. Economic growth allows greater spending on health, housing, nutrition and education, which is important in lowering mortality and in turn reducing fertility. Also, the nature of employment can have an impact on fertility. Many companies, particularly in HICs, do not want to lose valuable female workers and therefore may provide workplace childcare and offer the opportunity of flexible working time.

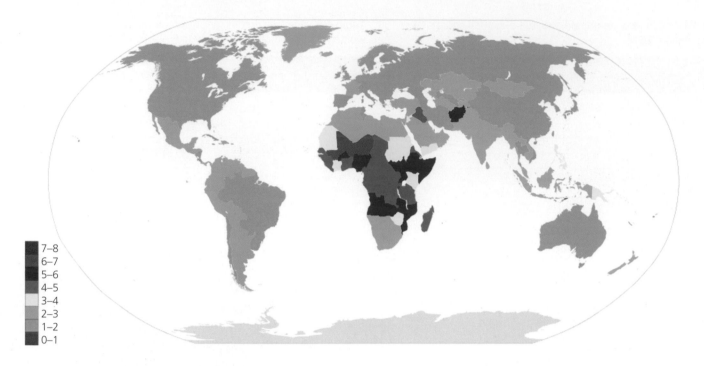

Key (Total fertility rate):
- 7–8
- 6–7
- 5–6
- 4–5
- 3–4
- 2–3
- 1–2
- 0–1

Figure 4.7 Total fertility rate by country, 2014

■ **Political** – There are many examples in the last century of governments attempting to change the rate of population growth for economic and strategic reasons. During the late 1930s, Germany, Italy and Japan all offered inducements and concessions to those with large families. In more recent years, Malaysia has adopted a similar policy. However, today, most governments that are interventionist in terms of fertility still want to reduce population growth.

Figure 4.8 The average age of marriage in a country is an important factor affecting fertility

Fertility can also be affected by general health factors such as being overweight or underweight, and using tobacco or alcohol. Being exposed to environmental hazards such as radiation, toxic chemicals or microwave emissions may reduce a woman's fertility.

The factors given above do not affect fertility directly; they influence another set of variables that determine the rate and level of childbearing. Figure 4.9 shows these 'intermediate variables'. These factors operate in every country, but their relative importance can vary from one country to another.

Fecundity
■ Ability to have a physical relationship
■ Ability to conceive
■ Ability to carry a pregnancy to term

Sexual unions*
■ The formation and dissolution of unions
■ Age at first physical relationship
■ Proportion of women who are married or in a union
■ Time spent outside a union (separated, divorced or widowed, for example)
■ Frequency of physical relationship
■ Sexual abstinence (religious or cultural customs, for example)
■ Temporary separations (military service, for example)

Birth control
■ Use of contraceptives
■ Contraceptive sterilisation
■ Induced abortion

*Includes marriage as well as long-term and casual relationships

Figure 4.9 The intermediate variables that affect fertility

Fertility decline

A study by the United Nations published in 2010 predicted that global population would peak at 10.1 billion in 2100, after reaching 9.3 billion by the middle of this century.

The global peak population has been continually revised downwards in recent decades. This is in sharp contrast to warnings in earlier decades of a population 'explosion'. The main reason for the slowdown in population growth is that fertility levels in most parts of the world have fallen faster than was previously expected.

In the second half of the 1960s, after a quarter-century of increasing growth, the rate of world population growth began to slow down. Since then, some LICs and MICs have seen the speediest falls in fertility ever known and thus earlier population projections did not materialise.

A fertility rate of 2.1 children per woman is replacement level fertility, below which populations eventually start falling. According to the 2014 World Population Data Sheet, there are almost 90 countries with total fertility rates at or below 2.1. This number is likely to increase. The movement to replacement level fertility is undoubtedly one of the most dramatic social changes in history, enabling many more women to work and children to be educated.

Figure 4.10 Lunch break at a school in Indonesia – increasing female literacy is an important factor In reducing fertility

Section 4.1 Activities

1 a Define the terms *fertility rate* and *total fertility rate*.
  b Why is the fertility rate a better measure of fertility than the crude birth rate?
2 a Describe and explain the differences in fertility by world region shown in Table 4.5.
  b Describe and attempt to explain the more detailed pattern of global fertility shown by Figure 4.7

3 How can **a** government policies and **b** religious philosophy influence fertility?
4 Why is replacement-level fertility an important concept?
5 Discuss the importance of three of the intermediate variables shown in Figure 4.9.

☐ The factors affecting mortality

Like crude birth rate, crude death rate is a very generalised measure of mortality. It is heavily influenced by the age structure of a population. For example, the crude death rate for the UK is 9/1000, compared with 6/1000 in Brazil. Yet life expectancy at birth in the UK is 81 years, compared with 75 years in Brazil. Brazil has a much younger population than the UK, but the average quality of life is significantly higher in the latter.

In 2014, the crude death rate varied around the world, from a high of 21/1000 in Lesotho to a low of 1/1000 in Qatar and the UAE. Table 4.6 shows variations by world region, and also includes data for infant mortality and life expectancy at birth. The infant mortality rate and life expectancy are much more accurate measures of mortality. The infant mortality rate is an age-specific rate; that is, it applies to one particular year of age.

Table 4.6 Death rate, infant mortality rate and life expectancy at birth by world region, 2014

Region	Crude death rate (per 1000)	Infant mortality rate (per 1000)	Life expectancy at birth (years)
World	8	38	71
HICs	10	5	79
MICs, LICs	7	42	60
Africa	10	62	59
Asia	7	34	71
Latin America/ Caribbean	6	18	75
North America	8	5	79
Oceania	7	21	77
Europe	11	6	78

Source: Population Reference Bureau, 2014 World Population Data Sheet

The causes of death vary significantly between the HICs and LICs (Figure 4.11; see also Figure 4.12). In LICs, infectious and parasitic diseases account for over 40 per cent of all deaths. They are also a major cause of disability and social and economic upheaval. In contrast, in HICs these diseases have a relatively low impact. In these countries, heart disease and cancer are the big killers. Epidemiology is the study of diseases. As countries develop, the ranking of major diseases tends to change from infectious to degenerative. This change is known as the 'epidemiological transition'.

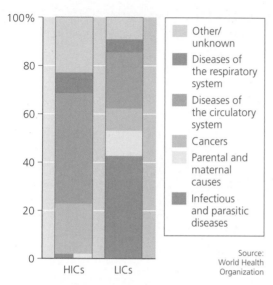

Figure 4.11 Contrasts in the causes of death between HICs and LICs

Figure 4.12 Drinking and dangerous driving warning to motorists in Mongolia – a significant cause of death among young men

Apart from the challenges of the physical environment in many LICs, a range of social and economic factors contribute to the high rates of infectious diseases. These include:

- poverty
- poor access to healthcare
- antibiotic resistance
- evolving human migration patterns
- new infectious agents.

When people live in overcrowded and insanitary conditions, communicable diseases such as tuberculosis and cholera can spread rapidly. Limited access to healthcare and medicines means that otherwise treatable conditions such as malaria and tuberculosis are often fatal to poor people. Poor nutrition and deficient immune systems are also key risk factors involved in many deaths from conditions such as lower respiratory infections, tuberculosis and measles.

Within most individual countries, variations in mortality occur due to:

- social class
- ethnicity
- place of residence
- occupation
- age structure of the population.

As Table 4.6 shows, there is a huge contrast in infant mortality by world region. Africa has the highest rate (62/1000), and North America (5/1000) the lowest rate. The variation among individual countries is even greater. In 2014, the highest infant mortality rates were in the Central African Republic (116/1000) and Congo D.R. (109/1000). In contrast, the lowest rate was in Hong Kong (1.7/1000). The infant mortality rate is frequently considered to be the most sensitive indicator of socio-economic progress, being heavily influenced by fundamental improvements in the quality of life, such as improvements in water supply, better nutrition and improved healthcare. Once children survive the crucial first year, their life chances improve substantially. Infant mortality in today's rich countries has changed considerably over time. In 1900, infant mortality in the USA was 120/1000. In 2013, it was down to 5.4/1000.

Table 4.6 shows that the lowest average life expectancy by world region is in Africa (59 years), with the highest average figure in northern America (79 years). Rates of life expectancy at birth have converged significantly between HICs and LICs during the last 50 years or so, in spite of a widening wealth gap. These increases in life expectancy have to a certain extent offset the widening disparity between per person incomes in HICs and LICs. However, it must not be forgotten that the ravages of AIDS in particular have caused recent decreases in life expectancy in some countries in Sub-Saharan Africa. It is likely that the life expectancy gap between HICs and LICs will continue to narrow in the future.

Section 4.1 Activities

1 Define the terms **a** *crude death rate*, **b** *infant mortality rate* and **c** *life expectancy*.
2 Why is crude death rate a very limited measure of mortality?
3 Using Figure 4.11, describe and explain the contrast in the causes of death between HICs and LICs.
4 To what extent does infant mortality vary around the world?
5 Discuss the main reasons for such large variations in infant mortality.
6 Describe the global variations in life expectancy.

☐ The interpretation of age/sex structure diagrams

The most studied aspects of **population structure** are age and gender (Figure 4.13). Other aspects of population structure that can be studied include race, language, religion and social/occupational groups.

Figure 4.13 Young children outside a *ger* in central Asia

Age and gender structures are conventionally illustrated by the use of **age/sex structure diagrams**. Diagrams can be used to portray either absolute or relative data. Absolute data show the figures in thousands or millions, while relative data show the numbers involved in percentages. The latter is most frequently used as it allows for easier comparison of countries of different population sizes. Each bar represents a five-year age-group, apart from the uppermost bar, which illustrates the population of a certain age and older, such as 85 or 100. The male population is represented to the left of the vertical axis, with females to the right.

Age/sex structure diagrams change significantly in shape as a country progresses through demographic transition (Figure 4.14).

■ The wide base in Niger's diagram reflects extremely high fertility. The birth rate in Niger is 50/1000, the highest in the world. The marked decrease in width of each successive bar indicates relatively high mortality and limited life expectancy. The death rate at 11/1000 is high,

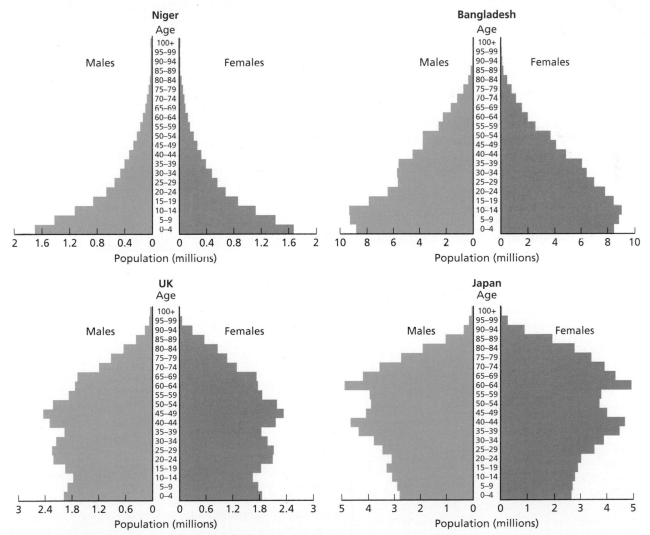

Source: *IGCSE Geography* 2nd edition, P. Guinness & G. Nagle (Hodder Education, 2014), p.27

Figure 4.14 Four age/sex structure diagrams

particularly considering how young the population is. The infant mortality rate has fallen steeply in recent decades to 54/1000. Life expectancy in Niger is 58 years; 50 per cent of the population is under 15, with only 3 per cent aged 65 or more. Niger is in stage 2 of demographic transition.

- The base of the second age/sex diagram, showing the population structure of Bangladesh, is narrower than that of Niger, reflecting a considerable fall in fertility after decades of government-promoted birth-control programmes. The reduced width of the youngest two bars compared with the 10–14 bar is evidence of recent falls in fertility. The birth rate is currently 20/1000. Falling mortality and lengthening life expectancy is reflected in the relatively wide bars in the teenage and young adult age groups. The death rate at 6/1000 is almost half that of Niger. The infant mortality rate is 33/1000. Life expectancy in Bangladesh is 70 years; 29 per cent of the population is under 15, while 5 per cent is 65 or over. Bangladesh is an example of a country in stage 3 of demographic transition.

- In the diagram for the UK, much lower fertility still is illustrated by narrowing of the base. The birth rate in the UK is only 12/1000. The reduced narrowing of each successive bar indicates a further decline in mortality and greater life expectancy compared with Bangladesh. The death rate in the UK is 9/1000, with an infant mortality rate of 3.9/1000. Life expectancy is 81 years; 18 per cent of the population is under 15, while 17 per cent is 65 or over. The UK is in stage 4 of demographic transition.

- The final diagram (Japan) has a distinctly inverted base, reflecting the lowest fertility of all four countries. The birth rate is 8/1000. The width of the rest of the diagram is a consequence of the lowest mortality and highest life expectancy of all four countries. The death rate is 10/1000, with infant mortality at 1.9/1000. Life expectancy is 83 years. Japan has only 13 per cent of its population under 15, with 26 per cent at 65 or over. With the birth rate lower than the death rate, Japan has entered stage 5 of demographic transition.

Figure 4.15 provides some useful tips for understanding age/sex structure diagrams. A good starting point is to divide the diagram into three sections:

- the young dependent population
- the economically active population
- the elderly dependent population.

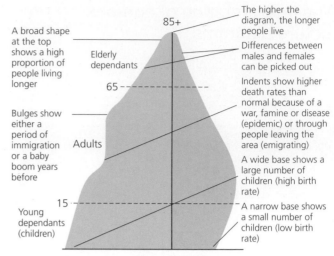

Figure 4.15 An annotated age/sex structure diagram

□ Population structure: differences within countries

In countries where there is strong rural-to-urban migration, the population structures of the areas affected can be markedly different. These differences show up clearly on age/sex diagrams. Out-migration from rural areas is age-selective, with single young adults and young adults with children dominating this process. Thus, the bars for these age groups in rural areas affected by out-migration will indicate fewer people than expected in these age groups. In contrast, the diagrams for urban areas attracting migrants will show age-selective in-migration, with substantially more people in these age groups than expected. Such migrations may also be sex-selective. If this is the case, it should be apparent on the diagrams.

Sex structure

The **sex ratio** is the number of males per 100 females in a population. In 2014, the global sex ratio at birth was estimated at 107 boys to 100 girls. Male births consistently exceed female births, for a combination of biological and social reasons. For example, more couples decide to complete their family on the birth of a boy than on the birth of a girl. However, after birth the gap generally begins to narrow until eventually females outnumber males, as at every age male mortality is higher than female mortality. This process happens most rapidly in the poorest countries, where infant mortality is markedly higher among males than females. Here the gap may be closed in less than a year.

However, there are anomalies to the picture just presented. In countries where the position of women is markedly subordinate and deprived, the overall sex ratio may show an excess of males. Such countries often exhibit high mortality rates in childbirth. For example, in India there are 107 males per 100 females for the population as a whole.

A recent report published in China recorded 118 male births for every 100 female births in 2010 due to the significant number of female foetuses aborted by parents intent on having a male child. Even within countries there can be significant differences in the sex ratio.

Figure 4.16 The women's section at a public event in India

The dependency ratio

Dependants are people who are too young or too old to work. The **dependency ratio** is the relationship between the working or economically active population and the non-working population. The formula for calculating the dependency ratio is:

$$\text{Dependency ratio} = \frac{\% \text{ population aged 0-14\% + population aged 65 and over}}{\% \text{ population aged 15-64}} \times 100$$

A dependency ratio of 60 means that for every 100 people in the economically active population there are 60 people dependent on them. The dependency ratio in HICs is usually between 50 and 75. In contrast, LICs typically have higher ratios, which may reach over 100. In LICs, children form the great majority of the dependent population. In contrast, in HICs and **MICs (middle-income countries)** there is a more equal balance between young and old dependants. Calculations of the **youth dependency ratio** and the **elderly dependency ratio** can show these contrasts more clearly.

$$\text{Youth dependency ratio} = \frac{\% \text{ population aged 0-14}}{\% \text{ population aged 15-64}} \times 100$$

$$\text{Elderly dependency ratio} = \frac{\% \text{ population aged 65 and over}}{\% \text{ population aged 15-64}} \times 100$$

For any country or region, the dependency ratio is equal to the sum of the youth dependency ratio and the elderly dependency ratio.

The dependency ratio is important because the economically active population will in general contribute more to the economy in income tax, VAT and corporation tax. In contrast, the dependent population tend to be bigger recipients of government funding, particularly for education, healthcare and public pensions. An increase in the dependency ratio can cause significant financial problems for governments if they do not have the financial reserves to cope with such a change.

The dependency ratio is an internationally agreed measure. Partly because of this it is a very crude indicator. For example:

- In HICs, few people leave education before the age of 18 and a significant number will go on to university and not get a job before the age of 21. In addition, while some people will retire before the age of 65, others will go on working beyond this age. Also, a significant number of people in the economically active age group do not work for various reasons, such as parents staying at home to look after children (Figure 4.17). The number of people in this situation can vary considerably from one country to another.
- In LICs, a significant proportion of children are working full or part-time before the age of 15. In some LICs, there is very high unemployment and underemployment within the economically active age group.

Figure 4.17 Early Learning Centre – this shop caters for the needs of young dependants, creating economic demand and jobs

However, despite its limitations, the dependency ratio does allow reasonable comparisons between countries (Table 4.7). It is also useful to see how individual countries change over time. Once an analysis using the dependency ratio has been made, more detailed research can look into any apparent anomalies.

Table 4.7 Dependency ratio calculations

Country	% population		Dependency ratio
	<15 years	65 and over	
USA	19	14	
Japan	13	26	
Germany	13	21	
UK	18	17	
Russia	16	13	
Brazil	24	7	
India	31	5	
China	16	10	
Nigeria	44	3	
Bangladesh	29	5	
Egypt	32	6	
Bolivia	35	5	

Source: Population Reference Bureau,
2014 World Population Data Sheet

4.2 Demographic transition

□ Changes in birth rate and death rate over time: the demographic transition model

Birth and death rates change over time. Although the birth and death rates of no two countries have changed in exactly the same way, some broad generalisations can be made about population growth since the middle of the eighteenth century. These generalisations are illustrated by the model of **demographic transition** (Figure 4.18), which is based on the experience of north-west Europe, the first part of the world to undergo such changes as a result of the significant industrial and agrarian advances that occurred during the eighteenth and nineteenth centuries.

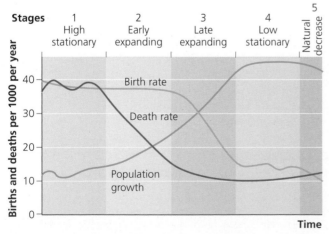

Source: *IGCSE Geography* by P. Guinness & G. Nagle (Hodder Education, 2003), p.7

Figure 4.18 Model of demographic transition

No country as a whole retains the characteristics of stage 1, which applies only to the most remote societies on Earth, such as isolated tribes in New Guinea and the Amazon who have little or no contact at all with the outside world. All the HICs of the world are now in stages 4 or 5, most having experienced all of the previous stages at different times. The poorest of the LICs (for example Bangladesh, Niger, Bolivia) are in stage 2, but are joined in this stage by a number of oil-rich Middle East nations where increasing affluence has not been accompanied by a significant fall in fertility. Most LICs that have registered significant social and economic advances are in stage 3 (for example Brazil, China, India), while some of the **newly industrialised countries (NICs)** such as South Korea and Taiwan have entered stage 4. With the passage of time, there can be little doubt that more countries will attain the demographic characteristics of the final stages of the model. The basic characteristics of each stage are as follows:

- **The high fluctuating stage (stage 1):** The crude birth rate is high and stable, while the crude death rate is high and fluctuating due to the sporadic incidence of famine, disease and war. In this stage, population growth is very slow and there may be periods of considerable decline. Infant mortality is high and life expectancy low (Figure 4.19). A high proportion of the population is under the age of 15. Society is pre-industrial, with most people living in rural areas, dependent on subsistence agriculture.

Figure 4.19 A graveyard dating from the eighteenth century in the UK – inscriptions show that life expectancy at that time was very low

Figure 4.20 Children on horses in Mongolia – a country in stage 3 of demographic transition

- **The early expanding stage (stage 2):** The death rate declines significantly. The birth rate remains at its previous level as the social norms governing fertility take time to change. As the gap between the two vital rates widens, the **rate of natural change** increases to a peak at the end of this stage. Infant mortality falls and life expectancy increases. The proportion of the population under 15 increases. Although the reasons for the decline in mortality vary somewhat in intensity and sequence from one country to another, the essential causal factors are: better nutrition; improved public health, particularly in terms of clean water supply and efficient sewerage systems; and medical advance. Considerable rural-to-urban migration occurs during this stage. However, for LICs in recent decades urbanisation has often not been accompanied by the industrialisation that was characteristic of the HICs during the nineteenth century.
- **The late expanding stage (stage 3):** After a period of time, social norms adjust to the lower level of mortality and the birth rate begins to decline. Urbanisation generally slows and average age increases. Life expectancy continues to increase and infant mortality to decrease. Countries in this stage usually experience lower death rates than nations in the final stage, due to their relatively young population structures (Figure 4.20).
- **The low fluctuating stage (stage 4):** Both birth and death rates are low. The former is generally slightly higher, fluctuating somewhat due to changing economic conditions. Population growth is slow. Death rates rise slightly as the average age of the population increases. However, life expectancy still improves as age-specific mortality rates continue to fall.
- **The natural decrease stage (stage 5):** In an increasing number of countries, the birth rate has fallen below the death rate, resulting in **natural decrease**. In the absence of net migration inflows, these populations are declining. Most countries in this stage are in eastern or southern Europe.

☐ Criticisms of the model

Critics of the demographic transition model see it as too Eurocentric as it was based on the experience of Western Europe. It is therefore not necessarily relevant to the experience of other countries. Critics argue that many LICs may not follow the sequence set out in the model. It has also been criticised for its failure to take into account changes due to migration.

The model presumes that all countries will eventually pass through all stages of the transition, just as the HICs have done. Because these countries have achieved economic success and enjoy generally high standards of living, completion of the demographic transition has come to be associated with socio-economic progress. This raises two major questions:

- Can LICs today hope to achieve either the demographic transition or the economic progress enjoyed by the HICs that passed through the transition at a different time and under different circumstances?

- Is the socio-economic change experienced by HICs a prerequisite or a consequence of demographic transition?

Demographic transition in LICs

There are a number of important differences in the way that LICs have undergone population change compared with the experiences of most HICs. In LICs:

- Birth rates in stages 1 and 2 were generally higher. About a dozen African countries currently have birth rates of 45/1000 or over. Twenty years ago, many more African countries were in this situation.
- The death rate fell much more steeply and for different reasons. For example, the rapid introduction of Western medicine, particularly in the form of inoculation against major diseases, has had a huge impact in reducing mortality. However, AIDS has caused the death rate to rise significantly in some countries, particularly in Sub-Saharan Africa.
- Some countries had much larger base populations and thus the impact of high growth in stage 2 and the early part of stage 3 has been far greater. No countries that are now classed as HICs had populations anywhere near the size of India and China when they entered stage 2 of demographic transition.
- For those countries in stage 3, the fall in fertility has also been steeper. This has been due mainly to the relatively widespread availability of modern contraception with high levels of reliability.
- The relationship between population change and economic development has been much more tenuous.

Different models of demographic transition

Although most countries followed the classical or UK model of demographic transition illustrated above, some countries did not. The Czech demographer Pavlik recognised two alternative types of population change, shown in Figure 4.21. In France, the birth rate fell at about the same time as the death rate and there was no intermediate period of high natural increase. In Japan and Mexico, the birth rate actually increased in stage 2 due mainly to the improved health of women in the reproductive age range.

✺ □ Changes in demographic indices over time

Fertility and mortality

Figure 4.22 illustrates change in birth and death rates in England and Wales between 1700 and 2000. The birth and death rates in stages 1, 2 and 3 broadly correspond to those in many poorer societies today. For example, infant mortality in England fell from 200/1000 in 1770 to just over 100/1000 in

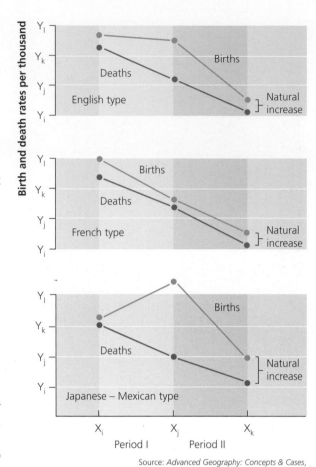

Source: *Advanced Geography: Concepts & Cases*, P. Guinness & G. Nagle (Hodder Education, 1999), p.3

Figure 4.21 Types of demographic transition

Section 4.2 Activities

1. What is a geographical model (such as the model of demographic transition)?
2. Explain the reasons for declining mortality in stage 2 of demographic transition.
3. Why does it take some time before the fall in fertility follows the fall in mortality (stage 3)?
4. Suggest why the birth rate is lower than the death rate in some countries (stage 5).
5. Discuss the merits and limitations of the model of demographic transition.
6. Why has the death rate in MICs and LICs fallen much more steeply over the last 50 years, compared with the fall in the death rate in earlier times in HICs?

1870. Today, the average infant mortality for the developing world as a whole is 50/1000. Only the very poorest countries in the world today have infant mortality rates over 100/1000. Infant mortality is regarded as a key measure of socio-economic development. Figure 4.22 identifies a range of important factors that influenced birth and death rates in England and Wales during the time period concerned.

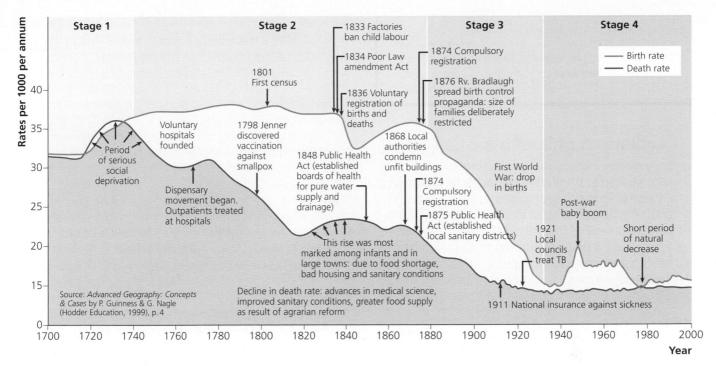

Figure 4.22 England and Wales – changes in birth and death rates, 1700–2000

☐ Issues of ageing populations

Different patterns of population growth can bring both benefits and problems to the countries concerned. This is particularly the case when countries have a very high percentage of either young or old people in their populations.

According to the United Nations, 'Population ageing is unprecedented, without parallel in human history and the twenty-first century will witness even more rapid ageing than did the century just past.' In western Europe in 1800, less than 25 per cent of men would live to the age of 60. Today, more than 90 per cent do.

The world's population is ageing significantly. **Ageing of population** (demographic ageing) is a rise in the median age of a population. It occurs when fertility declines while life expectancy remains constant or increases.

The following factors have been highlighted by the United Nations:

- The global average for life expectancy increased from 46 years in 1950 to 71 in 2014. It is projected to reach 76 years by 2050.
- In LICs, the population aged 60 years and over is expected to quadruple between 2000 and 2050.
- In HICs, the number of older people was greater than that of children for the first time in 1998. By 2050, older people in HICs will outnumber children by more than two to one.
- The population aged 80 years and over numbered 69 million in 2000. This was the fastest-growing section of the global population, which is projected to increase to 375 million by 2050.

The progressive demographic ageing of the older population itself – globally, the average annual growth rate of people aged 80 and over (3.8 per cent) is twice as high as the growth of population 60 and over (1.9 per cent).

- Europe is the 'oldest' region in the world, but Japan is the oldest nation with a median age of 45 years.
- Africa is the 'youngest' region in the world, with the proportion of children (under 15) accounting for 41 per cent of the population today. However, this is expected to decline to 28 per cent by 2050. In contrast, the proportion of older people is projected to increase from 5 per cent to 10 per cent over the same time period.

Table 4.8 shows that 8 per cent of the world's population are aged 65 years and over. On a continental scale, this varies from only 4 per cent in Africa to 17 per cent in Europe. Population projections show that the world population in the age group 65 years and over will rise to 10 per cent in 2025, and to 16 per cent by 2050.

Table 4.8 The percentage of total population aged 65 years and over, 2014

Region	Population aged 65 and over (%)
World	**8**
Africa	4
North America	14
Latin America/Caribbean	7
Asia	7
Europe	17
Oceania	11

Source: Population Reference Bureau, 2014 World Population Data Sheet

4.2 Demographic transition   97

The problem of demographic ageing has been a concern of HICs for some time, but it is now also beginning to alarm MICs and LICs. Although ageing has begun later in MICs and LICs, it is progressing at a faster rate. This follows the pattern of previous demographic change, such as declining mortality and falling fertility, where change in MICs and LICs was much faster than that previously experienced by HICs.

Demographic ageing will put healthcare systems, public pensions and government budgets in general under increasing pressure (Figure 4.23). Four per cent of the USA's population was 65 years of age and older in 1900. By 1995, this had risen to 12.8 per cent and by 2030 it is likely that one in five Americans will be senior citizens. The fastest-growing segment of the population is the so-called 'oldest-old': those who are 85 years or more. It is this age group that is most likely to need expensive residential care. The situation is similar in other HICs.

Figure 4.23 An elderly woman – demographic ageing is a worldwide phenomenon

Some countries have made relatively good pension provision by investing wisely over a long period of time. However, others have more or less adopted a pay-as-you-go system, as the elderly dependent population rises. It is this latter group who will be faced with the biggest problems in the future.

For much of the post-1950 period, the dominant demographic problem has been generally perceived as the 'population explosion', a result of very high fertility in LICs. However, greater concern is now being expressed about demographic ageing in many countries where difficult decisions about the reallocation of resources are having

to be made. At present, very few countries are generous in looking after their elderly. Poverty amongst the elderly is a considerable problem, but technological advance might provide a solution by improving living standards for everyone. If not, other less popular solutions, such as increased taxation, will have to be examined.

However, some demographers argue that there needs to be a certain rethinking of age and ageing, with older people adopting healthier and more adventurous lifestyles than people of the same age only one or two generations ago. Sayings such as '50 is the new 40' have become fairly commonplace. It is argued that we should not just think of chronological age, but also of prospective age – the remaining years of life expectancy people have (Table 4.9).

Table 4.9 Remaining life expectancy among French women, 1952 and 2005

Year	Years lived	Remaining life expectancy (years)
1952	30	44.7
2005	30	54.4

Source: **Population Bulletin** *Vol.63 No.4 2008*

It is of course easy to underestimate the positive aspects of ageing:

- Many older people make a big contribution to childcare by looking after their grandchildren.
- Large numbers of older people work as volunteers, for example in charity shops.

Section 4.2 Activities

1 Describe the variations shown in Table 4.8.
2 Why is a large elderly dependent population generally viewed as a problem?
3 Discuss the possible benefits of a large elderly population.
4 Briefly explain the data shown in Table 4.9.

Case Study: Population ageing in Japan

Japan has the most rapidly ageing population in the history of the world:

- 33 per cent of Japan's people are over the age of 60.
- Japan has the world's oldest population, with a median age of 46 years.
- The country's population peaked between 2005 and 2010 at 128 million (Figure 4.24). The most extreme population projection predicts a decline of 50 million by the end of this century.
- Fertility has declined substantially and the total fertility rate is an extremely low 1.4.
- No other country has a lower percentage of its population under 15 (13 per cent).

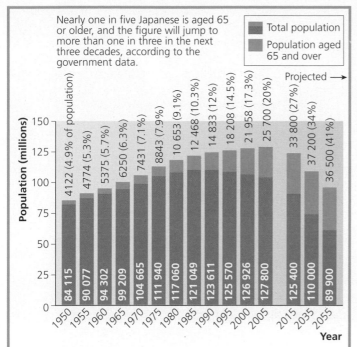

Nearly one in five Japanese is aged 65 or older, and the figure will jump to more than one in three in the next three decades, according to the government data.

Legend:
- Total population
- Population aged 65 and over

Projected →

Figure 4.24 Population trends in Japan

A high elderly dependency ratio presents considerable economic and social challenges to the country, not least in terms of pensions, healthcare and long-term care. Japan's workforce peaked at 67.9 million in 1998 and has been in decline since. This presents an increasing economic burden on the existing workforce. However, it must be noted that there is a high labour-force participation rate among the elderly. Japanese men work an average of five years after mandatory retirement.

Japan has a long tradition of positive attitudes towards older people. Every year, National Respect the Aged Day is a public holiday. However, while there is a strong tradition of elderly people being looked after by their families, the number of old people living in care homes or other welfare facilities is steadily rising. The cost of care is shared between the elderly person, their family and the government. As the number of people in this situation increases, more pressure is placed on the country's economy. Social changes are also occurring, for example the emergence of ageing as a theme in films and books.

Younger workers are at a premium and there is considerable competition to recruit them. One solution is for manufacturers to set up affiliated companies in China or other countries, but past results have been mixed. The possibility of expanding immigration to help reduce the rising dependency ratio appears to be politically unacceptable in Japan. Foreigners make up only 1 per cent of Japan's labour force. Legal immigration is practically impossible (except for highly skilled ethnic Japanese workers) and illegal immigration is strictly suppressed.

The UN predicts that by 2045 for every four Japanese aged 20–64 there will be three people aged 65 or over. The key question is: what is a socially acceptable level of provision for the elderly in terms of the proportion of the country's total GDP? This is a question many other countries are going to have to ask themselves as well. Pension reforms have been implemented, with later retirement and higher contributions from employers. However, it is likely that further changes will be required as the cost of ageing rises.

Section 4.2 Activities

1 Describe the changes in Japan's population shown in Figure 4.24.
2 Why does Japan have a rapidly ageing population?
3 Suggest why this trend may contribute to changing attitudes towards the elderly.

☐ Issues of youthful populations

Rapid population growth results in a large young dependent population. The young dependent population is defined as the population under 15 years of age. Table 4.10 shows the huge variation around the world average of 26 per cent. The 41 per cent for Africa is over two and a half times higher than the figure for Europe. The highest figures for individual countries are in Niger (50 per cent), Chad (49 per cent), Somalia, Uganda, Angola and Mali (all 48 per cent).

Table 4.10 The percentage of total population under 15 years of age (2013)

Country	% population under 15 years of age
World	26
Africa	41
North America	19
Latin America/Caribbean	27
Asia	25
Europe	16
Oceania	24

Source: Selected data from the 2014 World Population Data Sheet

Countries with large young populations (Figure 4.25) have to allocate a substantial proportion of their national resources to look after them. Young people require resources for health, education, food, water and housing above all. The money required to cover such needs may mean there is little left to invest in agriculture, industry and other aspects of the economy. A LIC government might see this as being too large a demand on the country's resources and as a result may introduce family planning policies to reduce the birth rate. However, individual parents may have a different view, where they see a large family as valuable in terms of the work children can do on the land. Alongside this, people in poor countries often have to rely on their children in old age because of the lack of state welfare benefits.

Figure 4.25 Young adults at a gathering point in an urban area

As a large young population moves up the age ladder over time, it will provide a large working population when it enters the economically active age group (15–64). This will be an advantage if a country can attract sufficient investment to create enough jobs for a large working population. Then, the large working population will contribute a lot of money in taxes to the country, which can be invested in many different ways to improve quality of life and to attract more foreign investment. Such a situation can create an upward spiral of economic growth. On the other hand, if there are few employment opportunities for a large working population, the unemployment rate will be high. The government and most individuals will have little money to spend and the quality of life will be low. Many young adults may seek to emigrate because of the lack of opportunities in their own country.

Eventually, the large number of people in this age group will reach old age. If most of them enter old age in poverty, this creates even more problems for the government.

☐ The link between population and development

Development, or improvement in the quality of life, is a wide-ranging concept. It includes wealth, but it also includes other important aspects of our lives. For example, many people would consider good health to be more important than wealth. People who live in countries that are not democracies

Case Study: The Gambia

A country with a high young dependent population

■ The Gambia has a young and fast-growing population, which has placed big demands on the resources of the country.

■ 95 per cent of the country's population is Muslim, and until recently religious leaders were against the use of contraception. In addition, cultural tradition meant that women had little influence on family size.

■ Children were viewed as an economic asset because of their help with crop production and tending animals. One in three children aged 10–14 are working.

■ In 2012, the infant mortality rate was 70/1000. With 44 per cent of the population classed as young dependents and only 2 per cent elderly dependents, the dependency ratio is 85.

■ Many parents in the Gambia struggle to provide basic housing for their families. There is huge overcrowding and a lack of sanitation, with many children sharing the same bed.

■ Rates of unemployment and underemployment are high and wages are low, with parents struggling to provide even the basics for large families.

■ Many schools operate a two-shift system, with one group of pupils attending in the morning and a different group attending in the afternoon.

■ Another sign of population pressure is the large number of trees being chopped down for firewood. As a result, desertification is increasing at a rapid rate.

■ In recent years, the government has introduced a family-planning campaign, which has been accepted by religious leaders.

■ Family planning programmes have had limited success to date, with the total fertility rate falling from 6.1 in 1970 to just 5.6 in 2013.

and where freedom of speech cannot be taken for granted often envy those who do live in democratic countries. Development occurs when there are improvements to the individual factors making up the quality of life. For example, development occurs in a LIC when:

- the local food supply improves due to investment in machinery and fertilisers
- the electricity grid extends outwards from the main urban areas to rural areas
- a new road or railway improves the **accessibility** of a remote province
- levels of literacy improve throughout the country
- average incomes increase above the level of inflation.

There has been much debate about the causes of development. Detailed studies have shown that variations between countries are due to a variety of factors:

Physical geography

- Landlocked countries have generally developed more slowly than coastal ones.
- Small island countries face considerable disadvantages in development.
- Tropical countries have grown more slowly than those in temperate latitudes, reflecting the cost of poor health and unproductive farming in the former. However, richer non-agricultural tropical countries, such as Singapore, do not suffer a geographical deficit of this kind.
- A generous allocation of natural resources has spurred economic growth in a number of countries.

Economic policies

- Open economies that welcomed and encouraged foreign investment have developed faster than closed economies.
- Fast-growing countries tend to have high rates of saving and low spending relative to GDP.
- Institutional quality in terms of good government, law and order and lack of corruption generally result in a high rate of growth.

Demography

Progress through demographic transition is a significant factor, with the highest rates of economic growth experienced by those nations where the birth rate has fallen the most.

The Human Development Index

In 1990, the **Human Development Index (HDI)** was devised by the United Nations to indicate levels of development in countries. The HDI contains four variables:

- life expectancy at birth
- mean years of schooling for adults aged 25 years

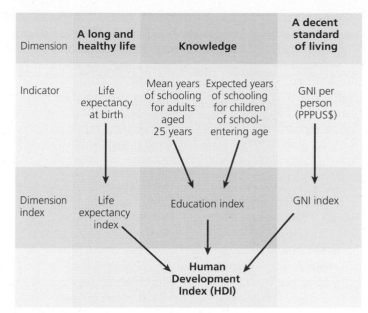

Figure 4.26 Constructing the Human Development Index

- expected years of schooling for children of school-entering age
- GNI per person (PPP).

One of the four variables used in the HDI is therefore a key demographic factor. The actual figures for each of these four measures are converted into an index (Figure 4.26) that has a maximum value of 1.0 in each case. The four index values are then combined and averaged to give an overall Human Development Index value. This also has a maximum value of 1.0. Every year, the United Nations publishes the *Human Development Report*, which uses the HDI to rank all the countries of the world in terms of their level of development.

Various academic studies have concluded that there is no straightforward relationship between population and economic growth. Thus some economies with a low level of economic growth may not be hugely affected by population growth, but are more affected by other factors such as political instability and lack of investment. On the other hand, some economies that achieve a high level of economic growth may not have done so mainly because of declining population growth, but due to other factors.

Table 4.11 shows the top 25 countries listed in the *Human Development Report* for 2014. All 25 countries are in stage 4 (or stage 5) of demographic transition, suggesting a very strong link between the rate of population growth and the level of economic development (Figure 4.27). Two first-generation NICs are on the list – Singapore and Hong Kong. In both of these countries, the rate of natural increase declined as economic growth progressed. Of course, the debate is – which comes first? Does economic growth lead to lower natural increase, or vice versa? Or is there a more complex relationship between the two variables?

Table 4.11 Top 25 countries in the *Human Development Report 2014*

HDI rank	Country	HDI value 2013	Life expectancy at birth (years) 2013	Mean years of schooling (years) 2012	Expected years of schooling (years) 2012	GNI per person (2011 PPP$) 2013	HDI value 2013
1	Norway	0.944	81.5	12.6	17.6	63909	0.943
2	Australia	0.933	82.5	12.8	19.9	41524	0.931
3	Switzerland	0.917	82.6	12.2	15.7	53762	0.916
4	Netherlands	0.915	81.0	11.9	17.9	42397	0.915
5	United States	0.914	78.9	12.9	16.5	52308	0.912
6	Germany	0.911	80.7	12.9	16.3	43049	0.911
7	New Zealand	0.910	81.1	12.5	19.4	32569	0.908
8	Canada	0.902	81.5	12.3	15.9	41887	0.901
9	Singapore	0.901	82.3	10.2	15.4	72371	0.899
10	Denmark	0.900	79.4	12.1	16.9	42880	0.900
11	Ireland	0.899	80.7	11.6	18.6	33414	0.901
12	Sweden	0.898	81.8	11.7	15.8	43201	0.897
13	Iceland	0.895	82.1	10.4	18.7	35116	0.893
14	UK	0.892	80.5	12.3	16.2	35002	0.890
15	Hong Kong , China (SAR)	0.891	83.4	10.0	15.6	52383	0.889
15	Korea (Republic of)	0.891	81.5	11.8	17.0	30345	0.888
17	Japan	0.890	83.6	11.5	15.3	36747	0.888
18	Liechtenstein	0.889	79.9	10.3	15.1	87085	0.888
19	Israel	0.888	81.8	12.5	15.7	29966	0.886
20	France	0.884	81.8	11.1	16.0	36629	0.884
21	Austria	0.881	81.1	10.8	15.6	42930	0.880
21	Belgium	0.881	80.5	10.9	16.2	39471	0.880
21	Luxembourg	0.881	80.5	11.3	13.9	58695	0.880
24	Finland	0.879	80.5	10.3	17.0	37366	0.879
25	Slovenia	0.874	79.6	11.9	16.8	26809	0.874

Source: United Nations Human Development Report 2014

Figure 4.27 The Waterfront, San Francisco – the USA has a very high level of human development

Figure 4.28 shows people living in a country at a low level of human development. Table 4.12 shows the human development index values for each category: very high, high, medium and low human development. The countries with low human development invariably have high rates of population growth and most are in stage 2 of demographic transition. The more advanced LICs are generally in stage 3 of demographic transition. This includes countries such as Brazil, Mexico, India and Malaysia. However, again it must be stated that the development process is complex and is the result of the interaction of a wide range of factors.

Figure 4.28 A fishing village in Papua New Guinea – a low human development country

Table 4.12 Human Development Index values

Level of human development	HDI value	Number of countries 2013
Very high	0.8 and over	49
High	0.7–0.799	53
Medium	0.55–0.699	42
Low	Below 0.55	43

Source: Human Development Report 2014

 Infant mortality: changes over time

The infant mortality rate is a significant measure of the general health of a population. It has a big influence on life expectancy at birth. The most recent data for infant mortality by world region is shown in Table 4.6 on page 89. The global average (2014) is now down to 38/1000 after a very considerable decline in recent decades. The first edition of this book gave 2010 data, which showed the global average for infant mortality at 46/1000. A fall of 8/1000 is a major success for global health improvement in such a short time period. Figure 4.29 shows how much infant mortality has fallen since 1950 and the UN projections to 2050.

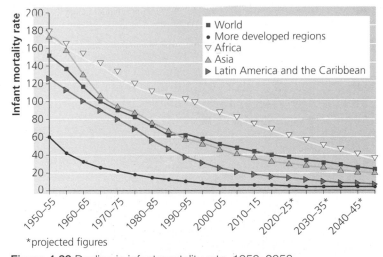

*projected figures

Figure 4.29 Decline in infant mortality rate, 1950–2050

To put the current figures in context, a review of infant mortality at the beginning of the twentieth century found that only one country in the world, Sweden, had an infant mortality rate of less than 100/1000 (or 10 per cent). This means that for the world as a whole, more than 10 per cent of all live births did not survive until their first birthday. In many countries, the figure was above 20 per cent, or 200/1000!

The factors responsible for the considerable decline in global infant mortality include:

- better nutrition
- improvements in public health, particularly with regard to water supply and sanitation
- significant medical advances, particularly in the field of paediatrics
- improvements in housing and other human environmental conditions
- the introduction of better maternity conditions (maternity leave, and so on) for new mothers.

A child's risk of dying is greatest in the first 28 days of life, known as the **neonatal period**. There have been major advances in caring for infants during this period worldwide. The number of deaths globally of children under 28 days fell from 4.7 million in 1990 to 2.8 million in 2013. In the same period, deaths of children under 1 year of age declined from 8.9 million in 1990 to 4.6 million in 2013.

Changes in life expectancy

The decline in levels of mortality and the increase in life expectancy has been the most important reward of economic and social development. On a global scale, 75 per cent of the total improvement in longevity has been achieved in the twentieth century and the early years of the twenty-first century. In 1900, the world average life expectancy is estimated to have been about 30 years, but by 1950–55 it had risen to 46 years. By 1980–85 it had reached a fraction under 60 years. The current global average is 70 years. Here there is a four-year gap between males and females (69 and 73 years). The gender gap is wider in HICs (75 and 82 years) than in LICs (67 and 71 years). It is likely that the life expectancy gap between HICs and LICs/MICs will continue to narrow in the future (Figures 4.30 and 4.31).

Figure 4.30 The provision of water pumps in this village in Nepal has improved life expectancy

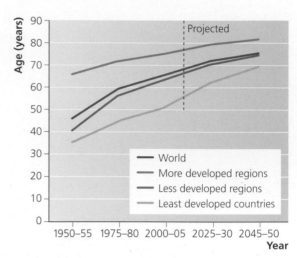

Figure 4.31 Life expectancy at birth – world and development regions, 1950–2050

4.3 Population–resource relationships

☐ Food security

The definition of **food security** is when all people at all times have access to sufficient, safe, nutritious food to maintain a healthy and active life. The total amount of food produced around the world today is enough to provide everyone with a healthy diet. The problem is that while some countries produce a food surplus or have enough money to buy it elsewhere, other countries are in food deficit and lack the financial resources to buy enough food abroad. About one in nine of the world's population remains chronically undernourished. The global distribution of undernourishment has changed significantly in recent decades.

Forecasts of famine tend to appear every few decades or so. In 1974, a world food summit held in Rome met against a background of rapidly rising food prices and a high rate of global population growth. The major concern was that the surge in population would overwhelm humankind's ability to produce food in the early twenty-first century. The next world food summit, again hosted by Rome, was held in 1996. It too met against a background of rising prices and falling stocks. But new concerns, unknown in 1974, had appeared. Global warming threatened to reduce the productivity of substantial areas of land and many scientists were worried about the long-term consequences of genetic engineering.

Rapidly rising food prices (Figure 4.32) in recent years and a range of other problems associated with food production have resulted in the frequent use of the term 'global food crisis'. At the World Summit on Food Security held in Rome in November 2009, the FAO Director-General Jacques Diouf referred to the over 1 billion hungry people in the world as 'our tragic achievement in these modern days'. He stressed the need to produce food where poor and hungry people live and to boost agricultural investment in these regions.

There is a huge geographical imbalance between food production and food consumption, resulting in a lack of food security in many countries. The three main strands of food security are:

- **food availability** – sufficient quantities of food available on a consistent basis
- **food access** – having sufficient resources to obtain appropriate foods for a nutritious diet
- **food use** – appropriate use based on knowledge of basic nutrition and care, as well as adequate water and sanitation.

Figure 4.32 Selling local produce in a Vietnamese food market

Figure 4.33 shows the Food Security Risk index for 2013. The greatest degree of risk is in Africa, the Middle East and Asia.

The current food crisis presents three fundamental threats, which are:

- pushing more people into poverty
- eroding the development gains that have been achieved in many countries in recent decades
- presenting a strategic threat by endangering political stability in some countries; a significant number of countries have experienced food-related riots and unrest in recent years.

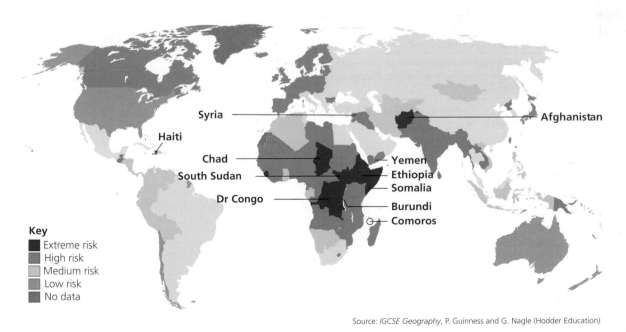

Key
- ■ Extreme risk
- ■ High risk
- ■ Medium risk
- ■ Low risk
- ■ No data

Source: *IGCSE Geography*, P. Guinness and G. Nagle (Hodder Education)

Figure 4.33 World map – Food Security Risk Index, 2013

Table 4.13 summarises some of the current adverse influences on food supply and distribution. LICs have long complained about the subsidies that the European Union and other HICs give to their farmers and the import tariffs they impose on food products coming from elsewhere. This denies valuable markets to many LICs. On the other side of the coin, production for local markets has declined in some LICs because of increasing production for export markets. Agricultural transnational corporations (TNCs) are often the driving force behind this trend, but governments in LICs are also anxious to increase exports to obtain foreign currency.

Table 4.13 Adverse influences on global food production and distribution

Nature of adverse influence	Effect of adverse influence
Economic	Demand for cereal grains has outstripped supply in recent years Rising energy prices and agricultural production and transport costs have pushed up costs all along the farm-to-market chain Serious underinvestment in agricultural production and technology in LICs has resulted in poor productivity and underdeveloped rural infrastructure The production of food for local markets has declined in many LICs as more food has been produced for export
Ecological	Significant periods of poor weather and a number of severe weather events have had a major impact on harvests in key food-exporting countries Increasing problems of soil degradation in both HICs and LICs Declining biodiversity may impact on food production in the future
Socio-political	The global agricultural production and trading system, built on import tariffs and subsidies, creates great distortions, favouring production in HICs and disadvantaging producers in LICs An inadequate international system of monitoring and deploying food relief Disagreements over the use of trans-boundary resources such as river systems and aquifers

☐ The causes and consequences of food shortages

About 800 million people in the world suffer from hunger. The problem is mainly concentrated in Africa but also affects a number of Asian and Latin American countries. Food shortages can occur because of both natural and human problems. The natural problems that can lead to food shortages include:

- soil exhaustion
- drought
- floods
- tropical cyclones
- pests
- disease.

However, economic and political factors can also contribute to food shortages. Such factors include:

- low capital investment
- rapidly rising population
- poor distribution/transport difficulties
- conflict situations.

The impact of such problems has been felt most intensely in LICs, where adequate food stocks to cover emergencies affecting food supply usually do not exist. However, HICs have not been without their problems. Thus HICs are not immune from the physical problems that can cause food shortages. However, they invariably have the human resources to cope with such problems, so actual food shortages do not generally occur.

The effects of food shortages are both short term and longer term. **Malnutrition** can affect a considerable number of people, particularly children, within a relatively short period when food supplies are significantly reduced. With malnutrition people are less resistant to disease and more likely to fall ill. Such diseases include beri-beri (vitamin B1 deficiency), rickets (vitamin D deficiency) and kwashiorkor (protein deficiency). People who are continually starved of nutrients never fulfil their physical or intellectual potential. Malnutrition reduces people's capacity to work, so the land may not be properly tended, and other economic activities may not be pursued to their full potential. This is threatening to lock parts of the developing world into an endless cycle of ill-health, low productivity and underdevelopment.

☐ The role of technology and innovation in the development of food production

The world has passed through many stages of food production, each marked by significant technological advance. Never before has the world's production-to-consumption food system been more complex. Contemporary food science and technology have made major contributions to the success of modern food systems by integrating key elements of many different academic disciplines to solve difficult problems, such as improving nutritional deficiencies and food safety.

Another huge challenge is the large and growing food-security gap around the world. As much as half of the food grown and harvested in LICs is not consumed, partly because proper handling, processing, packaging and distribution methods are lacking. Starvation and nutritional deficiencies in vitamins, minerals, protein and calories are still prevalent in many parts of the world. Science-based improvements in agricultural production,

Case Study: Sudan and South Sudan

The countries of Sudan and South Sudan (Figure 4.34), which were the single country of Sudan until 2011, have suffered food shortages for decades. The long civil war and drought have been the main reasons for famine in Sudan, but there are many associated factors as well (Figure 4.35). The civil war, which lasted for over 20 years, was between the government in Khartoum and rebel forces in the western region of Darfur and in the south (now South Sudan). A Christian Aid document in 2004 described the Sudan as 'A country still gripped by a civil war that has been fuelled, prolonged and part-financed by oil'. One of the big issues between the two sides in the civil war was the sharing of oil wealth between the government-controlled north and the south of the country where much of the oil is found. The United Nations has estimated that up to 2 million people were displaced by the civil war and more than 70 000 people died from hunger and associated diseases. At times, the UN World Food Programme stopped deliveries of vital food supplies because the situation was considered too dangerous for the drivers and aid workers.

Figure 4.34 Sudan and South Sudan

Figure 4.36 The fertile banks of the River Nile in Sudan with desert beyond

Physical factors	Social factors	Agricultural factors	Economic/political factors
• Long-term decline of rainfall in southern Sudan • Increased rainfall variability • Increased use of marginal land leading to degradation • Flooding	• High population growth (3%) linked to use of marginal land (overgrazing, erosion) • High female illiteracy rates (65%) • Poor infant health • Increased threat of AIDS	• Highly variable per person food production; long-term the trend is static • Static (cereals and pulses) or falling (roots and tubers) crop yields • Low and falling fertiliser use (compounded by falling export receipts) • Lack of a food surplus for use in crisis	• High dependency on farming (70% of labour force; 37% of GDP) • Dependency on food imports (13% of consumption 1998–2000) whilst exporting non-food goods, e.g. cotton • Limited access to markets to buy food or infrastructure to distribute it • Debt and debt repayments limit social and economic spending • High military spending

Drought in southern Sudan compounds low food intake; any remaining surpluses quickly used

Shorter-term factors leading to increased Sudanese food insecurity and famine

Conflict in Darfur reduces food production and distribution

Both reduce food availability in Sudan and inflate food prices

Situation compounded by:
• Lack of government political will
• Slow donor response
• Limited access to famine areas
• Regional food shortages

Figure 4.35 Summary of causes of famine in Sudan and South Sudan

food science and technology, and food distribution systems are vitally important in tackling this problem.

Agricultural technology covers a wide range of activities and includes:

- the development of high-yielding seeds
- genetic engineering, which remains a controversial issue although its use has spread significantly around the world in the last decade
- precision agriculture – the integration of information to improve agricultural knowledge in addressing site-specific production targets
- environmental modelling – the optimal use of genetics on specific soils within known weather profiles
- continued advances in the 'classical' agricultural technologies
- employing advanced techniques to remediate land that has been damaged by poor agricultural practices
- integrated pest management (IPM), which considers the site-specific conditions, but also the values and business considerations of the food producers.

Food production: the Green Revolution

Innovation in food production has been essential to feeding a rising global population. The package of agricultural improvements generally known as the **Green Revolution** was seen as the answer to the food problem in many LICs and MICs. India was one of the first countries to benefit when a high-yielding variety seed programme (HVP) started in 1966–67. In terms of production, it was a turning point for Indian agriculture, which had virtually reached stagnation. The HVP introduced new hybrid varieties of five cereals: wheat, rice, maize, sorghum and millet. All were drought-resistant, with the exception of rice; were very responsive to the application of fertilisers; and they had a shorter growing season than the traditional varieties they replaced. Although the benefits of the Green Revolution are clear, serious criticisms have also been made. The two sides of the story can be summarised as follows:

Advantages

- Yields are twice to four times greater than of traditional varieties.
- The shorter growing season has allowed the introduction of an extra crop in some areas.
- Farming incomes have increased, allowing the purchase of machinery, better seeds, fertilisers and pesticides.
- The diet of rural communities is now more varied.
- Local infrastructure has been upgraded to accommodate a stronger market approach.
- Employment has been created in industries supplying farms with inputs.
- Higher returns have justified a significant increase in irrigation.

Disadvantages

- High inputs of fertiliser and pesticide are required to optimise production. This is costly in both economic and environmental terms. In some areas, rural indebtedness has risen sharply.
- High-yielding varieties (HYVs) require more weed control and are often more susceptible to pests and disease.
- Middle- and higher-income farmers have often benefited much more than the majority on low incomes, thus widening the income gap in rural communities. Increased rural-to-urban migration has often been the result.
- Mechanisation has increased rural unemployment.
- Some HYVs have an inferior taste.
- The problem of salinisation has increased, along with the expansion of the irrigated areas.
- HYVs can be low in minerals and vitamins. Because the new crops have displaced the local fruits, vegetables and legumes that traditionally supplied important vitamins and minerals, the diet of many people in LICs and MICs is now extremely low in zinc, iron, vitamin A and other micronutrients.

Perennial crops: the next agricultural revolution?

The answer to many of the world's current agricultural problems may lie in the development of **perennial crops**. Today's annual crops die off once they are harvested and new seeds have to be planted before the cycle of production can begin again. The soil is most vulnerable to erosion in the period between harvesting and the next planting. Perennial crops would protect the soil from erosion and also offer other advantages. Over the next few years, plant biologists hope to breed plants that closely resemble domestic crops but retain their perennial habit. Classical crossing methods have been proved to work in the search for perennial crop plants, but the process is slow. Some plant breeders aim to speed up the process by using genetic engineering. The objective is to find the genes that are linked to domestication and then insert these into wild plants.

☐ The role of constraints in sustaining populations

There are a significant number of potential constraints in developing resources to sustain changing populations. Figure 4.37 illustrates the factors affecting the development of a particular resource body. The factors included in the diagram are those that operate in normal economic conditions and thus do not include war or other types of conflict, which can greatly increase the constraints operating on resource development.

Figure 4.37 Factors affecting the development of a particular resource body

War is a major issue for development. It significantly retards development and the ability of a country to sustain its population. Major conflict can set back the process of development by decades. In many conflicts, water, food and other resources are deliberately destroyed to make life as difficult as possible for the opposing population. Conversely, where development succeeds, countries become progressively safer from violent conflict, making subsequent development easier.

Trade barriers are another significant constraint. Many LICs complain that the trade barriers (tariffs, quotas and regulations) imposed by many HICs are too stringent. This reduces the export potential of poorer countries and hinders their development.

Climatic and other hazards in the short term, and climate change in the medium and long term, have a serious impact on the utilisation of resources. For example:

- Tropical storms are a major hazard and an impediment to development in a number of LICs such as Bangladesh and the countries of Central America and the Caribbean.
- Regions at significant risk of flooding, due to tropical storms and other factors, are often deprived of investment in agriculture and other aspects of development because of the potential losses involved.
- Drought has a considerable impact on the ability to sustain changing populations in many parts of the world. Desertification is reducing the agricultural potential of many countries, for example those in the Sahel region in Africa.

- Volcanic eruptions can devastate large areas, covering farmland in lava, burying settlements and destroying infrastructure. A major eruption on the island of Montserrat in the Caribbean in 1995 has had a huge impact on the development of the island. The southern third of the island had to be evacuated and all public services had to be removed to the north of the island.
- Earthquakes can have a significant impact on resource development, adding considerably to the costs of development because of the expensive construction techniques required to mitigate the consequences of this hazard.

Climate change has the potential to increase the frequency of extreme events in many parts of the world. In some regions, there will be wide-ranging implications for human health.

Section 4.3 Activities

1 Give three examples of the modern role of agricultural technology.
2 Discuss the advantages and disadvantages of the Green Revolution.
3 How might perennial crops lead to a new agricultural revolution?
4 Explain the role of constraints in sustaining changing populations.

☐ Carrying capacity

Carrying capacity is the largest population that the resources of a given environment can support. Carrying capacity is not a fixed concept as advances in technology can significantly increase the carrying capacity of individual regions and the world as a whole. For example, Abbé Raynal (*Révolution de l'Amérique*, 1781) said of the USA: 'If ten million men ever manage to support themselves in these provinces it will be a great deal'. Yet today the population of the USA is over 300 million and hardly anyone would consider the country to be overpopulated.

Resources can be classed as either natural or human. The traditional distinction is between renewable or flow resources and non-renewable or stock resources. However, the importance of aesthetic resources is being increasingly recognised. Further subdivision of the non-renewable category is particularly relevant to both fuel and non-fuel minerals. Renewable resources can be viewed as either critical or non-critical. The former are sustainable if prudent resource management is employed, while the latter can be seen as everlasting.

Figure 4.38 Water tower in a reservoir: Lake Vyrnwy, Wales – even in temperate countries like the UK, sufficient water supply is becoming of increasing concern

The relationship between population and resources has concerned those with an understanding of the subject for thousands of years. However, the assumptions made by earlier writers were based on very limited evidence, as few statistical records existed more than two centuries ago.

The enormous growth of the global economy in recent decades has had a phenomenal impact on the planet's resources and natural environment (Figure 4.38). Many resources are running out and waste sinks are becoming full. The remaining natural world can no longer support the existing global economy, much less one that continues to expand. The main responsibility lies with the rich countries of the world. The world's richest 20 per cent of the population accounted for over 94 per cent of the world's wealth in 2014, while the world's poorest 80 per cent currently own just 5.5 per cent!

Climate change will have an impact on a number of essential resources for human survival, increasing the competition between countries for such resources. An article in the British newspaper *The Times* (9 March 2009) about this global situation was entitled 'World heading for a War of the Resources'. In the same month, an article appeared in *The Guardian* newspaper (20 March 2009) entitled 'Deadly crop fungus brings famine threat to developing world'. It reported that leading crop scientists had issued a warning that a deadly airborne fungus could devastate wheat harvests in poor countries and lead to famines and civil unrest over significant regions of central Asia and Africa. A further article in *The Times* (14 May 2009) was entitled 'Russia warns of war within decade over hunt for oil and gas'.

The **ecological footprint** has arguably become the world's foremost measure of humanity's demands on the natural environment. It was conceived in 1990 by M. Wackernagel and W. Rees at the University of British Columbia. The concept of ecological footprints has been used to measure natural resource consumption, how it varies from country to country and how it has changed over time. The ecological footprint (Figure 4.39) for a country has been defined as 'the sum of all the cropland, grazing land, forest and fishing grounds required to produce the food, fibre and timber it consumes, to absorb the wastes emitted when it uses energy, and to provide

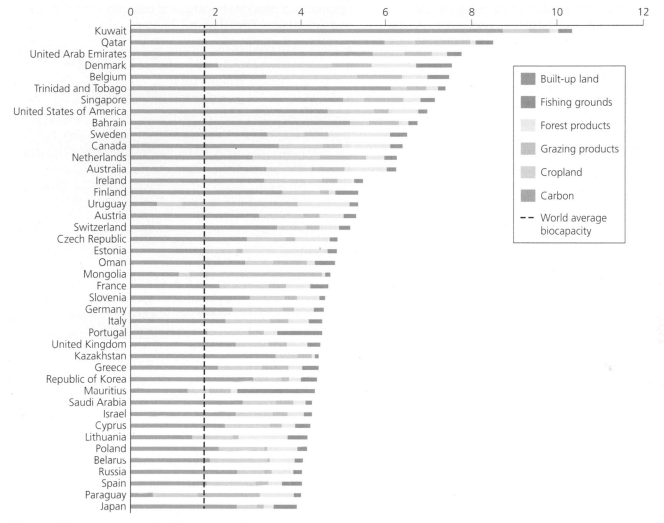

Figure 4.39 Per person ecological footprint (global hectares demanded per person), 2010

space for its infrastructure' (*Living Planet Report 2008*). Thus the ecological footprint, calculated for each country and the world as a whole, has six components (Figure 4.39):

- built-up land
- fishing grounds
- forest
- grazing land
- cropland
- carbon footprint.

In previous years, an additional component reflecting the electricity generated by nuclear power plants was included in ecological footprint accounts. This component is no longer used because the risks and demands of nuclear power are not easily expressed in terms of **biocapacity**.

The ecological footprint is measured in **global hectares**. A global hectare is a hectare with world-average ability to produce resources and absorb wastes. In 2005 the global ecological footprint was 17.5 billion global hectares (gha) or

2.7 gha per person. This can be viewed as the demand side of the equation. On the supply side, the total productive area, or biocapacity of the planet, was 13.6 billion gha, or 2.1 gha per person. With demand greater than supply, the Earth is living beyond its environmental means.

Figure 4.39 shows the ecological footprint of countries with the highest per person figures and how the footprint of each country is made up. Kuwait, Qatar, United Arab Emirates, Denmark and Belgium have the highest ecological footprints per person in the world. A total of 13 countries have figures above 6 gha per person. Nations at different income levels show considerable disparities in the extent of their ecological footprint. The lowest per person figures were attributed to Bangladesh, Pakistan, Afghanistan, Haiti, Eritrea, and Timor-Leste. All these countries have an ecological footprint of about 1.0 gha per person. Footprint and biocapacity figures for individual countries are calculated annually by Global Footprint Network.

population tends to increase in geometrical progression (1 – 2 – 4 – 8 – 16 – 32), multiplying itself by a constant amount each time. In time, population would outstrip food supply until a catastrophe occurred in the form of famine, disease or war. War would occur as human groups fought over increasingly scarce resources. These limiting factors maintained a balance between population and resources in the long term. In a later paper, Malthus placed significant emphasis on 'moral restraint' as an important factor in controlling population.

Clearly, Malthus was influenced by events in and before the eighteenth century and could not have foreseen the great advances that were to unfold in the following two centuries that have allowed population to grow at an unprecedented rate, alongside a huge rise in the exploitation and use of resources. There have been many advances in agriculture since the time of Malthus that have contributed to huge increases in agricultural production. These advances include: the development of artificial fertilisers and pesticides, new irrigation techniques, high-yielding varieties of crops, cross-breeding of cattle, greenhouse farming and the reclamation of land from the sea.

However, nearly all of the world's productive land is already exploited. Most of the unexploited land is either too steep, too wet, too dry or too cold for agriculture (Figure 4.45). In Asia, nearly 80 per cent of potentially arable land is now under cultivation.

Figure 4.45 The Gobi desert in central Asia – the process of desertification has been spreading in recent decades

Figure 4.46 summarises the opposing views of the **neo-Malthusians** and the resource optimists such as Esther Boserup (1910–99). Neo-Malthusians argue that an expanding population will lead to unsustainable pressure on food and other resources. In recent years, neo-Malthusians have highlighted:

There are two opposing views of the effects of population growth:

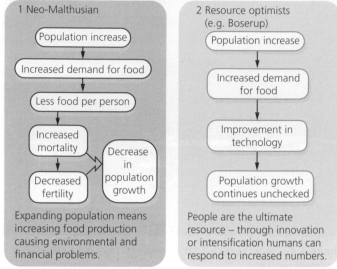

Source: *Advanced Geography: Concepts & Cases* by P. Guinness & G. Nagle (Hodder Education, 1999), p.35

Figure 4.46 The opposing views of the neo-Malthusians and the anti-Malthusians

- the steady global decline in the area of farmland per person
- the steep rise in the cost of many food products in recent years
- the growing scarcity of fish in many parts of the world
- the already apparent impact of climate change on agriculture in some world regions
- the switchover of large areas of land from food production to the production of biofuels, helping to create a food crisis in order to reduce the energy crisis
- the continuing increase in the world's population
- the global increase in the level of meat consumption as incomes rise in NICs in particular.

The **anti-Malthusians** or resource optimists believe that human ingenuity will continue to conquer resource problems, pointing to many examples in human history where, through innovation or intensification, humans have responded to increased numbers. Resource optimists highlight a number of continuing advances, which include:

- the development of new resources
- the replacement of less efficient with more efficient resources
- the rapid development of green technology, with increasing research and development (R&D) in this growing economic sector
- important advances in agricultural research
- stabilising levels of consumption in some HICs.

4.4 The management of natural increase

Population policy encompasses all of the measures explicitly or implicitly taken by a government aimed at influencing population size, growth, distribution or composition. Population policies generally evolve over time and are clearly documented in writing.

Such policies may promote large families (**pro-natalist policies**) or immigration to increase population size, or encourage limitation of births (**anti-natalist policies**) to decrease it. A population policy may also aim to modify the distribution of the population over the country by encouraging migration or by displacing populations. Population policies that narrow people's choices are generally very controversial.

A significant number of governments have officially stated positions on the level of the national birth rate. However, forming an opinion on demographic issues is one thing, but establishing a policy to do something about it is much further along the line. Thus not all nations stating an opinion on population have gone as far as establishing a formal policy.

Most countries that have tried to control fertility have sought to curtail it. For example, in 1952 India became the first LIC to introduce a policy designed to reduce fertility and to aid development, with a government-backed family planning programme. Rural and urban birth-control clinics rapidly increased in number. Financial and other incentives were offered in some states for those participating in programmes, especially for **sterilisation**. In the mid-1970s, the sterilisation campaign became increasingly coercive, reaching a peak of 8.3 million operations in 1977. **Abortion** was legalised in 1972 and in 1978 the minimum age of marriage was increased to 18 years for females and 21 years for males. The birth rate fell from 45/1000 in 1951–61 to 41/1000 in 1961–71. By 1987, it was down to 33/1000, falling further to 29/1000 in 1995. By 2014, it had dropped to 22/1000. It was not long before many other LICs followed India's policy of government investment to reduce fertility. The most severe anti-natalist policy ever introduced has been in operation in China since 1979.

What is perhaps surprising is the number of countries that now see their fertility as too low. Such countries are concerned about:

- the socio-economic implications of population ageing
- the decrease in the supply of labour
- the long-term prospect of population decline.

Russia has seen its population drop considerably since 1991. Alcoholism, AIDS, pollution and poverty are among the factors reducing life expectancy and discouraging births. In 2008, Russia began honouring families with four or more children with a Paternal Glory medal. The government has urged Russians to have more children, sometimes suggesting it is a matter of public duty.

Section 4.4 Activities

1 Define the term *population policy*.
2 What is the difference between a pro-natalist policy and an anti-natalist policy?
3 Suggest why the governments of some countries want to reduce their fertility while others want to increase it.
4 Why does the management of natural increase focus on fertility as opposed to mortality?

Case Study: Managing natural increase in China

China, with a population in excess of 1.3 billion, has been operating the world's strictest **family-planning programme** since 1979. Known as the one-child policy, it has drastically reduced population growth, but also brought about a number of adverse consequences, including:

- demographic ageing
- an unbalanced sex ratio
- a generation of 'spoiled' only children
- a social divide as an increasing number of wealthy couples 'buy their way round' the legislation.

Chinese demographers say that the one-child policy has been successful in preventing at least 300 million births, and has played a significant role in the country's economic growth.

Although it is the third largest country in the world in terms of land area, 25 per cent of China is infertile desert or mountain and only 10 per cent of the total area can be used for arable farming. Most of the best land is in the east and south, reflected in the extremely high population densities found in these regions. Table 4.14 ranks China's administrative regions by population size and shows a comparable country in terms of total population for each region. For example, Guangdong, with over 95 million people, equivalent to the population

Figure 4.47 The central business district of Beijing

of Mexico, has the largest population in China. Anhui has a population similar to that of the UK, and seven Chinese provinces have populations higher than the UK.

Table 4.14 China's administrative regions by population

Rank	Administrative division, China	Population	Comparable country (country rank worldwide)
	CHINA	**1 358 650 000**	**India (1.2 billion), or combined populations of western Europe, North America and South America**
1	Guangdong	95 440 000	Mexico (11)
2	Henan	94 290 000	Ethiopia (15) + Guatemala (66)
3	Shandong	94 170 000	Vietnam (13) + Sierra Leone (107)
4	Sichuan	81 380 000	Germany (14)
5	Jiangsu	76 770 000	Egypt (16)
6	Hebei	69 890 000	Iran (18)
7	Hunan	63 800 000	Thailand (21)
8	Anhui	61 350 000	UK (22)
9	Hubei	57 110 000	Italy (23)
10	Zhejiang	51 200 000	Myanmar (24)
11	Guangxi	48 160 000	South Korea (26)
12	Yunnan	45 430 000	Spain (28)
13	Jiangxi	44 000 000	Colombia (29)
14	Liaoning	43 150 000	Sudan (31)
15	Heilongjiang	38 250 000	Argentina (33)
16	Guizhou	37 930 000	Kenya (32)
17	Shaanxi	37 620 000	Poland (34)
18	Fujian	36 040 000	Algeria (35)
19	Shanxi	34 110 000	Canada (36)
20	Chongqing	28 390 000	Nepal (40)
21	Jilin	27 340 000	Uzbekistan (45)
22	Gansu	26 280 000	Saudi Arabia (45)
23	Inner Mongolia	24 140 000	North Korea (47)

Rank	Administrative division, China	Population	Comparable country (country rank worldwide)
23	Inner Mongolia	24 140 000	North Korea (47)
	Taiwan Province (Republic of China)	22 980 000	Texas state, USA
24	Xinjiang	21 310 000	Mozambique (51)
25	Shanghai	18 880 000	Cameroon (58)
26	Beijing	16 950 000	Netherlands (61)
27	Tianjin	11 760 000	Greece (73)
28	Hainan	8 540 000	Austria (92)
29	Hong Kong	7 000 000	Tajikistan (98)
30	Ningxia	6 180 000	Paraguay (102)
31	Qinghai	5 540 000	Denmark (108)
32	Tibet	2 870 000	Kuwait (136)
33	Macau	540 000	Solomon Islands (164)

The balance between population and resources has been a major cause of concern for much of the latter part of the twentieth century, although debate about this issue can be traced as far back in Chinese history as Confucius (Chinese philosopher and teacher of ethics, 551–479 BCE). Confucius said that excessive population growth reduced output per worker, depressed the level of living and produced strife. He discussed the concept of optimum numbers, arguing that an ideal proportion between land and numbers existed and any major deviation from this created poverty. When imbalance occurred, he believed the government should move people from overpopulated to underpopulated areas.

Between 1950 and 2014, the crude birth rate fell from 43.8/1000 to 12/1000. China's birth rate is now at the level of many HICs such as the UK. The impact of the one-child policy is very clear to see in Figure 4.48.

From 2000 to 2010, China's population increased by 0.57 per cent a year. This was only about half the level of the previous decade and only one-fifth of the level of 1970. The total fertility rate, which was 6.1 between 1965 and 1970, fell to 2.6 between 1980 and 1985, 1.9 between 1990 and 1995, 1.7 between

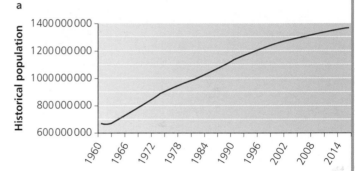

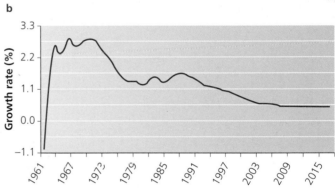

Figure 4.49 a China's historical population, 1960–2015 and **b** Growth rate, 1961–2015

2000 and 2005, and was down to 1.6 in 2013. Figure 4.49a shows the growth in China's population between 1960 and 2015 and Figure 4.49b shows the growth rate for the same period.

For people in the West, it is often difficult to understand the all-pervading influence over society that a government can have in a **centrally planned economy**. In the aftermath of the communist revolution in 1949, population growth was encouraged for economic, military and strategic reasons. Sterilisation and abortion were banned and families received a benefit payment for every child. However, by 1954 China's population had reached 600 million and the government was now worried about the pressure on food supplies and other resources. Consequently, the country's first birth-control programme was introduced in 1956. This was to prove short-lived, for in 1958 the 'Great Leap Forward' began. The objective was rapid industrialisation and modernisation.

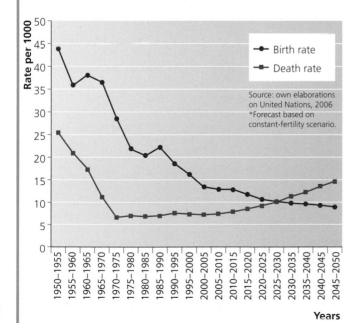

Figure 4.48 Birth and death rates, 1950–2050

Figure 4.50 Crowds at the Forbidden City, Beijing – China's population was only about 75 million when the Forbidden City was built in the early fifteenth century

The government was now concerned that progress might be hindered by labour shortages and so births were again encouraged. But by 1962, the government had changed its mind, heavily influenced by a catastrophic famine due in large part to the relative neglect of agriculture during the pursuit of industrialisation. An estimated 20 million people died during the famine. Thus a new phase of birth control ensued in 1964. Just as the new programme was beginning to have some effect, a new social upheaval – the Cultural Revolution – got underway. This period, during which the birth rate peaked at 45/1000, lasted from 1966 to 1971.

With order restored, a third family planning campaign was launched in the early 1970s with the slogan 'Late, sparse, few'. However, towards the end of the decade the government felt that its impact might falter and in 1979 the controversial one-child policy was imposed. The Chinese demographer Liu Zeng calculated that China's optimum population was 700 million, and he looked for this figure to be achieved by 2080. Some organisations, including the UN Fund for Population Activities, have praised China's policy on birth control. Many others see it as a fundamental violation of **civil liberties**.

Ethnic minorities were exempt from parts of the policy, which applied mainly to the Han ethnic majority, which makes up more than 90 per cent of the total population. China's policy is based on a reward and penalty approach. Rural households that obey family-planning rules get priority for loans, materials, technical assistance and social welfare. The slogan in China is *shao sheng kuai fu* – 'fewer births, quickly richer'. The one-child policy has been most effective in urban areas where the traditional bias of couples wanting a son has been significantly eroded. However, the story is different in rural areas where the strong desire for a male heir remains the norm. In most provincial rural areas, government policy has relaxed so that couples can now have two children without penalties.

The policy has had a considerable impact on the sex ratio, which at birth in China is currently 119 boys to 100 girls. This compares with the natural rate of 106:100. In some provinces, it is estimated the figure may be as high as 140. A paper published in 2008 estimated that China had 32 million more men aged under 20 than women. The imbalance is greatest in rural areas because women are 'marrying out' into cities. This is already causing social problems, which are likely to multiply in the future. **Selective abortion** after pre-natal screening is a major cause of the wide gap between the actual rate and the natural rate. But even if a female child is born, her lifespan may be sharply curtailed by infanticide or deliberate neglect. Feminist writers in China see 'son preference' as a blatant form of gender discrimination and gender-based violence.

Figure 4.51 The metro in Beijing

However, this is an issue that affects other countries as well as China.

The significant gender imbalance means that a very large number of males will never find a female partner, which could result in serious social problems as significant numbers of males are unable to conform to the basic **social norms** of society, which revolve around marriage and parenthood. Such unmarried men are known as 'bare branches'!

In recent years, reference has been made to the 'Four-Two-One' problem, whereby one adult child is left with having to provide support for two parents and four grandparents. Care for the elderly is clearly going to become a major problem for the Chinese authorities, since the only social security system for most of the country's poor is their family. China's ageing process is happening more quickly than in most other countries, mainly due to the speed of its demographic transition.

In July 2009, newspapers in the UK and elsewhere reported that dozens of babies had been taken from parents who had breached China's one-child policy and sold for adoption abroad. In the cities, the fines for having a second child can be up to 200 000 yuan (£20 000). This is meant to reflect the schooling and healthcare costs of additional children. However, an increasing number of affluent parents are prepared either to pay these fines outright or to travel to Hong Kong where no permit for a second child is needed.

In recent years, there has been a certain level of debate within China about the one-child policy. A substantial decline in the supply of young labour has been a major factor in pushing up wages. China is no longer the substantial source of cheap labour that it once was. Another consequence of the one-child rule has been the creation of a generation of so-called 'little emperors' – indulged and cosseted boy children who are often overweight, arrogant and lacking in social skills.

Since late 2013, 29 of the 31 provincial regions on the mainland have enacted policies that allow couples to have a second baby if either partner is a single child. However, although many Chinese couples would undoubtedly have more children if allowed by the government, in urban areas a new class of city workers has arisen with a Western-style reluctance to have more than one child, because they want to preserve their rising standard of living.

In September 2015, the Chinese government announced it would relax the rules to allow all couples to have two children from March 2016.

Section 4.4 Activities

1 Write a brief bullet-point summary of the main changes in Chinese fertility policy since 1949.
2 How successful has the one-child policy been in reducing fertility in China?
3 Discuss some of the disadvantages of the one-child policy.

5 Migration

5.1 Migration as a component of population change

☐ Movements of populations: definitions

Migration is more volatile than fertility and mortality, the other two basic demographic variables. It can react very quickly indeed to changing economic, political and social circumstances. However, the desire to migrate may not be achieved if the constraints imposed on it are too great.

Migration is defined as the movement of people across a specified boundary, national or international, to establish a new permanent place of residence (Figure 5.1). The United Nations defines 'permanent' as a change of residence lasting more than one year. Movements with a time scale of less than a year are termed 'circulatory movements'.

Figure 5.1 Chinatown in San Francisco – the Chinese community is long established in this city

It is customary to subdivide the field of migration into two areas: **internal migration** and **international migration**. International migrants cross international boundaries; internal migrants move within the frontiers of one nation. The terms **immigration** and **emigration**

are used with reference to international migration. The corresponding terms for internal movements are **in-migration** and **out-migration**. Internal migration streams are usually on a larger scale than their international counterparts. **Net migration** is the number of migrants entering a region or country less the number of migrants who leave the same region or country. The balance may be either positive or negative.

Migrations are embarked upon from an area of **origin** and are completed at an area of **destination**. Migrants sharing a common origin and destination form a **migration stream**. For every migration stream, a **counterstream** or reverse flow at a lower volume usually results as some migrants dissatisfied with their destination return home. Push and pull factors (Figures 5.2 and 5.3) encourage people to migrate. **Push factors** are the observations that are negative about an area in which the individual is presently living, while **pull factors** are the perceived better conditions in the place to which the migrant wishes to go. Once strong links between a rural and an urban area are established, the phenomenon of **chain migration** frequently results. After one or a small number of pioneering migrants have led the way, others from the same rural community follow. In some communities, the process of **relay migration** has been identified, whereby at different stages in a family's life cycle different people take responsibility for migration in order to improve the financial position of the family. Another recognisable process is **stepped migration**, whereby the rural migrant initially heads for a familiar small **town** and then after a period of time moves on to a larger urban settlement. Over many years, the migrant may take a number of steps up the **urban hierarchy**.

+ Positive factors   − Negative factors

O Factors perceived as unimportant to the individual

Source: *IGCSE Geography* by P. Guinness & G. Nagle (Hodder Education, 2009), p.23

Figure 5.2 Push and pull factors

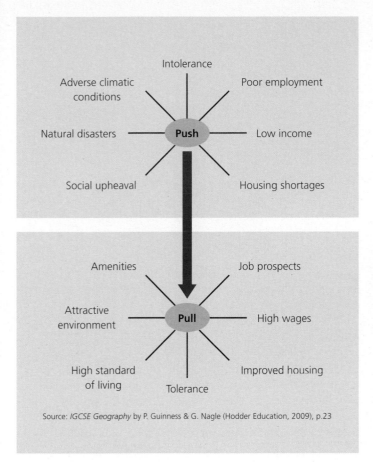

Figure 5.3 Push and pull factors

The most basic distinction drawn by demographers is between voluntary and forced migration (Figure 5.4). **Voluntary migration** is where the individual or household has a free choice about whether or not to move. **Forced**

migration occurs when the individual or household has little or no choice but to move. This may be due to environmental or human factors. Figure 5.4 shows that there are barriers to migration. In earlier times, the physical dangers of the journey and the costs involved were major obstacles. However, the low real cost of modern transportation and the high level of safety have reduced these barriers considerably. In the modern world, it is the legal restrictions that countries place on migration that are the main barriers to international migration. Most countries now have very strict rules on immigration, and some countries restrict emigration.

Section 5.1 Activities

1 Define *migration*.
2 Distinguish between **a** *immigration* and *emigration* and **b** *in-migration* and *out-migration*.
3 Explain the terms **a** *origin* and *destination* and **b** *stream* and *counterstream*.
4 Briefly describe each of the following:
   a chain migration
   b relay migration
   c stepped migration.
5 Discuss three push factors and three pull factors shown in Figure 5.3.
6 Write a brief summary to explain Figure 5.4.

☐ Causes of migration

Various attempts to classify migration have helped improve understanding of its causes. In 1958, W. Peterson noted the following five migratory types: primitive, forced, impelled, free and mass.

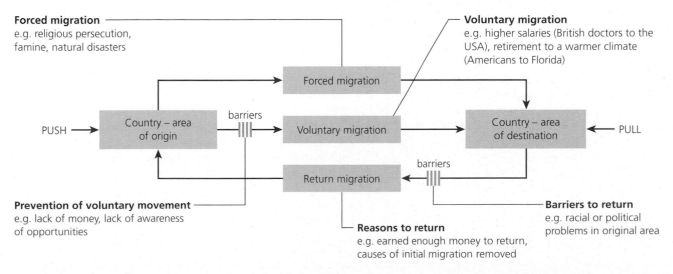

Figure 5.4 Voluntary and forced migration

- The nomadic pastoralism and shifting cultivation practised by the world's most traditional societies are examples of primitive migration. Physical factors such as seasonal rainfall and the limits of soil fertility govern such migratory practices.
- The abduction and transport of Africans to the Americas as slaves was the largest forced migration in history. In the seventeenth and eighteenth centuries, 15 million people were shipped across the Atlantic Ocean as slaves. The expulsion of Asians from Uganda in the 1970s, when the country was under the dictatorship of Idi Amin, and the forcible movement of people from parts of the former Yugoslavia under the policy of 'ethnic cleansing', are much more recent examples. Migrations may also be forced by natural disasters (volcanic eruptions, floods, drought, and so on) or by environmental catastrophe such as nuclear contamination in Chernobyl.
- Impelled migrations take place under perceived threat, either human or physical, but an element of choice lacking in forced migrations remains. Arguably the largest migration under duress in modern times occurred after the partition of India in 1947, when 7 million Muslims fled India for the new state of Pakistan and 7 million Hindus moved with equal speed in the opposite direction. Both groups were in fear of their lives but they were not forced to move by government, and minority groups remained in each country.
- The distinction between free and mass migration is one of magnitude only. The movement of Europeans to North America was the largest mass migration in history.

Within each category, Peterson classed a particular migration as either innovating or conservative. In the former, the objective of the move was to achieve improved living standards, while in the latter the aim was just to maintain present standards.

E.S. Lee (1966) produced a series of Principles of Migration in an attempt to bring together all aspects of migration theory at that time. Of particular note was his **origin-intervening obstacles-destination** model, which emphasised the role of push and pull factors (Figures 5.2 and 5.3). Here, he suggests there are four classes of factors that influence the decision to migrate:

1 those associated with the place of origin (Figure 5.5)
2 those associated with the place of destination
3 intervening obstacles that lie between the places of origin and destination
4 a variety of personal factors that moderate 1, 2 and 3.

Each place of origin and destination has numerous positive, negative and neutral factors for the individual. What may constitute a negative factor at destination

Figure 5.5 A severe winter in Mongolia caused great loss to animal herds, forcing farmers to leave the countryside for the capital city, Ulaanbaatar

for one individual – a very hot climate, say – may be a positive factor for another person. Lee suggested that there is a difference in the operation of these factors at origin and destination, as the latter will always be less well known: 'There is always an element of ignorance or even mystery about the area of destination, and there must always be some uncertainty with regard to the reception of a migrant in a new area'. This is particularly so with international migration. Another important difference noted by Lee between the factors associated with area of origin and area of destination related to stages of the life cycle. Most migrants spend their formative years in the area of origin enjoying the good health of youth with often only limited social and economic responsibilities. This frequently results in an overvaluation of the positive elements in the environment and an undervaluation of the negative elements. Conversely, the difficulties associated with **assimilation** into a new environment may create in the newly arrived a contrary but equally erroneous evaluation of the positive and negative factors at destination. The **intervening obstacles** between origin and destination include distance, the means and cost of transport and legal restraints (mainly in the form of immigration laws).

Akin Mabogunje, in his analysis of rural–urban migration in Africa, attempted to set the phenomenon in its economic and social context as part of a system of interrelated elements (Figure 5.6). The systems approach does not see migration in over-simplified terms of cause and effect, but as a circular, interdependent and self-modifying system.

In Mabogunje's framework, the African rural–urban migration system is operating in an environment of change. The system and the environment act and react upon each other continuously. For example, expansion in

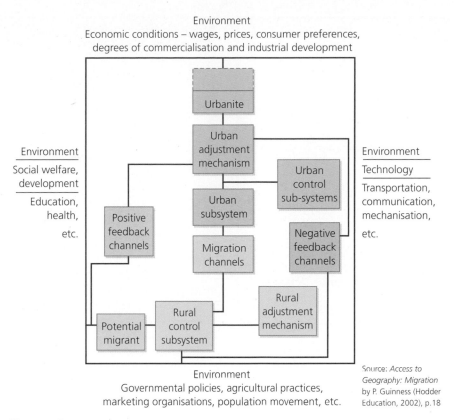

Environment
Economic conditions – wages, prices, consumer preferences,
degrees of commercialisation and industrial development

Environment

Social welfare,
development

Education,
health,
etc.

Environment
Technology

Transportation,
communication,
mechanisation,
etc.

Environment
Governmental policies, agricultural practices,
marketing organisations, population movement, etc.

Source: *Access to
Geography: Migration*
by P. Guinness (Hodder
Education, 2002), p.18

Figure 5.6 A systems approach to migration

the urban economy will stimulate migration from rural areas, while deteriorating economic conditions in the larger urban areas will result in a reduction of migration flows from rural areas.

If the potential migrant is stimulated to move to an urban area by the positive nature of the environment, he/she then comes under the influence of the 'rural control subsystem'. Here the attitudes of the potential migrant's family and local community come into strong play, either encouraging or restraining movement. If movement occurs, the migrant then comes under the influence of the 'urban control subsystem'. The latter will determine, by means of the employment and housing opportunities it offers, the degree to which migrants assimilate.

In addition, there are adjustment mechanisms. For example, at the rural point of origin a positive adjustment resulting from out-migration might be increased income per head for the remaining villagers. The most likely negative adjustment will be the reduced level of social interaction between the out-migrants and their families. At the urban destination, the in-migrant may benefit from the receipt of regular wages for the first time, but as a result may be drawn into the negative aspects of lower-income urban life such as gambling, excessive drinking and prostitution.

The flow of information between out-migrants and their rural origin is an important component of the system. Favourable reports from the new urban

dwellers will generally increase the migration flow, while negative perceptions will slow down the rate of movement. The trans-Siberian railway (Figure 5.7) is an important routeway for people moving between the Asiatic and European parts of Russia. Many small communities in Asiatic Russia have been abandoned because of high out-migration.

Figure 5.7 The trans-Siberian railway

☐ Recent approaches to migration

Figure 5.8 summarises the main differences in the most recent approaches to migration, each of which is briefly discussed below.

Determinants of migration	Effects	Unit of analysis		
		Individual	Household/family	Institutions
Economic	Positive	Todaro Push–pull	Stark and others; 'new economics' of migration	
	Negative			Marxism Structuralism
Sociological/ anthropological		←———— Structuration theory ————→ ←———— Gender analyses ————→		

Figure 5.8 Recent approaches to migration studies

The Todaro model: the cost–benefit approach

In the post-1950 period, there has been a huge movement of population from rural to urban areas in LICs. For many migrants, it appeared that they had just swapped rural poverty for urban poverty. The simplistic explanation put forward was that many rural dwellers had been attracted by the 'bright lights' of the large urban areas without any clear understanding of the real **deprivation** of urban life for those at or near the bottom of the socio-economic scale. They had migrated due to false perceptions picked up from the media and other sources. The American economist Michael Todaro challenged this view, arguing that migrants' perceptions of urban life were realistic, being strongly based on an accurate flow of information from earlier migrants from their rural

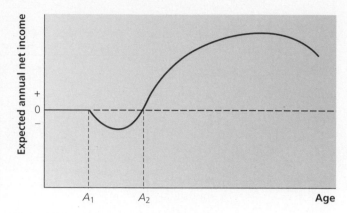

Figure 5.9 A typical net-income stream

community. Potential migrants carefully weighed up the costs and benefits of moving to urban areas, including the 'anticipated income differential'. They were very aware that in the short term they might not be better off but, weighing up the odds, the likelihood was that their socio-economic standing would improve in the long term. Thus people were willing to endure short-term difficulties in the hope of better prospects eventually, if not for themselves then for their children. Expected wages were discounted against the prospects of remaining unemployed for any length of time.

Figure 5.9 summarises the typical net-income stream of a young rural–urban migrant. While at school, the young rural dweller's net income is zero. At A_1 he migrates to a large urban area but is initially unable to find work because of the intense competition for employment and the limited nature of his contacts. His net income is negative as he has no option but to live on savings or borrowed money. However, in time, as his knowledge of the city improves and his contact base widens, he finds employment and his net income becomes positive (A_2), rising to a peak and then decreasing with age as his productivity begins to fall.

Stark's 'new economics of migration'

Stark, in what is often referred to as the 'new economics of migration', has extended the Todaro model by replacing the individual with the household as the unit of analysis. Stark, along with others, argued that insufficient attention had been paid to the institutions that determine migration. For example, in the Todaro model it is assumed that migrants act individually according to a rationality of economic self-interest. However, migration, according to Stark, is seen as a form of economic diversification by families whereby the costs and rewards are shared. It is a form of risk spreading. She asserts that 'even though the entities that engage in migration are often individual agents, there is more to labour migration than an individualistic optimising behaviour. Migration by one person can be due to, fully consistent with, or undertaken by a group of persons, such as the family.'

So often the initial cost of establishing the rural migrant in an urban area is carried by the family in the expectation of returns in the form of remittances. The migrant also has expectations in maintaining the link, for example in the form of inheritance. A number of studies have described how families invest in the education of one member of the family, usually the firstborn son, for migration to the urban formal sector. The expectation is that the remittances received will be crucial to the up-bringing of the remaining children and have an important effect on the general standard of living of the family.

The Stark model also takes account of: incomplete and imperfect information; imperfections in rural capital markets and transaction costs; and stresses the importance in migration decisions of relative deprivation in the local income distribution rather than absolute deprivation.

Marxist/structuralist theory

Some writers, often in the tradition of Marxist analyses, see labour migration as inevitable in the transition to capitalism (Figure 5.10). Migration is the only option for survival after alienation from the land. Structuralist theory draws attention to the advantages of migrant labour for capitalist production and emphasises the control that capitalism has over migrant labour. For example, employers in destinations do not bear the cost of their workers' reproduction as the latter maintain ties with their rural communities, and employers use migrant labour to reduce the bargaining power of local labour. In the international arena, migration is seen as a global movement in which labour is manipulated in the interest of HICs to the detriment of LICs. According to Rubenstein, remittances are 'a minor component of surplus labour extraction, a small charge to capital in a grossly unequal process of exchange between core and peripheral societies'.

Figure 5.10 Eastern European food shop in London – the population of Eastern Europeans in the UK has increased rapidly since Poland and other Eastern European countries joined the EU in 2004

Structuration theory

Structuration theory incorporates both individual motives for migration and the structural factors in which the migrants operate. It stresses that rules designed to regulate behaviour also provide opportunity and room for manoeuvre for those they seek to constrain. This approach also builds in an awareness of cultural factors.

Gender analyses

In recent decades, gender has come to occupy a significant place in migration literature. According to Arjan de Haan, 'There is now much more emphasis on the different migration responses by men and women, which themselves are context dependent, and on gender discrimination in returns to migrant labour.'

Case Study: Push and pull factors in Brazil

While recognising that individuals can react differently to similar circumstances, it is still important to consider the negative factors that act to 'push' people from rural areas of origin, and the positive influences that 'pull' them towards towns and cities. In Brazil, the push factors responsible for rural–urban migration can be summarised as follows:

- The mechanisation of agriculture has reduced the demand for farm labour in most parts of the country.
- Farms and estates have been amalgamated, particularly by agricultural production companies. In Brazil, as elsewhere in Latin America, the high incidence of landlessness has led to a much greater level of rural–urban migration than in most parts of Africa and Asia.
- Conditions of rural employment are generally poor. Employers often ignore laws relating to minimum wages and other employee rights.

- There is desertification in the north-east and deforestation in the north.
- Unemployment and **underemployment** are significant.
- Social conditions are poor, particularly in terms of housing, health and education.

The pull factors for internal migrants in Brazil revolve around individuals wanting to better their own and their children's lives. Within the larger urban areas such as São Paulo, Rio de Janeiro, Belo Horizonte and Brasilia, migrants hope to find particular advantages:

- A greater likelihood of paid employment – many people will be unable to find work in the formal sector, but opportunities in the informal sector, even if only part-time, may be available. Developing skills in the informal sector may open the way to work in the formal sector at a later date. Paid

⇨

- employment provides the opportunity to save money, even if the amounts initially are very small.
- Greater proximity to health and education services – this factor is particularly important for migrants with children. There is a clear urban/rural divide in standards for both health and education.
- Most migrants end up in *favelas* or *corticos* (deteriorating formal inner-city housing). However, even *favela* housing may be better than that found in some rural areas. Many *favelas* show substantial signs of upgrading over time and develop an important sense of community.
- Greater access to retail services than in rural areas – Competition in the urban retail services sector can result in lower prices, enabling the individual/household to purchase a wider range of goods.
- The cultural and social attractions of large cities may be viewed as important factors in the quality of life.
- Access to internet services is often lacking in rural areas. This is often an important factor for younger migrants.

□ The role of constraints, obstacles and barriers

Brief reference has already been made to factors that can either prevent migration or make it a difficult process. Here a distinction has to be made between internal and international migration. In most countries, there are no legal restrictions on internal migration. Thus the main constraints are distance and cost. In contrast, immigration laws present the major barrier in international migration where national borders have to be crossed.

The cost of migration can be viewed in three parts.

1 'Closing up' at the point of origin – this will vary considerably according to the assets owned by an individual or household. In LICs, the monetary value may be small, although the personal value may be high. In HICs, costs such as those of estate agents and legal fees for selling a house and selling possessions that cannot be transported at below market value, and other associated costs, can be substantial.
2 The actual cost of movement itself will depend on the mode of transport used and the time taken on the journey. Costs may involve both personal transport costs and the freight costs of transporting possessions.
3 The costs of 'opening up' at the point of destination – many HICs impose a 'stamp duty' on the purchase of a house above a certain value. This is in addition to estate agents' and legal fees. Other legal costs may also be required to begin life at the destination. If the migration is linked to employment, costs may be paid by an employer. In poorer countries, such costs may appear low in monetary value, but may be substantial for the individuals concerned because of their very low income.

The consideration of distance usually involves the dangers associated with the journey. Such dangers can be subdivided into physical factors and human factors. Physical factors include risks such as flood, drought, landslide and crossing water bodies (Figure 5.11). Human factors centre around any hostility from other people that may be encountered on the journey, and the chances of an accident while travelling. For example, in recent years people fleeing Zimbabwe for South Africa have

Figure 5.11 Iguaçu Falls, Brazil – the physical environment is much less of a barrier to migration than it once was

encountered bandits on both sides of the border, waiting at these locations to rob them. Ethnic tensions along a migration route may also result in significant danger.

In terms of international migration, government attitudes in the form of immigration laws usually present the most formidable barrier to prospective migrants. A number of reasonably distinct periods can be recognised in terms of government attitudes to immigration:

- Prior to 1914, government controls on international migration were almost non-existent. For example, the USA allowed the entry of anybody who was not a prostitute, a convict, a lunatic and, after 1882, Chinese. Thus the obstacles to migration at the time were cost and any physical dangers that might be associated with the journey.
- Partly reflecting security concerns, migration was curtailed between 1914 and 1945. During this period, many countries pursued immigration policies that would now be classed as overtly racist.
- After 1945, many European countries, facing labour shortages, encouraged migrants from abroad. In general, legislation was not repealed but interpreted very liberally. The Caribbean was a major source of labour for the UK during this period. The former West Germany

attracted 'guest workers' from many countries but particularly from Turkey.

- In the 1970s, slow economic growth and rising unemployment in HICs led to a tightening of policy that, by and large, has remained in force. However, in some countries immigration did increase again in the 1980s and early 1990s, spurring the introduction of new restrictions.

Thus over time the legal barriers to immigration have generally become more formidable. Most countries favour immigration applications from people with skills that are in short supply and from people who intend to set up businesses and create employment.

☐ Migration data

There are three principal sources of migration data: censuses, population registers and social surveys. For all three, moves are recorded as migration when an official boundary used for data collection is crossed. Moves that do not cross a boundary may go unrecorded even though they may cover longer distances. This is one of the major problems encountered by the researcher in the study of migration.

Population censuses are important sources of information because they are taken at regular intervals and cover whole countries. The two sorts of data generally provided are:

- birthplaces of the population
- period migration figures (movement over a particular period of time).

Birthplace data tells us a great deal about the broad picture of migration but it is not without its deficiencies. For example, there is no information about the number of residential moves between place of birth and present residence. In terms of period migration, recent British censuses have asked for place of residence a year before as well as place of birth. When these are compared with the present addresses of people at the time of the census, we can begin to trace **migration patterns**. However, again, intervening moves during the one-year period and between censuses (every ten years in the UK) will go unrecorded.

Japan and a number of European countries (including Norway, Sweden and Switzerland) collect 'continuous data' on migration through **population registers**. Inhabitants are required to register an address with the police or a civic authority and to notify all changes of residence. Population registers aim to record every move, rather than just those caught by the rather arbitrary administrative and period framework of the census. In the UK and many other countries, only partial registers exist to record movements for some parts of the population. Examples are electoral rolls, tax registers and school rolls. Social

researchers have argued for the introduction of population registers in countries like the UK but strong opposition has focused on possible infringements of individual liberties. Thus it was only under the exceptional circumstances of the Second World War and its immediate aftermath that a national register operated in the UK.

Specific **social surveys** can do much to supplement the sources of data discussed above. An example from the UK is the International Passenger Survey, a sample survey carried out at seaports and airports. It was established to provide information on tourism and the effect of travel expenditure on the balance of payments, but it also provides useful information on international migration. The annual General Household Survey of 15 000 households also provides useful information, as does the quarterly Labour Force Survey. Questionnaire-based surveys are perhaps the only means by which the relationship between attitudes and behaviour in the migration process can be fully analysed.

Even when all the available sources of information are used to analyse migration patterns, the investigator can be left in no doubt that a large proportion of population movements go entirely unrecorded; and even in those countries with the most advanced administrative systems, there is only partial recording of migrants and their characteristics.

☐ Conclusion

Migration has been a major process in shaping the world as it is today. Its impact has been economic, social, cultural, political and environmental. Few people now go through life without changing residence several times. Through the detailed research of geographers, demographers and others, we have a good understanding of the causes and consequences of the significant migrations of the past, which should make us better prepared for those of the future whose impact may be every bit as great. We can only speculate about the locations and causes of future migrations. Causal factors may include the following: continuing socio-economic disparity between rich and poor nations; global warming and all its implications; nuclear catastrophe; civil wars; and pandemics due to current and new diseases.

Section 5.1 Activities

1 Briefly discuss the cost–benefit approach of the Todaro model.
2 What are the main elements of Stark's new economics of migration?
3 Discuss the principal sources of migration data.

5.2 Internal migration (within a country)

☐ Distance, direction and patterns

Figure 5.12 provides a comprehensive classification of population movements in LICs and MICs, covering distance, direction and patterns. The 'distance continuum' ranges from relatively limited local movements to very long-distance movements, often crossing international frontiers. The majority of the movements shown in Figure 5.12 are internal migrations. In terms of settlement size, the following movements are included:

- rural–rural
- rural–urban
- urban–rural
- urban–urban.

Figure 5.13 Rural depopulation in northern Spain as a result of out-migration

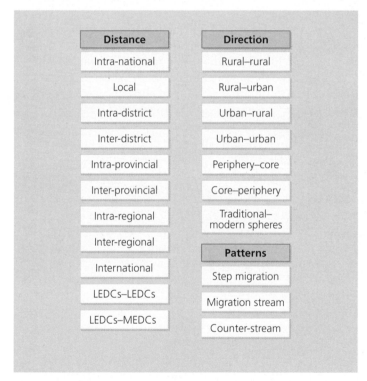

Distance	Direction
Intra-national	Rural–rural
Local	Rural–urban
Intra-district	Urban–rural
Inter-district	Urban–urban
Intra-provincial	Periphery–core
Inter-provincial	Core–periphery
Intra-regional	Traditional–modern spheres
Inter-regional	**Patterns**
International	Step migration
LEDCs–LEDCs	Migration stream
LEDCs–MEDCs	Counter-stream

Figure 5.12 Spatial dimensions of population movements in LICs and MICs

In this section, particular consideration will be given to **rural–urban migration** in LICs and urban–rural movements in HICs (Figure 5.13).

As Parnwell states in relation to Figure 5.12, 'Distance provides a useful basis for differentiating between types of movement and types of mover, because the distance over which a person travels can also be used as a proxy for other important variables'. As cost is a significant factor in the distance over which migration takes place, the relative distance of movements may have a filtering effect upon the kinds of people who are moving between different areas. There is also a broad relationship between social/cultural change and distance. A change of dialect or differences in the social organisation of groups may make the migrant seem an obvious 'outsider'. To avoid such changes, the prospective migrant may decide on a shorter-distance movement. Long-distance movement may also involve entry into areas with different ethnicity, colour or religion, which may all hinder the process of assimilation.

In terms of direction, the most prevalent forms of migration are from rural to urban environments and from peripheral regions to economic **core** regions. Thus the main migration streams are from culturally traditional areas to areas where rapid change, in all its manifestations, is taking place. In LICs and MICs, the socio-economic differences between rural and urban areas are generally of a much greater magnitude than in HICs. This may necessitate some quite fundamental forms of adaptation by rural–urban migrants in the poorer nations of the world.

Although of a lesser magnitude, rural–rural migration is common in LICs and MICs for a variety of reasons, including employment, family reunion and marriage. In some instances, governments have encouraged the agricultural development of frontier areas such as the Amazon basin in Brazil.

Movements between urban areas consist in part of stepped migration up the urban hierarchy as migrants improve their knowledge base and financial position, adding to a range of other urban–urban migrations for reasons such as employment and education. Urban–rural migration is dominated by counterstream movement; that is, urbanites who are returning to their rural origins. Very few people, apart from the likes of government officials, teachers and doctors, move to the countryside for the first time to live or work. Apart from perhaps Brazil and a few other more affluent developing nations, **counterurbanisation** has yet to gain any kind of foothold in LICs.

Section 5.2 Activities

1 What is *internal migration*?
2 Provide a brief explanation of Figure 5.12.
3 Define **a** *stepped migration* and **b** *counterurbanisation*.

□ The causes of internal migration

The reasons why people change their place of permanent residence can be viewed at three dimensions of scale: **macro-level**, **meso-level** and **micro-level**.

The macro-level

This dimension highlights socio-economic differences at the national scale, focusing particularly on the **core–periphery** concept. The development of core regions in many LICs had its origins in the colonial era, which was characterised by the selective and incomplete opening-up of territories, supporting development in a restricted range of economic sectors. At this time, migration was encouraged to supply labour for new colonial enterprises and infrastructural projects, such as the development of ports and the construction of transport links between areas of raw-material exploitation and the ports through which export would take place.

The introduction of capitalism, through colonialism, into previously non-capitalist societies had a huge influence on movement patterns. The demand for labour in mines, plantations and other activities was satisfied to a considerable extent by restricting native access to land and by coercing people into migration to work either directly through forced labour systems or indirectly through taxation. The spread of a cash economy at the expense of barter into peripheral areas further increased the need for paid employment that, on the whole, could only be found in the economic core region (Figure 5.14).

In the post-colonial era, most LICs and MICs have looked to industrialisation as their path to a better world, resulting in disproportionate investment in the urban-industrial sector and the relative neglect of the rural economy. Even where investment in agriculture has been considerable, either the objective or the end result was to replace labour with machinery, adding further to rural out-migration.

The macro-level perspective provides a general explanation of migration patterns in LICs and MICs. However, this approach has two weaknesses:

■ it fails to explain why some people migrate and others stay put when faced with very similar circumstances in peripheral areas
■ it offers no explanation as to why not all forms of migration occur in the direction of economic core regions.

Figure 5.14 The Ger district in Ulaanbaatar, Mongolia, which is expanding rapidly due to high levels of rural–urban migration

The meso-level

The meso-level dimension includes more detailed consideration of the factors in the origin and destination that influence people's migration decisions. E.S. Lee's origin-intervening obstacles-destination model, which is discussed in the previous section, is a useful starting point in understanding this level of approach, which looks well beyond economic factors and recognises the vital role of the perception of the individual in the decision-making process.

Lee argues that migration occurs in response to the prevailing set of factors both in the migrant's place of origin and in one or a number of potential destinations. However, what is perceived as positive and what is viewed as negative at origin and destination may vary considerably between individuals, as may the intervening obstacles. As Lee states, 'It is not so much the actual factors at origin and destination as the perception of these factors which results in migration'. Lee stressed the point that the factors in favour of migration would generally have to outweigh considerably those against, due to the natural reluctance of people to uproot themselves from established communities.

High population growth is often cited as the major cause of rural–urban migration. However, in itself population growth is not the main cause of out-migration. Its effects have to be seen in conjunction with the failure of other processes to provide adequately for the needs of growing rural communities. Even when governments focus resources on rural development, the volume of out-migration may not be reduced. The irony in many LICs and MICs is that people are being displaced from the countryside because in some areas change is too slow to accommodate the growing size and needs of the population, or because in other areas change is too quick to enable redundant rural workers to find alternative employment in their home areas. In such circumstances, out-migration does indeed provide an essential 'safety valve'.

The evidence in Table 5.1 and in other similar studies is that the economic motive underpins the majority of rural–urban movements. During the 1960s, most demographers cited the higher wages and more varied employment opportunities of the cities as the prime reasons for internal migration. It was also widely held that the level of migration was strongly related to the rate of urban unemployment. However, while rural/urban income differentials are easy to quantify, they do not take into account the lower cost of living in the countryside and the fact that non-cash income often forms a significant proportion of rural incomes.

Table 5.1 Reasons for migration from rural areas in Peru and Thailand

PERU	
Reason	% respondents citing reason
To earn more money	39
To join kin already working	25
No work in the villages	12
Work opportunities presented themselves	11
Dislike of village life	11
To be near the village and family	11
To support nuclear and/or extended family	9
Poor	8
To pay for education	7

Source: J. Laite 'The migrant response in central Peru', in J. Gugler (ed.) The Urbanization of the Third World, OUP 1988

NORTH-EAST THAILAND		
Principal reason	No. respondents citing reason	% respondents citing reason
To earn more money for the household	138	52.9
To earn more money for self	57	21.8
To earn more money for parents	31	11.9
To further education	12	4.6
To earn money to build a house	10	3.8
To earn money to invest in farming	4	1.5
For fun	3	1.1
To earn money to purchase land/land title	2	0.8
To earn money to repay a debt	1	0.4
To earn money to pay for hired labour	1	0.4
To see Bangkok	1	0.4
To earn money to get married	1	0.4
Total	261	100.0

Source: M. Parnwell, Population Movements and the Third World, Routledge, 1993

In the 1970s, as more and more cities in LICs experienced large-scale in-migration in spite of high unemployment, demographers began to reappraise the situation. Michael Todaro was one of the first to recognise that the paradox of urban deprivation on the one hand and migration in pursuit of higher wages on the other could be explained by taking a long-term view of why people move to urban areas. As the more detailed consideration of the Todaro model in the previous section explains, people are prepared to ensure urban hardship in the short term in the likelihood that their long-term prospects will be much better in the city than in the countryside. Apart from employment prospects, the other perceived advantages of the cities are a higher standard of accommodation, a better education for migrants' children, improved medical facilities, the conditions of infrastructure often lacking in rural areas and a wider range of consumer services. The most fortunate migrants find jobs in the formal sector. A regular wage then gives some access to the other advantages of urban life. However, as the demand for jobs greatly outstrips supply, many can do no better than the uncertainty of the informal sector.

Of all the factors that migrants take into account before arriving at a decision, the economic perspective invariably dominates the decision to leave the countryside. However, all the evidence shows that other factors, particularly the social environment, have a very strong influence on the direction that the movement takes. This largely explains why capital cities, with their wide range of social opportunities, attract so many rural migrants.

The micro-level

The main criticisms of the macro- and meso-level explanations of migration are that:

■ they view migration as a passive response to a variety of stimuli
■ they tend to view rural source areas as an undifferentiated entity.

The specific circumstances of individual families and communities in terms of urban contact are of crucial importance in the decision to move, particularly when long distances are involved. The alienation experienced by the unknown new migrant to an urban area should not be underestimated and is something that will be avoided if at all possible. The evidence comes from a significant number of sample surveys and of course from the high incidence of 'area of origin' communities found in cities. For example:

■ A sample survey of rural migrants in Mumbai found that more than 75 per cent already had one or more relatives living in the city, from whom 90 per cent had received some form of assistance upon arrival.
■ A survey of migration from the Peruvian Highlands to Lima found that 90 per cent of migrants could rely

on short-term accommodation on arrival in the city, and that for about half their contacts had managed to arrange a job for them.

The importance of established links between urban and rural areas frequently results in the phenomenon of 'chain migration'. After one or a small number of pioneering migrants have led the way, subsequent waves of migration from the same rural community follow. The more established a migrant community becomes in the city, the easier it appears to be for others in the rural community to take the decision to move and for them to assimilate into urban society.

Apart from contact with, and knowledge of, urban locations, differentiation between rural households takes the following forms:

- level of income
- size of land holding
- size of household
- stage in the life cycle
- level of education
- cohesiveness of the family unit.

All of these factors have an influence on the decision to migrate (Figure 5.15). Family ties and commitments may determine whether or not someone is able to migrate, and may also influence who from a family unit is most likely to take on the responsibility of seeking employment in the city. Here the stage in the life cycle is crucial and it is not surprising that the great majority of migrants in LICs and MICs are aged between 15 and 25 years. In some communities, the phenomenon of 'relay migration' has been identified, whereby at different stages in a family's life cycle, different people take responsibility for migration.

It is only by examining all three dimensions – macro, meso and micro – that the complexity of the migration process can be fully understood. As elsewhere in geographical analysis, there is a tendency to over-simplify.

Figure 5.15 Migrants from north-east Brazil farming a smallholding in the Amazon basin

This is often useful in the early stages of enquiry, but unless we are careful the understandable generalisation may mask essential detail.

Section 5.2 Activities

1 Why is it important to consider different dimensions of scale when examining internal migration?
2 Produce a brief summary of the information in Table 5.1.

☐ The impacts of internal migration

Socio-economic impact

Figure 5.16 provides a useful framework for understanding the costs and benefits of migration. It highlights the main factors that determine how rural areas are affected by migration; namely, the two-way transfers of labour, money, skills and attitudes. However, while all of the linkages seem fairly obvious, none is easy to quantify. Therefore, apart from very clear-cut cases, it is often difficult to decide which is greater – the costs or benefits of migration.

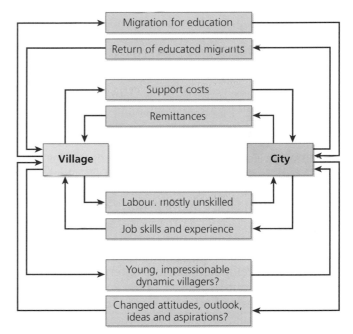

Figure 5.16 The costs and benefits of migration

Remittances from internal migration are even more difficult to estimate than those arising from international migration. Thus it is not surprising that research has produced a fairly wide range of conclusions, of which the following are but a sample:

- Williamson (1988) put urban–rural remittances at 10–13 per cent of urban incomes in Africa.
- Reardon (1997) noted that in rural areas in Africa not close to major cities, migrant earnings accounted for

Figure 5.19 Cairo has expanded rapidly due both to high in-migration and high natural increase

associated with the need for firewood and building materials. Increased pressure on the land can result in serious soil degradation.

A study of high in-migration into the coastal areas of Palawan in the Philippines found that the historical social processes that helped maintain reasonable patterns of environmental use had been overwhelmed by the rapid influx of migrants. The newcomers brought in new resource extraction techniques that were more efficient but also more destructive than those previously employed by the established community. The study concluded that high in-migration had caused severe environmental damage to the coastal environment.

☐ Impact on population structures

The age-selective (and often gender-selective) nature of migration can have a very significant impact on both areas of origin and destination. This is no more so than in rural areas of heavy out-migration and urban areas where heavy in-migration is evident.

Age/sex structure diagrams for rural areas in LICs and MICs frequently show the loss of young adults (and their children) and may also show a distinct difference between the number of males and females in the young-adult age group, due to a higher number of males than females leaving rural areas for urban destinations. However, in some rural areas female out-migration may be at a higher level than male out-migration, as Figure 5.20 illustrates. In contrast, urban population diagrams show the reverse impact, with age-selective in-migration.

In Figure 5.20, women aged 20 to 35 years in Grant County, USA comprise just 4.3 per cent of the population. This is a mainly rural area. The county's ageing population lowers the birth rate and increases the death rate. Here, out-migration has caused depopulation – an actual fall in the population. In contrast, in Orange County, Florida, 12 per cent of the population are women aged 20 to 35 years. Orange County is a predominantly urban area.

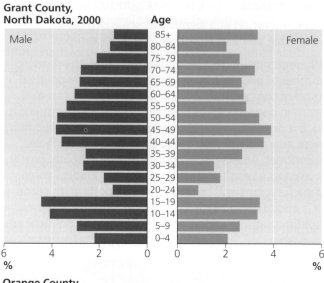

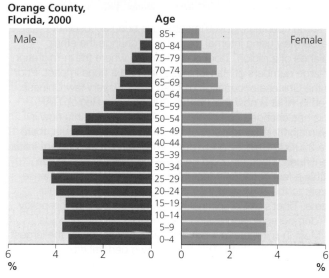

Source: *OCR A2 Geography* by M. Raw (Philip Allan Updates, 2009), p.135

Figure 5.20 Age/sex structure diagrams for Grant County, North Dakota and Orange County, Florida in the USA

Section 5.2 Activities

1 With reference to Figure 5.16:
  a Give two reasons for rural–urban migration.
  b To what extent and why is rural–urban migration selective?
  c Discuss the 'support costs' flowing from village to city.
  d What are *remittances*? Suggest how remittances are used in rural areas.
2 In what ways can internal migration have a political impact?
3 Describe how internal migration can have an impact on the environment.
4 Explain how rural–urban migration can have an impact on population structures.

Stepped migration and urban–urban movements

A number of analyses of internal migration, for example in Nigeria, have recognised a stepped structure to such movements, with migrants from rural areas often moving to a local town before later making a move further up the urban hierarchy. Figure 5.21 shows three ways stepped migration might occur in a LIC.

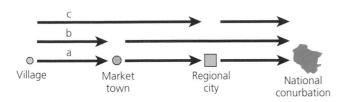

Figure 5.21 Stepped migration

During the initial move from a rural environment to a relatively small urban area, migrants may develop skills and increase their knowledge of and confidence in urban environments. They may become aware of better employment opportunities in larger urban areas and develop the personal contacts that can be so important in the migration process. For those working in the formal sector, a move up the urban hierarchy may be linked to a promotion within the company in which they work, or a transfer linked to public-sector employment.

Another important form of urban–urban migration is from towns and cities in economic **periphery** areas to urban areas in the economic core. An example is Brazil, with significant movement in the last 50 years from urban areas in the relatively poor north east such as Fortaleza, Natal, Recife and Salvador to the more prosperous cities of the south east, such as São Paulo, Rio de Janeiro and Belo Horizonte. Greater employment opportunities and higher average wages have been the main reason for such movements, but many of the other push and pull factors discussed earlier have also been significant.

Causes and impacts of intra-urban movements

Demographic analysis shows that movements of population within cities are closely related to stages in the **family life cycle**, with the available housing stock being a major determinant of where people live at different stages in their life. Studies in Toronto show a broad **concentric zone** pattern (Figure 5.22). Young adults frequently choose housing close to the **central business district (CBD)**, while older families occupy the next ring out. Middle-aged families are more likely to reside at a greater distance from the central area; and farther out still, in the newest suburban areas, young families dominate. This simplified model applies particularly well to a rapidly growing metropolis like Toronto where an invasion and succession process evolves over time.

Toronto's inner city has a much higher percentage of rented and small-unit accommodation than the outer regions, which, along with the stimulus of employment and the social attractions of the central area, has attracted young adults to the area. Most housing units built in the inner area in recent decades have been in the form of apartments.

Studies in the UK have highlighted the spatial contrasts in life cycle between middle- and low-income groups (Figure 5.23). With life cycle and income being the major determinants of where people live, residential patterns are also influenced by a range of organisations, foremost of which are local authorities, housing associations, building societies and landowners. On top of this is the range of choice available to the household. For those on low income this is frequently very restricted indeed. As income rises, the range of choice in terms of housing type and location increases.

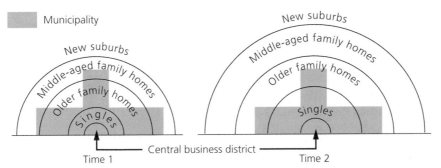

Source: *Toronto in Transition*, City of Toronto Planning and Development Dept. Policy and Research Division, April 1980, p. 21

Figure 5.22 Toronto – changing social structure in a growing city

Section 5.3 Activities

1 In terms of voluntary migration, distinguish between independent and dependent movements.
2 Describe and comment on the information illustrated in Figures 5.24 and 5.25.

☐ Forced migration

In the historical writings on migration in LICs, there is an emphasis on the forced recruitment of labour. The abduction and transport of Africans to the Americas as slaves was the largest forced migration in history. In the seventeenth and eighteenth centuries, 15 million people were shipped across the Atlantic Ocean as slaves.

Even in recent times, the scale of involuntary movement in LICs is considerably higher than most people think. However, giving due consideration to such movements should not blind us to the increasing scale of free labour migration that has occurred in recent decades. Here the focal points have been the most dynamic economies of the LICs, which have sucked in labour from more laggard neighbouring countries.

In the latter part of the twentieth century and at the beginning of the twenty-first century, some of the world's most violent and protracted conflicts have been in the LICs, particularly in Africa and Asia. These troubles have led to numerous population movements of a significant scale. Not all have crossed international frontiers to merit the term **refugee** movements. Instead, many have involved **internal displacement**. This is a major global problem, which is showing little sign of abatement.

A number of trends appear to have contributed to the growing scale and speed of forced displacement:

■ the emergence of new forms of warfare involving the destruction of whole social, economic and political systems
■ the spread of light weapons and land mines, available at prices that enable whole populations to be armed
■ the use of mass evictions and expulsions as a weapon of war and as a means of establishing culturally and ethnically homogeneous societies – the term 'ethnic cleansing' is commonly used to describe this process.

In a number of locations around the world, whole neighbourhoods of states have become affected by interlocking and mutually reinforcing patterns of armed conflict and forced displacement, for example in the Caucasus and Central Africa. The United Nations High Commission for Refugees (UNHCR) is responsible for guaranteeing the security of refugees in the countries where they seek asylum and aiding the governments of these nations in this task. The UNHCR has noted a growing number of situations in which people are repeatedly uprooted, expelled or relocated within and across state borders, forcing them to live a desperately insecure and

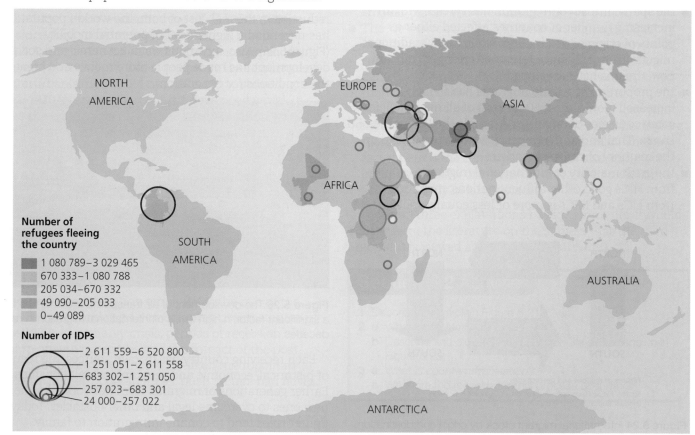

Figure 5.27 Map of world refugees and internally displaced people (IDPs), 2014

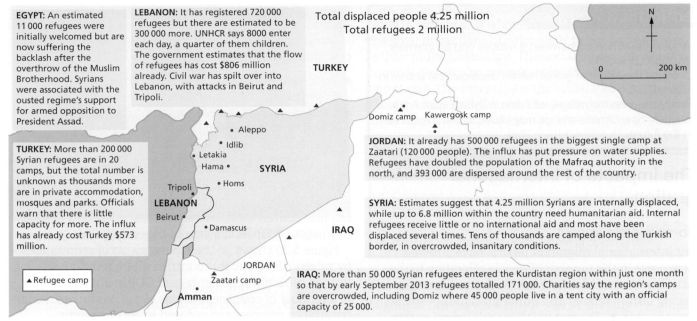

EGYPT: An estimated 11 000 refugees were initially welcomed but are now suffering the backlash after the overthrow of the Muslim Brotherhood. Syrians were associated with the ousted regime's support for armed opposition to President Assad.

LEBANON: It has registered 720 000 refugees but there are estimated to be 300 000 more. UNHCR says 8000 enter each day, a quarter of them children. The government estimates that the flow of refugees has cost $806 million already. Civil war has spilt over into Lebanon, with attacks in Beirut and Tripoli.

Total displaced people 4.25 million
Total refugees 2 million

TURKEY: More than 200 000 Syrian refugees are in 20 camps, but the total number is unknown as thousands more are in private accommodation, mosques and parks. Officials warn that there is little capacity for more. The influx has already cost Turkey $573 million.

JORDAN: It already has 500 000 refugees in the biggest single camp at Zaatari (120 000 people). The influx has put pressure on water supplies. Refugees have doubled the population of the Mafraq authority in the north, and 393 000 are dispersed around the rest of the country.

SYRIA: Estimates suggest that 4.25 million Syrians are internally displaced, while up to 6.8 million within the country need humanitarian aid. Internal refugees receive little or no international aid and most have been displaced several times. Tens of thousands are camped along the Turkish border, in overcrowded, insanitary conditions.

IRAQ: More than 50 000 Syrian refugees entered the Kurdistan region within just one month so that by early September 2013 refugees totalled 171 000. Charities say the region's camps are overcrowded, including Domiz where 45 000 people live in a tent city with an official capacity of 25 000.

▲ Refugee camp

Source: *IGCSE Geography* 2nd edition, P. Guinness & G. Nagle (Hodder Education, 2014) p.20

Figure 5.28 Syria – refugees and internally displaced people, September 2013

nomadic existence. The UNHCR has observed that 'the forced displacement of minorities, including depopulation and repopulation tactics in support of territorial claims and self-determination, has become an abominable characteristic of the contemporary world'. Figure 5.27 shows world refugees and internally displaced persons (IDPs) as of mid-2014. The UNHCR put the number of forcibly displaced people worldwide at 42.5 million. This includes 15.4 million refugees, the remainder being internally displaced people. The current conflict in Syria has produced large numbers of both refugees and internally displaced people (Figure 5.28). An increasing number of people have fled the conflict in Syria and other conflict situations such as in Eritrea, Iraq and Afghanistan, many to seek sanctuary in Europe.

Many LICs are prone to natural disasters. Because poor nations do not possess the funds to minimise the consequences of natural disaster, forced migration is often the result. Some areas have been devastated time and time again, often eliciting only a minimal response from the outside world. Ecological and environmental change are a common cause of human displacement. Much of central Asia is affected by problems such as soil degradation and desertification, a situation created by decades of agricultural exploitation, industrial pollution and overgrazing. The worst situation is in and around the Aral Sea, a large lake located between Kazakhstan and Uzbekistan. In a large-scale effort to increase cotton production in the region, most of the river water flowing into the Aral Sea was siphoned off for irrigation. Since 1960, the surface area of the sea has been reduced by half. Dust from the dried-up bed of the sea, containing significant amounts of agricultural and industrial chemicals, is carried long distances by the wind, adding

further to the pollution, salinisation and desertification of the land. Agricultural production has fallen sharply and food has increased in price, the fishing industry has been almost totally destroyed and local people are plagued by significant health problems. It has been estimated that more than 100 000 people have left the Aral Sea area since 1992 because of these problems.

Semipalatinsk in Kazakhstan, where almost 500 nuclear bombs were exploded between 1949 and 1989, 150 of them above ground, is another environmental disaster zone. Here, 160 000 people decided to leave, due to concerns about the consequences of nuclear radiation. Around half of these people moved to other parts of Kazakhstan, with the remainder going to a number of other former Soviet states. Tackling environmental degradation in this region will not be an easy task. The problem is so deep-rooted and was kept hidden for so long under Soviet rule that it may in some instances be too late for effective remedial action to be taken.

Increasingly large numbers of people have been displaced by major infrastructural projects and by the commercial sector's huge appetite for land. In LICs, the protests of communities in the way of 'progress' are invariably ignored for reasons of 'national interest' or pure greed. The World Bank and other international organisations have been heavily criticised in recent decades for financing numerous large-scale projects without giving sufficient consideration to those people directly affected.

It is predicted that climate change will force mass migrations in the future. In 2009, the International Organization for Migration estimated that worsening tropical storms, desert droughts and rising sea levels will displace 200 million people by 2050.

In the USA, the large inflow of migrants from Latin America has resulted in a substantial increase in the proportion of Spanish speakers in the country. Many areas in the southern part of the USA, in states such as California, New Mexico, Texas and Florida, are effectively bilingual. Many other traits of Latin American culture are also evident in the region. In turn, the contact that migrant workers have with their families and communities elicits a certain reverse flow of cultural traits, as workers relate their experiences and send money home.

The political impact

Significant levels of international migration can have a considerable political impact, both within and between countries. In many countries, there is a clear trend of immigrants being more likely to vote for parties of the centre and the left as opposed to political parties to the right of centre. In HICs, immigrants tend to head for economic core regions and to inner-city areas within these regions. Such concentrations can have a big impact on voting patterns.

Over time, immigrants gradually assimilate into host societies. In general, economic assimilation comes first, followed by social assimilation and then political assimilation. When immigrant groups reach a certain size and standing they begin to develop their own politicians, as opposed to voting for politicians from the host society. This process is more likely to happen in mature democracies where there is a long history of immigration. The UK and the USA are examples of countries where this process has been evident.

High levels of international migration between one country and another can lead to political tension. The high level of Mexican migration into the USA, both legal and illegal, has created tensions between the US and Mexican governments. In recent years, the USA has greatly increased the size of its Border Patrol. Critics refer to the 'militarisation of the Mexican border', which is costing $3 billion a year.

In a number of EU countries in recent decades, high levels of immigration have created sizeable immigrant populations. Such populations have been assimilated more successfully in some countries than others. Where and how people are housed is a big factor in assimilation.

Many LICs and MICs are looking to HICs to adopt a more favourable attitude to international migration. The subject is brought up regularly at international conferences. This political pressure is known as 'the pro-migration agenda of developing nations'.

Living within a new political system can also affect the attitudes of immigrant communities to what goes on back in their home country. The harshest critics of authoritarian governments in the Middle East and Asia are invariably exiles living in other countries.

The environmental impact

In an article entitled 'The Environmental Argument for Reducing Immigration to the United States', Winthrop Staples and Philip Cafaro argue that 'a serious commitment to environmentalism entails ending America's population growth by implementing a more restrictive immigration policy. The need to limit immigration necessarily follows when we combine a clear statement of our main environmental goals – living sustainably and sharing the landscape generously with other species – with uncontroversial accounts of our current demographic trajectory and of the negative environmental effects of U.S. population growth, nationally and globally.'

Staples and Cafaro explain how population growth contributes significantly to a host of environmental problems in the USA. They also argue that a growing population increases America's large environmental footprint beyond its borders and creates a disproportionate role in stressing global environmental systems.

There have been growing environmental concerns about immigration in other countries too, as the concept of sustainability has become understood in a more detailed way. However, some critics see such arguments as a disingenuous way of attempting to curtail immigration.

Figure 5.31 is a summary of the possible impacts of international migration. Many of these factors are also relevant to internal migration. Because migration can be such an emotive issue, you may not agree with all of these statements, and you may consider that some important factors have been omitted.

Section 5.3 Activities

1 With brief reference to one country, describe the cultural impact of international migration.
2 Give two examples of the way international migration can have a political impact.
3 How can international migration have an impact on the environment?

The impact of international migration		
Impact on countries of origin	**Impact on countries of destination**	**Impact on migrants themselves**
Positive		
• Remittances are a major source of income in some countries. • Emigration can ease the levels of unemployment and underemployment. • Reduces pressure on health and education services and on housing. • Return migrants can bring new skills, ideas and money into a community.	• Increase in the pool of available labour may reduce the cost of labour to businesses and help reduce inflation. • Migrants may bring important skills to their destination. • Increasing cultural diversity can enrich receiving communities. • An influx of young migrants can reduce the rate of population ageing.	• Wages are higher than in the country of origin. • There is a wider choice of job opportunities. • A greater opportunity to develop new skills. • They have the ability to support family members in the country of origin through remittances. • Some migrants have the opportunity to learn a new language.
Negative		
• Loss of young adult workers who may have vital skills, e.g. doctors, nurses, teachers, engineers (the 'brain-drain' effect). • An ageing population in communities with a large outflow of (young) migrants. • Agricultural output may suffer if the labour force falls below a certain level. • Migrants returning on a temporary or permanent basis may question traditional values, causing divisions in the community.	• Migrants may be perceived as taking jobs from people in the long-established population. • Increased pressure on housing stock and on services such as health and education. • A significant change in the ethnic balance of a country or region may cause tension. • A larger population can have a negative impact on the environment.	• The financial cost of migration can be high. • Migration means separation from family and friends in the country of origin. • There may be problems settling into a new culture (assimilation). • Migrants can be exploited by unscrupulous employers. • Some migrations, particularly those that are illegal, can involve hazardous journeys.

Source: *IGCSE Geography* 2nd edition, P. Guinness & G. Nagle (Hodder Education, 2014) p.23

Figure 5.31 Matrix showing the impact of migration

Case Study: Diasporas in London

London is undoubtedly the most cosmopolitan city in Europe (Figures 5.32 and 5.33). Some commentators go further and view London as the most multiracial city in the world. The diverse **ethnicity** of the capital is exemplified by the fact that over 200 languages are spoken within its boundaries. The lobby group Migration Watch estimates that two-thirds of immigration into the UK since the mid-1990s has been into London. Within the UK, the process of **racial assimilation** is much more advanced in London than anywhere else. Almost 30 per cent of people in London were born outside the UK, compared with 2.9 per cent in north-east England. London has the highest proportion of each ethnic minority group apart from Pakistanis, of whom there is a higher proportion in Yorkshire.

Figure 5.32 An Indian pub in Southall – the largest Indian community in the UK

Table 5.2 Factors encouraging migration from Mexico, by type of migrant

Type of migrant	Demand–pull	Supply–push	Network/other
Economic	Labour recruitment (guest workers)	Unemployment or underemployment; low wages (farmers whose crops fail)	Job and wage information flows
Non-economic	Family unification (family members join spouse)	Low income, poor quality of life, lack of opportunity	Communications; transport; assistance organisations; desire for new experience/adventure
Note: All three factors may encourage a person to migrate. The relative importance of pull, push and network factors can change over time.			

Source: P. Martin and J. Widgren, International Migration: Facing the Challenge *(2002) Vol. 57, No. 1 (page 8 table 1), quoted in* Population Bulletin *Vol.63 No.1 2008*

religious groups were against the programmes. Congress agreed with what was then a common view in the USA – that the inflow of Mexican workers was holding down the wages of US farm workers – and ended the programme.

The end of the *bracero* programme saw farm wages rise, along with the increasing mechanisation of US agriculture. Re-adjusting the labour market in America after several decades of significant dependence on Mexican workers was not easy. On the other side, the loss of US jobs and wages was a difficult adjustment for many Mexican workers. Under the *bracero* programme, American farmers were required to pay for the transportation of Mexican workers from the US/Mexican border. This was an incentive for many Mexicans to move to the border area in the hope of being selected for work in the USA. When the programme ended they returned to border communities in Mexico where unemployment was extremely high.

☐ The establishment of *maquiladoras*

The US and Mexican governments made changes to their trade laws to allow the establishment of *maquiladoras*. These were factories in Mexico that imported components and used Mexican labour to assemble them into goods such as televisions for export to the USA. The logical location for the *maquiladoras* was in towns just over the border in Mexico so that they were as close to their US markets as possible. As the number of factories grew, more Mexicans migrated from other parts of the country to the border towns, putting them in competition with returning *braceros* for jobs. The establishment of *maquiladoras* only solved the returning *bracero* problem to a certain extent, as many of the jobs in the factories went to women.

☐ The increase in illegal migration

Although many rural Mexicans had become dependent on US employment, there was very little illegal migration from Mexico to the USA in the 1960s and 1970s. However, high population growth and the economic crisis in the early 1980s resulted in a considerable increase in illegal migration across the border. Networks were soon established between Mexican communities and US employers. At this time, there were no penalties placed on American employers who knowingly hired illegal migrants. During this period, Mexican workers spread out more widely in the USA than ever before. They were employed mainly in agriculture, construction, various manufacturing industries and in low-paid services jobs. The US Border Patrol was responsible for apprehending illegal workers, but their numbers were limited and they only had a modest impact on the spread of illegal workers

Figure 5.36 US Border Patrol

(Figure 5.36).

As attitudes in America again hardened against illegal workers, Congress passed the Immigration Reform and Control Act (IRCA) of 1986. This imposed penalties on American employers who knowingly hired illegal workers.

The objective was to discourage Mexicans from illegal entry. Much of the opposition of the unions to guest workers was because they saw the process creating 'bonded workers' with very limited rights.

However, the Act also legalised 2.7 million unauthorised foreigners. Of this number, 85 per cent were Mexican. The legalisation substantially expanded network links between Mexican workers and US employers.

The formation of the North American Free Trade Agreement (NAFTA) lowered barriers to trade and investment flow between Mexico, the USA and Canada. At the time, the Mexican government expected Mexico's export trade to increase and Mexico–USA migration to fall due to NAFTA. However, this proved not to be the case and migration from Mexico to the USA increased. Labour migration continued at a high rate even after economic and employment growth in Mexico improved in the late 1990s.

Since 1980, Mexicans have been the largest immigrant group in the USA. In 2013, approximately 11.6 million Mexican immigrants lived in the USA, up from 2.2 million in 1980 (Figure 5.37).

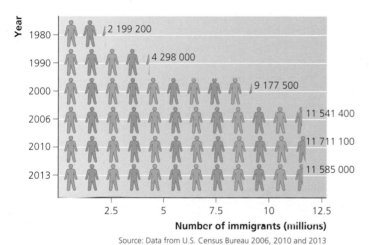

Source: Data from U.S. Census Bureau 2006, 2010 and 2013

Figure 5.37 Mexican immigration to the USA, 1980–2013

The US Census in 2000 found an estimated 8.4 million, mostly Mexican, unauthorised foreigners. This stimulated new attempts to regulate migration between the two countries. George Bush, elected President in 2000, stated that he favoured a guest-worker programme to permit more Mexicans to work in America. In 2001, Mexican President Vicente Fox pressed the US government to endorse what was known as the 'whole *enchilada*'. This would involve legalisation for unauthorised Mexicans in the USA, a new guest-worker programme, improved conditions along the border and exempting Mexico from immigrant visa ceilings. These discussions were halted by the 11 September 2001 terrorist attacks.

Legal and illegal migration from Mexico continued as before. By 2006, there were an estimated 11.5 million

Mexican-born people living in the USA. This amounted to around 11 per cent of living people born in Mexico. With their children also taken into account, the figure increased to more than 20 million. This was equivalent to almost a fifth of the population of Mexico. The next four leading countries of origin were the Philippines, India, China and Vietnam, with between 1.1 and 1.6 million people each. This illustrates the size and impact of Mexican immigration into the USA.

Figure 5.38 shows the distribution of the Mexican population in the USA by county. Counties are subdivisions of states in the USA. There is a very strong concentration of the US Mexican population in the four states along the Mexican border – California, Arizona, New Mexico and Texas. The concentration is particularly strong in California and Texas. Other western states, including Washington, Oregon, Colorado, Nevada and Idaho, also have above-average concentrations. The main reasons for this spatial distribution are:

- proximity to the border
- the location of demand for immigrant farm workers
- urban areas where the Mexican community is long-established.

Figure 5.39 illustrates the distribution of the Mexican population in the Los Angeles region. Within the urban area itself, the Mexican population is concentrated in areas of poor housing and low average income. In more peripheral areas, the Mexican population is concentrated in low-cost housing areas where proximity to farm employment is an important factor.

Mexican culture has had a sustained impact on many areas in the USA, particularly urban areas close to the border. As a result, many Mexican migrants find reassuring similarities between the two countries. One study on labour migration from Mexico to the USA stated: 'Many Mexicans find adapting to Los Angeles as easy as navigating Mexico City.'

There is no doubt that the Mexican population in the USA has undergone a process of assimilation over time. There are three facets to assimilation:

- economic
- social
- political.

Assimilation tends to occur in the order presented above, with economic assimilation occurring first. While most migrants from Mexico would be in the low skills category, their children and grandchildren usually aspire to, and gain, higher qualifications and skills. Such economic mobility inevitably results in greater social contact with the mainstream population. Eventually, more people from migrant populations get involved in politics and the migrant community gains better political representation.

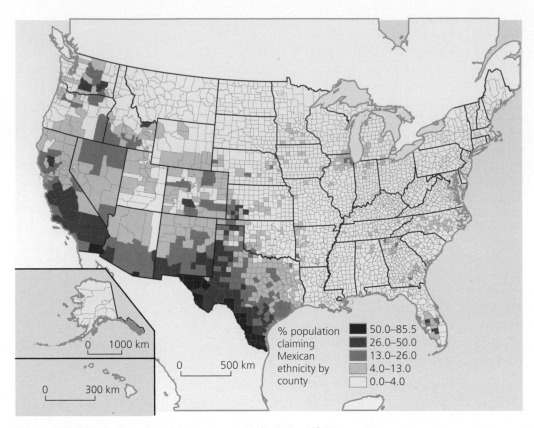

Figure 5.38 Distribution of the Mexican population in the USA by county

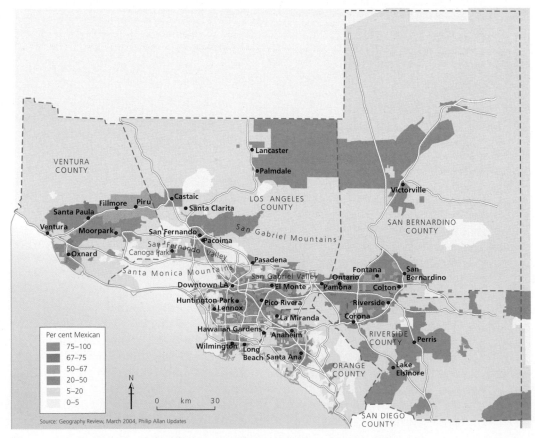

Figure 5.39 Distribution of the Mexican population in the Los Angeles region

5 Migration

The demography of Mexican migration to the USA

In an article entitled 'The demography of Mexican migration to the US', G.H. Hanson and C. McIntosh highlight the fact that with the US baby boom peaking in 1960, the number of US native-born people coming of working age actually declined in the 1980s. In contrast, high levels of fertility continued in Mexico in the 1960s and 1970s. The sharp increase in Mexico–USA relative labour supply coincided with the stagnation of Mexico's economy in the 1980s, after significant economic progress in the 1960s and 1970s. This created ideal conditions for an emigration surge.

However, the conditions behind recent emigration from Mexico are unlikely to be sustained. Today, Mexico's labour supply growth is converging to US levels. Between 1965 and 2000, Mexico's total fertility rate fell from 7.0 to 2.5, close to the US rate of 2.1. Thus, labour supply pressures for emigration from Mexico peaked in the late 1990s and are likely to fall in coming years.

Figure 5.40 is a simulation of migration from Mexico to the USA based on differences in labour supply and wage differentials between the two countries. Population projections are used to estimate future labour supply.

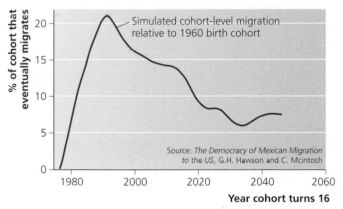

Effect in differences from the 1960 birth cohort, which turned 16 in 1976.

Figure 5.40 Labour supply pressures for Mexican migration to the USA

Opposition to Mexican migration into the USA

In the USA, the Federation for American Immigration Reform (FAIR) argues that unskilled newcomers:

- undermine the employment opportunities of low-skilled US workers
- have negative environmental effects
- threaten established US cultural values.

The recent global economic crisis saw unemployment in the USA rise to about 10 per cent, the worst job situation for 25 years. Immigration always becomes a more sensitive issue in times of high unemployment. FAIR has also highlighted the costs to local taxpayers of illegal workers in terms of education, emergency medical care, detention and other costs that have to be borne.

Those opposed to FAIR see its actions as uncharitable and arguably racist. Such individuals and groups highlight the advantages that Mexican and other migrant groups have brought to the country.

An ethnographic case study

A. Mountz and R. Wright (1996) presented an interesting **ethnographic** account of the transnational migrant community of San Agustín, a village in the Mexican state of Oaxaca, and Poughkeepsie, a city in New York state. The link between the two communities began with the migration of a lone Oaxacan to Poughkeepsie in the early 1980s. In classic network fashion, the Mexican population of Poughkeepsie, predominantly male, grew to well over a thousand over the next decade. Most Oaxacans found employment as undocumented workers in hotels, restaurants and shops and as building workers and landscapers. Their remittances transformed village life in their home community.

What struck Mountz and Wright most was the high level of connectedness between San Agustín and Poughkeepsie, with the migrant community keeping in daily contact with family and friends via telephone, fax, camcorders, videotape and VCRs – communications technology that was rapidly being introduced to San Agustín. Rapid migration between the two communities was facilitated by jet travel and systems of wiring payments. In effect, the community of San Agustín had been geographically extended to encompass the Oaxacan enclave in Poughkeepsie. This is a classic example of **time–space distanciation** – the stretching of social systems across space and time.

Migrant remittances were used not only to support the basic needs of families but also for home construction, the purchase of consumer goods and financing *fiestas*. The last provided an important opportunity for migrants to display continued village membership. However, as out-migration became more established, tensions began to develop between some migrants and the home community. The main point of conflict was over the traditional system of communal welfare that requires males to provide service and support to the village. Where this could not be done in terms of time, a payment could be substituted. This was increasingly resented by some migrants who saw 'their money as their own'. The traditionalists in the village cited migration as the major cause of the decline of established values and attitudes.

The researchers found that a **migrant culture** had now become established in San Agustín, as it had in so many other Mexican communities, for four main reasons:

- economic survival
- rite of passage for young male adults

- the growing taste for consumer goods and modern styles of living
- the enhanced status enjoyed by migrants in the home community.

What started out as an exception was now well on the way to becoming the rule for San Agustín's young males.

☐ The impact on Mexico

Sustained large-scale labour migration has had a range of impacts on Mexico, some of them clear and others debatable. Significant impacts include:

- the high value of remittances, which totalled $22 billion in 2013 – as a national source of income, this is only exceeded by oil exports; it represents about 2 per cent of the country's GDP – remittances from the USA to Mexico have increased 14-fold since 1985!
- reduced unemployment pressure as migrants tend to leave areas where unemployment is particularly high
- lower pressure on housing stock and public services as significant numbers of people leave for the USA
- changes in population structure with emigration of young adults, particularly males

- loss of skilled and enterprising people
- migrants returning to Mexico with changed values and attitudes.

Women and children often assume the agricultural labour previously performed by now-absent men. Sometimes, if no-one is able to work the land, agricultural plots are either sold or abandoned. In general, women and children's psychological health has been greatly affected by family members' migration.

Section 5.4 Activities

1 With reference to Table 5.2, discuss the factors that encourage migration from Mexico by type of migrant.
2 Comment on the information presented in Figure 5.37.
3 Describe the distribution of the Mexican population in the USA shown in Figure 5.38.
4 What impact has such a high rate of emigration had on Mexico?

6 Settlement dynamics

6.1 Changes in rural settlements

Rural settlements form an essential part of the human landscape (Figure 6.1). However, such settlements in HICs, MICs and LICs have undergone considerable changes in recent decades. This has happened for a number of reasons, which include:

- rural–urban migration
- urban–rural migration
- the consequences of urban growth
- technological change
- rural planning policies
- the balance of government funding between urban and rural areas.

Figure 6.1 Rural settlement in Nepal

☐ Changing rural environments in the UK

In the past, **rural** society was perceived to be distinctly different from urban society. The characteristics upon which this idea was based are shown in Figure 6.2. However, rapid rural change over the last 50 years or so in the UK and other HICs has seen the idea of a rural–

urban divide superseded by the notion of a rural–urban continuum. The latter is a wide spectrum that runs from the most remote type of rural settlement to the most highly urbanised. A number of the intermediate positions exhibit both rural and urban characteristics. Paul Cloke (1979) used 16 variables, including population density, land use and remoteness, to produce an 'index of rurality' for England and Wales (Figure 6.3). Urban areas now make substantial demands on the countryside, the evidence of which can be found in even the most remote areas.

Rural areas are dynamic spatial entities. They constantly change in response to a range of economic, social, political and environmental factors. In recent years, the pace of change has been more rapid than ever before. The UK reflects many of the changes occurring in rural areas in other HICs.

1 Close-knit community with everybody knowing and interacting with everyone else.

2 Considerable homogeneity in social traits: language, beliefs, opinions, mores, and patterns of behaviour.

3 Family ties, particularly those of the extended family, are much stronger than in urban society.

4 Religion is given more importance than in urban society.

5 Class differences are less pronounced than in urban society. Although occupational differentiation does exist, it is not as pronounced as in towns and cities. Also the small settlement size results in much greater mixing which in turn weakens the effects of social differentiation.

6 There is less mobility than in urban society, both in a spatial sense (people do not move house so frequently) and in a social sense (it is more difficult for a farm labourer to become a farmer or farm manager than for a factory worker to become a manager).

Source: *The Geography of Rural Resources*
by C. Bull, P. Daniel and M. Hopkinson, Oliver & Boyd, 1984

Figure 6.2 Principal characteristics of traditional rural society

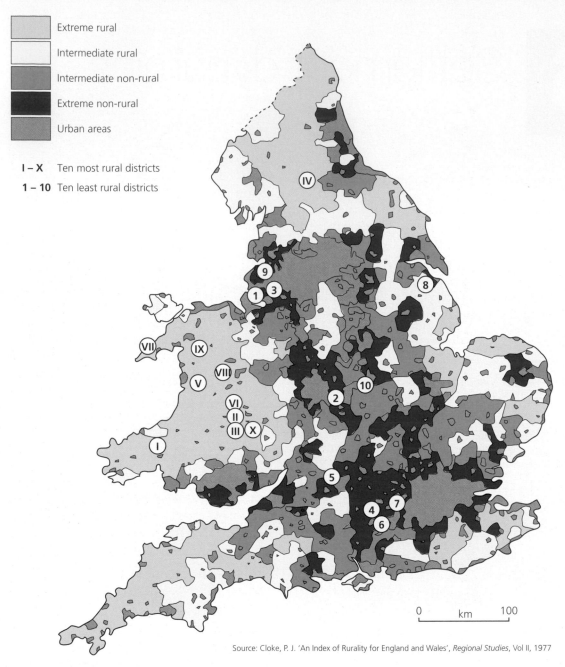

Figure 6.3 An index of rurality for England and Wales

Source: Cloke, P. J. 'An Index of Rurality for England and Wales', *Regional Studies*, Vol II, 1977

Legend:
- Extreme rural
- Intermediate rural
- Intermediate non-rural
- Extreme non-rural
- Urban areas

I – X   Ten most rural districts
1 – 10   Ten least rural districts

The economy of rural areas is no longer dominated by farmers and landowners. As agricultural jobs have been lost, new employers have actively sought to locate in the countryside. Manufacturing, high technology and the service sector have led this trend. Most of these firms are classed as SMEs – small and medium-sized enterprises. In fact, in recent decades employment has been growing faster in rural than in urban areas. Other significant new users of rural space are recreation, tourism and environmental conservation. The **rural landscape** has evolved into a complex multiple-use

resource and as this has happened the **rural population** has changed in character.

These economic changes have fuelled social change in the countryside with the in-migration of particular groups of people. To quote Brian Ilbery, a leading authority on rural geography, 'The countryside has been repopulated, especially by middle-class groups … who took advantage of relatively cheap housing in the 1960s and 1970s to colonize the countryside'. Once they are significant in number, the affluent newcomers exert a strong influence over the social and physical nature of rural space. In many areas, newcomers have dominated the housing market, to the detriment of the established population in the locality. Increased demand has pushed up house prices to a level beyond the means of many original families who then have no option but to move elsewhere.

Gentrification is every bit as evident in the countryside as it is in selected inner-city areas. However, the increasing mobility of people, goods and information has eroded local communities. A transformation that has been good for newcomers has been deeply resented by much of the established population.

In the post-war period, the government attempted to contain expansion into the countryside by creating **green belts** and by the allocation of housing to urban areas or to large **key villages**. Rural England has witnessed rising owner-occupation and low levels of local-authority housing. The low level of new housing development in smaller rural communities has been reflected in higher house prices and greater social exclusivity.

Such social and economic changes have increased the pressure on rural resources so that government has had to re-evaluate policies for the countryside. Regulation has become an important element in some areas, notably in relation to sustainability and environmental conservation.

Changing agriculture

The countryside in the UK and other HICs has been affected by major structural changes in agricultural production. Although agricultural land forms 73 per cent of the total land area of the UK, less than 2 per cent of the total workforce are now employed in agriculture. This is down from 6.1 per cent in 1950 and 2.9 per cent in 1970. Even in the most rural of areas, agriculture and related industries rarely account for more than 15 per cent of the employed population.

At the same time, the size of farms has steadily increased (Figure 6.4). Such changes have resulted in a significant loss of hedgerows, which provide important ecological networks. Agricultural wages are significantly below the national average and as a result farmers are among the poorest of the working poor. As many farmers have struggled to make a living from traditional agricultural practices, a growing number have sought to diversify both within and outside agriculture (Figure 6.5). However, while diversification may initially halt job losses, if too many farmers in an area opt for the same type of diversification, a situation of over-supply can result in a further round of rural decline.

Figure 6.4 Large-scale cereal farming in the Paris Basin, France

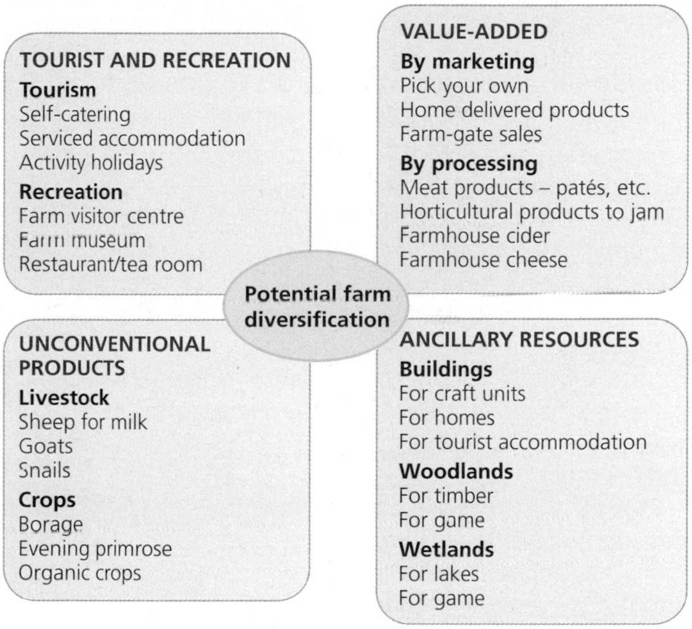

TOURIST AND RECREATION

Tourism
Self-catering
Serviced accommodation
Activity holidays

Recreation
Farm visitor centre
Farm museum
Restaurant/tea room

VALUE-ADDED

By marketing
Pick your own
Home delivered products
Farm-gate sales

By processing
Meat products – patés, etc.
Horticultural products to jam
Farmhouse cider
Farmhouse cheese

Potential farm diversification

UNCONVENTIONAL PRODUCTS

Livestock
Sheep for milk
Goats
Snails

Crops
Borage
Evening primrose
Organic crops

ANCILLARY RESOURCES

Buildings
For craft units
For homes
For tourist accommodation

Woodlands
For timber
For game

Wetlands
For lakes
For game

Source: Slee, 1987

Figure 6.5 Areas of potential farm diversification

Section 6.1 Activities

1 With reference to Figure 6.2, outline the principal characteristics of traditional rural society.
2 Briefly describe the pattern of rural areas shown in Figure 6.3.
3 What impact has agricultural change had on the rural landscape?
4 Why does the potential for farm diversification vary from region to region?

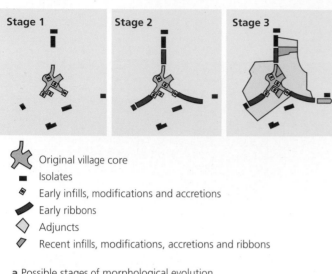

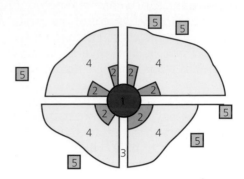

Key (Figure 6.6a):

- ⬡ Original village core
- ▪ Isolates
- ◈ Early infills, modifications and accretions
- ◣ Early ribbons
- ◇ Adjuncts
- ◢ Recent infills, modifications, accretions and ribbons

a Possible stages of morphological evolution of a suburbanised village

Key (Figure 6.6b):

1 Original village core
2 Infills, modifications and accretions
3 Ribbon development
4 Adjuncts
5 Isolates

Note: This model diagram indicates all the morphological elements likely to be present in a metropolitan village. The arrangement of these elements is likely to vary considerably between villages.

b Metropolitan village: morphological features

Source: *Advanced Geography: Concepts & Cases*, P. Guinness & G. Nagle (Hodder Education, 1999), p.75

Figure 6.6 Morphology of metropolitan villages

Counterurbanisation and the rural landscape

In recent decades, **counterurbanisation** has replaced urbanisation as the dominant force shaping settlement patterns. It is a complex and multifaceted process that has resulted in a 'rural population turnaround' in many areas where depopulation had been in progress. Green-belt restrictions have limited the impact of counterurbanisation in many areas adjacent to cities. But, not surprisingly, the greatest impact of counterurbanisation has been just beyond green belts where commuting is clearly viable. Here, rural settlements have grown substantially and been altered in character considerably.

Figure 6.6 shows the changing morphology of **metropolitan villages** identified by Hudson (1977). Stage 1 is characterised by the conversion of working buildings into houses with new building mainly in the form of infill. However, some new building might occur at the edge of the village. The major morphological change in stage 2 is ribbon development along roads leading out of the village. Stage 3 of the model shows planned additions on a much larger scale of either council or private housing estates at the edge of villages. Clearly, not all metropolitan villages will have evolved in the same way as the model, particularly those where green-belt restrictions are in place. Nevertheless, the model provides a useful framework for reference.

Rural depopulation

Because of the geographical spread of counterurbanisation since the 1960s or so, the areas affected by **rural depopulation** have diminished. Depopulation is now generally confined to the most isolated areas of the country, but exceptions can be found in other areas where economic conditions are particularly dire. Figure 6.7 is a simple model of the depopulation process.

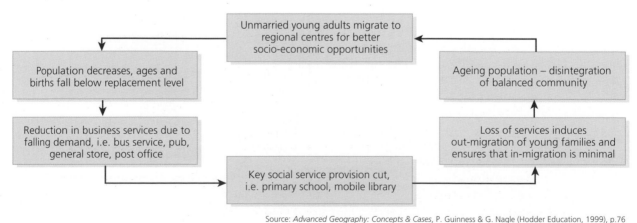

Source: *Advanced Geography: Concepts & Cases*, P. Guinness & G. Nagle (Hodder Education, 1999), p.76

Figure 6.7 Model of rural depopulation

The issue of rural services

Services – access to shops and post offices, healthcare, activities – are the basis for any community, creating and enhancing a feeling of belonging and a sustainable future for the area. However, rural services have been in decline for a number of decades, with a significant impact on the quality of life of many people, particularly those without a car. A major report published in 2008 revealed that nearly half of communities have seen the loss of key local services in the previous four years. The Oxford University study warned that poorer people in the countryside 'form a forgotten city of disadvantage'. It found that residents of the village of Bridestowe on Dartmoor had the fewest amenities, while the village of Wrotham, in Kent, had suffered the greatest loss of services since 2004 and was the most excluded community in the south-east of the UK.

Critics accused the government of masterminding the 'near certain death of the village post office' with its plans to close 2500 branches by the end of the year. One in 13 rural primary schools has closed since 1997, and more are under threat as new Whitehall rules mean schools could lose funding by failing to fill their places. Existing village GP surgeries are also at risk as the government promotes its new 'polyclinics'.

The Commission for Rural Committees warned that 233 000 people are living in 'financial service deserts' – areas with no post office within 1.25 miles, or no bank, building society or cashpoint for 2.5 miles.

ACRE (Action with Communities in Rural England) highlights the following reasons for rural service decline:

- the effect of market forces and, in some cases, the arrival of supermarkets in local areas, making local services no longer competitive
- the changing pattern of rural population, with more mobile residents with different shopping and consumer patterns becoming a greater part of the rural pattern of life
- a change in expectations of rural residents themselves, no longer prepared to make do with relatively poor and expensive services and, in many cases, with the means and opportunity to access better services.

Key villages

Between the 1950s and 1970s, the concept of key settlements was central to rural settlement policy in many parts of Britain, particularly where depopulation was occurring (Figure 6.8). The concept relates to central place theory and assumes that focusing services, facilities and employment in one selected settlement will satisfy the essential needs of the surrounding villages and **hamlets**. The argument was that with falling demand, dispersed services would decline rapidly in vulnerable areas. The only way to maintain a reasonable level of service provision in such an area was to focus on those locations

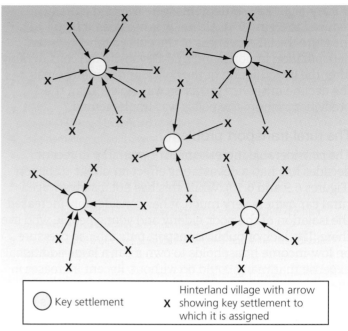

| ◯ Key settlement | Hinterland village with arrow |
| | X showing key settlement to which it is assigned |

Source: *Advanced Geography: Concepts & Cases*, P. Guinness & G. Nagle (Hodder Education, 1999), p.79

Figure 6.8 Key settlement concept

with the greatest accessibility and the best combination of other advantages. In this way, threshold populations could be assured and hopefully the downward spiral of service decline would be halted.

Devon introduced a key settlement policy in 1964 to counter the impact of:

- rural depopulation
- the changing function of the village in relation to urban centres
- the decline in agricultural employment
- the contraction of public transport.

The selection of key settlements in Devon was part of a wider settlement policy involving sub-regional centres, suburban towns and coastal resorts. The criteria used for selecting key settlements were as follows:

- existing services
- existing employment other than agriculture in or near the village
- accessibility by road
- location in relation to current bus (and possibly rail) services
- location in relation to other villages that would rely on them for some services
- the availability of public utilities capable of extension for new development
- the availability and agricultural value of land capable of development
- proximity to urban centres (key settlements would not flourish too close to competing urban areas).

6.1 Changes in rural settlements   155

Section 6.1 Activities

1 Describe the location of the Isle of Purbeck.
2 With reference to Figure 6.17, describe and explain the differences between the population structures of the Isle of Purbeck and England as a whole.
3 Discuss the main issues affecting the rural population on the Isle of Purbeck.

6.2 Urban trends and issues of urbanisation

 ## □ The development of the urban environment

The first cities

Gordon Childe used the term **urban revolution** to describe the change in society marked by the emergence of the first cities some 5500 years ago. The areas that first witnessed this profound social-economic change were:

■ Mesopotamia – the valleys of the Tigris and Euphrates rivers
■ the lower Nile valley
■ the plains of the river Indus.

Later, urban civilisations developed around the Mediterranean, in the Yellow River valley of China, in South East Asia and in the Americas. Thus the first cities mainly emerged in areas that are now considered to be LICs.

The catalyst for this period of rapid change was the Neolithic Revolution, which occurred about 8000 BCE. This was when sedentary agriculture, based on the domestication of animals and cereal farming, steadily replaced a nomadic way of life. As farming advanced, irrigation techniques were developed. Other major advances that followed were the ox-drawn plough, the wheeled cart, the sailing boat and metallurgy. However, arguably the most important development was the invention of writing in about 4000 BCE, for it was in the millennium after this that some of the villages on the alluvial plains between the Tigris and Euphrates rivers increased in size and changed in function so as to merit the classification of urban.

Considerably later than the first cities, trading centres began to develop. The Minoan civilisation cities of Knossos and Phaistos, which flourished in Crete during the first half of the second millennium BCE, derived their wealth from maritime trade. Next it was the turn of the Greeks and then the Romans to develop urban and trading systems on a scale larger than ever

before. For example, the population of Athens in the fifth century BCE has been estimated at a minimum of 100 000. The fall of the Roman Empire in the fifth century CE (Figure 6.19) led to a major recession in urban life in Europe, which did not really revive until medieval times.

The medieval revival was the product of population growth and the resurgence of trade, with the main urban settlements of this period located at points of greatest accessibility. While there were many interesting developments in urban life during the medieval period, it required another major technological advance to set in train the next urban revolution.

The urban industrial revolution

The second 'urban revolution', based on the introduction of mass production in factories, began in Britain in the late eighteenth century. This was the era of the Industrial Revolution when industrialisation and urbanisation proceeded hand in hand. The key invention, among many, was the steam engine, which in Britain was applied to industry first and only later to transport. The huge demand for labour in the rapidly growing coalfield towns and cities was satisfied by the freeing of labour in agriculture through a series of major advances. The so-called 'Agricultural Revolution' had in fact begun in the early seventeenth century.

By 1801, nearly one-tenth of the population of England and Wales was living in cities of over 100 000 people. This proportion doubled in 40 years and doubled again in another 60 years. The 1801 census recorded London's population at 1 million, the first city in the world to reach this figure. By 1851, London's population had doubled to 2 million. However, at the global scale less than 3 per cent of the population lived in urban places at the beginning of the nineteenth century.

As the processes of the Industrial Revolution spread to other countries, the pace of urbanisation quickened. The change from a population of 10–30 per cent living in urban areas of 100 000 people or more took about 80 years in England and Wales, 66 years in the USA, 48 years in Germany, 36 years in Japan and 26 years in Australia.

The initial urbanisation of many LICs was restricted to concentrations of population around points of supply of raw materials for the affluent HICs. For example, the growth of São Paulo was firmly based on coffee; Buenos Aires on mutton, wool and cereals; and Kolkata on jute.

By the beginning of the most recent stage of urban development in 1950, 27 per cent of people lived in towns and cities, with the vast majority of urbanites still living in HICs. In fact, in HICs the cycle of urbanisation was nearing completion.

Figure 6.19 Remains of the Roman city of Pompeii, with Mt Vesuvius in the background

The post-1945 urban 'explosion' in LICs and MICs

Throughout history, **urbanisation** and significant economic progress have tended to occur together. In contrast, the rapid urban growth of LICs and MICs in the latter part of the twentieth century in general far outpaced economic development, creating huge problems for planners and politicians (Figure 6.20). Because urban areas in LICs and MICs have been growing much more quickly than did the cities of HICs in the nineteenth century, the term 'urban explosion' has been used to describe contemporary trends.

However, the clear distinction between urbanisation and urban growth should be kept in mind, as some of the least urbanised countries, such as China and India, contain many of the world's largest cities and are recording the fastest rates of growth.

An approach known as 'dependency theory' has been used by a number of writers to explain the urbanisation of LICs and MICs, particularly the most recent post-1950 phase. According to this approach, urbanisation in LICs and MICs has been a response to the absorption of countries and regions into the global economy. The capitalist global economy induces urbanisation by concentrating production and consumption in locations that:

- offer the best economies of scale and agglomeration
- provide the greatest opportunities for industrial linkage
- give maximum effectiveness and least cost in terms of control over sources of supply.

Figure 6.20 Street market in Nabul, Tunisia

☐ The cycle of urbanisation

The development of urban settlement in the modern period can be seen as a sequence of processes known as the cycle of urbanisation (Figure 6.23). The key processes and their landscape implications are: **suburbanisation**, counterurbanisation and **reurbanisation**. In the UK, suburbanisation was the dominant process until the 1960s. From this decade, counterurbanisation increasingly had an impact on the landscape. Reurbanisation of some of the largest cities, beginning in the 1990s, is the most recent phenomenon.

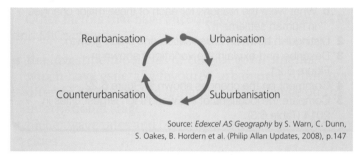

Source: *Edexcel AS Geography* by S. Warn, C. Dunn, S. Oakes, B. Hordern et al. (Philip Allan Updates, 2008), p.147

Figure 6.23 The cycle of urbanisation

Suburbanisation

Although the urban Industrial Revolution in Britain began in the late eighteenth century, it was not really until the 1860s that urban areas began to spread out significantly. The main factor in this development was the construction of suburban railway lines. Each railway development spurred a rapid period of house building. Initially, the process of suburbanisation was an almost entirely middle-class phenomenon. It was not until after the First World War, with the growth of public housing, that working-class suburbs began to appear.

In the interwar period, about 4.3 million houses were built in the UK, mainly in the new suburbs. Just over 30 per cent were built by local authorities (councils). The reasons for such a rapid rate of suburban growth were:

- government support for house-building
- the willingness of local authorities to provide piped water and sewerage systems, and gas and electricity
- the expansion of building societies
- low interest rates
- development of public transport routes
- improvements to the road network.

Figure 6.24 describes the development of Stoneleigh, an outer suburb in south-west London (Figure 6.25). In the latter half of the twentieth century, suburbanisation was limited by the creation of Green Belts and the introduction of general planning controls.

STONELEIGH: A RAILWAY SUBURB

By the end of the 1930s developments were taking place on the rural–urban fringe. Stoneleigh acquired a railway station in 1932 and witnessed spectacular growth thereafter. The Stoneleigh Estate consisted of three farms. These had been offered for building development in the early 1900s but by the end of the 1920s only a few dozen houses had been built. However, following the arrival of the railway, development intensified. By 1933 a 3500 acre site for 3000 homes existed, and the area had a complete set of drains and sewers. By 1937 all farmland and woodland within a 1 mile radius of the railway station had been destroyed.

The housing density at Stoneleigh was low at eight houses per acre. As well as the railway there was a good bus service to Epsom, Surbiton and Kingston. Further developments followed quickly:

- a block of 18 shops (by 1933)
- a sub-post office (1933) and a bank (1934)
- Stoneleigh's first public house (1934)
- a cinema (1937)
- a variety of churches (1935 onwards)
- schools (from 1934)
- recreational grounds at Nonsuch Park and Cuddington.

Stoneleigh benefited from a strong and dynamic residents' association. The residents were aggrieved that nearby working-class areas in Sutton and Cheam were reducing their own land values. They canvassed successfully for boundaries to be redrawn, raising the values of their properties. There were many social activities too, including dances, whist drives, cricket, children's parties, choral societies, cycling and tennis. This went a long way to creating a sense of community. The chairman of the residents' association was also the editor of a local newspaper, which helped the residents in their aims.

By 1939 Stoneleigh was a model railway suburb. Over 3000 people used the railway each day for commuting to work and it was also useful for reaching the south coast. However, the railway also split the community in two. There were problems for buses and cars trying to move from one side of the town to the other. Socially, it also split the community.

The development of Stoneleigh shows many similarities with other suburbs:

- a variety of housing styles, reflecting the different building companies
- a somewhat chaotic road layout
- complete destruction of the former farming landscape
- ponderous shopping parades
- the claim by some that it is dull and soulless.

Yet because of its poor road layout, in particular the lack of railway crossings, and its housing developments right up to the railway line, it does not have the worst trappings of modern suburban development.

Source: *Geography Review*, September 1998, Philip Allan Updates

Figure 6.24 Stoneleigh – a railway suburb

Figure 6.25 Stoneleigh – an outer suburb in south-west London

Counterurbanisation

Urban deconcentration is the most consistent and dominant feature of population movement in most cities in HICs today, in which each level of the settlement hierarchy is gaining people from the more urban tiers above it but losing population to those below it. However, it must be remembered that the net figures hide the reasonable numbers of people moving in the opposite direction. There has been a consistent loss of population for metropolitan England in terms of net within-UK migration. It does not, however, mean an overall population decline of this magnitude, because population change is also affected by natural increase and international migration. London is the prime example of the counterbalancing effect of these last two processes.

Around London, where central rents are particularly high, much office employment has diffused very widely across south-east England. Between 20 and 30 decentralisation centres can be identified in the Outer Metropolitan Area, between 20 and 80 kilometres from central London, especially along the major road and rail corridors. Examples include Dorking, Guildford and Reigate.

Reurbanisation

In very recent years, British cities have, to a limited extent so far, reversed the population decline that has dominated the post-war period. Central government finance, for example the millions of pounds of subsidies poured into London's Docklands, Manchester's Hulme wastelands and Sheffield's light railway, has been an important factor in the revival. New urban design is also playing a role. The rebuilding of part of Manchester's city centre after a massive IRA bomb has allowed the planners to add new pedestrian areas, green spaces and residential accommodation. A recent example is Birmingham's Big City Plan, set out in 2010, which plans for radical change in the city centre. The Big City Plan will co-ordinate the redevelopment of the area over the next 20 years.

The reduction in urban street crime due to the installation of automated closed-circuit surveillance cameras has significantly improved public perception of central areas (Figure 6.26). Rather than displacing crime to nearby areas, as some critics have claimed, a Home Office study found that, on the contrary, the installation of cameras had a halo effect, causing a reduction in crime in surrounding areas.

Is the recent reurbanisation just a short-term blip or the beginning of a significant trend, at least in the medium term? Perhaps the most important factor favouring the latter is the government's prediction in the late 1990s of the formation of 4.4 million extra households over the next two decades, 60 per cent of which will have to be housed in existing urban areas because there is such fierce opposition to the relaxation of planning restrictions in the countryside. Also, as many of the new households will be single-person units, the existing urban areas may well be where most of them would prefer to live.

Figure 6.26 Reurbanisation in the central area of Reading, UK

Case Study: The rejuvenation of inner London

For the first time in about 30 years, London stopped losing population in the mid-1980s and has been gaining people ever since, due to net immigration from overseas and natural increase. Perhaps the most surprising aspect of this trend is the rejuvenation of inner London, where the population peaked at 5 million in 1900 but then steadily dropped to a low of 2.5 million by 1983. The 2011 census recorded a population of 3.23 million in inner London, the highest since the 1961 census (3.34 million). Inner London has benefited from a number of regeneration projects (Figure 6.27), some of them very large in scale. The overall effect has been to improve housing, services, employment and the environment. ⇨

Figure 6.27 Regeneration in inner London

Young adults now form the predominant population group in inner London, whereas in the 1960s all the inner London boroughs exhibited a mature population structure. Inner London is seen as a vibrant and attractive destination by young migrants from both the UK and abroad.

Gentrification has been an important part of change in inner London. The term 'gentrification' was first coined in 1963 by the sociologist Ruth Glass to describe the changes occurring in the social structure and housing market in parts of inner London. The process involved:

- the physical improvement of the housing stock (Figure 6.28)
- a change in housing tenure from renting to owning
- an increase in house prices
- the displacement or replacement of the working class by the new middle class.

Figure 6.28 Gentrification of terraced housing in inner London

Section 6.2 Activities

1 What is the cycle of urbanisation?
2 With reference to Figure 6.24, describe the process of suburbanisation.
3 What is counterurbanisation and when did it begin?
4 a Define reurbanisation.
  b Explain the reasons for the occurrence of this process.

☐ Competition for land

All urban areas exhibit competition for land to varying degrees. Such competition varies according to location, and the level of competition can change over time. The best measures of competition are the price of land and the rents charged for floorspace in buildings. However, planning measures such as **land-use zoning** and other restrictions can complicate the free market process to a considerable degree. Bid-rent theory does much to explain how competition for land can result in **functional zonation** – this is discussed in more detail in the next section. Space does not usually stay idle for long in the sought-after parts of urban areas. However, there are areas of some cities where dereliction has been long-standing. Here, the land may be unattractive for both residential and business purposes and it may require substantial investment from government to bring the area back into active use again.

Renewal and redevelopment

Urban redevelopment involves complete clearance of existing buildings and site infrastructure and constructing new buildings, often for a different purpose, from scratch. In contrast, **urban renewal** keeps the best elements of the existing urban environment (often because they are safeguarded by planning regulations) and adapts them to new usages. Simple examples are where a bank has been turned into a restaurant, keeping the former's façade but altering the inside of the building to suit its new purpose. Urban renewal helps to maintain some of the historic character of urban areas.

In cities in various countries where damage was extensive as a result of the Second World War, large-scale redevelopment took place in the subsequent decades. The general model was to completely clear the land (redevelopment) and build anew. However, from the 1970s renewal gained increasing acceptance and importance in planning circles. In more recent years, the term **urban regeneration** has become increasingly popular. This involves both redevelopment and renewal.

In the UK, urban development corporations were formed in the 1980s and early 1990s to tackle large areas of urban blight in major cities around the country. The establishment of the London Docklands Development Corporation in 1981 set in train one of the largest urban regeneration projects ever undertaken in Europe. An important part of this development was the construction of Canary Wharf, which extended London's CBD towards the east. The regeneration of the Lower Lea Valley is a more recent development, stimulated by the granting of the 2012 Olympic Games to London.

The Lower Lea Valley was one of the most deprived communities in the country and was seen as the largest remaining regeneration opportunity in inner London. Unemployment was high and the public health record poor. This run-down environment with an industrial history suffered from a lack of **infrastructure**. Most of the existing industry provided only low-density employment. Flytipping had been a major problem here for many years. The area is one of the most ethnically diverse in the UK. It had a negative image both within East London and in the capital city as a whole. It is hoped that the Olympic Games have transformed the Lower Lea Valley and much of the surrounding area, bringing permanent prosperity through the process of **cumulative causation**. The development that the Olympics brought was dovetailed with the existing regeneration framework. The total investment in the area exceeded £9 billion, much of which went into transport improvement. Plans to develop the Lower Lea Valley had been around for some time – the development role of the Olympic Games has been to speed up this process.

☐ Global (world) cities

A **global (world) city** is one that is judged to be an important nodal point in the global economic system. The term 'global city' was first introduced by Saskia Sassen in her book *The Global City* published in 1991. Initially referring to New York, London and Tokyo, Sassen described global cities as ones that play a major role in global affairs in terms of politics, economics and culture. The number of global cities has increased significantly in recent decades as the process of globalisation has deepened. Global cities are defined by influence rather than size. Which large cities in terms of population do not appear on Figure 6.29? For example, in the USA, Los Angeles is larger in population size than Chicago, but while Chicago has Alpha status, Los Angeles does not merit an Alpha ranking.

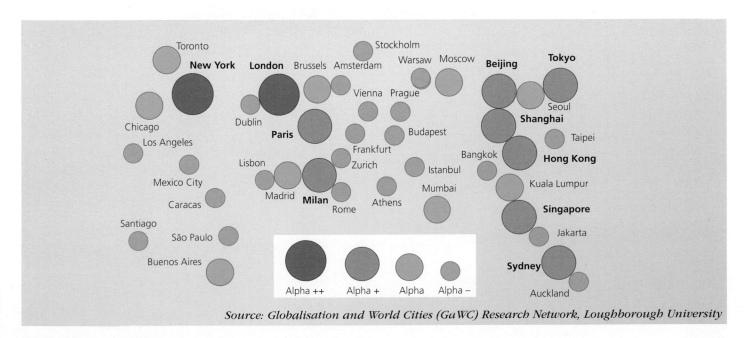

Source: Globalisation and World Cities (GaWC) Research Network, Loughborough University

Figure 6.29 Alpha global cities

Hierarchy of world cities

The Globalisation and World Cities (GaWC) Research Network at Loughborough University has identified various levels of global city. Figure 6.29 shows those cities termed the 'Alpha' cities in 2008, which are subdivided into four categories. Only New York and London are placed in the highest Alpha++ category under this classification. Beijing is in the Alpha+ category, along with Shanghai, Hong Kong and Tokyo in the East Asia geographical region. The remaining cities in this category are Paris, Singapore (Figure 6.30) and Sydney. The GaWC analysis also recognises four lower levels of urban area around the world. The next two levels in the global city hierarchy, the Beta and Gamma levels, are shown in Table 6.4. The results are based upon the office networks of 175 advanced producer service firms in 526 cities in 2008.

Figure 6.30 Singapore – a world city

Table 6.4 Beta and Gamma global cities

Beta +	Beta	Beta –	Gamma +	Gamma	Gamma –
Washington	Oslo	Munich	Montreal	Ljubljana	Detroit
Melbourne	Berlin	Jeddah	Nairobi	Shenzhen	Manchester
Johannesburg	Helsinki	Miami	Bratislava	Perth	Wellington
Atlanta	Geneva	Lima	Panamá City	Kolkata	Riga
Barcelona	Copenhagen	Kiev	Chennai	Guadalajara	Guayaquil
San Francisco	Riyadh	Houston	Brisbane	Antwerp	Edinburgh
Manila	Hamburg	Guangzhou	Casablanca	Philadelphia	Porto
Bogatá	Cairo	Beirut	Denver	Rotterdam	San Salvador
Tel Aviv	Luxembourg	Karachi	Quito	Amman	St Petersburg
New Delhi	Bangalore	Düsseldorf	Stuttgart	Portland	Tallinn
Dubai	Dallas	Sofia	Vancouver	Lagos	Port Louis
Bucharest	Kuwait	Montevideo	Zagreb		San Diego
	Boston	Nicosia	Manama		Islamabad
		Rio de Janeiro	Guatemala City		Birmingham (UK)
		Ho Chi Minh City	Cape Town		Doha
			San José (CR)		Calgary
			Minneapolis		Almaty
			Santo Domingo		Columbus
			Seattle		

In 2008, the American journal *Foreign Policy* published its Global Cities Index. The rankings are based on 24 measures over five areas:

- business activity
- human capital
- information exchange
- cultural experience
- political engagement.

Foreign Policy noted that 'the world's biggest, most interconnected cities help set global agendas, weather transnational dangers, and serve as the hubs of global integration. They are the engines of growth for their countries and the gateways to the resources of their regions.'

Causes of the growth of world cities

The growth of global cities has been due to:

- **demographic trends** – significant rates of natural increase and in-migration at different points in time for cities in HICs, MICs and LICs; large population clusters offer potential in terms of both workforce and markets
- **economic development** – the emergence of major manufacturing and service centres in national and continental space, along with the development of key transport nodes in the global trading system
- **cultural/social status** – the cultural facilities of large cities are an important element of their overall attraction to FDI and tourism
- **political importance** – many global cities are capital cities, benefiting from particularly high levels of investment in infrastructure.

There will undoubtedly be many changes in the hierarchy of global cities as the years unfold. The rapid development of many NICs will have a significant impact on the rankings. Africa is so far unrepresented on the Alpha list, but cities such as Johannesburg, Cairo and Lagos may well get there in the not-too-distant future. In contrast, other established global cities may decline in importance.

Section 6.2 Activities

1 What is a *global city*?
2 Describe the levels and distribution of global cities shown in Figure 6.29.
3 On an outline map of the world, show the locations of the Beta global cities shown in Table 6.4.
4 Suggest how global cities can rise and fall in terms of their level or grading.

6.3 The changing structure of urban settlements

☐ Functional zonation

The patterns evident and the processes at work in large urban areas are complex, but by the beginning of the twentieth century geographers and others interested in urban form were beginning to see more clearly than before the similarities between cities, as opposed to laying stress on the uniqueness of each urban entity. The first generalisation about urban land use to gain widespread recognition was the concentric zone model emanating from the so-called 'Chicago School'.

The concentric zone model

Published in 1925, and based on American Mid-Western cities, particularly Chicago (Figure 6.31), E.W. Burgess's model (Figure 6.32) has survived much longer than perhaps its attributes merit as it has only limited applicability to modern cities. However, it did serve as a theoretical foundation for others to investigate further.

Figure 6.31 The CBD of Chicago

The main assumptions upon which the model was based are:

- a uniform land surface
- free competition for space
- universal access to a single-centred city
- continuing in-migration to the city, with development taking place outward from the central core.

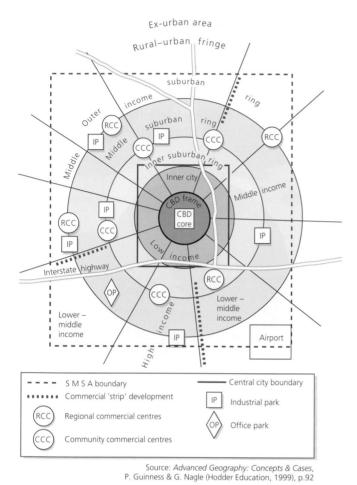

Figure 6.36 The spatial structure of the post-industrial American city

Source: *Advanced Geography: Concepts & Cases*, P. Guinness & G. Nagle (Hodder Education, 1999), p.92

Models of cities in LICs and MICs

Although the development of urban land-use models has favoured Western cities, some interesting contributions relating to cities in LICs, MICs and socialist cities have appeared at various points in time.

Griffin and Ford's model, upon which Figure 6.37 is based, summarises many of the characteristics that they noted in modern Latin American cities:

■ Central areas, which had changed radically from the colonial period to now, exhibit most of the characteristics of modern Western CBDs.

■ The development of a commercial spine, extending outwards from the CBD, is enveloped by an elite residential sector.

■ There is a tendency for industries, with their need for urban services such as power and water, to be near the central area.

■ The model includes a 'zone of maturity' with a full range of services containing both older, traditional-style housing and more recent residential development. The traditional housing, once occupied by higher-income families who now reside in the elite sector, has generally undergone subdivision and deterioration. A significant

proportion of recent housing is self-built of permanent materials and of reasonable quality.

■ Also included is a zone of 'in situ accretion', with a wide variety of housing types and quality but with much still in the process of extension or improvement. Urban services tend to be patchy in this zone, with typically only the main streets having a good surface. Government housing projects are often a feature of this zone (Figure 6.38).

■ There is a zone of squatter settlements, which are the place of residence of most recent in-migrants. Services in this zone are at their most sparse, with open trenches serving as sewers and communal taps providing water. Most housing is of the **shanty** type, constructed of wood, flattened oil-cans, polythene and any other materials available at the time of construction. The situation is dynamic and there is evidence of housing at various stages of improvement.

Source: *OCR AS Geography* by M. Raw (Philip Allan Updates, 2008), p.205

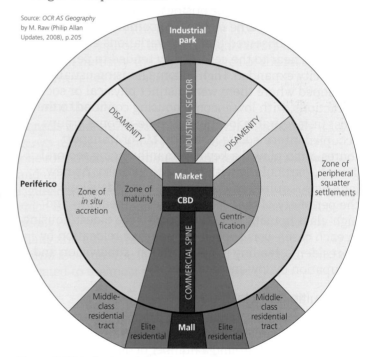

Figure 6.37 Latin American city model

Figure 6.38 Low-cost government housing – Manaus, Brazil

Figure 6.39 High-rise apartment blocks in the inner area of São Paulo

Urban density gradients

Contrasting functional zones within urban areas characteristically vary in residential population density. Examination of population density gradients, termed 'gradient analysis', shows that for most cities densities fall with increasing distance from the centre (Figure 6.39).

Gradient analysis of cities in HICs over time (Figure 6.40a) shows the following trends:

- the initial rise and later decline in density of the central area
- the outward spread of population and the consequent reduction in overall density gradient over time.

In contrast, analysis of density gradients in LICs and MICs shows:

- a continuing increase in central area densities (Figure 6.40b)
- the consequent maintenance of fairly stable density gradients as the urban area expands.

In cities in LICs and MICs, both personal mobility and the sophistication of the transport infrastructure operate at a considerably lower level. Also, central areas tend to retain an important residential function. Both of these factors result in a more compact central area and the transport factor in particular has restricted urban sprawl to levels below that of cities in HICs. The presence of extensive areas of informal settlement in the outer areas also results in higher suburban densities. However, in the more advanced LICs and MICs, where car ownership is rising rapidly, significant sprawl is now occurring.

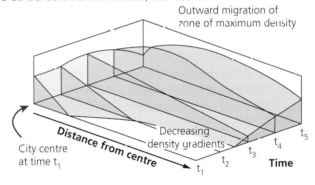

a Central densities first increase, later decline

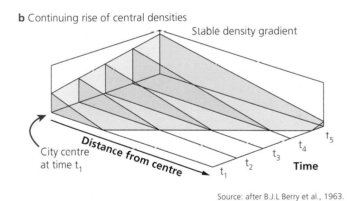

b Continuing rise of central densities

Source: after B.J.L Berry et al., 1963.

Figure 6.40 Changes in urban density gradients through time

Section 6.3 Activities

1 With reference to Figure 6.32, briefly explain the differences between the three models illustrated.
2 Identify the main features of Mann's land-use model for a typical British city.
3 Briefly describe and explain the main elements of the model illustrated in Figure 6.36.
4 Comment on the main characteristics of the model of Latin American cities shown in Figure 6.37.
5 **a** Define the term *urban density gradient*.
   b How and why do urban density gradients differ between cities in HICs and LICs/MICs?

☐ Factors affecting the location of urban activities

A range of factors affects the location of urban activities such as retailing, manufacturing, office functions, education, health, leisure and open space. Most, if not all, of these factors can be placed under two general headings:

- **Market forces** – the demand and supply of land in various locations dictates its price.
- **Local or central government planning decisions** – planners can overrule market forces where they consider it necessary for the public good. Government may be able to decide, within certain constraints, where the locations of public housing, open spaces, schools, hospitals and public buildings should be.

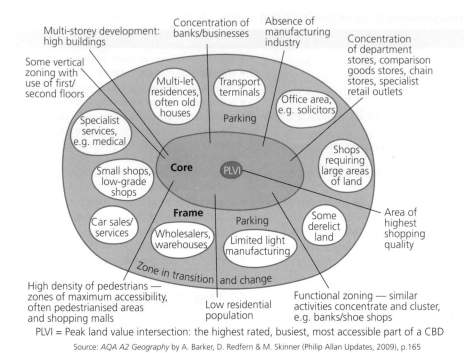

PLVI = Peak land value intersection: the highest rated, busiest, most accessible part of a CBD

Source: *AQA A2 Geography* by A. Barker, D. Redfern & M. Skinner (Philip Allan Updates, 2009), p.165

Figure 6.43 The key features of the CBD

Major retailing and office functions dominate the core, alongside theatres, cinemas, restaurants, bars, hotels and key public buildings. Vertical zoning is often apparent, with retailing occupying lower floors and offices above. Similar functions often locate together, for example department stores and theatres. The high land values of the CBD result in extremely low residential populations. This contrasts with the very high pedestrian flows recorded in CBDs – a combination of a large number of people attracted to the CBD to purchase goods and services and the very significant number of people who work there.

Traffic congestion is a universal problem in CBDs and thus it is not surprising that this is the urban zone with the greatest traffic restrictions. In London, a congestion charge zone covers much of the CBD. At the time of writing, motorists have to pay £11.50 a day to enter the zone.

CBDs change over time. Common changes in many HICs and an increasing number of MICs and LICs have been:

- pedestrianised zones
- indoor shopping centres
- environmental improvements
- greater public transport coordination
- ring roads around the CBD with multi-storey car parks.

Some parts of the CBD may expand into the adjoining inner city (a zone of assimilation), while other parts of the CBD may be in decline (a zone of discard). The CBD is a major factor in the economic health of any urban area. Its prosperity can be threatened by a number of factors (Figure 6.44). CBDs are often in competition with their nearest neighbours and are constantly having to upgrade their facilities to remain attractive to their catchment populations.

Source: *AQA A2 Geography* by A. Barker, D. Redfern & M. Skinner (Philip Allan Updates, 2009), p.166

Figure 6.44 Factors influencing CBD decline

Urban redevelopment can be a major factor in CBD change (Figure 6.45). The redevelopment of London Docklands changed London's CBD from a bi-nuclear entity (the West End and the City) to its current tri-nuclear form (West End, City, Canary Wharf). In the West End retailing is the dominant function, whereas in the City offices dominate, for example the latter area contains the Bank of England, the Stock Exchange and Lloyd's of London (insurance). Canary Wharf was planned to have a good mixture of both offices and retailing. It has been an important development in maintaining London's position as a major global city (Figure 6.46).

Figure 6.45 Times Square – part of the CBD of New York

Figure 6.46 London's CBD

☐ Residential segregation

Residential segregation is very apparent in cities in HICs, MICs and LICs. The main causes of residential segregation are income and race/ethnicity. The processes that result in residential segregation include:

- the operation of the housing market
- planning
- culture
- the influence of family and friends.

The way the housing market operates in an urban area or a country as a whole significantly determines the number of housing units built, the type of housing, the availability of mortgages and where housing construction takes place. The last is of course also heavily influenced by planning. In terms of the number of housing units built, the ideal situation is where the supply of housing matches the demand. In such a situation, in theory, housing should be reasonably affordable with little overcrowding. However, in so many urban areas housing is in short supply, resulting in high prices and overcrowding at the lower end of the housing market in particular. Residential segregation tends to become more intense when housing is in short supply, with people on lower incomes gradually pushed out of desirable areas and into what is sometimes termed the 'urban periphery'.

Access to finance, mainly in terms of the availability of mortgages, is an important factor in the efficient operation of the housing market. Where access to housing finance is generally good, the level of residential segregation is likely to be less intense compared to a situation where housing finance is difficult to access.

The tenure of housing is also a major issue, with the proportion of housing units that are classed as 'social housing' a major factor here. Where the proportion of social housing is significant, the way it is distributed in an urban area has a major impact on residential segregation. The situation where planners aim for a good social mix in an urban area is very different from the grouping of social housing in distinct areas, which may result in 'urban ghettos'.

Culture can be a strong determinant of where people want to live in an urban area. Income may allow people to live in certain areas, but if they do not feel 'comfortable' in an area they will tend to avoid it. This factor is strongly linked to the influence of family and friends.

London provides a prime example of residential segregation. On all socio-economic measures, the contrast between the relative deprivation of inner London and the affluence of outer London is striking. London is made up of the City of London and the 32 boroughs, of which 13 are in inner London and 19 in outer London (Figure 6.47).

The Cairo Traffic Congestion Study 2014 produced by the World Bank highlighted the problems of chronic congestion in many parts of Cairo. Total daily commute time has been estimated to be an average of about 90 minutes. Congestion is a by-product of:

- population growth
- high and rising levels of car ownership
- lack of sufficient off-street parking
- insufficient capacity of the public transport system.

Twenty per cent of the total population are private car owners, and private cars make up over 80 per cent of traffic congestion. There are rising concerns over greenhouse gas emissions, deteriorating air quality and noise. About half of all motorised vehicles in Egypt operate in Cairo. Poor traffic management and the inadequate supply of mass transport add to the problems of congestion. No Bus Rapid Transit (BRT) system currently exists. The high ridership on buses and the metro is evidence of the strong demand for public transport in Cairo.

The 2014 study also highlighted the poor environment in the city for both pedestrians and cyclists. There is a very high accident rate especially for pedestrians, with more than 1000 deaths each year on Cairo's roads.

The standard full-size bus service is run by the Cairo Transport Authority (CTA). There are about 450 formal bus routes in the city. There are also minibuses run by companies sub-contracted by the government. In addition, taxis and micro-buses privately run by individuals are an important component of road transport.

The tram system in Cairo has been running since 1896 but has now largely been shut down. The focus of the rail system is the centrally located Ramses Station, which links with major commuter stations and beyond to the national urban network. Trains are run by Egyptian National Railways. Nile ferries also play a role in the daily movement of people. Popular among tourists are the *feluccas*, comparable to Venice's *gondolas*, which operate along the Nile to and from the pyramids at Giza.

Cairo's metro (Figure 6.56) carries an average of about 2.2 million passenger rides a day. The metro underground system was developed in an attempt to cut the number of vehicles on Cairo's roads. Line 1 was opened in 1987 and Line 2 in 1996. Line 3 was opened in 2012, although the extension to connect with Cairo airport will not be completed until 2019. The fourth line is about to start construction in 2016, and two more lines are planned. By 2014, there were 61 stations with a total track length of 78 kilometres. The ticket price in 2013 was EGP1.0 ($0.14), regardless of distance.

Cairo's airport is the second busiest in Africa after Johannesburg International Airport. It has had to expand to keep pace with demand. The airport is located about 15 kilometres from the heart of the city's business area. It is a large site covering an area of about 37 km². There are three main terminals and a Seasonal Flights Terminal opened in 2011. The purpose of the latter is to ease the strain on the existing terminals during pilgrim seasons. The airport handled 14.7 million passengers in 2012. The expansion of the airport has been vital to maintain Cairo's global city status.

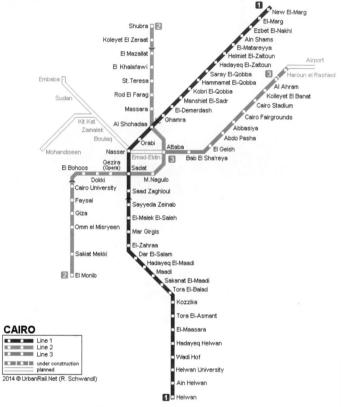

Source: www.urbanrail.net/af/cairo/cairo-map.gif

Figure 6.56 Map of the Cairo metro

Section 6.4 Activities

1 With reference to an atlas and Figure 6.52, describe the location of Cairo.
2 Describe and explain the growth of Cairo's population since 1960.
3 Distinguish between *hard infrastructure* and *soft infrastructure*.
4 a Describe the main elements of Cairo's transport infrastructure.
  b Why is transportation so important to the successful development of a city?

7 Tropical environments

7.1 Tropical climates

The tropical environment is the area between 23.5°N and 23.5°S (Figure 7.1). This area covers about 50 million km² of land, almost half of it in Africa. According to the climatologist Köppen, there are three types of tropical climate (A). These include rainforest climates (Af), **monsoon** climates (Am) and savanna climates (Aw). These are shown on Figure 7.1.

Section 7.1 Activities

1 Describe the distribution of areas with a rainforest climate (Af), as shown on Figure 7.1.
2 Comment on the distribution of areas with a monsoon climate, as shown on Figure 7.1.
3 Describe the main features of a savanna climate, as shown by the climate graph for Banjul in Figure 7.1.

   You should refer to maximum and minimum temperatures, seasonal variations in temperatures, total rainfall, seasonal variations in rainfall and any links between temperatures and rainfall.

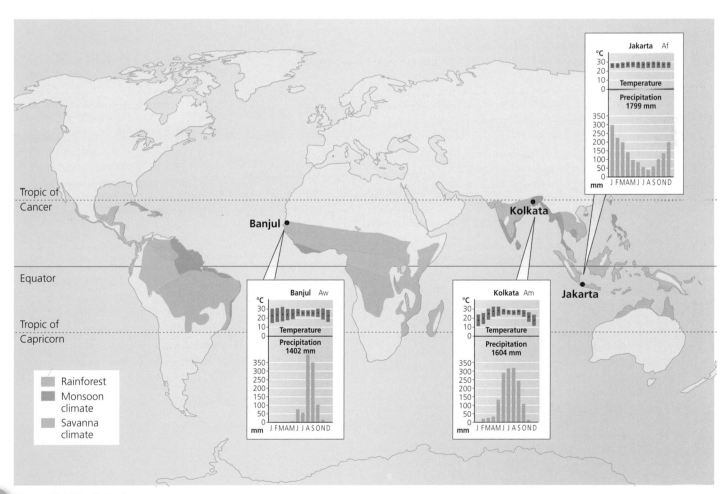

Figure 7.1 Tropical climates

Air masses

The original concept of an **air mass** was that it was a body of air whose physical properties, especially temperature and humidity, were uniform over a large area. By contrast, it is now redefined as a large body of air in which the horizontal gradients (variation) of the main physical properties are fairly slack. It is generally applied only to the lower layers of the atmosphere, although air masses can cover areas of tens of thousands of km² (Figure 7.2).

Air masses derive their temperature and humidity from the regions over which they lie. These regions are known as source regions. The principal ones are:

- areas of relative calm, such as semi-permanent high-pressure areas
- where the surface is relatively uniform, including deserts, oceans and ice-fields.

Air masses can be modified when they leave their sources, as Figure 7.3 illustrates.

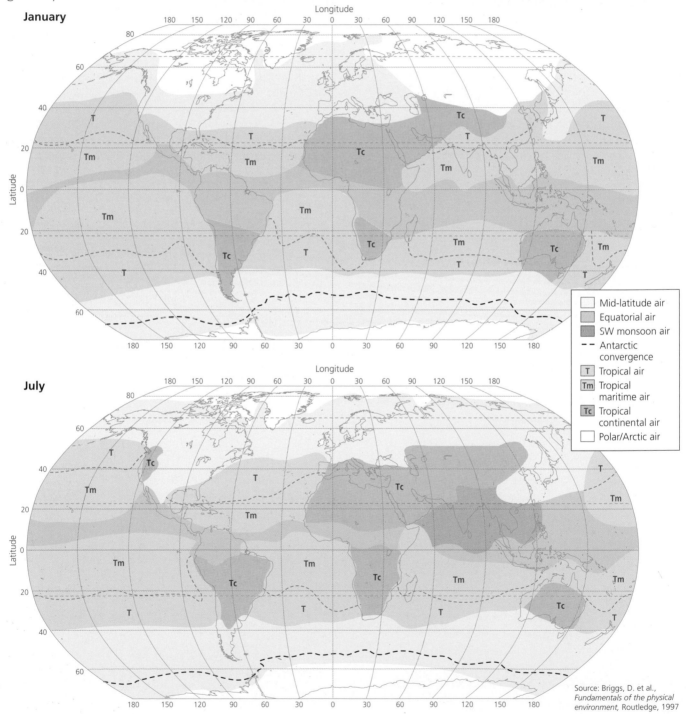

Source: Briggs, D. et al., *Fundamentals of the physical environment*, Routledge, 1997

Figure 7.2 Air masses

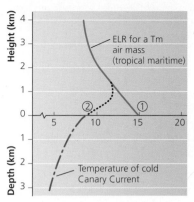

① Original temperature
② Temperature after being cooled by cold water

Temperature (°C)

Because it is chilled it becomes denser and therefore more stable. It is often associated with fog.

Source: *Advanced Geography: Concepts and Cases* by P. Guiness & G. Nagle (Hodder Education, 1999), p.417

Figure 7.3 Temperature profiles showing the modification of air masses

The initial classification of air masses was made by Tor Bergeron in 1928. Primarily, they are classified first by the latitude of the source area (which largely controls its temperature) and secondly by whether the source area is continental (dry) or maritime (moist) (Table 7.1). A maritime tropical (mT) air mass is one that is warm and moist. A third subdivision refers to the stability of the air mass – that is, whether it has cooled and become more stable, or whether it has become warmer and less stable (Figure 7.3).

Table 7.1 Dominant air masses in tropical environments

Approx. latitude of source	Label	Temp. (°C) winter/summer	Stability winter/summer
Equatorial (0–10°)	cE (continental equatorial)	25/25	Moist/neutral
	mE (maritime equatorial)	25/25	Moist/neutral
Subtropical (30°)	cT (continental tropical)	15/25	Stable, dry/neutral
	mT (maritime tropical)	18/22	Stable/stable

As air masses move from their source regions, they may be changed due to:

■ internal changes and
■ the effects of the surface over which they move.

These changes create **secondary air masses** (Table 7.1). For example, a warm air mass that travels over a cold surface is cooled and becomes more stable. Hence it may form low cloud or fog but is unlikely to produce much rain. By contrast, a cold air mass that passes over a warm surface is warmed and becomes less stable. The rising air is likely to produce more rain. Air masses that have been warmed are given the suffix 'w' and those that have been cooled are given the suffix 'k' (*kalt*).

Section 7.1 Activities

1 Define the term *air mass*.
2 Outline the main characteristics of mE and cT air masses.
3 Compare the seasonal distribution of tropical air masses as shown in Figure 7.2.

☐ Intertropical convergence zone (ITCZ)

Winds between the tropics converge on a line known as the **ITCZ** or 'equatorial trough'. It is actually a band a few hundred kilometres wide, enclosing places where wind flows are inwards and subsequently rise convectively (Figure 7.4).

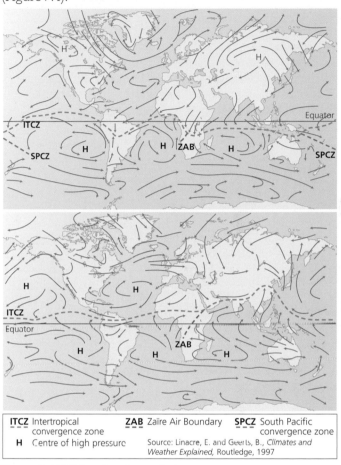

ITCZ Intertropical convergence zone   **ZAB** Zaïre Air Boundary   **SPCZ** South Pacific convergence zone
H Centre of high pressure

Source: Linacre, E. and Geerts, B., *Climates and Weather Explained*, Routledge, 1997

Figure 7.4 ITCZ and surface winds

The ITCZ lies at about 5°N on average. This is known as the 'meteorological equator'. It wanders seasonally, lagging about two months behind the change in the overhead Sun. The latitudinal variation is most pronounced over the Indian Ocean because of the large Asian continental land mass to the north. Over the eastern Atlantic and eastern Pacific Oceans, the ITCZ moves seasonally due to the cold Benguela and Peru **ocean currents**.

The movement of the ITCZ over South Africa, for example, is complicated by the land's shape, elevation and location. There is a southerly spur of the ITCZ known as the 'Zaire Air Boundary' (ZAB) (Figure 7.4). The largest and most prominent spur in the South Pacific is the South Pacific Convergence Zone (SPCZ), which is related to the 'warm pool' near Papua New Guinea, and is most pronounced in summer. It lies mostly over water. There is a convergence of:

■ moist northerlies from the semi-permanent high pressure in the south-east Pacific, and

- south-easterlies from mobile highs moving across the south-west Pacific in summer.

Winds at the ITCZ are commonly light or non-existent, creating calm conditions called the doldrums. There are, though, occasional bursts of strong westerlies, known as a 'westerly wind burst'.

☐ Subtropical anticyclones

Centres or ridges of high pressure imply subsiding air and a relatively cold atmosphere. They tend to be found over continents, especially in winter. There is a high-pressure belt over south-east Australia in winter, whereas high pressure is located south of the continent in summer, when the sea is colder (Figure 7.5).

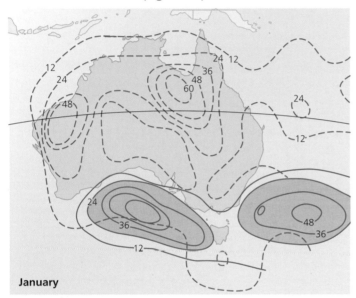

January

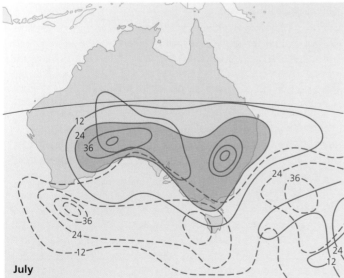

July

The continuous lines and shaded areas show the number of hours in a month when high pressure is centred over an area. The dotted lines show the number of hours that a centre of low pressure is positioned over a location. The shaded areas have high pressure centred there for at least 24 hours per month.

Figure 7.5 Seasonal high pressure over Australia

The subtropical high or warm anticyclone is caused by cold air descending at the tropopause. Two rings of high pressure lie around 30–35° north and south. The position of the high pressure coincides with the subsiding part of the Hadley cell, so it alters in response to the seasonal drift of the ITCZ, but only over 5–10°. On average, highs cross the east coast of Australia south of Sydney in summer, whereas in winter they pass further north.

The subtropical high-pressure belt tends to lie over the ocean, especially in summer, when there are low pressures over the continents caused by heating. One subtropical high is anchored over the eastern Pacific by an anticyclonic swirl induced by the Andes. The high there shifts from 32° in January to 23° in July. The high is particularly strong due to the cold ocean surface, except during **El Niño** events. Another semi-permanent high lies over the Indian Ocean, moving nearer to Australia in summer and Africa in winter.

Highs tend to be larger than low-pressure systems, reaching up to 4000 kilometres in width and 2000 kilometres north–south. Therefore smaller pressure gradients are involved and so winds are lighter.

The subtropical high-pressure belt is intersected by cold fronts. The subtropical high-pressure areas generally move eastwards at speeds of 30–50 kilometres/hour – hence a 4000 kilometre system moving at an average of 40 kilometres/hour would take about four days to pass over – if it keeps moving.

An anticyclone's movement may sometimes stall, and it may travel a distance equivalent to less than 20° latitude in a week. This is known as a 'blocking anticyclone', or 'blocking high'. It is not so common in the southern hemisphere because there are fewer land masses and mountain ranges to disturb the airflow.

Effects

Highs at the surface are associated with subsidence. A temperature inversion may occur, especially where there is a cold high in winter over a continent. Where there is moisture at low level and air pollution, low-level stratus clouds may form, causing an anticyclonic gloom, as found in Santiago and Melbourne in winter, for example.

Arid climates result from a prevalence of high pressure. North-east Brazil is arid, even at a latitude of 8°S, because it protrudes far enough into the South Atlantic to be dominated by high pressure. The same is true for the Galapagos Islands, dominated by the South Pacific High, even though they are located at the equator.

Section 7.1 Activities

1 Outline the main cause of the subtropical high-pressure belt.
2 Explain why it varies seasonally.
3 Describe the seasonal changes that occur to the ITCZ and wind patterns, as shown in Figure 7.4.

☐ Ocean currents

The oceanic **gyre** (swirl of currents) explains why east coasts in the southern hemisphere and west coasts in the northern hemisphere are usually warm and wet, because warm currents carry water polewards and raise the air temperature of maritime areas. In contrast, cold currents carry water towards the equator and so lower the temperatures of coastal areas. West coasts tend to be cool and dry due to:

- the advection of cold water from the poles and
- cold upwelling currents.

For instance, in South Africa the east coast is 3–8 °C warmer than the west coast. An exception is Australia where the southward Leeuwin Current brings warmth to Perth. Continental east coasts in the subtropics are humid and west coasts arid – this is mainly due to the easterly winds around the tropics.

Table 7.2 shows the mean monthly temperatures and rainfall of some coastal cities on the west and east sides of three continents.

Table 7.2 A comparison of mean monthly temperatures (°C) and rainfall of some coastal cities

Continent	Coast	Place	Temperature (°C) January	Temperature (°C) July	Rainfall (mm) January	Rainfall (mm) July
Around 23 °S						
South America	West	Antofagasta	21	14	0	5
	East	Rio de Janeiro	26	21	125	41
South Africa	West	Walvis Bay	19	15	0	0
	East	Maputo	25	18	130	13
Australia	West	Carnarvon	27	17	20	46
	East	Brisbane	25	15	163	56
Around 34 °S						
South America	West	Santiago	21	9	3	76
	East	Buenos Aires	23	10	79	56
South Africa	West	Cape Town	21	12	79	56
	East	Durban	24	17	127	85
Australia	West	Perth	23	13	8	170
	East	Sydney	22	12	89	117

Section 7.1 Activities

1 Using examples, compare the temperatures of west-coast locations with those of east-coast locations for the same latitude.
2 How do you explain this pattern?

Wind

The temperature of the wind is determined by the area where the wind originates and by the characteristics of the surface over which it subsequently blows. A wind blowing from the sea tends to be warmer in winter but cooler in summer than the corresponding wind blowing from the land.

Monsoon

The word 'monsoon' is used to describe wind patterns that experience a pronounced seasonal reversal. The best-known monsoon is that experienced in India, but there are also monsoons in East Africa, Arabia, Australia and China. The basic cause is the difference in heating of land and sea on a continental scale.

In India, two main seasons can be observed:

- the north-east monsoon (Figure 7.6a), consisting of **a** the winter season (January and February) and **b** a hot, dry season between March and May
- the south-west monsoon (Figure 7.6b), consisting of **a** the rainy season of June–September and **b** the post-monsoon season of October–December; most of India's rain falls during the south-west monsoon.

During the winter season, winds generally blow outwards since high pressure is centred over the land. Nevertheless, parts of southern India and Sri Lanka receive some rain, while parts of north-west India receive rainfall as a result of depressions. These winter rains are important as they allow the growth of cereals during the winter. Mean temperatures in winter range from 26 °C in Sri Lanka to about 10 °C in the Punjab (these differences are largely the result of latitude). Northern regions and interior areas have a much larger temperature range than in coastal areas. In the north, daytime temperatures may reach over 26 °C, while frosts at night are common.

The hot, dry season occurs between March and May. It gradually spreads northwards throughout India. Daytime temperatures in the north may exceed 49 °C, while coastal areas remain hot and humid. Vegetation growth is prevented by these conditions and many rivers dry up.

In spring, the high-pressure system over India is gradually replaced by a low-pressure system. As there is low pressure over the equator, there is little regional air circulation. However, there are many storms and dust storms. Increased humidity near the coast leads to rain. Parts of Sri Lanka, the southern part of India and the Bay of Bengal receive rain and this allows the growth of rice and tea. For most of India, however, there is continued **drought**.

The rainy season occurs as the low-pressure system intensifies. Once pressure is low enough, it allows for air from the equatorial low and the southern hemisphere to be sucked in, bringing moist air. As it passes over the ocean, it picks up more moisture and causes heavy rain when it passes over India.

The south-west monsoon in southern India generally occurs in early June, and by the end of the month it affects most of the country, reaching its peak in July or early August (Figure 7.7). Rainfall is varied, especially between windward and leeward sites. Ironically, the low-pressure system over north-west India is one of the driest parts of India, as the monsoon has shed its moisture en route. The monsoon weakens after mid-September but its retreat is slow and may take up to three months. Temperature and rainfall fall gradually, although the Bay of Bengal and Sri Lanka still receive some rainfall.

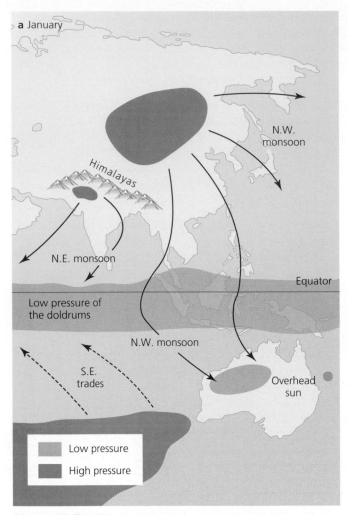

Figure 7.6 The Asian monsoon

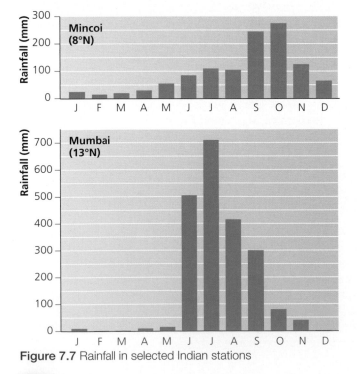
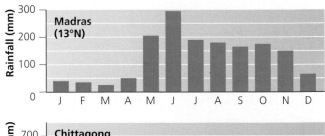
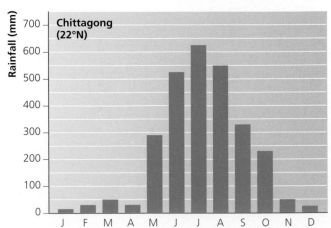

Figure 7.7 Rainfall in selected Indian stations

The simplest explanation for the monsoon is that it is a giant land–sea breeze. The great heating of the Asian continent and the high mountain barrier of the Himalayas, barring winds from the north, allow the equatorial rain systems to move as far north as 30°N in summer. At this time, central Asia becomes very hot, warm air rises and a centre of low pressure develops. The air over the Indian Ocean and Australia is colder, and therefore denser, and sets up an area of high pressure. As air moves from high pressure to low pressure, air is drawn into Asia from over the oceans. This moist air is responsible for the large amount of rainfall that occurs in the summer months. In the winter months, the Sun is overhead in the southern hemisphere. Australia is heated (forming an area of low pressure), whereas the intense cold over central Asia and Tibet causes high pressure. Thus in winter, air flows outwards from Asia, bringing moist conditions to Australia. The mechanism described here, a giant land–sea breeze, is, however, only part of the explanation.

Between December and February, the north-east monsoon blows air outwards from Asia. The upper airflow is westerly, and this flow splits into two branches north and south of the Tibetan plateau. The Tibetan plateau – over 4000 metres in height – is a major source of cold air in winter, especially when it is covered in snow. Air sinking down from the plateau, or sinking beneath the upper westerly winds, generally produces cold, dry winds. During March and April, the upper airflows change and begin to push further north (in association with changes in the position of the overhead Sun). The more

northerly jet stream (upper wind) intensifies and extends across India and China to Japan, while the southern branch of the jet stream remains south of Tibet and loses strength.

There are corresponding changes in the weather. Northern India is hot and dry with squally winds, while southern India may receive some rain from warm, humid air coming in over the ocean. The southern branch of the jet stream generally breaks down around the end of May, and then shifts north over the Tibetan plateau. It is only when the southern jet stream has reached its summer position, over the Tibetan plateau, that the south-west monsoon arrives. By mid-July, the monsoon accounts for over three-quarters of India's rainfall. The temporal and spatial pattern is varied: parts of the north-west attract little rainfall, whereas the Bay of Bengal and the Ganges receive large amounts of summer rainfall. The monsoon rains are highly variable each year, and droughts are not uncommon in India. In autumn, the overhead Sun migrates southwards, as too does the zone of maximum insolation and convection. This leads to a withdrawal of the monsoon winds and rain from the region.

Section 7.1 Activities

1 Describe the seasonal variations in the monsoon, as shown in Figure 7.6.
2 Account for the variations in rainfall as shown in Figure 7.7.
3 Briefly explain the formation of the Asian monsoon.

Case Study: Pakistan's floods, August 2010

The heaviest monsoon rain to affect Pakistan for 80 years destroyed the homes of over 110 000 people, killing an estimated 1600, leaving 2 million homeless and affecting over 20 million people (Figure 7.8). Cases of the deadly disease cholera were reported following the flood.

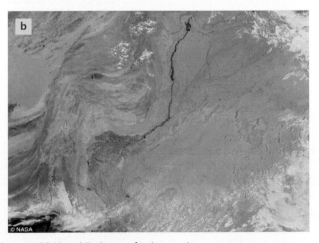

Figure 7.8 Flooding in Pakistan, 2010 – image **a** was taken on 15 August 2010, while image **b** shows the same area a year previously; the blue patches show the extent of the flooding, which left 20 million people homeless

The heavy rainfall, which was many times the usual expected during the monsoon and fell further north and west than usual, exposed the lack of investment in water infrastructure, including big dams, much of which was built in the 1960s. Over 270 millimetres of rain fell in Peshawar in one 24-hour period. The removal of forest cover may also have allowed rainwater to drain faster into the rivers. Further flooding in mid-August prevented vital repairs of embankments, allowing water to reach previously unaffected areas. The floods were the worst since at least 1929. Water levels in the River Indus, which flows through the middle of Pakistan and has most of the population huddled around it, were said to be the highest in 110 years. For example, between 27 and 30 July, 373 millimetres of rain fell at Murree, 394 millimetres at Islamabad and 415 millimetres at Risalpur. Discharge at Guddu peaked at 1.18 million cusecs, compared with a normal discharge of 327 000 cusecs.

One of the hardest-hit areas was the scenic Swat valley, further north, where the population was only just recovering from the Taliban takeover and a military operation in the previous year to drive out insurgents. One factor that may have contributed to the extreme flooding in Swat is the deforestation that accompanied the Taliban takeover. When the landowners fled after being targeted by the Taliban, the timber smugglers joined up with the insurgents to chop down as many trees as possible.

Agriculture was badly affected, causing spiralling food prices and shortages. Floods submerged 69 000 km² of Pakistan's most fertile crop area. The World Bank estimated that crops worth $1 billion (£640 million) were ruined and that the disaster would cut the country's growth in half. More than 200 000 animals also died.

Many roads, bridges and irrigation canals were destroyed, along with the electricity supply infrastructure. Also affected was Mohenjo-Daro, one of the largest settlements of the ancient Indus valley, built around 2500 BCE.

The Asian Development Bank estimated the cost of reconstruction at US$6.8–9.7 billion. This would force the Pakistani government to divert funds from other badly needed development programmes.

UN aid agencies and their partners requested almost $460 million (£295 million) to help Pakistan. Ten days after the start of the floods, just $157 million had been pledged. Based on the estimate of 14 million people affected, the UN said this meant only $4.11 had been committed for each affected person. After the Kashmir earthquake in 2005, which left 2.8 million people needing shelter, $247 million was committed in the first ten days – $70 per person. Ten days after the Haiti earthquake, $495 had been committed for each person affected. Following criticism from the UN, Saudi Arabia pledged $100 million to the Pakistan flood appeal.

Section 7.1 Activities

Study Figure 7.8, which shows the Indus valley before and during the 2010 floods. Compare the river during flood with that of normal flow.

☐ World climates: a classification

The most widely used classification of climate is that of W. Köppen (Figure 7.9). His classification first appeared in 1900, and he made many modifications to it before his death in 1936. Although it has been refined since, the current version bears many resemblances to its early patterns.

Köppen classified climate with respect to two main criteria: temperature and seasonality of rainfall. Indeed, five of the six main climatic types are based on mean monthly temperature:

A **Tropical rainy climate** – coldest month >18 °C
B **Dry** (desert)
C **Warm temperate rainy climate** – coldest month –3 °C–18 °C; warmest month >10 °C
D **Cold boreal forest climate** – coldest month ≤–3 °C; warmest month ≥10 °C
E **Tundra** – warmest month 0–10 °C
F **Perpetual frost climate** – warmest month <0 °C.

The choice of the specific figures is as follows: 18 °C is the critical winter temperature for tropical forests; 10 °C is the poleward limit of forest growth; –3 °C is generally associated with two to three weeks of snow annually.

There are subdivisions that relate to rainfall:

- **f** – no dry season
- **m** – monsoonal (short dry season and heavy rains in the rest of the year)
- **s** – summer dry season
- **w** – winter dry season

Tropical humid climates and seasonally humid climates have a maximum temperature of >20 °C and a minimum monthly temperature of >13 °C. Tropical humid climates have a mean monthly rainfall of over 50 millimetres for between eight and twelve months. In contrast, seasonally humid climates have a mean monthly rainfall of over 50 millimetres for between one and seven months.

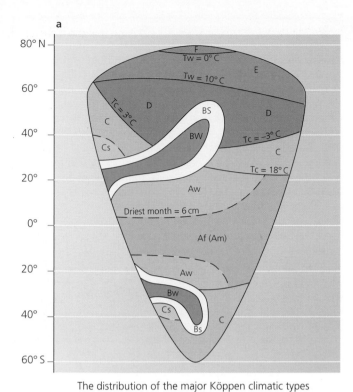

a

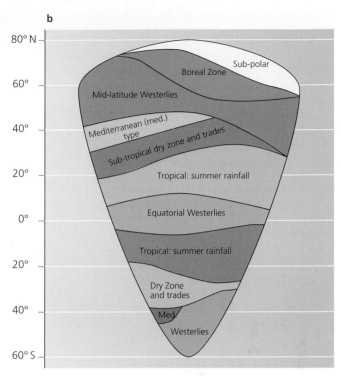

b

The distribution of the major Köppen climatic types on a hypothetical continent of low and uniform elevation.

Tw = mean temperature of warmest month	Tc = mean temperature of coldest month

The distribution of Fohn's climatic types on a hypothetical continent of low and uniform elevation.

Source: Barry, R. and Chorley, R., *Atmosphere, Weather and Climate*, Routledge, 1998

Figure 7.9 Köppen's climate classification

Tropical humid climates (Af) are generally located within 5–10° of the equator. Some higher latitudes may receive high levels of rainfall from unstable tropical easterlies. In Af climates, the midday Sun is always high in the sky – but high humidity and cloud cover keep temperatures from soaring. Some months, such as April and October, may be wetter due to movements of the ITCZ.

In contrast, seasonally humid climates (Aw) have a dry season, which generally increases with latitude. Rainfall varies from the moist low latitudes to semi-desert margins. As the midday Sun reaches its highest point (zenith), temperatures increase, air pressure falls, and strong convection causes thundery storms. However, as the angle of the noon Sun decreases, the rains gradually cease and drier air is re-established.

A variation of the Aw is the tropical wet monsoon (Am) climate. Winters are dry with high temperatures. They reach a maximum just before the monsoon, and then fall during the cloudy wet period when inflows of tropical maritime air bring high rainfall to windward slopes

Section 7.1 Activities

1 Describe the main climate characteristics of Jakarta (Af) and Kolkata (Am), as shown on Figure 7.1.
2 How do you account for the differences between them?

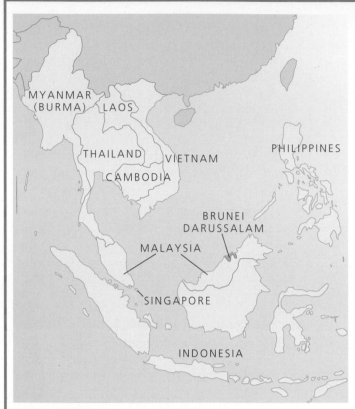

Figure 7.10 Location of Brunei Darussalam

The climate of Brunei is influenced by its location on the north-west coast of Borneo within the equatorial tropics (Figure 7.10), and by the wind systems of South East Asia.

Brunei is located in an area of low pressure at the equator, sandwiched between two areas of high pressure over the subtropics. The low-pressure 'trough' at the equator (the ITCZ) is the area where air masses from the northern and southern hemispheres converge. The annual movements of the ITCZ and the associated trade winds produce two main seasons in Brunei, separated by two transitional periods.

Between December and March, the north-east monsoon winds affect the South China Sea and Borneo. The average position of the ITCZ is between the latitudes of 5 °S and 10 °S, having moved southwards across Borneo and Brunei during late December. From June to September, the ITCZ is situated at a latitude of around 15 °N to the east of the Philippines, but to the west the ITCZ becomes a monsoon trough. The first transitional period occurs in April and May and the second one in October and November.

On a longer time scale (three to seven years), the climate of Brunei is influenced by the El Niño Southern Oscillation (ENSO). The warm episode or El Niño is normally associated with prolonged dry conditions in Brunei Darussalam. In contrast, La Niña episodes are cold and wetter than normal.

The annual rainfall total exceeds 2300 millimetres. There are clear seasonal patterns, with two maxima and two

minima. The first maximum is from late October to early January, with December being the wettest month (Figure 7.11). The second minor maximum is from May to July, with May being relatively wetter. This seasonality is a reflection of the two monsoon seasons in conjunction with the related movements of the ITCZ and the influence of the localised land–sea circulations. The lowest minimum occurs from late January to March, and the next minor minimum is from June to August. The concept of a dry month or dry season in Brunei is relative!

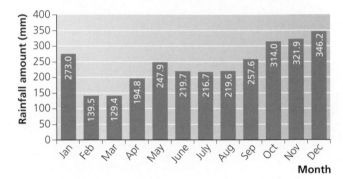

Figure 7.11 Mean monthly rainfall in Brunei, 1966–2006

The orographic effect on rainfall in Brunei is notable, particularly in Temburong District. The stations of Semabat and Selangan in Temburong have mean annual rainfall totals of over 4000 millimetres, compared with stations nearer the coast such as Puni and Bangar with annual means of around 3600 millimetres. Rainfall in Brunei is characterised by high intensities (measured in millimetres per hour), with very large amounts falling over sharply delimited areas at short time intervals, in contrast to prolonged rainfalls associated with large-scale systems such as **tropical cyclones** (Figure 7.12).

The probability of thunderstorms shows that two peaks are evident: the higher peak in April and May from late afternoon to the early hours of the morning, and a secondary peak from September to November, mostly from late afternoon to just around midnight.

The temperature regime is notable for its uniformity, with only small variations both seasonally and in different parts of the country. Higher temperatures are generally recorded during the months of March to September, with higher solar heating and less cloudiness and rainfall than in other months. Cold air surges originating from the Siberia/China area during the north-east monsoon season affect Brunei, resulting in lower minimum temperatures.

Section 7.1 Activities

1 Describe and explain the main climate characteristics for Brunei.
2 Describe and suggest reasons for spatial variations in Brunei's climate.

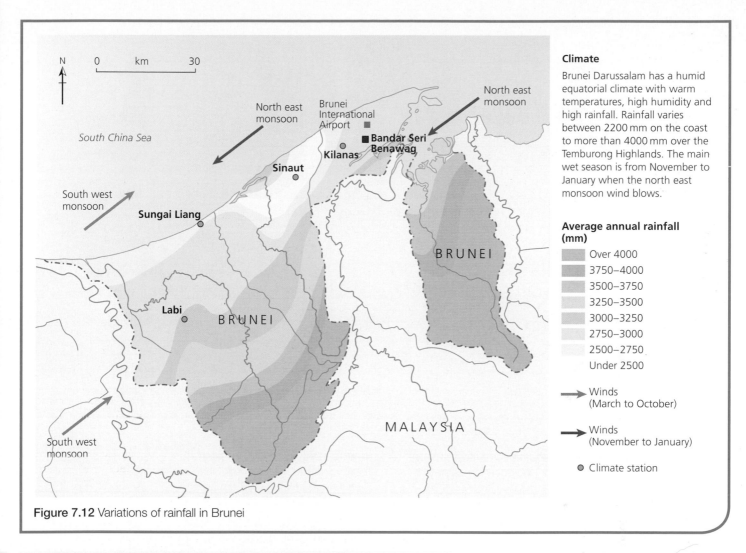

Climate

Brunei Darussalam has a humid equatorial climate with warm temperatures, high humidity and high rainfall. Rainfall varies between 2200 mm on the coast to more than 4000 mm over the Temburong Highlands. The main wet season is from November to January when the north east monsoon wind blows.

Average annual rainfall (mm)

- Over 4000
- 3750–4000
- 3500–3750
- 3250–3500
- 3000–3250
- 2750–3000
- 2500–2750
- Under 2500

→ Winds (March to October)

→ Winds (November to January)

● Climate station

Figure 7.12 Variations of rainfall in Brunei

Case Study: The El Niño Southern Oscillation

This case study shows that there are major interruptions to the normal functioning of tropical climates. One such interruption is the El Niño Southern Oscillation (ENSO). El Niño, which means 'the Christ Child', is an irregular occurrence of warm surface water in the Pacific off the coast of South America that affects global wind and rainfall patterns (Figure 7.13). In July 1997, the sea surface temperature in the eastern tropical Pacific was 2.0–2.5 °C above normal, breaking all previous climate records. El Niño's peak continued into early 1998 before weather conditions returned to normal.

During the 1920s, Sir Gilbert Walker identified a characteristic of the Southern Oscillation that consisted of a sequence of surface pressure changes within a regular time period of three to seven years, and was most easily observed in the Pacific Ocean and around Indonesia. When eastern Pacific pressures are high and Indonesian pressures are low relative to the long-term average, the situation is described as having a high Southern Oscillation Index (SOI). By contrast, when Pacific pressures are low and Indonesian pressures are high, the SOI is described as low. In 1972, J.N. Walker

identified a cell-like circulation in the tropics that operates from an east to west (interzonal) direction, rather than a north–south (meridional) direction. The cell works by convection of air to high altitudes caused by intense heating followed by movement within the subtropical easterly jet stream and its subsequent descent.

El Niño is a phase of the Southern Oscillation when the trade winds are weak and the sea surface temperatures in the equatorial Pacific increase by between 1 and 4°C. The impacts of the ENSO, which occurs every three to seven years and is the most prominent signal in year-to-year natural climate variability, are felt worldwide. During the 1982–83 ENSO, the most destructive event of the last century, damages amounted to about $13 billion. The event has been blamed for droughts (India, Australia), floods (Ecuador, New Zealand) and fires (West Africa, Brazil). Scientists are capable of forecasting the onset of El Niño up to one year in advance through sea-surface temperature signals.

El Niño has been studied since the early 1900s. In 1904, Sir Gilbert Walker investigated annual variability of the monsoon.

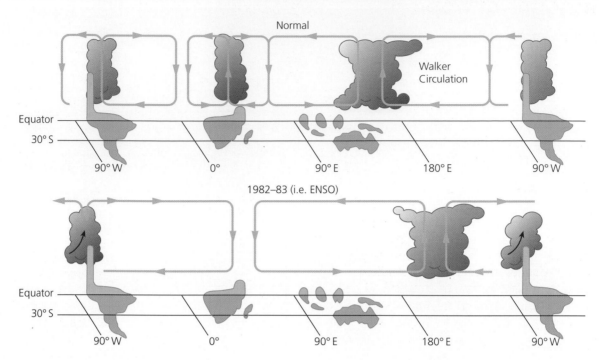

Normal

Walker Circulation

Equator
30° S

90° W   0°   90° E   180° E   90° W

1982–83 (i.e. ENSO)

Equator
30° S

90° W   0°   90° E   180° E   90° W

Source: Linacre, E. and Geerts, B., *Climates and Weather Explained*, Routledge, 1997

Figure 7.13 El Niño events

He discovered a correlation in the patterns of atmospheric pressure at sea level in the tropics, the ocean temperature and rainfall fluctuations across the Pacific Ocean, which he named the Southern Oscillation. He showed that the primary characteristic of the Southern Oscillation is a seesaw in the atmospheric pressure at sea level between the south-eastern subtropical Pacific and the Indian Ocean. During normal conditions, dry air sinks over the cold waters of the eastern tropical Pacific and flows westwards along the equator as part of the trade winds. The air is moistened as it moves towards the warm waters of the western tropical Pacific. The sea surface temperature gradients between the cold waters along the Peruvian coast and the warm waters in the western tropical Pacific are necessary for the atmospheric gradients that drive circulation.

Climatic anomalies induced by El Niño have been responsible for severe damage worldwide. Among the effects of the 1997–98 El Niño were:

- a stormy winter in California (and the 1982–83 event took 160 lives and caused $2 billion of damage in floods and **mudslides**)
- worsening drought in Australia, Indonesia, the Philippines, southern Africa and north-east Brazil

- drought and floods in China
- increased risk of malaria in South America
- lower rainfall in northern Europe
- higher rainfall in southern Europe.

Perception of the El Niño hazard has developed in a series of stages. Until the 1972–73 event, it was perceived as affecting local communities and industries along the eastern Pacific coast near Peru; then between 1972/73 and 1982/83, El Niño was recognised as a cause of natural disasters worldwide. However, since 1982–83, countries have begun to realise that there is a need for national programmes that will use scientific information in policy planning and that an integrated approach from a number of countries is required to reduce the effects of El Niño.

Section 7.1 Activities

1 Describe the climatic processes that occur during El Niño events.
2 What were the possible effects of the 1997–98 El Niño season?

7.2 Landforms of tropical environments

Tropical landforms are diverse and complex. They are the result of many interrelated factors, including climate, rock type, tectonics, scale, time, vegetation and, increasingly, human impact. In the early part of the twentieth century, the development of tropical landforms was thought to be largely the result of climate. Climatic geomorphology suggested that in a certain climate, a distinct set of processes and landforms would be produced. Critics of this idea showed that as the scale of the investigation became increasingly small and local, site-specific conditions, such as drainage and topography, become more important and the role of climate less important. In addition, many of the early studies of tropical landforms were in parts of tropical Africa, India and South America – regions that are tectonically stable.

☐ Weathering

Mechanical and chemical weathering occur widely in the tropics. Much attention has been given to the processes of hydrolysis and exfoliation. **Hydrolysis** is a form of chemical weathering. Hydrolysis occurs on rocks with orthoclase feldspar, notably granite. Feldspar reacts with an acid water to produce kaolin (or kaolinite/china clay), silicic acid and potassium hydroxyl. The acid and hydroxyl are removed in solution, leaving kaolin as the end product. In the **humid tropics**, the availability of water and the consistently high temperatures maximise the efficiency of chemical reactions, and in the oldest part of the tropics these have been operating for a very long period. In contrast, in many savanna areas where there is less moisture, **exfoliation** or disintegration occurs. This is a form of mechanical weathering. Owing to large-scale diurnal (daytime/night-time) differences in the heating and cooling of rocks, rocks expand by day and contract by night. As rock is a poor conductor of heat, these stresses only occur in the outer layers of the rock, causing peeling or exfoliation.

In many regions, the depth of the weathering profile is very deep. As the depth increases, slopes may become less stable. Rapid mass movements are likely to take place in a cyclical pattern, once a certain amount of weathering has occurred (Figure 7.14).

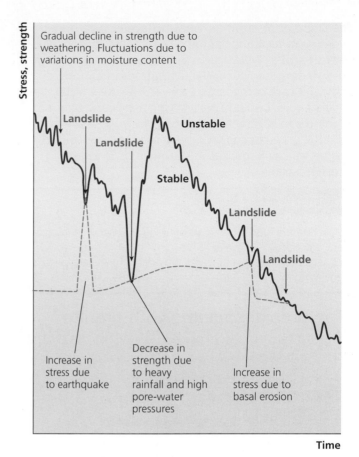

Figure 7.14 Cycles of weathering and landslides

Weathering profiles vary widely. The idealised weathering profile has three zones: residual soil, weathered rock and relatively unweathered bedrock. The residual soil is a zone of eluviation caused by the infiltration of water. Weathering alters the structure and texture of the parent rock. Soluble components may be removed, leaving behind clay-sized materials, especially those containing iron, aluminium and silica. In the lower part of the residual horizon (the C horizon), features of the original rock such as structure, joints and faults may be recognisable. Weathered rock is also known as **saprolite**.

In the weathered zone, at least 10 per cent of the rock is unweathered corestones. This zone is typically highly permeable, especially in the upper sections, and contains minerals in a wide range of weathering stages. In contrast, in the zone of unweathered bedrock, there is little alteration of feldspars and micas. The greatest depth at which unweathered rocks is reached occurs in areas where humid tropical conditions have existed for a prolonged time.

The **weathering front** or **basal surface of weathering** between solid rock and saprolite (weathered rock) may be very irregular. Typically, deep weathering occurs down to 30–60 metres, but because of variations in jointing density and rock composition, the depth varies widely over short distances. Depth of weathering has been recorded at 90 metres in Nigeria and Uganda, while in Australia a depth of 275 metres has been measured in New South Wales, as well as 80 metres in Victoria and 35 metres in Western Australia.

Section 7.2 Activities

1 Describe the main types of weathering that occur in tropical environments.
2 Explain why 'deep weathering' may occur in some tropical environments.

☐ Features associated with granite

The word 'inselberg' describes any isolated hill or hills that stand prominently over a level surface. Inselbergs include:

- laterite-capped masses of saprolite
- hills of sedimentary rocks
- castle kopjes
- **tors** of residual core stones
- massive rock domes with near-vertical sides, called 'bornhardts' or 'domed inselbergs'.

Tors

Most tors and castle kopjes are found in strongly jointed rock. Tors are ridges or piles of spheroidally (rounded) weathered boulders (Figure 7.15) that have their bases in the bedrock and are surrounded by weathered debris. They vary in height from 20 to 35 metres and have core stones up to 8 metres in diameter.

Figure 7.15 Spheroidal weathering

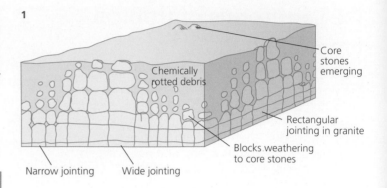

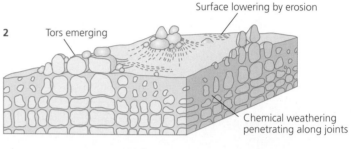

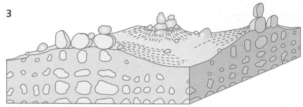

Figure 7.16 Tor formation

Tors are formed by chemical weathering of the rock along joints and bedding planes beneath the surface (Figure 7.16). If the joints are widely spaced the core stones are large, whereas if the joints are close together the amount of weathering increases and the core stones are much smaller. If denudation of the surface exceeds the rate of chemical weathering at the weathering front, the blocks will eventually be exposed at ground level. Gradually, the rocks below the surface will be chemically weathered and cause the collapse of the tor.

Good examples of tors are found on the Jos Plateau of Nigeria and in the Matopas region of Zimbabwe and around Harare.

Inselbergs

A striking feature of tropical plains is the rock hills known as 'inselbergs' (Figure 7.17). Conditions especially favourable for residual hills occur in the seasonal tropics. Residual hills are best developed on volcanic materials, especially granite and gneiss, with widely spaced joints and a high potassium content. However,

Figure 7.17 Inselberg and low plains

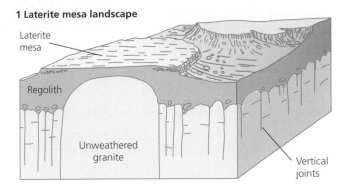

1 Laterite mesa landscape

Laterite mesa

Regolith

Unweathered granite

Vertical joints

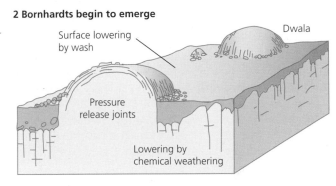

2 Bornhardts begin to emerge

Dwala

Surface lowering by wash

Pressure release joints

Lowering by chemical weathering

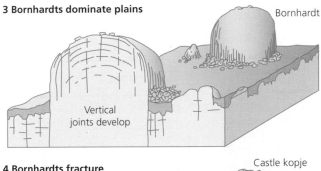

3 Bornhardts dominate plains

Bornhardt

Vertical joints develop

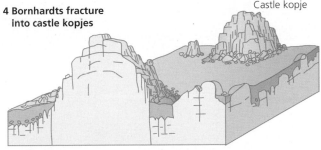

4 Bornhardts fracture into castle kopjes

Castle kopje

Figure 7.18 Formation of bornhardts

they can also be found on sedimentary rocks, such as sandstone.

Residual hills occur in a variety of sizes. The term 'inselberg' usually describes an almost abrupt rise from the plains. **Duricrust**-topped hills are the major exception. The terms 'tor' or 'boulder inselberg' are used to describe spheroidally weathered boulders rooted in bedrock.

Residual hills are the result of stripping weathered regolith from a differentially weathered surface. The major debate is whether deep weathering is needed for hill formation. The two-stage model requires the development of a mass of weathered material beneath the ground and its subsequent removal. Alternatively, weathering and erosion could occur simultaneously. The diversity of residual hills suggests that both mechanisms operate simultaneously.

Bornhardts

The monolithic domed inselberg or 'bornhardt' is a characteristic landform of granite plateaux of the African savanna, but can also be found in tropical humid regions. They are characterised by steep slopes and a convex upper slope (Figure 7.18). Bornhardts are eventually broken down into residual castle kopjes.

Bornhardts occur in igneous and metamorphic rocks. Granite, an igneous rock, develops joints up to 35 metres below the surface during the process of pressure release. Vertical jointing in granite is responsible for the formation of castle kopjes.

The two main theories for the formation of bornhardts are:

- the stripping or exhumation theory – increased removal of regolith occurs so that unweathered rocks beneath the surface are revealed
- Lester King's parallel retreat theory, which states that the valley sides retreat until only remnant inselbergs are left.

In the case of bornhardts, it seems that the major theory is that of etchplanation. There is a gradual evolution of the domed rock (ruware) through the bornhardt to an inselberg and then the residual castle kopje. As the theory invokes evolution through a series of periods of deep weathering, changing climatic conditions and uplift, this evolutionary approach might be the most appropriate.

Classic examples of bornhardts include Mount Hora in the Mzimba District of Malawi, and Zuma Rock near Abuja in Nigeria. Bornhardts are an example of **equifinality**; that is, different processes leading to the same end product.

1 Explain why tors may be considered 'joint-controlled'.
2 To what extent are inselbergs 'relict' features?
3 Describe the development of bornhardts and castle kopjes as shown in Figure 7.18.

☐ Features associated with limestone

Tropical karst

There are two major landform features associated with tropical karst. **Polygonal** or **cockpit karst** (Figure 7.19) is a landscape pitted with smooth-sided soil-covered depressions and cone-like hills. Cone karst is a landscape with closely spaced conical hills – it is a type of cockpit karst. The Cockpit Country in Jamaica is a classic example of cone karst. The terms 'cockpit' and 'cone' karst are used almost interchangeably. **Tower karst** is a landscape characterised by upstanding rounded blocks set in a region of low relief. Although water is less able to dissolve carbon dioxide in tropical areas, the higher temperatures and the presence of large amounts of organic matter produce high amounts of carbon dioxide in the soil water.

Some geographers believe that there is an evolution of limestone landscapes that eventually leads to cockpit karst and tower karst (Figure 7.20). However, according to Marjorie Sweeting (1989), the distinction between cockpit karst and tower karst is fundamental, as the hydrological and tectonic conditions associated with each are quite different.

Polygonal or cockpit karst is characterised by groups of hills, fairly uniform in height (Figure 7.21). These can be up to 160 metres high in Jamaica, with a base of up to 300 metres. They develop mainly as a result of solution. They are as common to some tropical areas as dry valleys and dolines are to temperate areas. Polygonal karst tends to occur in areas:

- that have been subjected to high rates of tectonic uplift and
- where vertical erosion by rivers is intense.

Figure 7.19 Cockpit karst

Solution holes

The surface is broken up by many small solution holes but the overall surface remains generally level.

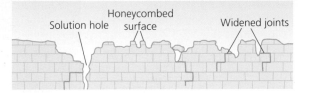

Cockpit karst

Cockpit karst is usually a hilly area in which many deep solution holes have developed to give it an 'eggbox' appearance.

Tower karst

The widening and deepening of the cockpits has destroyed much of the limestone above the water table. Only a few limestone towers remain, sticking up from a flat plain of sediments that have filled in the cockpits at a level just above the water table. Eventually the towers will be entirely eroded, and disappear.

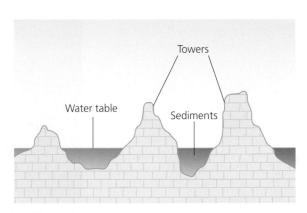

Figure 7.20 Cockpit karst and tower karst

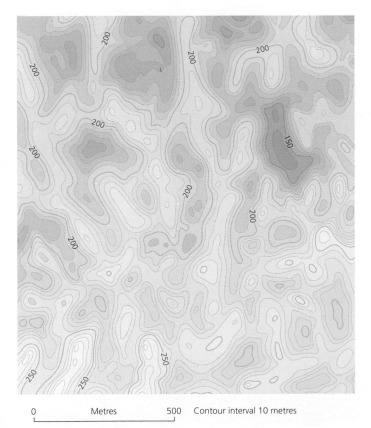

0   Metres   500   Contour interval 10 metres

Figure 7.21 Cockpit karst in northern Puerto Rico, from a US Geological Survey map

The spacing of the cones may be related to the original stream network. Concentrated solution along preferred routes, such as wider joints, leads to accelerated weathering of certain sections of the limestone, especially during times of high flow, such as during a storm. Water will continue to weather the limestone as far down as the water table. This creates closed depressions and dolines. Once the water table is reached, water will flow laterally rather than vertically, so developing a flat plain.

An alternative theory suggests that the formation and subsequent collapse of cave systems is the main mechanism for cockpit-karst formation. Caves in limestone migrate upwards through the hillside. Collapse of the ceiling is due to the solution by water percolating downwards. Every time the ceiling collapses, the ceiling gets higher and the floor is raised by the debris, so the whole chamber gets higher. Eventually, the cave roof collapses.

By contrast, tower karst is much more variable in size than the conical hills of cockpit karst, and ranges from just a few metres to over 150 metres in height in Sarawak. Other areas of tower karst include southern China, Malaysia, Indonesia and the Caribbean. They are characterised by steep sides, with cliffs and overhangs, and with caves and solution notches at their base. The steepest towers are found on massive, gently tilted limestone. According to Sweeting, tower karst develops in areas where:

- tectonic uplift is absent or limited
- limestone lies close to other rocks
- the water table is close to the surface.

In wet monsoonal areas, rivers will be able to maintain their flow over limestone, erode the surface and leave residual blocks set in a river plain. It is likely that there are other important processes. These include:

- differential erosion of rock with varying resistance
- differential solution along lines of weakness
- the retreat of cockpit karst slopes to produce isolated tower karst
- lateral erosion.

Why is China so important for tropical karst?

Southern China is one of the best areas in the world for the development of cockpit and tower karst (Figure 7.22). A number of conditions help explain this:

- large amounts of rainfall – over 2000 millimetres per annum (in the north of China where rainfall is low, limestone features include escarpments and dry valleys)
- long periods of slow uplift exposing broad, gently dipping plateaus
- thick beds of limestone, up to 3000 metres deep, allow spectacular landforms to develop.

Figure 7.23 Stalactite, Harrison's Cave, Barbados

Figure 7.22 Tower karst, China

☐ Underground features

Underground features include caves and tunnels formed by carbonation-solution and erosion by rivers. Carbonation is a reversible process. When calcium-rich water drips from the ceiling, it leaves behind calcium in the form of **stalactites** and **stalagmites**. These are cave deposits formed by the precipitation of dissolved calcium carbonate. Stalactites (Figure 7.23) develop from the top of the cave, whereas stalagmites (Figure 7.24) are formed on the base of the cave. Rates of **deposition** are slow; about 1 millimetre (the thickness of a coat of paint) per 100 years. The speed with which water drips from the cave ceiling appears to have some influence on whether stalactites (slow drip) or stalagmites (fast drip) are formed.

Figure 7.24 Stalagmite, Harrison's Cave, Barbados

Case Study: Blue holes in the Bahamas

Sea-level changes in the Caribbean have caused some limestone caves to be submerged, forming **blue holes** (Figure 7.25). These are a major tourist attraction in the Bahamas, for example. The Bahamas has a variety of blue holes (Figure 7.26). The islands' limestone has been carved out by carbonation-solution over millennia, during periods when the sea level fluctuated – the seas were some 130 metres lower 10 000 years ago, for example. As the sea level rose over the last 10 000 years, it submerged many of the limestone sinks, caves and tunnels. These drowned sinks became the blue holes. The classic form of a blue hole is circular, extending in a bell shape beneath the surface. However, some open into the edge of an oceanic wall or are simply openings in a shallow reef. Not all blue holes are oceanic – many are found inland. The deepest known blue hole in the Bahamas reaches down to over 200 metres, and many systems drop to around 100 metres and then extend into a network of caverns and caves at the bottom.

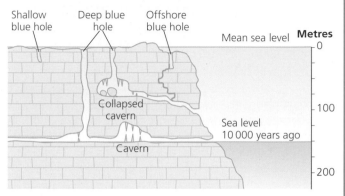

Figure 7.25 Formation of blue holes

Figure 7.26 Blue hole in the Bahamas

Section 7.2 Activities

1 How are blue holes formed?
2 Outline the differences between cockpit karst and tower karst.

7.3 Humid tropical (rainforest) ecosystems and seasonally humid tropical (savanna) ecosystems

An **ecosystem** is the interrelationship between plants and animals and their biotic and abiotic environment. In contrast, **ecology** is the study of organisms in relation to the environment, and **biogeography** is the geographical distribution of ecosystems, where and why they are found.

A **community** is a group of populations (animals and plants) living and interacting with each other in a common habitat (such as a savanna or a tropical rainforest). This contrasts with the term **population**, which refers to just one species, such as zebra.

The productivity of an ecosystem is how much organic matter is produced per annum. It is normal to refer to **net primary productivity**; that is, the amount of organic material produced by plants and made available to the herbivores (Table 7.3). Net primary productivity varies widely, and is affected by water availability, heat, nutrient availability and age and health of plant species. In contrast, **biomass** is a measure of stored energy. Forest ecosystems normally have much higher rates of biomass than grassland ecosystems, since much of the energy is stored as woody tissue.

Biodiversity refers to the variety of habitats, species and genetic diversity in an ecosystem. It is very high in tropical rainforests due to the high levels of productivity caused by year-round rainfall and high temperatures. It is also related to the long period of evolution without significant climate change. Biodiversity in savannas is also high due to the mosaic of habitats caused by physical and human factors.

Table 7.3 A comparison of mean net primary productivity and biomass for the world's major biomes

Ecosystem	Mean NPP (kg/m²/year)	Mean biomass (kg/m²)
Tropical rainforest	2.2	45
Tropical deciduous forest	1.6	35
Tropical scrub	0.37	3
Savanna	0.9	4
Mediterranean sclerophyll	0.5	6
Desert	0.003	0.002
Temperate grassland	0.6	1.6
Temperate forest	1.2	32.5
Boreal forest	0.8	20
Tundra and mountain	0.14	0.6
Open ocean	0.12	0.003
Continental shelf	0.36	0.001
Estuaries	1.5	1

☐ Succession in plant communities

Succession refers to the spatial and temporal changes in plant communities as they move towards a seral climax (Figure 7.27). Each **sere** or stage is an association or group of species, which alters the micro-environment and allows another group of species to dominate. At the start of succession (pioneer communities), there are few nutrients and limited organic matter. Organisms that can survive are small and biodiversity is low. In late succession, there is more organic matter, higher biodiversity and longer-living organisms (Figure 7.28). Nutrients may be held by organisms, especially trees, so nutrient availability may be low. A well-documented case of succession is that following the eruption of Krakatoa in 1883 (Table 7.4). The **climax community** is the group of species that are at a dynamic equilibrium with the prevailing environmental conditions. On a global scale,

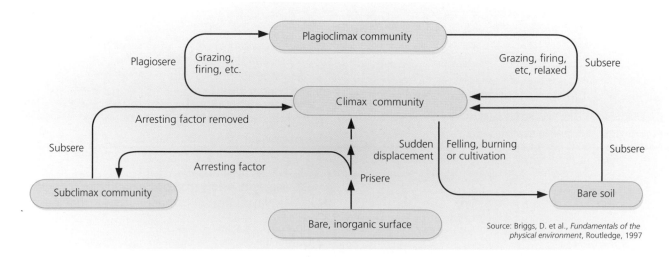

Source: Briggs, D. et al., *Fundamentals of the physical environment*, Routledge, 1997

Figure 7.27 A model of succession

climate is the most important factor in determining large ecosystems or **biomes** such as tropical rainforest and temperate woodland (Figure 7.29). In some areas, however, vegetation distribution may be determined by soils rather than by climate. This is known as **edaphic** control. For example, in savanna areas, forests are found on clay soils, whereas grassland occupies sandy soils. On a local scale, within a climatic region, soils may affect plant groupings.

Attribute	Early	Late
Organic matter	Small	Large
Nutrients	External	Internal
Nutrient cycles	Open	Closed
Role of detritus	Small	Large
Diversity	Low	High
Nutrient conservation	Poor	Good
Niches	Wide	Narrow
Size of organisms	Small	Large
Life-cycles	Simple	Complex
Growth form	r species	k species
Stability	Poor	Good

Source: Briggs, D.et al., *Fundamentals of the physical environment*, Routledge 1997

Figure 7.28 Community changes through succession

Table 7.4 Primary succession on Krakatoa after the 1883 volcanic eruption

Year	Total number of plant species	Vegetation on the coast	Vegetation on the lower slopes	Vegetation on the upper slopes
1883	Volcanic eruption kills all life on the island			
1884	No life survives			
1886	26	9 species of flowering plant	Ferns and scattered plants; blue-green algae beneath them on the ash surface	Ferns and scattered plants; blue-green algae beneath them on the ash surface
1897	64	Coastal woodland develops	Dense grass	Dense grasses with shrubs interspersed
1908	115	Wider belt of woodland with more species, shrubs and coconut palms	Dense grasses to 3m high, woodland in the larger gullies	Dense grasses to 3m high, woodland in the larger gullies
1919			Scattered trees in grassland, single or in groups with shade species beneath; thicket development in large gullies	Scattered trees in grassland, single or in groups with shade species beneath; thicket development in large gullies
1928	214			
1934	271		Mixed woodland largely taken over from savanna	Woodland with smaller trees, fewer species taking over
Early 1950s		Coastal woodland climax	Lowland rainforest climax	Submontane forest climax

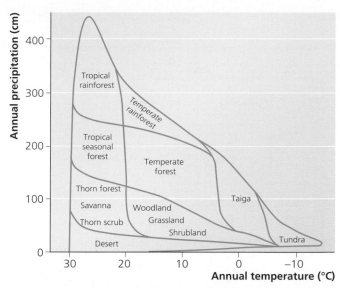

Source: *Advanced Geography: Concepts and Cases* by P. Guiness & G. Nagle (Hodder Education, 1999), p.441

Figure 7.29 The relationship between climate and vegetation

Sometimes a sub-climax occurs. This is often the case when a factor – such as a flood, **hurricane** or tsunami – prevents the climax community from forming or from remaining. It may, over time, form again, once the arresting factor has been removed. In contrast, a **plagioclimax** refers to a plant community permanently influenced by human activity. It is prevented from reaching climatic climax by burning, grazing, and so on. The maintenance of grasslands through burning is an example of plagioclimax.

Section 7.3 Activities

1 Briefly explain the meaning of the terms *succession* and *plagioclimax*.
2 Describe the main changes in plant species over time following the eruption of Krakatoa.
3 Suggest reasons for the differences between the vegetation on the coast, the lower slopes and the upper slopes.

Figure 7.30 shows a model of succession in a rainforest. A contemporary example of succession being affected by arresting factors is the rainforest of Chances Peak in Montserrat. In the early 1990s, Chances Peak had some of the finest tropical rainforest in the Caribbean region. It had a high biodiversity of plant life, insects, lizards, birds and bats. By January 1996, surveys indicated that vegetation loss from acid rain, gases, heat and dust on the top of Chances Peak and in the surrounding area was severe. The cloud forest had disappeared. Vegetation was gradually dying further down the mountain. On the east side, the lush forests of the Tar River valley were degraded from ash and gases, and were finally destroyed by lahars (mudflows) (Figure 7.31).

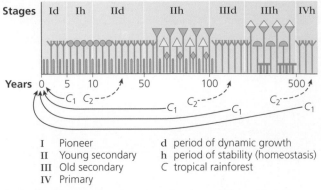

Stages	Id	Ih	IId	IIh	IIId	IIIh	IVh

Years 0   5   10   50   100   500

I   Pioneer
II   Young secondary
III Old secondary
IV   Primary

d   period of dynamic growth
h   period of stability (homeostasis)
C   tropical rainforest

Figure 7.30 A model of succession in a rainforest from bare soil to primary forest

Figure 7.31 The effect of a volcanic eruption – Chances Peak, Montserrat

Volcanoes emit sulphurous gases. On a global scale, the amount of sulphur from active volcanoes is minor compared with that from other sources. On a local scale, the impacts from volcanic sulphur emissions have important consequences for plant and animal life. Acid rain increases leaching of some nutrients and renders other nutrients unavailable for uptake by plants.

By January 1996, the pH of the lake at the top of Chances Peak was 2.0 (1000 times more acidic than a pH of 5.0). Other lakes and streams measured about 1.5 at the same time.

Continued volcanic activity and erosion prevent reforestation. Sustained recovery will not take place until volcanic activity greatly reduces or stops. Recovery of the forest will start in two ways: seeds in the ground will start to germinate, and new seeds will blow in from surrounding areas. The rate of recovery will depend in part on the availability of seeds, which in turn depends on the proximity of other forest species and the availability of animals (for example, birds) to disperse seeds. In Montserrat, there is currently cloud forest in the Centre Hills and this may act as a source of new seeds for the southern Soufrière Hills close to the volcano. Pioneers will colonise the soil and make it suitable for other forest

species. Early pioneer species must be able to tolerate high light intensities and high temperatures (because there is no longer any forest to provide shade). Species such as *Cecropia* are early pioneers: they are light tolerant, and their seeds are dispersed by a variety birds and bats (76 species of birds are known to feed on *Cecropia*). High light levels and high temperatures often stimulate the germination of these seeds. *Heliconia* and some palm species will colonise disturbed areas. Once these pioneer species have established, the shade they create often prevents other members of the same species from germinating and surviving. Thus new shade-tolerant species can now establish themselves. The forest therefore begins to increase in diversity. Studies in Puerto Rico have shown that some forests recover from hurricane destruction in about 40 years.

Section 7.3 Activities

Suggest **two** contrasting reasons for the death of the trees shown in Figure 7.31.

☐ Vegetation in tropical rainforests

There is a wide variety of ecosystem types in the humid tropics (Figure 7.32). The tropical rainforest is the most diverse ecosystem or biome in the world, yet it is also the most fragile. This stems from the conditions of temperature and humidity being so constant that species here specialise to a great extent. Their food sources are limited to only a few species. Thus when this biome is subjected to stress by human activity it often fails to return to its original state.

The net primary productivity (NPP) of this ecosystem is 2200 g/m²/year. This means that the solar energy fixed by the green plants gives 2200 grams of living matter per m² every year (Figure 7.33). This compares with the NPP for savanna of 900 g/m²/year, temperate deciduous forests of 1200 g/m²/year and agricultural land of 650 g/m²/year. Figure 7.33 shows the pattern of production and consumption in a typical undisturbed area of tropical rainforest. Only a minute amount passes through animals (and even less through hunter-gatherers). Most of the plant material falls to the forest floor to become litter. Although the forest contains many animals, it does not contain a large biomass of animals, in part because much of the plant material is inedible. The woody parts of plants are indigestible to animals. Leaves, fruits and seeds are digestible, but may contain chemicals that are poisonous to animals. Consequently, many animals specialise on particular plant species, upon which they feed successfully despite their toxicity. Many plants counter this with synchronised fruiting, sometimes as infrequent as once per decade, so that most of the time there is little food to support the animals that specialise in feeding on them.

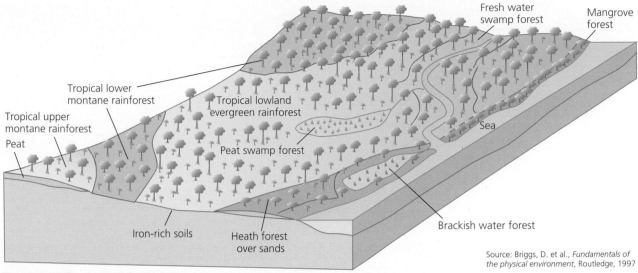

Figure 7.32 Tropical rainforest types

Source: Briggs, D. et al., *Fundamentals of the physical environment*, Routledge, 1997

The hot, humid climate gives ideal conditions for plant growth and there are no real seasonal changes. The plants are therefore aseasonal and the trees shed their leaves throughout the year rather than in one season. There is a great variety of plant species – in some parts of Brazil, there can be 300 tree species in 2 km² (Figure 7.34). The trees are tall and fast growing. The need for light means that only those trees that can grow rapidly and overshadow their competitors will succeed. Thus trees are notably tall and have long, thin trunks with a crown of leaves at the top; they also have buttress roots to support the tall trees.

The trophic structure of the rainforest is similar to other ecosystems (including the savannas). Primary producers (the first trophic layer – including trees such as teak) trap energy from sunlight and convert it to food energy. This in turn is consumed by herbivores (the second trophic layer, such as tree squirrels), which in turn are consumed by primary and secondary consumers. Eventually, detritivores decompose dead material and return the nutrients to the soil. Given the high level of biodiversity in rainforests, the food web is complex.

There are, broadly speaking, three main layers of tiers of trees (Figure 7.35):

- **emergent**, which extend up to 45–50 metres
- a **closed canopy** 25–30 metres high, which cuts out most of the light from the rest of the vegetation and restricts its growth
- a limited **understorey** of trees, denser where the canopy is weaker; when the canopy is broken by trees falling, by clearance or at rivers, there is a much denser understorey vegetation.

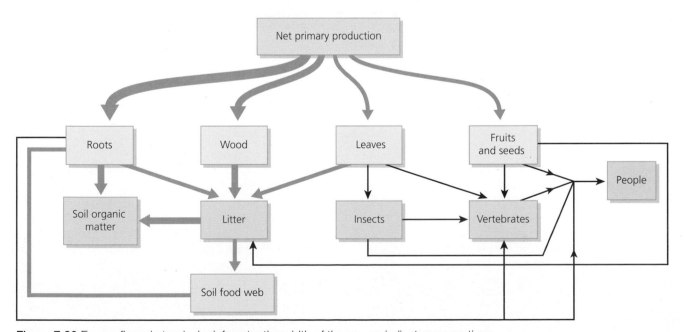

Figure 7.33 Energy flows in tropical rainforest – the width of the arrows indicates proportions

Figure 7.34 Tropical rainforests have vegetation in a number of different layers

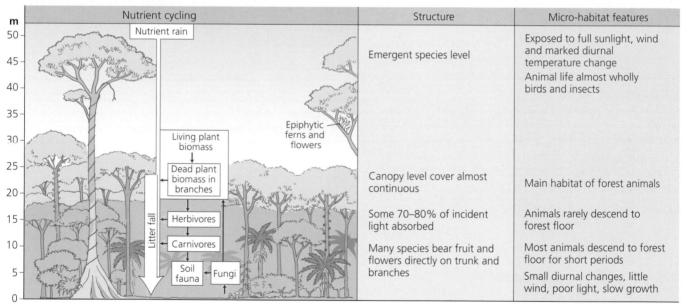

Nutrient cycling	Structure	Micro-habitat features
	Emergent species level	Exposed to full sunlight, wind and marked diurnal temperature change
		Animal life almost wholly birds and insects
	Canopy level cover almost continuous	Main habitat of forest animals
	Some 70–80% of incident light absorbed	Animals rarely descend to forest floor
	Many species bear fruit and flowers directly on trunk and branches	Most animals descend to forest floor for short periods
		Small diurnal changes, little wind, poor light, slow growth

Source: Briggs, D et al., *Fundamentals of the physical environment,* Routledge 1997

Figure 7.35 Vegetation structure in a rainforest

Trees are shallow-rooted as they do not have problems getting water. Other layers include lianas and epiphytes, and the final layer is an incomplete field layer limited by the lack of light. The floor of the rainforest is littered with decaying vegetation, rapidly decomposing in the hot, humid conditions. Tree species include the rubber tree, wild banana and cocoa; pollination is not normally by wind due to the species diversity – insects, birds and bats, which have restricted food sources.

Rainforest supports a large number of epiphytes, which are attached to the trees. Many of these are adapted to a system that requires only a small intake of nutrients. Some plants, such as the carnivorous pitcher plant (Figure 7.36), get their nutrients from insects and small mammals. There are also parasites taking nutrients from host plants, while those flora living on dead material are called 'saprophytes', an important part of the decomposing cycle. The fauna are as diverse as the flora.

Figure 7.36 A carnivorous pitcher plant, which takes its nutrients from insects and small mammals

1 Explain why tropical rainforests have very high rates of productivity.
2 Suggest reasons why there are different types of rainforest as shown in Figure 7.32.
3 Describe and explain the vegetation structure of the tropical rainforest as shown in Figure 7.35.

Nutrient cycles

Nutrients are circulated and reused frequently. All natural elements are capable of being absorbed by plants, either as gases or soluble salts. Only oxygen, carbon, hydrogen and nitrogen are needed in large quantities. These are known as **macronutrients**. The rest are **trace elements** or micronutrients, such as magnesium, sulfur and phosphorus. These are needed only in small doses. Nutrients are taken in by plants and built into new organic matter. When animals eat the plants, they take up the nutrients. These nutrients are eventually returned to the soil when the plants and animals die and are broken down by decomposers.

All nutrient cycles involve interaction between soil and the atmosphere and involve many food chains. Nevertheless, there is great variety between the cycles. Generally, gaseous cycles are more complete than sedimentary ones as the latter are more susceptible to disturbance, especially by human activity. Nutrient cycles can be sedimentary based, in which the source of the nutrient is from rocks, or they can be atmospheric based, as in the case of the nitrogen cycle.

Case Study: Civil war and the Rwandan rainforest

In 1925, the Albert National Park, an area of lush tropical rainforest, was established by Belgian colonial authorities. In 1979, UNESCO declared the park, renamed Virunga National Park, as Africa's first National Park Heritage Site, on account of its biodiversity. However, Virunga has been plagued by problems for decades and the park is being deprived of its ecological treasures:

- Political and economic breakdown in the Democratic Republic of Congo has starved the park of funds.
- In the 1960s, much of the big game was killed by poachers, rebels and government soldiers during the civil war in the Congo.
- Tourism in the park is limited, thereby reducing the park's revenue.

- Poaching has halved the number of hippopotamuses and buffalo in the area.
- Villagers in the park have depleted the forest and overfished the lake.
- The Rwandan civil war (1990–94) intensified pressures on the land.

The park was looted by Rwandan refugees and Hutu soldiers from around the town of Goma in eastern Congo. Up to 300 km² of forest was destroyed in less than six months in 1994. Nearly 900 000 refugees lived within or near Virunga and up to 40 000 people entered the park daily, notably for food and fuel. Soldiers, too, cut wood to sell to the refugees. The rivers were polluted by human and animal waste and medical products, leading to a high risk of disease transmission.

Nutrient cycles can be shown by means of simplified diagrams (Gersmehl's nutrient cycles), which indicate the stores and transfers of nutrients (Figure 7.37). The most important factors that determine these are availability of moisture, heat, fire (in grasslands), density of vegetation, competition and length of growing season.

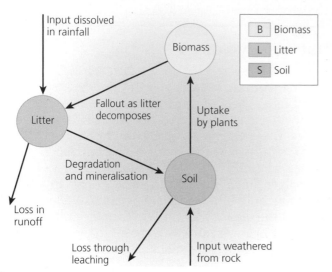

Figure 7.37 Gersmehl's nutrient cycle

The factors affecting the store of nutrients and their transfer are:

- the amount and type of weathering
- overland runoff and soil erosion
- eluviation
- the amount of rainfall
- rates of decomposition
- the nature of vegetation (woody species hold on to nutrients for much longer than annuals)
- the age and health of plants
- plant density
- fire.

Hence, explaining the differences between nutrient cycles in different ecosystems involves a consideration of many processes.

In the tropical rainforest, soil fertility is generally low. This can be explained by looking at the Gersmehl nutrient cycle model. The input of nutrients from weathering and precipitation is high (Figure 7.38) owing to the sustained warm wet conditions. However, most of the nutrients are held in the biomass due to the continual growing season. Breakdown of nutrients is rapid and there is a relatively small store in the soil. Where vegetation has been removed, the loss of nutrients is high due to high rates of leaching and overland flow.

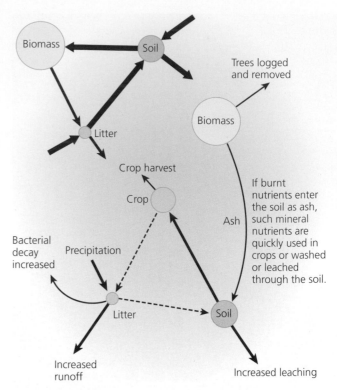

Figure 7.38 Nutrient cycle in a tropical rainforest

In contrast, in a savanna grassland ecosystem the biomass store is less than that of the tropical rainforest due to a shorter growing season. The litter store is also small due to fire. This means that the soil store is relatively large.

The savanna nutrient cycle differs from the tropical rainforest nutrient cycle because of the combined effects of a seasonal drought and the occurrence of fire. Consequently, there is:

- a lower nutrient availability
- a reduced biomass store
- a small litter store
- a relatively large soil store.

Section 7.3 Activities

1 Describe and explain the main characteristics of the nutrient cycle associated with tropical rainforests.
2 Outline the changes that occur as a result of human activity. Suggest reasons for the changes you have identified.

☐ Savanna ecosystems

Origin

Savannas are areas of tropical grassland that can occur with or without trees and shrubs. Savannas are widespread in low latitudes, covering approximately one-quarter of the world's land surface – 18 million km² (compared with 2.6 million km² covered by tropical rainforest). Their origin is partly related to natural conditions and partly to human activities, especially burning. They occur between the tropical rainforest and the hot deserts, but there is not a gradual transition from rainforest through savanna to desert – rather, the savanna is a mosaic of plant communities influenced by many factors: climate, soils, drainage, geomorphology, geology and human factors, such as burning and animal grazing (Figure 7.39). Savannas are under increasing pressure from human activities. Trying to protect them is not proving easy.

According to Hills (1965), the factors affecting the development of savannas can be divided into predisposing, causal, resulting and maintaining. For example, in South America, climate predisposes vegetation to grassland rather than forest because grass is better able to survive the dry period. Landscape may be a causal factor as it affects drainage; laterite may be a resulting factor (being caused by drainage, but then in turn limiting the vegetation that can survive on impoverished soils), and fire a maintaining factor – human activity regulates the use of fire to produce grassland for grazing at the expense of forest.

Others suggest that edaphic (soil) characteristics may be important. Infertile soils with low water retention may only support grassland, whereas more fertile, moist soils may support forest. There is also a close relationship between soils and landscape – for example, soils at the base of slopes are often more moist and receive nutrients from further upslope.

Others suggest that savannas are the result of climate change in the Pleistocene period, when conditions were more arid. Many savanna trees are fire-resistant. Frequency of fire is important, as it allows fire-resistant species to invade. Woody species that take water from deeper may invade an area following overgrazing since the latter removes annual grasses and results in overland flow.

Climate

The climate that characterises savanna areas is a tropical wet/dry climate. However, there is great variation in the climate between savanna areas. The wet season occurs in summer: heavy **convectional rain** (monsoonal) replenishes the parched vegetation and soil. Rainfall can range from as little as 500 millimetres to as much as 2000 millimetres (enough to support deciduous forest). However, all savanna areas have an annual drought: these can vary from as little as one month to as much as eight months. It is on account of the dry season that grasses predominate. Temperatures remain high throughout the year, ranging between 23 and 28 °C. The high temperatures, causing high evapotranspiration rates, and the seasonal nature of the rainfall cause a twofold division of the year into seasons of water surplus and water deficiency. This seasonal variation has a great effect on soil development.

Climate and soils

The link between climate and soil could hardly be closer (Figure 7.40). Soils in the savanna are often leached ferralitic soils (ferruginous soils, tropical red earths). These are similar to soils of the rainforest but not as intensely weathered, are less leached and exhibit a marked seasonal pattern in soil process. They may have a concentration of iron and aluminium oxides in them, so are sometimes

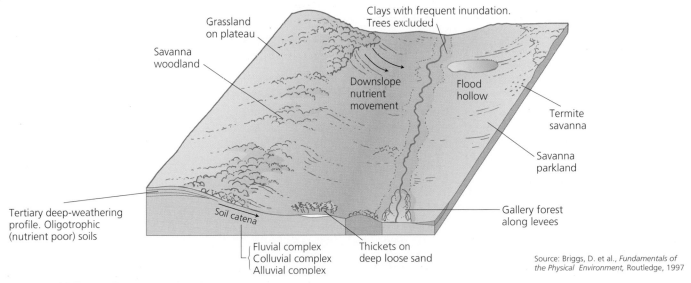

Figure 7.39 Types of savanna

Source: Briggs, D. et al., *Fundamentals of the Physical Environment*, Routledge, 1997

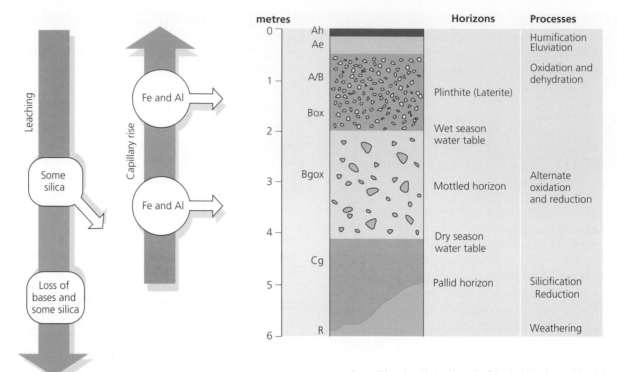

metres | Horizons | Processes

Horizons		Processes
Ah		Humification
Ae		Eluviation
A/B	Plinthite (Laterite)	Oxidation and dehydration
Box		
	Wet season water table	
Bgox	Mottled horizon	Alternate oxidation and reduction
	Dry season water table	
Cg	Pallid horizon	Silicification Reduction
R		Weathering

Source: Briggs, D. et al., *Fundamentals of the physical environment*, Routledge, 1997

Figure 7.40 Savanna soils

referred to as **red earths** and **tropical brown earths**. During the wet season, the excess of precipitation (P) over potential evapotranspiration (E) means that leaching of soluble minerals and small particles will take place down through the soil. These are deposited at considerable depth. By contrast, in the dry season E > P. Silica and iron compounds are carried up through the soil and precipitated close to the surface. Capillary action also brings soluble bases to the surface. During the following wet season, leaching removes silicates, thereby leaving an upper horizon of iron and aluminium oxides, often in such a hard, compact layer that it forms an impermeable crust, known as 'laterite'. The bedrock tends to weather into a type of clay. This may become very plastic in the wet season, but very hard during the dry season. The soils are not fertile, and are poor for agriculture. They are certainly more suited for pastoral than for agricultural purposes. Soils with impermeable laterite have been called a 'pedological leprosy' and are subject to wind erosion.

Geomorphology plays an important role too. Some areas, notably at the base of slopes and in river valleys, are enriched by clay, minerals and humus that is deposited there. By contrast, plateaus, plains and the tops of slopes may be depleted of nutrients by erosion. The local variation in soil leads to variations in vegetation: this control by the soil is known as edaphic control. For example, on the thicker clay-based soils there is frequently woodland, whereas on the leached sandy soils, with poor water retention, grassland predominates. Savanna areas are frequently found on tectonically stable geological **shields**: these have therefore been weathered and are

lacking in nutrients. Hence, even some river valleys may not be as fertile as their temperate counterparts.

A soil catena (Figure 7.41) can be observed in:

- thin, immature soils on the steep slopes close to the plateau top
- ferruginous soils on the freely drained slopes
- soils containing laterite where the lower part of the soil profile is affected by the water table
- **vertisols** (a vertical soil structure as a result of expansion of swelling clays) where the soil is largely beneath the water table.

Vegetation

Savanna vegetation is a mosaic including grasses, trees and scrub. All are **xerophytic** (adapted to drought) and therefore adapted to the savanna's dry season, and **pyrophytic** (adapted to fire). Adaptations to drought include deep tap roots to reach the water table, partial or total loss of leaves and sunken stomata on the leaves to reduce moisture loss. Those relating to fire include a very thick bark and thick budding that can resist burning, the bulk of the biomass being below ground level, and rapid regeneration after fire. Unlike shrubs, where growth occurs from the tips, the growth tissue in grasses is located at the base of the shoot close to the soil surface, so burning, and even grazing, encourage the growth of grass relative to other plants.

The warm wet summers allow much photosynthesis and there is a large net primary productivity of 900 g/m²/year. This varies from about 1000 g/m²/year where it borders rainforest areas to only about 200 g/m²/year

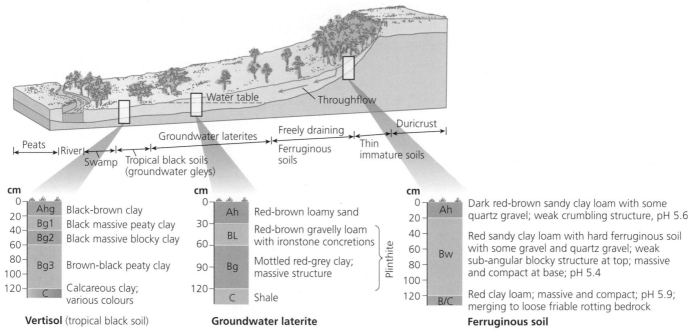

Figure 7.41 Soil catena in the savanna

Vertisol columns:
cm		
0	Ahg	Black-brown clay
20	Bg1	Black massive peaty clay
40	Bg2	Black massive blocky clay
60		
80	Bg3	Brown-black peaty clay
100		
120	C	Calcareous clay; various colours

Vertisol (tropical black soil)

Groundwater laterite:
cm		
0	Ah	Red-brown loamy sand
30	BL	Red-brown gravelly loam with ironstone concretions
60		
90	Bg	Mottled red-grey clay; massive structure
120	C	Shale

Plinthite

Groundwater laterite

Ferruginous soil:
cm		
0	Ah	Dark red-brown sandy clay loam with some quartz gravel; weak crumbling structure, pH 5.6
20		
40	Bw	Red sandy clay loam with hard ferruginous soil with some gravel and quartz gravel; weak sub-angular blocky structure at top; massive and compact at base; pH 5.4
60		
80		
100		
120	B/C	Red clay loam; massive and compact; pH 5.9; merging to loose friable rotting bedrock

Ferruginous soil

where it becomes savanna scrub. By contrast, the biomass varies considerably (depending on whether it is largely grass or wood) with an average of 4 kg/m². Typical species in Africa include the acacia, palm and baobab trees (Figure 7.42) and elephant grass, which can grow to a height of over 5 metres. Trees grow to a height of about 12 metres and are characterised by flattened crowns and strong roots.

The nutrient cycle also illustrates the relationship between climate, soils and vegetation (Figure 7.43). The store of nutrients in the biomass is less than that in the rainforest because of the shorter growing season. Similarly, the store in the litter is small because of fire. Owing to fire, many of the nutrients are stored in the soil so that they are not burnt and leached out of the system. The role of fire, whether natural or caused by people, is very important (Figure 7.44). It helps to maintain the savanna as a grass community; it mineralises the litter layer; it kills off weeds, competitors and diseases; and prevents any trees from colonising relatively wet areas.

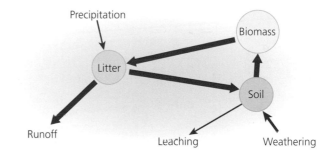

Figure 7.43 Savanna nutrient cycle

Figure 7.42 A baobab tree, Kruger National Park, South Africa

Figure 7.44 Fire is an important element of the savanna ecosystem

7.3 Humid tropical (rainforest) ecosystems and seasonally humid tropical (savanna) ecosystems

Other factors include the activities of animals. This includes locusts, for example, which can destroy large areas of grassland with devastating speed, and termites, which aerate the soil and break down up to 30 kilograms of cellulose per hectare each year. In some areas, up to 600 termite hills per hectare can be found, thus having a significant effect on the upper horizons on the soil.

The fauna associated with savannas is diverse. The African savanna has the largest variety of grazers (over 40), including giraffe, zebra, gazelle, elephants and wildebeest (Figure 7.45). Selective grazing allows a great variety of herbivores: for example, the giraffe feeds off the tops of the trees, the rhinoceros the lower twigs and gazelle the grass beneath the trees. These animals are largely migratory, searching out water and fresh pastures as the dry season sets in. A variety of carnivores including lions, cheetahs and hyenas are also supported.

Figure 7.45 Savanna fauna – zebra

Section 7.3 Activities

1 Outline the ways in which **a** savanna vegetation and **b** savanna fauna are adapted to seasonal drought.
2 Explain why fire is important in savanna ecosystems.

☐ Tropical rainforest soils

The soils of these tropical areas are usually heavily leached and **ferralitic**, with accumulations of residual insoluble minerals containing iron, aluminium and manganese (Figure 7.46). They are variously called **latosols**, **oxisols** and ferralitic soils. A distinction can be made between ferralitic soils (of the rainforest) and weathered ferralitic (ferruginous) soils that are found in savanna regions. The hot humid environment speeds up chemical weathering and decay of organic matter. This biome also covers ancient shield areas that have remained tectonically stable for a very long time and were unaffected by the Pleistocene glaciations. Not only are the soils well developed, but they have been weathered for a long time and are therefore lacking in nutrients and so are inherently infertile. More than 80 per cent of the soils have severe limitations of acidity, low nutrient status, shallowness or poor drainage. This is unusual given the richness of the vegetation that it supports. However, the nutrients are mainly stored in the biomass due to the rapid leaching of nutrients from the A horizon (see Figure 7.40). There is only a small store of nutrients in the litter or the soil itself.

Figure 7.46 Tropical red earth, Brunei

The rate of litter fall is high – 11 tonnes/hectare/year – and there is humus turnover of 1 per cent a day. At 25–30 °C, the breakdown and supply of litter is approximately equal. The rapid rates of decomposition and the rapid leaching of nutrients from the rooting area have led to an unusual adaptation in this ecosystem. The main agents of decay are fungi in **mycorrhizal** relationships with the tree roots. Nutrients are passed directly from the litter to the trees by the fungi (living on the tree roots). This bypasses the soil storage stage, when there is a strong chance that the nutrients will be lost from the nutrient cycle completely.

The rapid decay of litter gives a plentiful supply of bases. Clay minerals break down rapidly and the silica element is carried into the lower layers. Iron and aluminium sesquioxides, which are relatively insoluble, remain in the upper layers, as they require acidic water to mobilise them. These leached red or red–brown soils are termed ferralitic soils (Figure 7.40). Where it is wet, the iron may be hydrated and yellow soils develop. Deep weathering is a feature of these areas and the depth of the regolith may be up to 150 metres deep. Where a parent material allows free drainage and is poor in bases, such as a coarse sandstone, a tropical podsol will form. Catenas are very poorly developed under tropical rainforests.

These soils are not easy to manage. If they are ploughed, severe soil erosion may occur. Vegetation interrupts the nutrient cycle. In the rainforest, vegetation and soil are the major components in an almost closed nutrient cycle. The major store of plant nutrients is in the vegetation. The leaves and stems falling to the soil surface break down rapidly and nutrients are released during the processes of decomposition. These are almost immediately taken up by the plants. By contrast, the supply of nutrients from the underlying mineral soil is a small component. If the forest cover is removed, the bulk of the system's nutrient store is also removed. This leaves a well-weathered, heavily leached soil capable of supplying only low levels of nutrients.

Even when the forest is burnt, the nutrients held in the plant biomass store are often lost. During burning, there may be gaseous losses and afterwards rainfall may leach nutrients from the ash on the surface. In addition, the soils have a low cation-exchange capacity (CEC – see the following section). Unless a plant cover is rapidly established, most of the nutrients released from the plant biomass during burning will be lost within a short time. Thus, shifting cultivation can only take place for a few years before the overall fertility of the soil is reduced to such an extent that it is not worthwhile continuing cropping the plot. Indeed, farmers try to replicate the rainforest environment by intercropping. This provides shelter for the soil and protects it from the direct attack of intense rain (rainsplash erosion can otherwise be a serious problem). Compaction of the soil by heavy raindrops and the reduction of the infiltration capacity as a result will lead to overland flow and soil erosion even if there is only a slight slope.

Soils that are predominant in the region offer conditions only marginally suitable for most of these crops. They are often clayey textured, of low pH value, generally of less than medium fertility and offer only restricted rooting depths.

Changes to rainforest soils

Tropical rainforests are disappearing at an alarming rate and 'green jungles' are being changed into 'red deserts'. The loss of rainforest is up to 200 000 km²/year. By 2050, it is possible that there will be no extensive tracts of primary tropical rainforest, but simply isolated refuges of a few tens or hundreds of square kilometres. The tropical rainforest is a unique natural resource with a tremendous diversity of flora and fauna, much of which has still to be scientifically identified and studied.

To those who live within or close to the rainforest, the forest is a resource they are eager to exploit. To many economically marginal households, the land presently occupied by forest is seen as a way of improving their quality of their life, and to become self-sufficient in farming. Rainforests are areas of low population density, and in some areas are relatively unexploited. However, in some cases new farmers have little experience of the tropical environment. Some are the urban poor, while others are farmers familiar with very different environments. Table 7.5 shows the distribution of the different types of soil in the humid tropics.

Table 7.5 Distribution of the main types of soil in the humid tropics

	Million hectares
Acid, infertile soils	938
Moderately fertile, well-drained soils	223
Poorly drained soils	119
Very infertile sandy soils	104
Shallow soils	75
Total	**1459**

Source: S. Nortcliff, 'The clearance of the tropical rainforest', Teaching Geography, April 1987

Table 7.6 shows the characteristics of an oxisol (leached ferruginous soil) under tropical rainforest in Amazonas State, Brazil. The oxisol is deep and acidic (low pH). It has a low cation-exchange capacity and a low base saturation. The CEC is a measure of the soil's capacity to absorb and exchange positively charged ions (cations) such as potassium, calcium, magnesium, hydrogen and aluminium. The base saturation measures the proportion of the exchangeable cations that are bases (that is, not hydrogen and aluminium, which are acidic). A low CEC and low base saturation mean that the soil has a poor reserve of nutrients readily available to plants and that it also has an undesirable balance between ions for plant growth. Thus it is infertile. In the oxisol, the only horizon with a substantial CEC is the surface horizon, due to the presence of organic matter.

Table 7.6 Characteristics of an oxisol under tropical rainforest in Amazonas, Brazil

Depth (cm)	% organic matter	% sand	% silt	% clay	pH1 (H₂O)	CEC2 meq/100 g	BS3 %
0–8	3.4	10	15	75	3.4	18	2
8–18	0.6	11	11	78	3.7	8	4
18–50	0.5	7	8	85	4.2	4	5
50–90	0.3	7	4	89	4.5	3	5
90–150	0.2	7	3	90	4.7	3	5
150–170	0.2	5	3	92	4.9	2	5

Notes:
1   pH determined in a 1:2.5 soil/water ratio
2   CEC expressed as milli-equivalents per 100 g of soil
3   Percentage base saturation (sum of base cations/CEC 100).

Source: S. Nortcliff, 'The clearance of tropical rainforest', Teaching Geography, *April 1987*

Research near Manaus in Brazil investigated changes in the soil's physical characteristics that resulted when rainforest was cleared, first using traditional slash-and-burn techniques and second using a bulldozer (Table 7.7). Several observations of soil characteristics were made, including changes in soil surface, dry bulk density, moisture content and infiltration rate.

Table 7.7 Soil moisture contents – comparisons between uncleared forest, burned and bulldozed sites, Amazonas, Brazil

Depth below final soil surface (cm)	Virgin forest	Burnt	Bulldozed
	% soil moisture content		
0–10	52.1	40.8	40.8
10–20	44.4	40.4	39.4
20–40	41.4	40.4	38.6

On the site cleared by bulldozer, the change in soil surface height ranged from 2 to 9 centimetres, with an average of 5.7 centimetres. Thus much of the topsoil horizon was removed, leaving a denser subsurface horizon at or very close to the surface. The removal of the topsoil removes much of the soil organic matter, which is often the major store of plant nutrients within the soil.

There were increases in the soil dry bulk density resulting from the passage of heavy machinery over the surface, causing changes in the infiltration rate (the rate at which water can enter the soil) from over 200 centimetres/hour under uncleared forest, to 192 centimetres/hour at the slash-and-burn site and to 39 centimetres/hour at the bulldozed site.

The moisture content of the 0–10 centimetres and 10–20 centimetres layers of both burned and bulldozed sites were similar, and were significantly lower than the uncleared forest site. These differences reflect the removal of the organic matter during clearance. The organic matter acts both as a store of moisture and as a natural mulch restricting moisture loss.

Crop yields were higher in burnt plots because of the nutrient content of the ash (Table 7.8). The ash caused major changes in soil conditions. Soil acidity decreased and, compared with the bulldozed site, the organic matter was higher (although it was lower than that of the uncleared forest). The importance of ash to soil fertility and crop yield following clearance is substantial, especially in soils of low fertility.

Table 7.8 Crop response to forest clearance and different soil fertiliser applications, Vurimaguas in Peru

Fertility level	Slash and burn	Bulldozed	Bulldozed/ Burnt
	Yield tonnes per hectare		
Maize (grain yield)	0	0.1	0
NPK	0.4	0.04	10
NPKL	3.1	2.4	76
Soyabeans (grain yield)	0	0.7	0.2
NPK	1.0	0.3	34
NPKL	2.7	1.8	67
Cassava (fresh root yield)	0	15.4	6.4
NPK	18.9	14.9	78
NPKL	25.6	24.9	97

Fertiliser applications:
N = 50 kg nitrogen/ha
P = 172 kg phosphorus/ha
K = 40 kg potassium/ha
L = 4 tonnes lime/ha

Sites labelled 0 received none of the above fertiliser applications.
Sites labelled NPK received applications of nitrogen, phosphorus and potassium.
Sites labelled NPKL received all applications.

Forest clearance in the tropics will continue in order to satisfy the demands of the growing population (see Topic 4, Section 4.1). The priority therefore should be to slow down the rate of clearance. This can be done by making the most effective use of the cleared land and by limiting the need for forest clearance to replace land cleared at an earlier stage. These aims can be achieved by:

- increasing the productivity of land already cleared by selection of suitable crop and land-management combinations
- minimising the damage that results from forest clearing methods by adopting methods of clearance and timing of clearance that produce the least detrimental effects
- restoring eroded or degraded land by the establishment of more appropriate land-management systems.

1. Study Table 7.5. Explain why tropical rainforests have some of the world's most luxuriant vegetation and yet some of the world's least fertile soils.
2. Using examples, examine the effects of human activities on tropical soils.
3. With the use of an example, describe and explain the formation of a soil catena.

7.4 Sustainable management of tropical environments

Case Study: The Heart of Borneo

The forests of Borneo have suffered hugely in recent decades from rampant logging, slash-and-burn farming and cutting for oil palm and rubber plantations (Figure 7.47). The island's rich lowland forests have nearly vanished, with rates of forest loss still among the highest on Earth. There has also been widespread hunting. At Lambir Hills National Park in Sarawak, for example, half of the primate species, six of seven hornbill species and nearly all of the endangered mammal species have been hunted out. With the disappearance of its key seed dispersers, the fruits of many trees now just rot on the forest floor.

Figure 7.47a Rainforest converted to arable farming

Figure 7.47b Rainforest flooded to create HEP scheme

Figure 7.47c Cut trees choke a river

In 2007, the three governments of Borneo (Indonesia, Malaysia and Brunei) signed the Heart of Borneo (HoB) Declaration. The HoB was established in 2010 to protect the ecological, biological and cultural features of the Borneo rainforest. The Iban long house is part of the cultural tradition of Borneo (Figure 7.48). The HoB is largely in the central parts of Borneo (Figure 7.49), where the rainforest is mainly still intact. It covers 22 million hectares of rainforest, including some of the most biologically diverse habitats on Earth.

The forests of the HoB provide many benefits to the industries and communities of Borneo, such as clean water supplies, carbon sequestration, biodiversity benefits, economic revenues and important cultural services (Figure 7.50).

$\Rightarrow$

Figure 7.48 Iban long house

Figure 7.49 The location of the HoB

Figure 7.50 Food from the rainforest on sale at a market

Figure 7.51 Tourists boating – many rainforest sites are accessible only by river

Ecological services of the Heart of Borneo

Of the 20 major rivers in Borneo, 14 have their source in the mountainous forests of the HoB. The forests and peat lands of Borneo are particularly important because they are very effective carbon stores, with an average of 230 tonnes per hectare in above-ground biomass, and 2 400 tonnes per hectare in below-ground peat soils; most of this is released by deforestation and land degradation. Deforestation in Indonesia and Malaysia currently accounts for more than 80 per cent of their total carbon emissions, representing more than 2.5 Gt CO_2 per year; this is equivalent to almost four times the annual emissions from the global aviation industry.

Borneo has a rich biodiversity: more than 350 species of bird, 150 species of reptile and 15 000 flowering plants. Indeed, for every 10 hectares, there are over 700 species of trees.

The economic value of the HoB forests is equally significant. The HoB has rich natural resources in the form of timber and other forest products. If managed sustainably, these resources have the potential to provide continuous and long-term income. Sustainable forest management also maintains the flow of ecosystem services, such as water provision, pollination and local climate regulation.

Communities living within the forest depend on it for food, water, medicine and construction materials. Tourism also brings in a lot of money (Figure 7.51). In 2009, Sabah in Malaysia alone recorded more than 2 million arrivals, generating an estimated US$1.2 billion in tourism receipts.

Table 7.9 Extent of forestry, palm oil and coal mining concessions within the HoB

	Kalimantan (ha)	Malaysia (ha)	Brunei (ha)	Total (ha)
Palm oil	830 000	770 000	–	1 600 000
Palm oil RSPO certified	–	12 000	–	12 000
Forestry	2 600 000	3 200 000	138 000	5 938 000
FSC certified	424 000	215 000	–	639 000
Mining	1 100 000	–	–	1 100 000

Table 7.10 Potential environmental issues that can arise from poor management of forestry activities

Habitat loss	Logging natural forests	Clear-felling removes the whole forest ecosystem, resulting in severe reduction or complete loss of habitat and ecosystem values. The impacts associated with selective logging are less adverse, but can nevertheless be significant if managed poorly.
	Plantations	Plantations demand large areas, and their monocultures have very limited biodiversity. If plantations are built by clear-felling natural forest they lead to a severe loss of habitat and ecosystem values.
Carbon emissions		Borneo's forests are a carbon sink of global importance; however, deforestation releases this carbon. Fire is often used to clear forest, which can spread uncontrollably. The 1997–98 fires burnt 9.7 million ha of land, releasing huge quantities of carbon dioxide. Plantations on cleared land only sequester a small proportion of these emissions.
Land erosion	Logging natural forests	Clear-felling exposes the land to soil erosion, and poorly implemented selective logging can also result in serious soil degradation. Heavy rain and wind removes exposed and disturbed topsoil, rendering the land less productive for agriculture and severely impacting the prospects of forest regeneration in the area.
	Plantations	Without proper management, plantations suffer from soil erosion and land degradation, especially where large expanses of land are cleared and then not subsequently planted.
Degradation of watercourses	Logging natural forests	The clearance of watershed forest cover can degrade the quality of watercourses, leading to unpredictable and severe flash floods and mudslides that endanger downstream settlements.
	Plantations and processing operations	Plantations often use chemical fertilisers and pesticides. In addition, effluent waste from paper processing can contain high concentrations of bleaches that are toxic if not properly treated. This can leach into groundwater, affecting drinking supplies for the local communities and downstream urban areas. The overuse of water and diversion of watercourses can lead to shortages elsewhere.
Social issues		The allocation of logging and plantation concessions does not always take into consideration the traditional land rights of indigenous and other communities. These communities may use the land for crops and fruit trees, or for social activities. Logging and plantation expansion can result in conflict and displacement.

Source: WWF, Business solutions: delivering the Heart of Borneo declaration

However, growing populations and international demand have led to greater production from the palm oil, forestry and mining industries (amongst others), putting increasing pressure on Borneo's forests (Table 7.9). Moreover, the states of Borneo remain some of the poorest in the region, with an estimated 23 per cent of the population living below the poverty line in Sabah, for example.

Sustainable forestry

Changes in the production of Indonesia's and Malaysia's timber industry illustrate the need for an urgent shift towards a sustainable extraction model. The forestry sector manages the most land of any sector operating inside the HoB, and therefore has the greatest opportunities for sustainable use, but also the greatest risks in the absence of good practice (Table 7.10).

Sustainable logging of natural forests is a good example of how the standing forests can provide long-term revenues while maintaining a large proportion of the forests' values.

Plantation area is increasing across Borneo to meet the growing demand for timber and fibre for paper mills. It is essential that the expansion of mill capacity is matched by commensurate increases in sustainable plantations.

Sustainable forestry – the way forward:

- Logging activities should be avoided in high conservation-value forests; and elsewhere, reduced impact logging and sustainable forest management should be implemented to minimise environmental impacts.
- Plantations should not replace high conservation-value forests, and should instead be cultivated on the available idle land.

- Investors, traders and consumers should help drive **sustainable management** through financing and sourcing Forest Stewardship Council (FSC)-certified production.

Palm oil

The palm oil industry in Borneo has undergone rapid growth, and continues to expand to meet growing world demand. Indonesia's and Malaysia's palm oil production amounts to 85 per cent of the global supply, and production in Borneo in 2008 was 16.5 million tonnes, representing more than a third of this global supply. Oil palm plantations in Sabah grew from almost nothing in the mid 1980s to covering nearly 20 per cent of Sabah's land by 2010.

Palm oil plantations require the complete conversion of land use; if concessions are placed in high conservation-value areas, it can result in a significant loss of ecosystem value (Table 7.11). The challenge for the governments' vision enshrined in the HoB Declaration is to ensure that as the cultivated area increases, adequate protection is given to the HoB.

Palm oil – the way forward:

- Future revenues from the industry can be maintained and even increased by concentrating on increasing productivity, particularly amongst smallholders; by expanding plantations on idle lands; and by developing downstream processing industries. This adds value without increasing pressure to convert natural forests.
- Planners should ensure concessions are not allocated in high conservation-value areas of the HoB, but rather on idle land with low conservation values. ⇨

Table 7.11 Potential environmental issues that can arise due to inappropriate choice of locations or poor management of palm oil plantations

Habitat loss	Conversion of high conservation-value areas and their ultimate replacement with palm oil plantations results in reduced habitat and the loss of 80–90% of species, many of which may be endemic or threatened. Orang-utan habitat declined 39% in 1992–2002.
Carbon emissions	Borneo's forests are a carbon sink of global importance; deforestation releases this carbon, contributing to global climate change.
Fire	Despite its use being illegal across Borneo, fire is still used to clear forests. In 1997–98, fires burnt 9.7 million ha of land, releasing huge quantities of carbon dioxide; and fires can still be severe – most recently in May 2010.
Watershed degradation	Palm oil plantations often use chemical fertilisers and pesticides. Inappropriate use can result in polluting runoff entering watercourses and contaminating groundwater through leaching. This in turn can pollute drinking water for downstream communities and adversely affect aquatic wildlife and fishing yields. Inappropriate irrigation and diversion of watercourses can also lead to water shortages.
Land degradation	Deforestation, forest fires and peatland drainage expose the land to soil erosion. If land is left uncultivated, or is not effectively managed, soil erosion and land degradation can occur, particularly on sloped land. Heavy rain and wind remove topsoil, rendering the land less productive for agriculture and reducing the chance of forest regeneration. This can increase the frequency and severity of unpredictable flash floods, threatening lives, infrastructure and the environment.
Social issues	The allocation of palm oil concessions does not always take into consideration the traditional land rights of indigenous and other communities. These communities may use the land for crops and fruit trees, or for social activities. Plantation expansion can result in conflict and displacement.

- A shift to sustainable production, independently certified through the Roundtable for Sustainable Palm Oil (RSPO), will result in the improved environmental performance of existing and new plantations. Food suppliers and consumers are putting pressure on palm oil producers to guarantee that their palm oil is from sustainable sources. The RSPO was formed in 2004 to promote the growth and use of sustainable palm oil products through credible global standards and engagement of stakeholders. RSPO is a not-for-profit group that unites palm oil producers, processors and traders; consumer goods manufacturers; retailers; banks; investors; and environmental, social and development NGOs.
- Another approach has been adopted at the Yayasan Sabah forest. This was meant to be sustainably logged over an 80-year period. Instead, about 75 per cent of it was intensively logged. The director of forestry for the state of Sabah is trying to reverse the region's trend toward deforestation, moving forward with a plan to convert 10 per cent of Yayasan Sabah to palm oil plantations to generate revenue, and then gradually restore most of the rest of the tract so it can be sustainably logged. Palm oil plantations have been established on 100 000 hectares of land cleared for the pulp plantations. The revenue created by palm oil should generate the needed funds for Sabah's social programmes. The long-term plan is to restore the forests to a state of health where they can be logged again, albeit under the more stringent guidelines of the FSC.

Mining in Borneo

The Indonesian and Malaysian governments are both considering increasing their mining production, especially that of coal. In the short term, coal will remain an important and relatively low-cost source of energy for developing countries.

Legal and illegal coal and gold mining has significant social and environmental impacts (Table 7.12).

Table 7.12 Potential environmental issues that can arise from poor management of mining activities

Habitat loss	Open-cast mining is land intensive and requires the removal of large areas of terrestrial habitat and the loss of associated ecosystem value. There are also often secondary impacts that lead to habitat loss associated with both open-cast and underground mining activities. These include direct habitat removal and habitat fragmentation for construction of access roads or rail infrastructure.
Land removal and soil degradation	Large volumes of soil and overburden are extracted and processed in mining operations, and these can generate contaminated tailings as by-products. This can result in soil degradation, erosion and contamination, and also generate 'geohazards' such as subsidence and landslides.
Degradation of watercourses	Mine effluent can adversely affect water quality by increasing sedimentation in local watercourses and introducing contaminants. Even low levels of mercury and cyanide (used in gold processing) are toxic to most forms of wildlife and humans. Tailings (materials left over from mining and quarrying) can form acids through oxidisation that leach into the groundwater and enter watercourses.
Social conflict, health and displacement	Mining operations (both legal and illegal) attract large influxes of workers and associated temporary settlements and informal economies. This can encourage the spread of communicable diseases (e.g. HIV-AIDS) and diseases that thrive in poor-quality living quarters for workers (which can also spread to local communities). There is often high workforce turnover, caused in part by adverse health effects of mercury and cyanide where these are used. Mining activities can in some cases displace both indigenous and local communities, resulting in conflicts with mining companies over security and land rights.

Mining in Borneo – the way forward:

- Clear regulation and effective enforcement is needed across the region, for example of Environmental Impact Assessments and reclamation of land. Increased efforts are needed to control illegal mining and reduce the use of mercury by gold miners.
- Mining companies should identify high conservation-value forests before starting mining operations and ensure an adequate management plan is put in place to protect the value of the area during mining and after it is completed.
- Mine rehabilitation needs to be planned and implemented.

Conclusion

Success has been varied in terms of protecting areas of tropical rainforest. Sabah is still far ahead of Sarawak and Kalimantan, since surviving primary forest areas are being conserved; reforestation and forest restoration is happening; and encroachers have moved out of forest reserves. However, there is uncertainty as to whether a change in government may lead to a change in policies. Gains from conservation and reforestation could be reversed. Moreover, population growth and changes in standards of living will continue to put pressure on the forests of Borneo, and the land of Borneo, to provide for its people.

Source: The information in this Case Study has come from WWF, Business solutions: delivering the Heart of Borneo declaration

Section 7.4 Activities

1. Outline the consequences of the conversion of rainforest to palm-oil plantations.
2. Comment on the impacts of mining in rainforest areas.
3. Evaluate the attempts to manage Borneo's rainforest in the HoB.

Case Study: Sustainable agroforestry, Santa Rosa rainforest, Mexico

The Popoluca Indians of Santa Rosa in the Mexican rainforest practise a form of agriculture that resembles shifting cultivation, known as the **milpa system** (Table 7.13). This is a labour-intensive form of agriculture, using **fallow**. It is a diverse form of **polyculture** with over 200 species cultivated, including maize, beans, cucurbits, papaya, squash, water melon, tomatoes, oregano, coffee and chilli. The Popolucas have developed this system that mimics the natural rainforest. The variety of a natural rainforest is replicated in the form of shifting cultivation, for example lemon trees, peppervine and spearmint are **heliophytes** (light-seeking plants), and prefer open conditions, not shade. By contrast, coffee is a **sciophyte** (shade-tolerant plant), while the mango tree requires damp conditions.

The close associations that are found in natural conditions are also seen in the Popolucas' farming system, for example maize and beans grow well together, as maize extracts nutrients from the soil whereas beans return them to the soil. Tree trunks and small trees are left because they are useful for many purposes, such as returning nutrients to the soil and preventing soil erosion. They are also used as a source of material for housing and hunting spears, and for medicines.

As in a rainforest the crops are multi-layered, with tree, shrub and herb layers. This increases NPP per unit area, because photosynthesis is taking place on at least three levels, and soil erosion is reduced, as no soil is left bare. Most plants are self-seeded and this reduces the cost of inputs. The Popolucas show a huge amount of ecological knowledge and management. In all, 244 species of plant are used in their farming system. Animals include chickens and turkeys. These are used as a source of food, for barter and in exchange for money, and their waste is used as manure. Rivers and lakes are used for fishing and catching turtles. Thus it is not entirely a subsistence lifestyle, since wood, fruit, turtles and other animals are traded for some seeds, mainly maize.

Pressures on the Popolucas

About 90 per cent of Mexico's rainforest has been cut down in recent decades, largely for new forms of agriculture. This is partly a response to Mexico's huge international debt and attempts by the government to increase its agricultural exports and reduce its imports. The main new forms of farming are:

- **cattle ranching** for export, and
- **plantations** of cash crops, such as tobacco, also for export.

However, these new methods are not necessarily suited to the physical and economic environment. Tobacco needs protection from too much sunlight and excess moisture, and the soil needs to be very fertile. The cleared rainforest is frequently left bare and this leads to soil erosion. Unlike the milpa system, the new forms of agriculture are very labour intensive. Pineapple, sugar cane and tobacco plantations require large inputs of fertiliser and pesticides. Inputs are expensive and the costs are rising rapidly.

Ranching prevents the natural succession of vegetation, because there is a lack of seed from nearby forests and the cattle graze off young seedlings. Grasses and a few legumes become dominant. One hectare of rainforest supports about 200 species of trees and up to 10 000 individual plants. By contrast, 1 hectare of rangeland supports just one cow and one or two types of grass. However, it is profitable in the short term because land is available, and it is supported by the Mexican government. $\Rightarrow$

Extensive **monoculture** is increasingly mechanised, and uses large inputs of fertiliser, pesticides and insecticides. However, it is very costly and there are problems of soil deterioration and microclimatic change. Yet there is little pressure to improve efficiency because it is easy to clear new forest.

The Mexican rainforest can be described as a 'desert covered by trees'. Under natural conditions it is very dynamic, but its resilience depends on the level of disturbance. Sustainable development of the rainforest requires the management and use of the natural structure and diversity; that is, local species, local knowledge and skills, rather than a type of farming that has been developed elsewhere and then imported.

Table 7.13 A comparison between the milpa system and the new forms of agriculture

	Milpa system	Tobacco plantation or ranching
NPP	High, stable	Declining
Work (labour)	High	Higher and increasing
Inputs (clearing and seeding)	Few	Very high: 2.5–3 tonnes fertiliser/ha/year

Crops	Polyculture (244 species used)	Monoculture (risk of disease, poor yield, loss of demand and/or overproduction)
Yield (compared to inputs)	200%	140% (at best)
Reliability of farming system	Quite stable	High-risk operation
Economic organisation	Mainly subsistence	Commercial
Money	None/little	More
Carrying capacity (livestock)	Several families/4 ha	1 family on a plantation (200 ha)
Ranching	1 ha of good land = 1 cow	20 ha of poor land = 1 cow

Section 7.4 Activities

Compare the Popolucas' methods of farming with the ecosystem of the natural tropical rainforest. What lessons can be learnt from this?

Case Study: CAMPFIRE, Zimbabwe – sustainable management in savanna areas

Almost 5 million people live in communal lands that cover almost half of Zimbabwe. Most of this land can be described as savanna land. CAMPFIRE (Communal Areas Management Programme for Indigenous Resources) is a programme designed to assist rural development and conservation. It works with the people who live in communal lands, supporting the use of wildlife as an important natural resource. CAMPFIRE is helping people in these areas manage the environment in ways that are both sustainable and appropriate.

National Parks

Approximately 12 per cent of Zimbabwe is protected as National Parks or conservation areas, such as Hwange National Park and the Matopas. Many local people were evicted from their homes when the Parks were created. Most now live in the surrounding communal lands. They are no longer permitted to hunt the animals or harvest the plants now found inside protected areas. However, animals frequently roam outside Park boundaries, destroying crops and killing livestock and sometimes people. This has created much conflict between local people and National Park staff, often resulting in illegal hunting.

Raising awareness and raising money

The CAMPFIRE movement began in the mid-1980s. It encourages local communities to make their own decisions about wildlife management and control. It aims to help people manage natural resources so that plants, animals and people – the whole ecosystem – all benefit. It helps provide legal ways for such communities to raise money by using local, natural resources in a sustainable way. As a result, many communities now actively protect local wildlife, seeing it as a valuable asset. In some areas, local people have even provided animals with emergency food and water in times of shortage.

Five particular activities help provide extra income to local communities:

- **Trophy hunting** – About 90 per cent of CAMPFIRE's income comes from selling hunting concessions to professional hunters and safari operators working to set government quotas. Individual hunters pay high fees to shoot elephant (US$15 000) and buffalo and are strictly monitored, accompanied by local, licensed professionals. Cecil the lion, a 13-year-old male, was a major tourist attraction in Hwange National Park (Figure 7.52). He was shot and killed in 2015, by a North American dentist. The killing caused an international outrage. **Trophy hunting** is considered to be the ultimate form of ecotourism as hunters usually travel in small groups, demand few amenities and cause minimal damage to the local ecosystem, yet provide considerable income.
- **Selling live animals** – This is a fairly recent development. Some areas with large wildlife populations sell live animals to National Parks or game reserves. For example, Guruve district raised US$50 000 by selling ten roan antelope.
- **Harvesting natural resources** – A number of natural resources, for example crocodile eggs, caterpillars, river-sand and timber are harvested and sold by local communities. Skins and ivory can be sold from 'problem animals' (individual animals that persistently cause damage or are a threat can legally be killed).

- **Tourism** – In the past, most revenue from tourists has not gone to local communities. During the 1990s, pilot projects were set up and five districts now benefit from tourism. There has been development of specialist tourist areas, such as culture tourism, bird watching and visits to hot springs. Some local people are employed directly as guides, or run local facilities for tourists.
- **Selling wildlife meat** – Where species are plentiful, for example impala, the National Parks Department supervises killing and selling of skins and meat. However, this raises relatively small sums of money.

Figure 7.52 Cecil the lion

Organisation

Each village taking part in the CAMPFIRE project has a wildlife committee responsible for counting animals, anti-poaching activities, conflicts that arise through 'problem animals' and environmental education. Game scouts are trained to help stop poaching and manage wildlife.

Quotas

For hunting concessions to be granted and wildlife managed sustainably, local communities need to monitor their wildlife populations and manage their habitats, protecting them from poaching or alternative forms of land use, for example farming. Every year, the Department of National Parks helps to estimate the wildlife population totals so that sustainable quotas can be set.

Tour operators must, by law, keep detailed records of animals killed – their size, weight, length of certain animals and/or horns and tusks. This helps check that young animals are not being taken, which would put future numbers at risk. New quotas are not issued until operators produce these records for analysis by the Department for National Parks.

Up to 80 per cent of the money raised is given directly to local communities, which collectively decide how it should be spent. Money is used for the community, for example for building and equipping clinics and schools, constructing fences, drilling wells and building roads. In bad years, usually drought years, money may be used to buy maize and other foodstuffs. Since 1989, over 250 000 Zimbabweans have been involved in CAMPFIRE projects.

In 1980, Binga District in north-west Zimbabwe had just 13 primary schools, and most of its people lived in poverty. Money from hunting concessions, fishing and tourism was used by Sinkatenge village (near Matusadona National Park) to build a 12 kilometre length of electric fencing to enclose their fields, preventing animals from trampling their crops and providing full-time work for two local people to maintain it. Today, partly as a result of income from CAMPFIRE projects, the District has about 80 primary and around 40 secondary schools, and several health clinics and wells.

Masoka in the north-east was one of the first to join CAMPFIRE. Local people now receive more than four times their previous income via hunting concessions, using it to buy maize and other food in drought years, building a clinic, buying a tractor and funding their football team. For the first time here, local rural women were employed, working on CAMPFIRE projects. CAMPFIRE is also actively encouraging women to participate in community decision-making, something that has been traditionally dominated by men. Women have also been encouraged to attend workshops and take part in training schemes.

Nyaminyama District, on the southern edge of Lake Kariba, is introducing land-use zoning, with specific areas for wildlife conservation, tourism, crocodile breeding and hunting. A recent WWF report estimated that CAMPFIRE has increased incomes in communal areas by up to 25 per cent.

The future

There are many advantages of CAMPFIRE's activities in Zimbabwe:

- It creates jobs – local people are trained and become involved as environmental educators, game scouts, and so on.
- It prompts environmental education and promotes the benefits of wildlife conservation to communities.
- It provides an incentive for people to conserve wild species.
- It generates funds, which are used for community projects or to supplement household incomes.
- Communal lands can act as game corridors between existing National Parks, protecting the genetic diversity of wild species.

Rural communities benefit from secure land tenure and rights over their wildlife. The ability of CAMPFIRE to assist wildlife conservation in Zimbabwe depends on two broader factors:

- the acceptance of hunting as a wildlife management tool by the international community
- placing economic value on wild species.

Section 7.4 Activities

1 Briefly explain the difficulties in developing tropical environments.
2 With the use of examples, outline opportunities for sustainable development in tropical environments.

8 Coastal environments

☐ Introduction

Coastal environments are influenced by many factors, including physical and human processes. As a result, there is a great variety in coastal landscapes (Figure 8.1). For example, landscapes vary on account of:

- **lithology** (rock type) – hard rocks such as granite and basalt give rugged landscapes, for example the Giant's Causeway in Northern Ireland, whereas soft rocks such as sands and gravels produce low, flat landscapes, as around the Nile delta, for example
- **geological structure** – concordant (Pacific) coastlines occur where the geological strata lie parallel to the coastline, for example the south coast of Ireland or the Pacific coast of the USA, whereas discordant (Atlantic-type) coastlines occur where the geological strata are at right-angles to the shoreline, for example the south-west coastline of Ireland
- **processes** – erosional landscapes, for example the east coast of England, contain many rapidly retreating cliffs, whereas areas of rapid deposition, for example the Netherlands, contain many sand dunes and coastal flats
- **sea-level changes** – interaction with erosional and depositional processes produce advancing coasts (those growing, either due to deposition and/or a relative fall in sea level) or retreating coasts (those being eroded and/or drowned by a relative rise in sea level).
- **human impacts** (increasingly common) – some coasts, for example in Florida, are extensively modified, whereas others are more natural, for example Norway and Iceland
- **ecosystem types** – mangrove, coral, sand dune, saltmarsh and rocky shore, for example, add further variety to the coastline.

☐ Coastal zones

The coastal zone includes all areas from the deep ocean, which may lie beyond political jurisdiction (up to 320 kilometres offshore), to 60 kilometres inland. The inland areas may affect coastal areas through sediment supply and pollution sources, as well as being affected by coastal processes such as land–sea breezes. At the coast, there is the **upper beach** or **backshore** (backed by cliffs or sand dunes), the **foreshore** (periodically exposed by the tides) and the **offshore** area (covered by water).

The coastal zone is a dynamic area, with inputs and processes from land, sea and the atmosphere so is, geologically speaking, an area of very rapid change.

Figure 8.1 Boats converging in a cove, Capri, Italy

8.1 Coastal processes

☐ Wave generation and characteristics

Waves result from friction between wind and the sea surface. Waves in the deep, open sea (waves of oscillation) are different from those breaking on shore. Waves of oscillation are forward surges of energy. Although the surface wave shape appears to move, the water particles actually move in a roughly circular orbit within the wave (Figure 8.2).

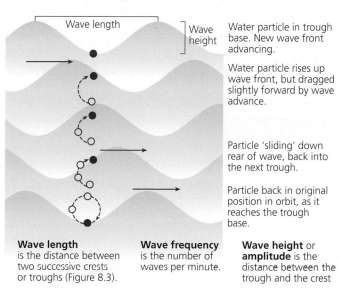

Water particle in trough base. New wave front advancing.

Water particle rises up wave front, but dragged slightly forward by wave advance.

Particle 'sliding' down rear of wave, back into the next trough.

Particle back in original position in orbit, as it reaches the trough base.

Wave length is the distance between two successive crests or troughs (Figure 8.3).

Wave frequency is the number of waves per minute.

Wave height or **amplitude** is the distance between the trough and the crest

Figure 8.2 Water movement

The **wave orbit** is the shape of the wave. It varies between circular and elliptical. The orbit diameter decreases with depth, to a depth roughly equal to wavelength (the distance between neighbouring crests or troughs), at which point there is no further movement related to wind energy – this point is called the **wave base**.

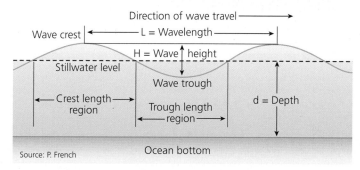

Source: P. French

Figure 8.3 Wave terminology

Wave height is an indication of **wave energy**. It is controlled by wind strength, **fetch** (the distance of open water a wave travels over) and the depth of the sea. Waves of up to 12–15 metres are formed in open sea and

can travel vast distances away from the generation area, reaching distant shores as **swell waves**, characterised by a lower height and a longer wavelength. In contrast, **storm waves** are characterised by a short wavelength, greater height and high frequency.

Waves reaching the shore are known as **waves of translation**. As waves move further onshore, the wave base comes into contact with the seabed. Friction slows down the wave advance, causing the wave fronts to crowd together. Wavelengths are reduced and the wave height increases. The shortening of the wave causes an increase in wave height – this process is known as **wave shoaling**. Thus a **breaker** is formed.

There are three main types of **breaking waves** (Figure 8.4):

- **Spilling breakers** are associated with gentle beach gradients and steep waves (wave height relative to wave length). They are characterised by a gradual peaking of the wave until the crest becomes unstable, resulting in a gentle spilling forward of the crest.
- **Plunging breakers** tend to occur on steeper beaches than spilling breakers, with waves of intermediate steepness. They are distinguished by the shore-ward face of the wave becoming vertical, curling over and plunging forward and downward as an intact mass of water.
- **Surging breakers** are found on steep beaches with low steepness waves. In surging breakers, the front face and crest of the wave remain relatively smooth and the wave slides directly up the beach without breaking. In surging breakers, a large proportion of the wave energy is reflected at the beach.

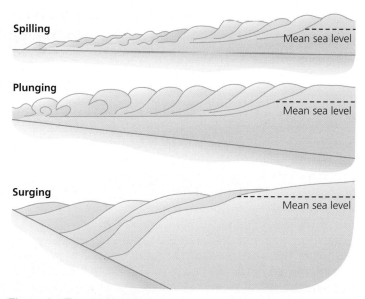

Figure 8.4 Types of breaker

Once the breaker has collapsed, the wave energy is transmitted onshore as a wave of translation. The **swash** will surge up the beach, with its speed gradually lessened by friction and the uphill gradient. Gravity will draw the water back as the **backwash**. There are two basic types of wave translation – **constructive** and **destructive waves**.

Constructive waves tend to occur when wave frequency is low (6 to 8 arriving onshore per minute), particularly when these waves advance over a gently shelving sea floor (formed, for example of fine material, such as sand; Figure 8.5). These waves have been generated far offshore. The gentle offshore slope creates a gradual increase in friction, which will cause a gradual steepening of the wave front. Thus a spilling breaker is formed, where water movement is elliptical. As this breaker collapses, the powerful constructive swash surges up the gentle gradient. Because of its low frequency, the backwash of each wave has time to return to the sea before the next wave breaks – the swash of each wave is not impeded and retains maximum energy.

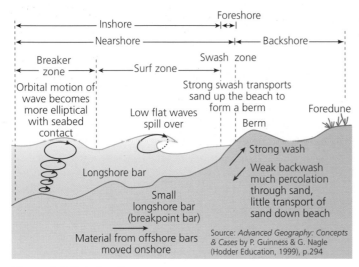

Figure 8.5 Constructive waves

Destructive waves are the result of locally generated winds, which create waves of high frequency (12 to 14 per minute; Figure 8.6). This rapid approach of the waves, particularly if they are moving onshore up a steeply shelving coastline (formed from coarse material such as gravel or shingle), creates a rapid increase in friction and thus a very steep, plunging breaker where water movement is circular. Due to the rapid steepening and curling of the wave breaker, the energy of the wave is transmitted down the beach (on breaker collapse), accelerated by the steeper gradient, and so the wave becomes destructive, breaking down the beach material.

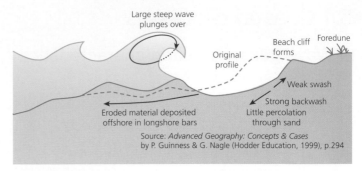

Figure 8.6 Destructive waves

Waves are dominant in some coastal environments, whereas in others it is the tide or winds. This has an important impact on the landforms that are found there (Table 8.1).

Table 8.1 Processes and landforms in coastal environments

Wave dominated	Tide dominated	Wind dominated
Shore platforms	Mudflats	Sand dunes
Cliffs	Sandflats	
Beaches	Saltmarshes	
Spits, tombolos	Mangroves	
Deltas	Deltas	
High energy	**Low energy**	**High energy**

Section 8.1 Activities

1 Using annotated diagrams, outline the main differences between constructive and destructive waves.
2 Explain the meaning of the terms *wave frequency*, *wave height* and *fetch*.

☐ Tides and the tidal cycle

Tides are regular movements in the sea's surface, caused by the gravitational pull of the Moon and Sun on the oceans. The Moon accounts for the larger share of the pull. **Low spring tides** occur just after a new Moon, whereas **high spring tides** occur after a full Moon. They occur when the Sun and the Moon are aligned. **Neap tides** occur when the Sun and Moon are at right-angles to the Earth. Tides are influenced by the size and shape of ocean basins, the characteristics of the shoreline, the Coriolis force and meteorological conditions. In general:

■ tides are greatest in **bays** and along funnel-shaped coastlines
■ in the northern hemisphere, water is deflected to the right of its path
■ during low pressure systems, water levels are raised 10 centimetres for every decrease of 10 millibars.

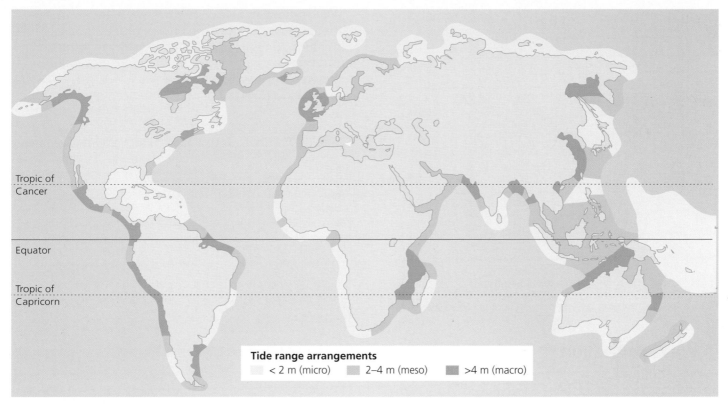

Figure 8.7 Tidal environments/ranges

The difference between high and low tide is called the **tidal range**. This varies from almost nothing in enclosed seas such as the Mediterranean to almost 15 metres in the Bay of Fundy, Canada. Tidal range varies with distance from the **amphidromic point** (place where there is no tidal range) and according to the shape of the coast; the strength of tidal currents varies enormously. If the coast is funnelled, as the tide advances it is concentrated in an ever narrowing space. Therefore its height rises rapidly, producing a **tidal bore**. A good example is the Severn Bore, which occurs in the Severn Estuary between Wales and England as a wave of up to 1 metre in height travelling at a speed of up to 30 kilometres/hour.

Coastal areas can be classified into **microtidal**, which have a very low tidal range (less than 2 metres), **mesotidal** (2–4 metres) and **macrotidal** (over 4 metres) (Figure 8.7). The tidal range has important influences on coastal processes:

- It controls the vertical range of erosion and deposition.
- Weathering and biological activity is affected by the time between tides.
- Velocity of tidal flow is influenced by the tidal range and has an important scouring effect.

Rip currents are important for transporting sediment. They can be caused by tidal motion or by waves breaking along a shore. A cellular circulation is caused by differences in wave height parallel to the shore. Water from the higher sections of the breaker travels further up the shore and returns back through the points where lower sections have broken. Once rip currents are formed, they modify the beach by creating cusps, which perpetuate the currents.

☐ Storm surges

Storm surges are changes in the sea level caused by intense low-pressure systems and high wind speeds. For every drop in air pressure of 10 millibars, sea water is raised 10 centimetres. During tropical cyclones, low pressure may be 100 millibars less than normal, raising sea level by 1 metre. In areas where the coastline is funnel-shaped, this rise in level is intensified. During high tides, the results can be devastating (Figure 8.8). Surges are common in the Bay of Bengal, on the south-east coast of the USA and in Japan. They are particularly hazardous in low-lying areas.

The Ganges delta experiences many storm surges. These may exceed 4 metres and the accompanying storm waves can add a further 4 metres to the wave height. The funnel shape of the Bay of Bengal forces water to build up. Seven of the nine worst storms during the last century affected Bangladesh. In 1970, over 300 000 people were killed in a surge. A further 225 000 people were killed in 1989, and in 1991 another 140 000 were killed in surges. In addition, millions were killed by the diseases and famines that followed, even more were made homeless and vast numbers of cattle were killed.

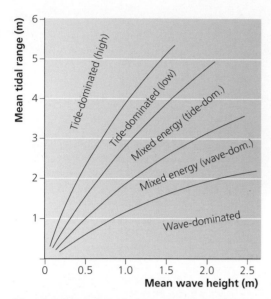

Figure 8.8 Relationship between mean tidal range and wave height

☐ Wave refraction

As wave fronts approach the shore, their speed of approach is reduced as the waves 'feel bottom'. Usually, due to the interaction between onshore wind direction (and therefore direction of wave advance) and the trend of the coast, the wave fronts approach the shore obliquely. This causes the wave fronts to bend and swing around in an attempt to break parallel to the shore. The change in speed and distortion of the wave front is called **wave refraction** (Figure 8.9). If refraction is completed, the fronts break parallel to the shore. However, due primarily to the complexities of coastline shape, refraction is not always totally achieved – this causes **longshore drift**, which is a major force in the transport of material along the coast.

Source: *Advanced Geography: Concepts & Cases* by P. Guinness & G. Nagle (Hodder Education, 1999), p.294

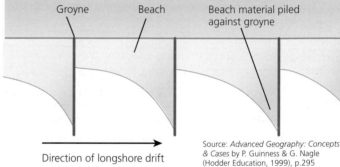

Figure 8.9 Wave refraction

Wave refraction also distributes wave energy along a stretch of coast. Along a complex transverse coast with alternating **headlands** and bays, wave refraction concentrates wave energy and therefore erosional activity on the headlands, while wave energy is dispersed in the bays – hence deposition tends to occur in the bays.

If refraction is not complete, longshore drift occurs (Figure 8.10). This leads to a gradual movement of sediment along the shore, as the swash moves in the direction of the **prevailing wind**, whereas the backwash moves straight down the beach following the steepest gradient.

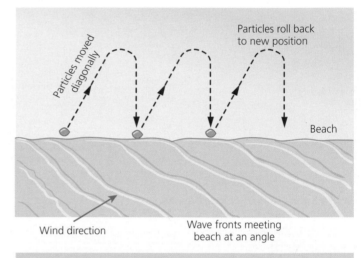

Source: *Advanced Geography: Concepts & Cases* by P. Guinness & G. Nagle (Hodder Education, 1999), p.295

Figure 8.10 Longshore drift

☐ Marine erosion

Waves perform a number of complex and interacting processes of **erosion** (Figure 8.11 and Table 8.2). **Hydraulic action** is an important process as waves break onto cliffs. As the waves break against the cliff face, any air trapped in cracks, joints and bedding planes is momentarily placed under very great pressure. As the wave retreats, this pressure is released with explosive force. This is known as **cavitation**. Stresses weaken the coherence of the rock, aiding erosion (comparable to cavitation in rivers). This is particularly obvious in well-bedded and well-jointed rocks such as limestones, sandstones, granite and chalk, as well as in rocks that are poorly consolidated such as clays and

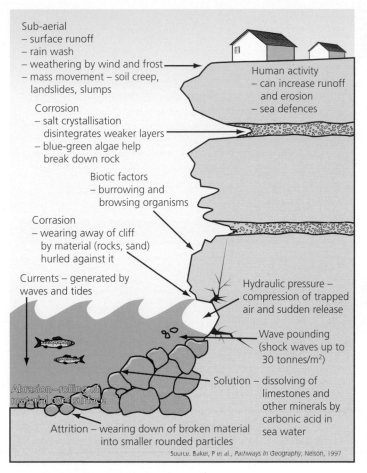

Sub-aerial
– surface runoff
– rain wash
– weathering by wind and frost
– mass movement – soil creep, landslides, slumps

Corrosion
– salt crystallisation disintegrates weaker layers
– blue-green algae help break down rock

Biotic factors
– burrowing and browsing organisms

Corrasion
– wearing away of cliff by material (rocks, sand) hurled against it

Currents – generated by waves and tides

Human activity
– can increase runoff and erosion
– sea defences

Hydraulic pressure – compression of trapped air and sudden release

Wave pounding (shock waves up to 30 tonnes/m²)

Solution – dissolving of limestones and other minerals by carbonic acid in sea water

Abrasion—rolling of material over surface

Attrition – wearing down of broken material into smaller rounded particles

Source: Baker, P et al., *Pathways in Geography*, Nelson, 1997

Figure 8.11 Types of erosion

Table 8.2 Factors affecting the rate of erosion

Energy factors	
Waves	Wave steepness – steep destructive waves formed locally have more erosive power than less steep constructive waves. Wave breaking point – waves breaking at a cliff base cause maximum erosion, whereas waves breaking off shore lose energy.
Tides	Neap and spring tides vary the zone of wave attack. Strong tidal currents can scour estuary channels.
Currents	Longshore and rip currents can move large quantities of material.
Winds	Onshore winds erode fine beach sand to form dunes. Offshore winds may erode dunes and nourish the beach. The longer the fetch, the greater the wave-energy potential.
Material factors	
Sediment supply	Continual supply is necessary for abrasion, whereas an oversupply can protect the coast.
Beach/rock platform width	Beaches/rock platforms influence wave energy by absorbing waves before they can attack cliffs.
Rock resistance	Rock type influences the rate of erosion, e.g. granites are very resistant, whereas unconsolidated volcanic ash has little resistance to wave attack. Erosion is rapid where rocks of different resistance overlie one another.
Rock structure and dip	Well-jointed or faulted rocks are very susceptible to erosion. Horizontal or vertical structures produce steep cliffs. Rocks dipping away from the sea produce gentle cliffs.
Shore geometry	
Offshore topography (bathymetry)	A steep seabed creates higher and steeper waves than one with a gentle gradient. Longshore bars cause waves to break off shore and lose energy.
Orientation of coast	Headlands with vertical cliffs tend to concentrate wave energy by refraction. Degree of exposure to waves influences erosion rates.
Direction of fetch	The longer the fetch, the greater the potential for erosion by waves.

glacial deposits. Hydraulic action is also notable during times of storm wave activity – for example, the average pressure of Atlantic storm waves is 11 tonnes/m². Another term used for this process is **wave pounding**.

Abrasion (also known as **corrasion**) is the process whereby a breaking wave can hurl pebbles and shingle against a coast, thereby abrading it. **Attrition** takes place as other forms of erosion continue. The eroded material is itself worn down by attrition, which partly explains the variety of sizes of beach material. **Solution** is a form of chemical erosion. In areas of calcareous (lime-rich) rock, waves remove material by acidic water. The source of the acidity is from organisms such as barnacles and limpets. These secrete organic acids that may make the water more acidic, especially in rock pools at low tide.

As wave action is constantly at work between high water mark (HWM) and low water mark (LWM), it causes undercutting of a cliff face, forming a notch and overhang. Breaking waves, especially during storms and spring tides, can erode the coast above HWM. As the undercutting continues, the notch becomes deeper and the overhang more pronounced. Ultimately, the overhang will collapse, causing the cliff line to retreat. The base

of the cliff will be left behind as a broadening **platform**, often covered with deposited material, with the coarsest near the cliff base, gradually becoming smaller towards the open sea.

☐ Sub-aerial processes

Sub-aerial, or cliff-face, processes also include different types of weathering and mass movement:

- **Salt weathering** is the process by which sodium and magnesium compounds expand in joints and cracks, thereby weakening rock structures.
- **Freeze–thaw weathering** is the process whereby water freezes, expands and degrades jointed rocks.

Lagos is located at a break in the coast, and the city developed rapidly in the nineteenth century and early twentieth century (Figure 8.12). Dredging started in 1907 and the harbour was begun in 1908. Breakwaters and a jetty provide a deepwater channel for large ships.

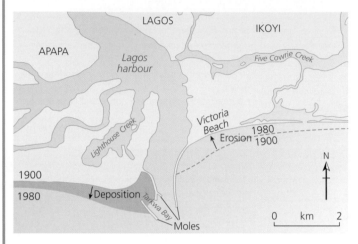

Figure 8.12 Erosion at Lagos, Nigeria

These developments interrupted the west–east longshore drift along the coast of West Africa. As a result, there has been an increase of deposition on Lighthouse Beach on the western updrift side of the jetty at Tarkwa Bay, which traps sediment, and an increase in erosion on the eastern downdrift side of the jetty and shipping channel. Victoria Beach has been eroded by almost 70 metres/year, and over 2 kilometres of beach has been lost since then (Table 8.3). This beach is much used as a recreational area for the people of Lagos. Hence, beach replenishment has been used here since 1976, but it continues to be eroded.

Table 8.3 Erosion of the Niger delta

	m/year
Western delta	18–24
Central delta	15–20
Eastern delta	10–24

From time to time, marine currents change direction, material is deposited in the channel leading to the harbour, and this has to be dredged to keep the channel clear. Continued deposition on Lighthouse Beach will eventually lead to deposition beyond the jetties of Tarkwa Bay, potentially cutting off the deep channel of Lagos Harbour.

- **Biological weathering** is carried out by molluscs, sponges and sea urchins (and is very important on low-energy coasts).
- **Solution weathering** is the chemical weathering of calcium by acidic water, which tends to occur in rock pools due to the presence of organisms secreting organic acids.
- **Slaking** is where materials disintegrate when exposed to water; this can be caused by hydration cycles.

Mass movements are also important in coastal areas, especially slumping and rockfalls (see Topic 3, Section 3.3).

☐ Marine transportation and deposition

Sediment sources are varied in marine transportation and deposition. They include reworked beach deposits, off-shore marine deposits, river deposits, glacial deposits, materials from mass movements on cliffs, wind-blown deposits and artificial beach nourishment. Some beaches are formed of volcanic ash – and may be black in colour – whereas others may be formed of shingle. The shingle beaches of southern Britain have sediment that is derived from glacial and periglacial deposits over 10 000 years old. The sandy beaches of the Caribbean, on the other hand, are largely the result of river sediment that has been reworked by waves. Other beaches are completely artificial, using sand imported from elsewhere.

Sediment transport is generally categorised into two modes:

- **Bedload** – Grains transported by bedload are moved through continuous contact (**traction** or dragging) or discontinuous contact (**saltation**) with the sea floor. In traction, grains slide or roll along – a slow form of transport. Weak currents may transport sand, or strong currents may transport pebbles and boulders. In saltation, the grains hop along the seabed. Moderate currents may transport sand, whereas strong currents transport pebbles and gravel.
- **Suspended load** – Grains are carried by turbulent flow and generally are held up by the water. Suspension occurs when moderate currents are transporting silts, or strong currents are transporting sands. Grains transported as **wash loads** are permanently in suspension, and typically consist of clays and dissolved material.

Deposition is governed by sediment size (mass) and shape. In some cases, sediments will flocculate (stick together), become heavier and fall out in deposition.

Sediment cells

The coastal sediment system, or **littoral cell system**, is a simplified model that examines coastal processes and patterns in a given area (Figure 8.13). It operates at a variety of scales, from a single bay, for example Turtle Bay in North Queensland, Australia, to a regional scale, for example the south California coast. Each littoral cell is a self-contained cell, in which inputs and outputs are balanced.

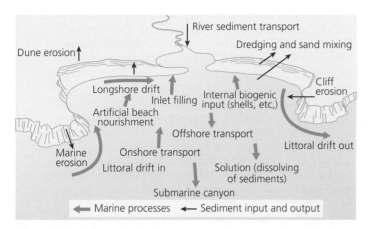

Figure 8.13 Sediment cells

The concept of dynamic equilibrium is important to littoral cells. This concept states that any system (or in this case, littoral cell) is the result of the inputs and processes that operate within it. Change to one of the inputs (for example an increase in sediment to the shoreline following cliff collapse) causes a knock-on effect on the processes (such as longshore drift, transport or beach protection) and a resulting change in the landforms (such as stabilisation of cliffs or downdrift beach enlargement). The balance changes – hence **dynamic equilibrium**.

Longshore drift

As discussed on page 230, longshore drift leads to sediment moving gradually along the shore; the swash moves sediment up the beach in the direction of the prevailing wind, while the backwash moves straight down the beach in the direction of the steepest gradient (see Figure 8.10). The net movement is downdrift.

Case Study: Changes in part of the Oceanside littoral cell, southern California

Inputs into the Californian sediment cell (Figure 8.14) include:

- river deposits
- sediments from cliffs
- materials for beach replenishment
- north–south longshore drift.

Irregular and variable river supplies have been further reduced by 33 per cent due to dam construction. Most of the material supplied for beach replenishment is fine-grained silt and sand.

The region is very active. Each year, rip currents and offshore currents move $100\,000\,m^3$ of sediment into the La Jolla submarine canyon and over $200\,000\,m^3$ of material drifts southwards. In addition, seasonal variations in constructive and destructive waves redistribute coastal sediments, and **sea levels** are rising 6–15 millimetres each year.

Human impacts

- Dams have reduced the supply of sediment to the beaches by 33 per cent.
- Buildings, houses, swimming pools, boats, private protection schemes and roads are destabilising the cliffs.
- Oceanside Harbour in the north is blocking the southward movement of sediment and most is now diverted to offshore currents and to the La Jolla submarine canyon.

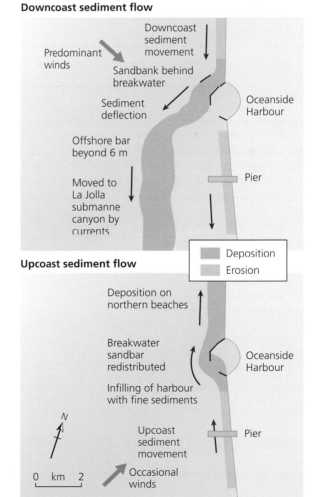

Figure 8.14 Sediment cells in southern California

Ocean currents along the coast of West Africa have removed huge amounts of beach material along the coast between Ghana and Nigeria. This has affected settlements, tourism and industry. The increase in coastal retreat has been blamed on the construction of the Akosombo Dam on the Volta River in Ghana. The Guinea Current is among the strongest in the world, and is removing approximately 1.5 million m³ of sand each year between the Ivory Coast and Nigeria (Figure 8.15). The effect on Ghana, Benin and Togo is potentially catastrophic.

Figure 8.15 Human activity and longshore drift along the coast of West Africa

The cause of the coastal retreat is traced to the building of the Akosombo Dam in 1961. It is just 110 kilometres from the coast and disrupts the flow of sediment from the River Volta, preventing about 40 per cent of it from reaching the coast. Thus there is less sand to replace that which has already been washed away, so the coastline retreats due to erosion by the Guinea Current. In addition, mining of sand and gravel from beaches and estuaries further reduces the natural **coastal protection**. Towns such as Keta, 30 kilometres east of the Volta estuary, have been destroyed as their protective beach has been removed. Other towns

in neighbouring Togo – Kpeme and Tropicana – are now threatened with destruction.

In Togo, the problem has been intensified by the use of artificial breakwaters. In the mid-1960s, a deepwater port was opened at Lomé, the country's capital, to improve trade with landlocked neighbouring countries such as Mali, Niger and Burkina Faso. Lomé is protected by a 1300 metre breakwater, which obstructs the natural flow of the Guinea Current from west to east. Sand carried by the current collects on the westward (updrift) side of the breakwater. Thus the east side (downdrift) is open to erosion. The result has been the erosion of the beach and local infrastructure. In 1984, a 100 metre stretch of the main Ghana–Benin highway was destroyed in just 24 hours. Erosion near the holiday resort of Tropicana caused the sea to advance 100 metres towards the holiday complex. Ironically, the erosion uncovered a bed of resistant sandstone, which now protects the resort, but is not as attractive for the tourist trade as the sandy beach that previously existed. Kpeme, 18 kilometres to the east of Tropicana, is a port from which most of Togo's processed phosphate is exported, accounting for more than half of Togo's foreign exchange. The jetty at Kpeme was threatened with erosion. To manage the risk, engineers have reinforced the foundations of the jetty with boulders. The boulders now trap sand and stop it from moving down the coastline. As a result, towns further east, such as Aneho, are now even more at risk from erosion. The coastline near Accra has been eroding at up to 6 metres a year, whereas other areas have retreated by as much as 30 metres a year. At a cost of between £1 million and £2 million to protect every kilometre of coastline, it is hard to imagine how the coast can be protected. If Togo were to protect its coastline by preventing the movement of sand eastwards, it might lead to an increase in erosion in Benin, where the foundations of oil wells may be threatened.

---

1 Explain what is meant by the term *littoral cell* (*sediment cell*).
2 Why is there more erosion on some coasts than others? Use examples to support your answer.
3 Outline the ways in which human activities can disrupt the operation of sediment cells.
4 With the use of an annotated diagram, explain what happens when wave refraction takes place.

8.2 Characteristics and formation of coastal landforms

☐ Erosional landforms

Cliff profiles are very variable and depend on a number of controlling factors. One major factor is the influence of bedding and jointing. The well-developed jointing and

bedding of certain harder limestones creates a geometric cliff profile with a steep, angular cliff face and a flat top (bedding plane). Wave erosion opens up these lines of weakness, causing complete blocks to fall away and the creation of angular overhangs and cave shapes. In other well-jointed and bedded rocks, a whole variety of features is created by wave erosion, such as caves, geos, **arches**, **stacks** and **stumps** (Figure 8.16). Wave refraction concentrates wave energy on the flanks of headlands. If there are lines of weakness, these may be eroded to form a **geo** (a widened crack or inlet). Geos may be eroded and enlarged to form caves, and if the caves on either side of a headland merge, an arch is formed. Further erosion and weathering of the arch may cause the roof of the arch to collapse, leaving an upstanding stack. The eventual erosion of the stack produces a stump. Good examples of arches and stacks are to be found at Etretat in northern France and at Dyrhólaey, southern Iceland (Figure 8.17).

Plan view / Side view

1 Two caves in a headland

2 A natural arch

3 A sea stack

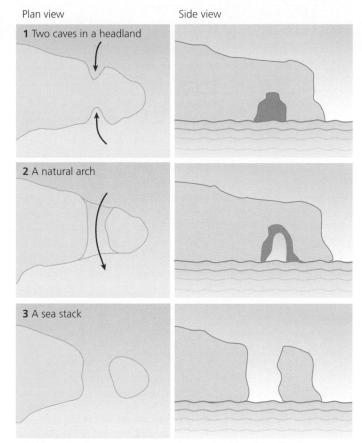

1 Wave refraction concentrates erosion on the sides of headlands. Weaknesses such as joints or cracks in the rock are exploited, forming caves.
2 Caves enlarge and are eroded further back into the headland until eventually the caves from each side meet and an arch is formed.
3 Continued erosion, weathering and mass movements enlarge the arch and cause the roof of the arch to collapse, forming a high standing stack.

Figure 8.16 Caves, arches, stacks and stumps

Figure 8.17 Cliffs and arches at Dyrhólaey, southern Iceland

Rocky shores can be divided into three main types (Figure 8.18): sloping shore platforms, sub-horizontal shore platforms and plunging cliffs. As with all classifications it is an over-simplification, but it is useful in illustrating the range of features associated with rocky shores (Table 8.4).

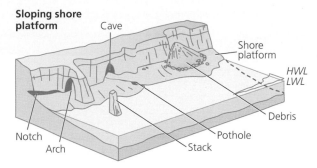

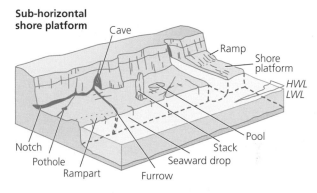

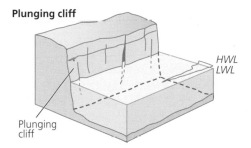

Figure 8.18 Types of rocky shoreline

The dip of the bedding alone will create varying cliff profiles. For example, if the beds dip vertically, then a sheer cliff face is formed. By contrast, if the beds dip steeply seaward, then steep, shelving cliffs with landslips are the result.

Each cliff profile is, to some extent, unique, but a model of cliff evolution or modification (Figure 8.19) takes into account not only wave activity but also sub-aerial weathering processes.

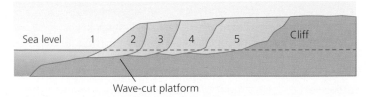

Source: Goudie, A, *The Nature of the Environment*, Blackwell, 1993

Figure 8.19 Evolution of wave-cut platforms

8.2 Characteristics and formation of coastal landforms 235

Table 8.4 Summary of the main erosional processes on rocky shores

Process	Description	Conditions conducive to the process
Mechanical wave erosion		
Erosion	Loose material is removed by waves.	Energetic wave conditions and microtidal range
Abrasion	Rock surfaces are scoured by wave-induced flow with a mixture of water and sediment.	'Soft' rocks, energetic wave conditions, a thin layer of sediment and microtidal range
Hydraulic action	Wave-induced pressure variations within the rock cause and widen rock cracks.	'Weak' rocks, energetic wave conditions and microtidal range
Weathering		
Physical (mechanical) weathering	Frost action and cycles of wetting/drying cause and widen rock cracks.	Sedimentary rocks in cool regions
Salt weathering	Volumetric growth of salt crystals in rocks widens cracks.	Sedimentary rocks in hot and dry regions
Chemical weathering	A number of chemical processes remove rock materials; these processes include hydrolysis, oxidation, hydration and solution.	Sedimentary rocks in hot and wet regions
Water-layer levelling	Physical, salt and chemical weathering working together along the edges of rock pools.	Sedimentary rocks in areas with high evaporation
Bio-erosion		
Biochemical	Chemical weathering by products of metabolism.	Limestone in tropical regions
Biophysical	Physical removal of rock by grazing and boring animals.	Limestone in tropical regions
Mass movements		
Rockfalls and toppling	Rocks fall straight down the face of the cliff.	Well-jointed rocks, undercutting of cliff by waves
Slides	Deep-seated failures.	Deeply weathered rock, undercutting of cliff by rock
Flows	Flow of loose material down a slope.	Unconsolidated material, undercutting of cliffs by waves

Many cliffs are composed of more than one rock type. These are known as **composite cliffs**. The exact shape and form of the cliff will depend on such factors as strength and structure of rock, relative hardness and the nature of the **waves** involved.

Figure 8.20 illustrates cliffs with rocks of different properties as well as varying wave energy. In Figure 8.20a, there is relatively strong rock of uniform resistance. Cliff retreat will be determined by rock strength – for granites this will be slow, while for glacial tills it will be rapid (Table 8.5). In Figure 8.20b, the rock strength is weaker (than in Figure 8.20a) and cliff retreat is faster. In addition, rock properties vary. Figure 8.20b shows that the form of the cliff depends on the relative position of the weaker rock: if it is at the base of the cliff, undercutting and collapse may occur; if it is near the top, the cliff is subject to sub-aerial processes and wave undercutting.

In Figure 8.21a, the beds are dipping landwards. Sliding is unlikely as the movement is landwards. However, in Figure 8.21b the movement is seawards, and the potential for sliding is great. Seaward-dipping rocks consequently pose greater management challenges than landward-dipping rocks.

Figures 8.21c and d show the impact of rock permeability on cliff development. If impermeable rock overlies permeable rock (Figure 8.21c), there is limited percolation and so the cliff is more stable. If permeable rock overlies impermeable rock, water may soak into the cliff, and slope failure is more likely where water builds up at the junction of the two rock types.

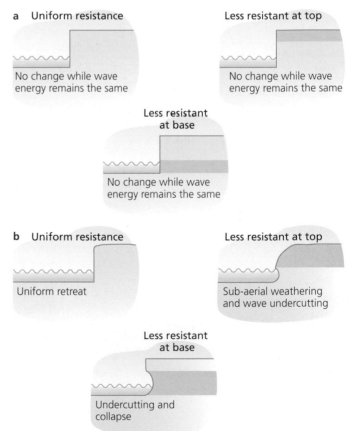

Figure 8.20 Effect of relationships between resistant and less resistant rocks in cliff morphology

Table 8.5 Mean cliff erosion rates

Rate	Rock type
< 0.001 m/year	Granitic
0.01–0.1 m/year	Limestone
0.01–0.1 m/year	Shale
0.1–1 m/year	Chalk
0.1–10 m/year	Quaternary deposits (glacial till)
10 m/year	Volcanic ash

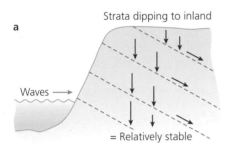

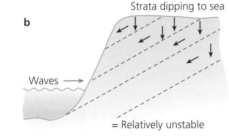

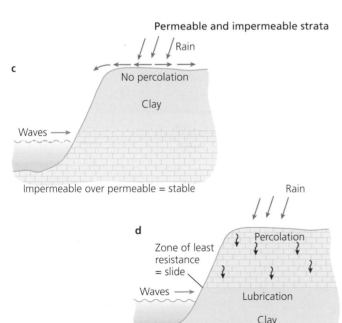

Figure 8.21 The controlling factors in cliff stability

Another type of composite cliff is a **bevelled cliff** (Figure 8.22). This was formed in a number of stages:

1 A vertical cliff was formed due to marine processes in the last interglacial (warm) period, when sea levels were higher than they are today.

2 During the subsequent glacial (cold) phase, sea levels dropped, and periglacial processes such as solifluction and freeze–thaw affected the former sea cliff, forming a bevelled edge.

3 When the sea levels rose again during the following interglacial, there was renewed wave erosion, which removes the debris and steepens the base of the cliff, leaving the upper part at a lower angle.

Cliff form can also be related to latitude. In the tropics, low wave-energy levels and high rates of chemical weathering generally produce low-gradient coasts. Coastal cliffs in high latitudes are also characterised by relatively low gradients since the periglacial processes produce large amounts of cliff-base materials. Temperate regions tend to have the steepest cliffs. The rapid removal of debris by high-energy waves prevents the build-up of material on the base, while active cliff development occurs as a result of undercutting.

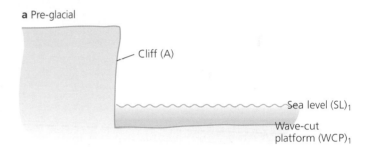

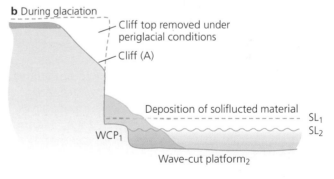

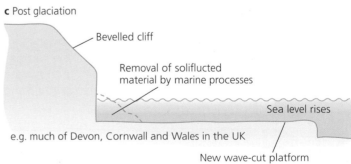

Figure 8.22 Bevelled cliffs

Coastal platforms

As a result of cliff retreat, a platform along the coast is normally created (Figure 8.23). Traditionally, up to very recent times, this feature was described as a **wave-cut platform** (or abrasion platform) because it was believed that it was created entirely by wave action (Figure 8.19). However, there is some controversy over the importance of other agents of weathering and erosion in the production of coastal platforms, especially the larger ones.

Figure 8.23 A shore platform in Jersey, Channel Islands, UK

In post-glacial times, sea level has not remained sufficiently constant to erode such platforms. However, some marine geomorphologists believe that these platforms could be **relict** or ancient features, originally cut long ago at a period when sea level was constant, and that the contemporary sea level is at about the same height and is just modifying slightly the ancient platform.

Secondly, in high latitudes, **frost action** could be important in supporting wave activity, particularly as these areas are now rising as a result of isostatic recovery (after intense glaciation). In other areas, solution weathering, **salt crystallisation** and slaking could also support wave activity, particularly in the tidal and splash zones. **Marine organisms**, especially algae, can accelerate weathering at low tide and in the area just above HWM. At night, carbon dioxide is released by algae because photosynthesis does not occur. This carbon dioxide combines with the cool sea water to create an acidic environment, causing 'rotting'.

Other organisms, such as limpets, secrete organic acids that can slowly rot the rock. Certain marine worms (polychaetes and annelids), molluscs and sea urchins can actually 'bore' into rock surfaces, particularly of chalk and limestone.

Section 8.2 Activities

1 Outline the main processes of coastal erosion.
2 Briefly explain the factors that affect the rate of erosion.
3 With the use of annotated diagrams, explain how stacks and stumps may be formed.
4 How are shore platforms formed?

Case Study: Shore platform at Kaikoura peninsula, New Zealand

Well-developed shore platforms can be found at Kaikoura on South Island, New Zealand (Figure 8.24). These platforms are periodically exposed to high-energy waves (>1.5 metres), but generally experience calm conditions (0.5 metre). The average rate of platform lowering is 1.4 millimetres per year^{-1}. Despite the high-energy waves, the amount of wave energy delivered to the platform is low. This is because most of the offshore waves break before they hit the platform. Sub-aerial weathering appears to be the dominant process here, in particular the number of cycles of wetting and drying.

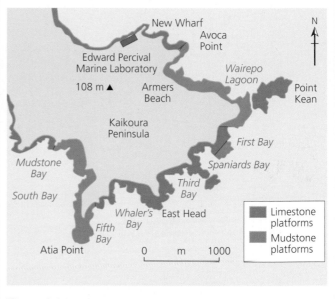

Figure 8.24 Kaikoura peninsula, New Zealand

☐ Depositional landforms

Beaches

A whole variety of materials can be moved along the coast by waves, fed by longshore drift. The coarse material is found deposited (and fallen from the backing cliffs) in the backshore and foreshore zones as **littoral deposits**. The finer material, worn down largely by attrition, is usually found in the offshore zone as **neritic** deposits.

The term **beach** refers to the accumulation of material deposited between low water mark (LWM) spring tides and the highest point reached by storm waves at high water mark (HWM) spring tides. A typical beach will have three zones: backshore, foreshore and offshore (Figure 8.25). The backshore is a cliff or is marked by a line of **dunes**. Above and at HWM there may be a **berm** or **shingle ridge**. This is coarse material pushed up the beach by spring tides and aided by storm waves that fling material well above the level of the waves themselves. These are often referred to as 'storm beaches'. The seaward edge of the berm is often scalloped and irregular due to the creation of beach **cusps** (Figure 8.26). Their origin is still controversial – they could be due to the edge of the swash itself, which is often scalloped, or to the action of two sets of wave fronts approaching the shore obliquely from opposite directions. Once initiated, the cusps are self-perpetuating: the swash is broken up by the cusp projection, concentrating energy onto the cusp (compare with refraction onto headlands), which excavates material. Cusps develop best in areas of high tidal range where waves approach the coast at right-angles. The spacing of cusps is related to wave height and swash strength.

The foreshore is exposed at low tide. The beach material may be undulating due to the creation of ridges, called **fulls**, running parallel to the water line, pushed up by constructive waves at varying heights of the tide. These are separated by troughs, called **swales**. Great stretches of sand, too, may comprise the foreshore. In areas of complex coast, sand beaches may only be exposed as small **bayhead beaches** in bays.

Offshore, the first material is deposited. In this zone, the waves touch the seabed and so the material is usually disturbed, sometimes being pushed up as **offshore bars**, when the offshore gradient is very shallow.

Excellent beach development occurs on a lowland coast (constructive waves), with a sheltered aspect/ trend, composed of 'soft' rocks, which provide a good supply of material, or where longshore drift supplies abundant material.

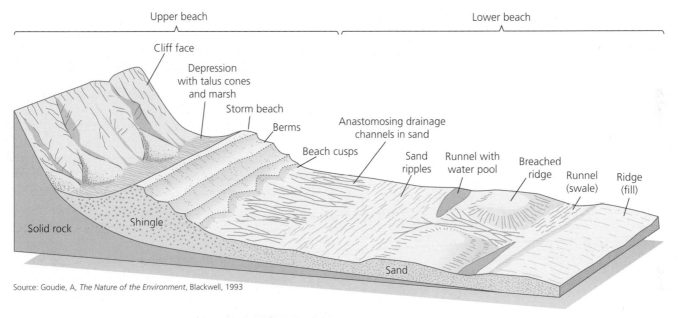

Source: Goudie, A, *The Nature of the Environment*, Blackwell, 1993

Figure 8.25 Beach deposits

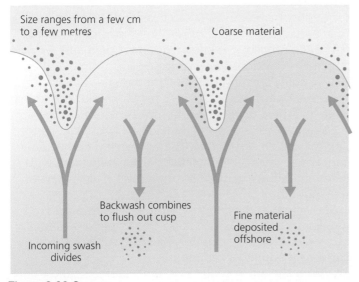

Figure 8.26 Cusps

Beach form is affected by the size, shape and composition of materials, the tidal range and wave characteristics. As storm waves are more frequent in winter and swell waves more important in summer, many beaches differ in their winter and summer profile (Figure 8.27). Thus the same beach may produce two very different profiles at different times of the year. For example, constructive waves in summer may build up the beach, but destructive waves in winter may change the size and shape of the beach. The relationship between wave steepness and beach angle is a two-way affair. Steep destructive waves reduce beach angle, whereas gentle constructive waves increase it. A low gradient produces shallow water, which in turn increases wave steepness. Hence plunging waves are associated with gentle beaches, whereas surging waves are associated with steeper beaches.

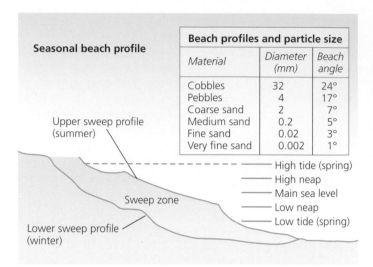

Figure 8.27 Seasonal changes on a beach

Beach profiles and particle size		
Material	Diameter (mm)	Beach angle
Cobbles	32	24°
Pebbles	4	17°
Coarse sand	2	7°
Medium sand	0.2	5°
Fine sand	0.02	3°
Very fine sand	0.002	1°

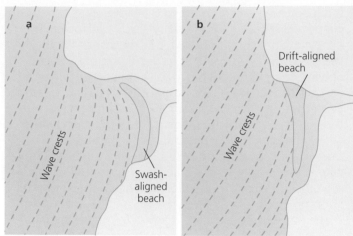

Figure 8.28 Swash-aligned and drift-aligned coasts

Sediment size affects the beach profile through its percolation rate (Figure 8.27). Shingle/pebbles allow rapid infiltration and percolation, so the impact of swash and backwash are reduced. As the backwash is reduced, it will not impede the next swash. If the swash is stronger than the backwash then deposition may occur. By contrast, sand produces a lower angle and allows less percolation. Backwash is likely to be greater than on a gravel beach. The pattern is made more complex because sediment size varies up a beach. The largest particles – the products of cliff recession – are found at the rear of a beach. Large, rounded material on the upper beach is probably supplied only during the highest spring tides and is unaffected by 'average' conditions. On the lower beach, wave action is more frequent, attrition is common and consequently particle size is smaller.

Swash-aligned and drift-aligned beaches

It is important to distinguish between two types of coastline:

- **Swash-aligned coasts** are oriented parallel to the crests of the prevailing waves (Figure 8.28a). They are closed systems in terms of longshore drift transport and the net littoral drift rates are zero.
- **Drift-aligned coasts** are oriented obliquely to the crest of the prevailing waves (Figure 8.28b). The shoreline of a drift-aligned coast is primarily controlled by longshore sediment-transport processes. Drift-aligned coasts are open systems in terms of longshore transport. **Spits**, **bars**, tombolos and cuspate forelands are all features of drift-aligned coasts.

Section 8.2 Activities

1 Describe the distinction between drift-aligned and swash-aligned coastlines.
2 Explain the origin of **a** beaches and **b** cusps.

Localised depositional features

Spits and other localised features develop where:

- abundant material is available, particularly shingle and sand
- the coastline is irregular due, for example, to local geological variety (transverse coast)
- deposition is increased by the presence of vegetation (reducing wave velocity and energy)
- there are estuaries and major rivers (Figure 8.29).

Spits are common along an indented coast. For example, along a transverse coast where bays are common, or near river mouths (estuaries and rias), wave energy is reduced. The long, narrow ridges of sand and shingle that form spits are always joined at one end to the mainland. The simplest spit is a linear spit (simple spit), but it may be curved at its distal (unattached) end.

Spits often become curved as waves undergo refraction. Cross-currents or occasional storm waves may assist this hooked formation. If the curved end is very pronounced, it is known as a **recurved spit**. Many spits have developed over long periods of time and have a complex morphology. For example, a **compound recurved spit** has a narrow proximal (joined) end and a wide, recurved distal end that often encloses a lagoon – for example Presque Isle, Lake Erie (Figure 8.30) and Sandy Hook in New Jersey. The wide distal end usually consists of several dune/beach ridges associated with older shorelines, demonstrating seaward migration of the shoreline.

Spits grow in the direction of the predominant longshore drift, and are a classic example of a drift-aligned feature that can only exist through the continued supply of sediment.

Within the curve of the spit, the water is shallow and a considerable area of mudflat and saltmarsh (**salting**)

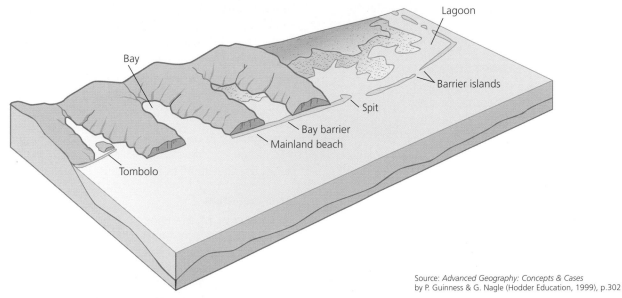

Figure 8.29 Localised depositional features

Source: *Advanced Geography: Concepts & Cases*
by P. Guinness & G. Nagle (Hodder Education, 1999), p.302

is exposed at low water. These saltmarshes continue to grow as mud is trapped by the marsh vegetation. The whole area of saltmarsh is intersected by a complex network of channels, or **creeks**, which contain water even at low tide.

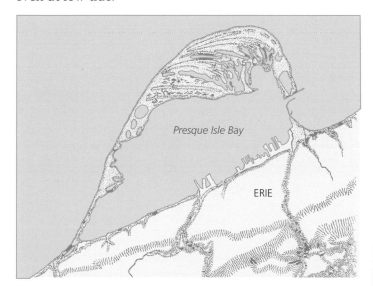

Figure 8.30 Compound recurved spit with lagoon – Presque Isle, Lake Erie

Section 8.2 Activities

Draw a labelled diagram to show the formation of a spit.

Bars

A bar is a ridge of material that is connected at both ends to the mainland. It is located above sea level. If a spit continues to grow lengthwise, it may ultimately link two headlands to form a **bay bar**. These are composed either of shingle, as in the case of the Low Bar in Cornwall, UK, or of sand, such as the *nehrung* of the Baltic coast, which pond back lagoons called *haff*. Bars may also be formed by the onshore movement of material.

Tombolo

If a ridge of material links an island with the mainland, this ridge is called a **tombolo**. A typical example is Chesil Beach on the south coast of England (Figure 8.31). Chesil Beach is 25 kilometres long, connecting the Isle of Portland with the mainland Dorset coast at Burton Bradstock, near Abbotsbury. At its eastern end, at Portland, the ridge is 13 metres above sea level and composed of flinty pebbles about the size of a potato. At its western end, near Abbotsbury, the ridge is lower, only 7 metres above sea level, and built of smaller flinty material about the size of a pea.

Figure 8.31 Chesil Beach tombolo

The height of the ridge and the sizing of material would suggest that dominant wave action occurred from east to west – the largest material is piled up at the eastern end, being the heaviest and most difficult to transport. Smaller, lighter material is carried further west before being deposited. However, the dominant wave action comes from the south-west, up the Channel from the Atlantic Ocean. In other words, the morphology of the ridge should be completely opposite from what it is.

The origin of Chesil Beach remains a problem. One theory to explain this situation is that Chesil is a very youthful feature and so is unstable in the present environment. Around 18 000–20 000 years ago, in the Pleistocene period, sea level fell to a level at least 100 metres below its present position. As a result, much of what is now the English Channel was dry. During the Ice Age, vast amounts of debris were produced on the nearby land surface by glacial and periglacial action. This debris could have been carried into the dry Channel area by meltwater at the close of the Ice Age. As the sea level rose in early post-glacial times, this material could have been pushed onshore and trapped by the Isle of Portland and Lyme Bay. Present-day wave action is gradually sorting this material.

Cuspate forelands

Cuspate forelands consist of shingle ridges deposited in a triangular shape, and are the result of two separate spits joining, or the combined effects of two distinct sets of regular storm waves. A fine example is at Dungeness near Dover in the UK, where the foreland forms the seaward edge of Romney Marsh (Figure 8.32). As recently as 900 CE, this marsh was a bay. Within the last 1000 years, it has silted up with mudflats and marshes as a direct result of the growth of the cuspate foreland. The shingle was deposited by longshore drift curling west from the North Sea and by the longshore drift flowing eastwards up the English Channel.

Figure 8.32 Dungeness, a cuspate foreland

Section 8.2 Activities

1 Distinguish between tombolos and cuspate forelands.
2 For either tombolos or cuspate forelands, explain how they are formed.

Offshore bars

Offshore bars are usually composed of coarse sand or shingle. They develop as bars offshore on a gently shelving seabed. Waves feel bottom far offshore. This causes disturbance in the water, which leads to deposition, forming an offshore bar below sea level. Between the bar and shore, lagoons (often called **sounds**) develop. If the water in the lagoon is calm – and fed by rivers – marshes and mudflats can develop. Bars can be moved onshore by storm winds and waves. A classic area for this is off the coast of the Carolinas in south-east USA.

Barrier beaches

Barrier beaches, or barrier islands, are natural sandy breakwaters that form parallel to a flat coastline. By far the world's longest series is that of roughly 300 islands along the east and southern coasts of the USA (Figure 8.33). The distance between barrier beaches and the shore is variable. The islands are generally 200–400 metres wide, but some are wider. Some Florida islands are so close to the shore that residents do not even realise they are on an island. By contrast, parts of Hatteras Island in North Carolina are 20 kilometres offshore.

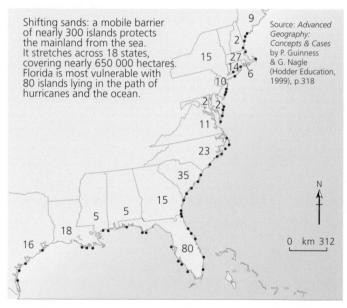

Figure 8.33 Barrier beaches along the east coast of the USA

Barrier beaches form only under certain conditions, and America's eastern seaboard provides the ideal conditions for barrier beaches (Figure 8.34). First, a gently sloping and low-lying coast unprotected by cliffs faces an ocean. Over the last 15 000 years, the sea level has risen by 120 metres as glaciers and icecaps have melted. Wind and waves have formed sand dunes at the edge of the continental shelf. As the rising sea breaks over the dunes, this forms lagoons behind the sandy ridge, which divides into islands. Waves wash sand from the islands, depositing it further inland and forming new islands. Currents, flowing parallel to the coast, scour sand from barrier beaches and deposit it further up or down the coast to form new islands. The island in Chatham Harbour appears to have migrated south over a period of 140 years.

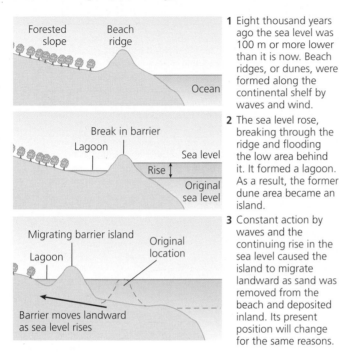

1 Eight thousand years ago the sea level was 100 m or more lower than it is now. Beach ridges, or dunes, were formed along the continental shelf by waves and wind.

2 The sea level rose, breaking through the ridge and flooding the low area behind it. It formed a lagoon. As a result, the former dune area became an island.

3 Constant action by waves and the continuing rise in the sea level caused the island to migrate landward as sand was removed from the beach and deposited inland. Its present position will change for the same reasons.

Figure 8.34 The formation of a barrier beach

Coastal dunes

Coastal sand dunes form where there is a reliable supply of sand, strong onshore winds, a large tidal range and vegetation to trap the sand. Extensive sandy beaches are almost always backed by sand dunes because strong onshore winds can easily transport inland the sand that has dried out and is exposed at low water. The sand grains are trapped and deposited against any obstacle on land, to form dunes. Vegetation causes the wind velocity to drop, especially in the lowest few centimetres above the ground, and the reduction in velocity reduces energy and increases the deposition of sand. Dunes can be blown inland and can therefore threaten coastal farmland and even villages. Special methods are used to slow down the migration of dunes:

- planting of special grasses, such as marram, which has a long and complex tap root system that binds the soil
- erecting brushwood fences to reduce sand movement
- planting of conifers that can stand the saline environment and poor soils, such as Scots and Corsican pines.

Sand-dune succession

Initially, sand is moved by the wind. However, wind speed varies with height above a surface. The belt of no wind is only 1 millimetre above the surface. As most grains protrude above this height, they are moved by saltation. The strength of the wind and the nature of the surface are important. Irregularities cause increased wind speed and

eddying, and more material is moved. On the leeward side of irregularities, wind speed is lower, transport decreases and deposition increases.

For dunes to become stable, vegetation is required. Plant succession and vegetation succession can be interpreted by the fact that the oldest dunes are furthest from the sea and the youngest ones are closest to the shore. On the shore, conditions are windy, arid and salty. The soil contains few nutrients and is mostly sand – hence the fore dunes are referred to as 'yellow dunes'. Few plants can survive, although sea couch and marram can tolerate these conditions. Once the vegetation is established, it reduces wind speed close to ground level. The belt of no wind may increase to a height of 10 millimetres. As grasses such as sea couch and marram need to be buried by fresh sand in order to grow, they keep pace with deposition. As the marram grows, it traps more sand. As it is covered it grows more, and so on. Once established, the dunes should continue to grow, as long as there is a supply of sand. However, once another younger dune, a fore dune, becomes established, the supply of sand – and so the growth of the dune – is reduced.

As the dune gets higher, the supply of fresh sand is reduced to dunes further back. Thus marram dies out. In addition, as wind speeds are reduced, evapotranspiration losses are less, and the soil is moister. The decaying marram adds some nutrients to the soil, which in turn becomes more acidic. In the slacks, the low points between the dunes, conditions are noticeably moister, and marsh vegetation may occur.

Towards the rear of the dune system, 'grey' dunes are formed – grey due to the presence of humus in the soil. The climax vegetation found here depends largely on the nature of the sand. If there is a high proportion of shells (providing calcium), grasslands are found. By contrast, acid dunes are found on old dunes where the calcium has been leached out, and on dunes based on outwash sands and gravels. Here, acid-loving plants such as heather and ling dominate. Pine trees favour acid soils, whereas oak can be found on more neutral soils. Thus the vegetation at the rear of the sand-dune complex is quite variable (Figure 8.35).

Tidal sedimentation in estuaries

In areas with a large tidal range (see Section 8.1), the characteristic shape of an estuary is triangular, with the base open to the sea (Figure 8.36). Within the estuary, there are shifting banks of sand, silt and clay. The rising tide flows inwards in the flood channel, while the falling ebb tide uses ebb channels. Where the tidal range is low, the effect of moving water is less important and wave action tends to block off the estuary from the sea.

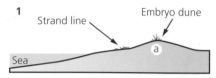

1
Strand line   Embryo dune

Sea

As the tide goes out, the sand dries out and is blown up the beach. At the top of the beach is a line of seaweed and litter called the strand line.
A small embryo dune forms in the shelter behind the strand line. This dune can be easily destroyed unless colonised by plants.

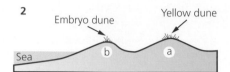

2
Embryo dune   Yellow dune

Sea

Sea couch grass colonises and helps bind the sand. Once the dune grows to over 1 m high, marram grass replaces the sea couch. A yellow dune forms at 10–20 m high with the long-rooted marram forming a good sand trap.

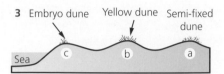

3 Embryo dune   Yellow dune   Semi-fixed dune

Sea

Once the yellow dune is over 10 m high, less sand builds up behind it and marram grass dies to form a thin humus layer. As soil begins to form, other plants are able to grow on the dune including dandelions. This kind of dune is called a semi-fixed dune. As the original dune (a) has developed, new embryo and yellow dunes have formed.

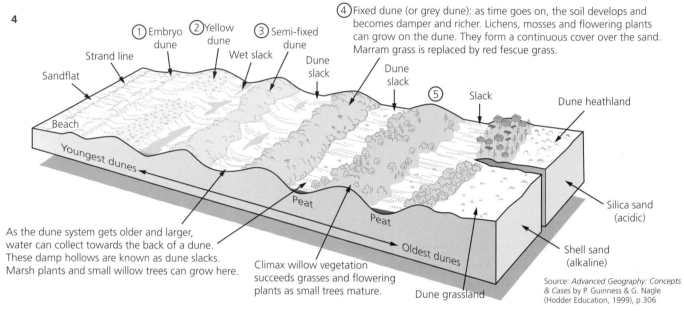

4

④ Fixed dune (or grey dune): as time goes on, the soil develops and becomes damper and richer. Lichens, mosses and flowering plants can grow on the dune. They form a continuous cover over the sand. Marram grass is replaced by red fescue grass.

As the dune system gets older and larger, water can collect towards the back of a dune. These damp hollows are known as dune slacks. Marsh plants and small willow trees can grow here.

Climax willow vegetation succeeds grasses and flowering plants as small trees mature.

Source: *Advanced Geography: Concepts & Cases* by P. Guinness & G. Nagle (Hodder Education, 1999), p.306

Figure 8.35 Sand-dune succession

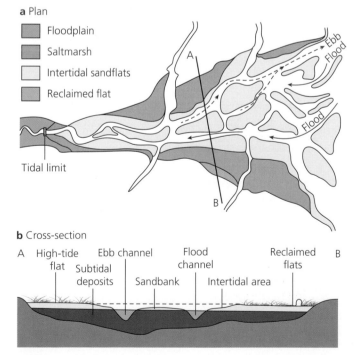

Figure 8.36 Tidal sedimentation

Sedimentation (deposition) in estuaries is increased by their relatively sheltered nature. The flood waves carry in a large volume of material in suspension. At high tide, there is little flow, and deposition occurs. Clay carried downstream by a river flocculates in salt water; that is, the particles stick together, become heavier and are deposited.

In addition, vegetation causes much deposition in an estuary. It slows down the speed of the water, thereby increasing rates of deposition. Over time, the surface of the mudflats may become higher than the surface of the estuary and be colonised by terrestrial species.

The intertidal zone – the zone between high tide and low tide – experiences severe environmental changes in salinity, tidal inundation and sediment composition. Halophytic (salt-tolerant) plants have adapted to the unstable, rapidly changing conditions (Figure 8.37).

Coastal saltmarshes

Saltmarshes are typically found in three locations: on low-energy coastlines; behind spits and barrier islands; and in estuaries and harbours.

Figure 8.37 Saltmarsh vegetation

Silt accumulates in these situations and on reaching sea level forms mudbanks. With the appearance of vegetation, saltmarsh is formed. The mudbanks are often intersected by creeks.

Saltmarshes can be found at Scolt Head Island, which is located on the north Norfolk coast of England (Figure 8.38). It is exposed to cold winds from the east and at high tide it is cut off by the sea, while at low tide it is joined to the mainland. The island developed from an extensive sand and shingle foreshore. Wave action during stormy seas sorted the shingle from the sand, forming shingle ridges near the high-water mark. The early ridges were unstable and mobile. However, as they became more stable, dunes developed and gradually moved the island westwards, in a series of stages. Most of the shingle came from offshore glacial deposits, while other shingle was drifted along the shore. Each of the former ends of the island are marked by curving lateral ridges of shingle, some with high and well-developed dunes.

The marshes change in age, and height, from east to west. The older marshes are higher, with more developed creek patterns. However, in some cases human activity has disrupted the pattern. Drainage may lead to settling and subsidence, so the oldest marshes may not always be the highest.

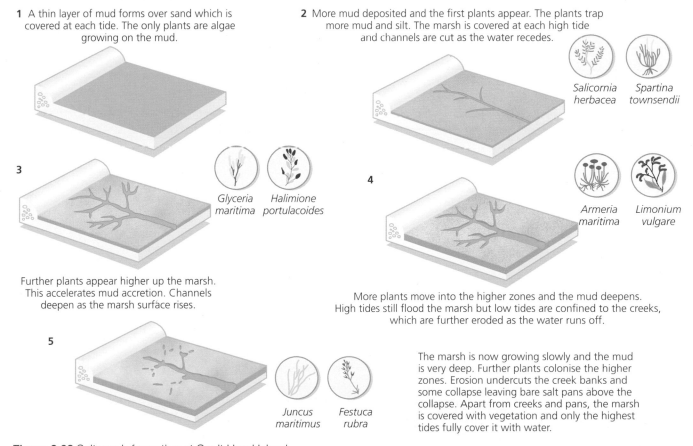

1 A thin layer of mud forms over sand which is covered at each tide. The only plants are algae growing on the mud.

2 More mud deposited and the first plants appear. The plants trap more mud and silt. The marsh is covered at each high tide and channels are cut as the water recedes.

Salicornia herbacea   *Spartina townsendii*

3 Further plants appear higher up the marsh. This accelerates mud accretion. Channels deepen as the marsh surface rises.

Glyceria maritima   *Halimione portulacoides*

4 More plants move into the higher zones and the mud deepens. High tides still flood the marsh but low tides are confined to the creeks, which are further eroded as the water runs off.

Armeria maritima   *Limonium vulgare*

5 The marsh is now growing slowly and the mud is very deep. Further plants colonise the higher zones. Erosion undercuts the creek banks and some collapse leaving bare salt pans above the collapse. Apart from creeks and pans, the marsh is covered with vegetation and only the highest tides fully cover it with water.

Juncus maritimus   *Festuca rubra*

Figure 8.38 Saltmarsh formation at Scolt Head Island

The marshes include small basins called pans. These can result from creeks being dammed by bank collapse. This impedes drainage, and the water in the pan slowly evaporates, leaving very salty water in the pan. High salinity will inhibit vegetation growth, and so the floor of the pan remains bare.

On Scolt Head Island, the vegetation is varied and natural. By contrast, many other marshes in southern England are dominated by the recently introduced cord grass (*Spartina anglica*). Once the bare marshflat is formed, the first plants, such as green algae (*Enteromorpha*), colonise the mudflat. The algae trap sediment from the sea and provide ideal conditions for the seeds of the salt-tolerant marsh samphire (*Salicornia*) and eel grass (*Zostera*), which then colonise the marsh. These plants increase the rate of deposition by slowing down the water as it passes over the vegetation. This is known as **bioconstruction**. Gradually, the clumps of vegetation become larger and the flow of tidal waters is restricted to specific channels, namely the creeks. The slightly increased height of the surface around plants leads to more favourable conditions. Here, plants are covered by sea water for shorter periods of time and this encourages other plants to colonise, such as sea aster, sea pea and seablite. These are even more efficient at trapping sediment and the height of the saltmarsh increases. New plants colonise as the marsh grows, including sea lavender, sea pink and sea purslane. As the height increases, tidal inundation of the marsh becomes less frequent and the rate of growth slows down. Sea rush (*Juncus*) and saltwort (*Salsola kali*) become the most common plants. It takes about 200 years on Scolt Head to progress from the marsh-samphire stage to the sea-rush stage.

Figure 8.39 Mangroves are adapted to rapidly changing conditions

Most mangroves do not tolerate high-energy localities. This may be because of the inability of seedlings to become established in such environments. Extensive mangroves may be found lining the estuaries of main rivers that flow into shallow seas, such as the Ganges-Brahmaputra, and the Mekong.

Mangroves inhabit a challenging environment. They act as filters, removing soil and terrestrial organic matter that flow into them from adjacent land. Mangroves provide an important habitat for fish, insects, birds and mammals.

Figure 8.40 Mangroves in northern Borneo

Mangrove root penetration is very shallow, but with extensive lateral spreading. Some mangroves have buttress roots to give them extra stability. The conditions in which mangrove trees grow are harsh: high salinity, low oxygen, poor nutrient availability, strong wind and wave action and unstable sediments. Luckily, they have little competition. Mangrove leaves exhibit xerophytic adaptation (drought adapted), such as a thick cuticle, sunken stomata, thick epidermis and leaf

---

Section 8.2 Activities

1 Outline the environmental conditions in which mudflats and saltmarshes occur.
2 Describe the typical succession associated with saltmarshes.

Mangroves

Mangroves occur between the sea and the land and line approximately 75 per cent of tropical coasts (Figures 8.39 and 8.40). Mangroves are essentially tidal forests. There are two main centres of mangrove: a more diverse eastern group (East Africa, India, South East Asia and Australasia) with around forty species, and a less diverse western group (West Africa, South America, the Caribbean) with only eight species. With increasing latitude, the number of species decreases. Mangrove forests are limited to where the temperature is greater than 24 °C in the warmest month and where rainfall exceeds 1250 millimetres.

hairs. Nevertheless, mangrove communities are highly productive, and they provide an important source of organic matter for food chains both within the forest and in adjacent areas.

Mangroves are used by humans to provide fuel, wood, charcoal, dyes, poisons, timber, thatching for housing and food. Consequently, overexploitation has led to a decline in mangrove forests in recent decades.

☐ The role of sea-level change in the formation of coastal landforms

Sea levels change in connection with the growth and decay of ice sheets. Eustatic change refers to a global change in sea level. At the height of glacial advance 18000 years ago, sea level was 100–150 metres below current sea level. The level of the land also varies in relation to the sea. Land may rise as a result of tectonic uplift or following the removal of an ice sheet. The change in the level of the land relative to the level of the sea is known as **isostatic adjustment** or **isostacy**. Parts of Scandinavia and Canada are continuing to rise at rates of up to 20 millimetres/year.

Figure 8.41 shows Valentin's classification of advancing and retreating coasts. These are, in part, the result of sea-level changes, and, in part, the balance of erosion and deposition. On the diagram, points A and X both experienced uplift of around 100 metres. However, erosion has taken place at A, reducing the effect of uplift. Similarly, B and Y have experienced about 100 metres of submergence. However, Y has experienced deposition, so the overall change in its position relative to sea level is negligible.

A simple sequence of sea-level change can be described:

1 Temperatures decrease, glaciers and ice sheets advance and sea levels fall, eustatically.
2 Ice thickness increases and the land is lowered isostatically.
3 Temperatures rise, ice melts and sea levels rise eustatically.
4 Continued melting releases pressure on the land and the land rises isostatically.

Features of emerged coastlines include:

■ **raised beaches**, such as the Portland raised beach
■ **coastal plains**
■ **relict cliffs**, such as those along the Fall Line in eastern USA
■ **raised mudflats**, for example the Carselands of the River Forth.

Submerged coastlines include:

■ **rias**, such as the River Fal – drowned river valleys caused by rising sea levels during the Flandarin Transgression or due to a sinking of the land
■ **fjords**, such as Milford Sound on the west coast of South Island, New Zealand, and the Oslo Fjord – glacial troughs occupied by the sea; common in uplifted mid-latitude coasts, notably Norway, Greenland and Chile (Figure 8.42) – an early view was that they were tectonic in origin; this has been rejected and replaced by drowning of U-shaped valleys.
■ **fjards** or 'drowned glacial lowlands'.

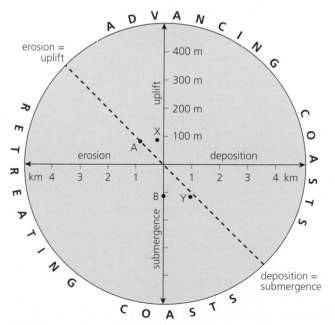

Figure 8.41 Valentin's classification of advancing and retreating coasts

Figure 8.42 A submerged glacial trough (fjord) in Norway

At the top of the sea cliff is a dark brown stratified soil, which is believed to contain wind-blown silt. Immediately below is a classic example of a periglacial head deposit. This is a mass (generally 1–1.5 metres thick) of angular, frost-shattered debris of local limestone formed during the periglacial period (Figure 8.43). Beneath this layer is a section of fine loam containing freshwater shells, indicating a warmer climate. These may have been laid down in a lagoon similar to the Fleet today (see Figure 8.31 on page 241, Chesil Beach, which shows the lagoon, named the 'Fleet', on the landward side of Chesil Beach between the beach and the road linking Chesil with the mainland). The loam lies on a deposit of rounded deposits slightly cemented by calcium carbonate, in which all of the features of a shingle beach can be seen. The pebbles are of limestone, flint and chert, similar to those at Chesil but at a higher altitude. Some pebbles have come from as far as Devon and Cornwall. The shingle deposit is about 3 metres thick.

There is argument among geomorphologists concerning the dating and environmental meaning of these deposits. The current view is that the beach represents a high shoreline of 15–16 metres above present sea level. This would have been formed during the high sea level of a warm interglacial period

125 000 years ago, while the overlying head deposits would have been formed during the last glaciation when sea level was much lower.

Figure 8.43 Portland raised beach

8.3 Coral reefs

☐ Characteristics

Coral reefs are calcium carbonate structures, made up of reef-building stony corals. Coral is limited to the depth of light penetration and so reefs occur in shallow water, ranging to depths of 60 metres. This dependence on light also means that reefs are only found where the surrounding waters contain relatively small amounts of suspended material. Although corals are found quite widely, reef-building corals live only in tropical seas, where temperature, salinity and a lack of turbid water are conducive to their existence.

Coral reefs occupy less than 0.25 per cent of the marine environment, yet they shelter more than 25 per cent of all known marine life, including polyps, fish, mammals, turtles, crustaceans and molluscs. There are as many as 800 different types of rock-forming corals. Some estimates put the total diversity of life found in, on and around all coral reefs at up to 2 million species.

☐ The development of coral

All tropical reefs begin life as polyps – tiny, soft animals, like sea anemones – which attach themselves to a hard surface in shallow seas where there is sufficient light for growth. As they grow, many of these polyps exude calcium

carbonate, which forms their skeleton. Then as they grow and die these 'rock'-forming corals create the reefs.

Polyps have small algae, **zooxanthellae**, growing inside them. There is a symbiotic relationship between the polyps and the algae; that is, both benefit from the relationship. The algae get shelter and food from the polyp, while the polyp also gets some food via photosynthesis. This photosynthesis (turning light energy from the Sun into food) means that algae need sunlight to live, so corals only grow where the sea is shallow and clear.

Rate of growth

Tropical reefs grow at rates ranging from less than 2.5 cm to 60 cm per year, forming huge structures over incredibly long periods of time – which makes them the largest and oldest living systems on Earth. The 2600 kilometre Great Barrier Reef off eastern Australia, for example, was formed over 5 million years.

Conditions required for coral growth

The distribution of coral (Figure 8.44) is controlled by seven main factors:

■ **Temperature** – no reefs develop where the mean annual temperature is below 20°C; optimal conditions for growth are between 23°C and 25°C.
■ **Depth of water** – most reefs grow in depths of water less than 25 metres, and so they are generally found on the margins of continents and islands.

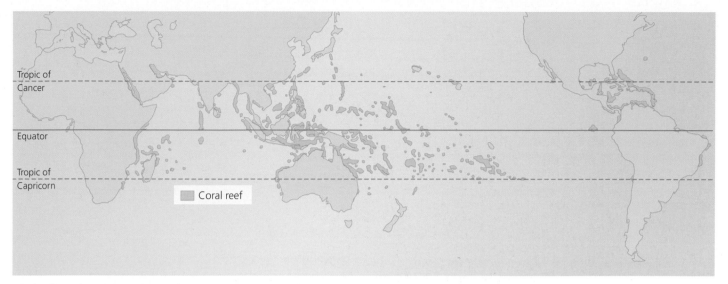

Figure 8.44 World distribution of coral reefs

■ **Light** – corals prefer shallow water because the tiny photosynthetic algae that live in the coral need light; in return, they supply the coral polyps with as much as 98 per cent of their food requirements.

■ **Salinity** – corals are marine organisms and are intolerant of water with salinity levels below 32 psu, although they can tolerate high salinity levels (> 42 psu), as found in the Red Sea or the Persian Gulf.

■ **Sediment** – this has a negative effect on coral; it clogs up their feeding structures and cleansing systems, and sediment-rich water reduces the light available for photosynthesis.

■ **Wave action** – coral reefs generally prefer strong wave action, which ensures oxygenated water, and where there is a stronger cleansing action, which helps remove any trapped sediment and also supplies microscopic plankton to the coral; however, in storm conditions such as the South Asian tsunami, the waves may be too destructive for the coral to survive.

■ **Exposure to the air** – coral die if they are exposed to the air for too long; they are therefore mostly found below the low tide mark.

□ Types of coral

■ **Fringing reefs** are those that fringe the coast of a land mass (Figure 8.45). They are usually characterised by an outer reef edge capped by an algal ridge, a broad reef flat and a sand-floored 'boat channel' close to the shore. Many fringing reefs grow along shores that are protected by barrier reefs and are thus characterised by organisms that are best adapted to low wave-energy conditions.

■ **Barrier reefs** occur at greater distances from the shore than fringing reefs and are commonly separated from it by a wide, deep lagoon. Barrier reefs tend to be broader, older and more continuous than fringing reefs; the Beqa barrier reef off Fiji stretches unbroken for more than 37 kilometres; that off Mayotte in the Indian Ocean for around 18 kilometres. The largest barrier-reef system in the world is the Great Barrier Reef, which extends 2600 kilometres along the east Australian coast, usually tens of kilometres offshore. Another long barrier reef is located in the Caribbean off the coast of Belize between Mexico and Guatemala.

■ **Atoll reefs** rise from submerged volcanic foundations and often support small islands of wave-borne detritus. Atoll reefs are essentially indistinguishable in form and species composition from barrier reefs except that they are confined to the flanks of submerged oceanic islands, whereas barrier reefs may also flank continents. There are over 300 atolls in the Indian and Pacific Oceans but only 10 are found in the western Atlantic.

■ **Patch reef** describes small circular or irregular reefs that rise from the sea floor of lagoons behind barrier reefs or within atolls.

Section 8.3 Activities

1 Outline the main factors that limit the distribution of coral reefs.
2 Describe the main types of coral reef.

Origin

The origin of fringing reefs is quite clear – they simply grow seaward from the land. Barrier reefs and atolls, however, seem to rise from considerable depth, far below the level at which coral can grow, and many atolls are isolated in deep water. The lagoons between the barrier and the coast are usually 45–100 metres in depth, and often many kilometres in width – and this requires some explanation.

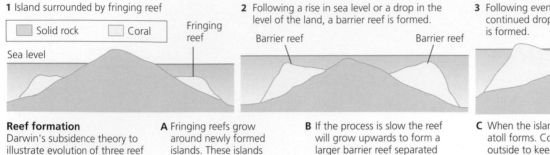

1 Island surrounded by fringing reef

Solid rock   Coral   Fringing reef

Sea level

2 Following a rise in sea level or a drop in the level of the land, a barrier reef is formed.

Barrier reef   Barrier reef

3 Following even more sea-level rises, and/or a continued drop in the level of the land, an atoll is formed.

Reef formation
Darwin's subsidence theory to illustrate evolution of three reef types (South Pacific model), linking the formation of the three types of reef together.

A Fringing reefs grow around newly formed islands. These islands subside, or sea level rises relative to land.

B If the process is slow the reef will grow upwards to form a larger barrier reef separated from the island by a deeper lagoon.

C When the island disappears beneath the sea an atoll forms. Corals can continue to grow on the outside to keep the reef on the surface. On the inside, where the land used to be, quiet water with increased sedimentation prevails.

Figure 8.45 Formation of fringing reefs, barrier reefs and atolls

In 1842, Charles Darwin, supported by Dana and others, explained the growth of barrier reefs and atolls as a gradual process, the main reason being subsidence. In his classic book *The Structure and Distribution of Coral Reefs*, Darwin outlines the way in which coral reefs could grow upwards from submerging foundations. From this, it became clear that fringing reefs might be succeeded by barrier reefs and then by atoll reefs. A fringing reef grows around an island, for example, and as the island slowly subsides, the coral continues to grow, keeping pace with the subsidence. Coral growth is more vigorous on the outer side of the reef, so it forms a higher rim, whereas the inner part forms an increasingly wide and deep lagoon (Figure 8.46). Eventually, the inner island is submerged, forming a ring of coral that is the atoll. Supporters of Darwin have shown that submergence has taken place, as in the case of drowned valleys along parts of Indonesia and along the Queensland coast of Australia. However, in other areas, such as the Caribbean, there is little evidence of submergence.

An alternative theory was that of Sir John Murray, who in 1872 suggested that the base of the reef consisted of a submarine hill or plateau rising from the ocean floor. These reached within 60 metres of the sea surface and consisted of either sub-surface volcanic peaks or wave-worn stumps. According to Murray, as a fringing reef grows, pounded by breaking waves, masses of coral fragments gradually accumulate on the seaward side, washed there by waves, and are cemented into a solid bank.

Yet another theory was that of Daly. He suggested that a rise in sea level might be responsible. A rise did take place in post-glacial times as ice sheets melted. He discovered traces of glaciation on the sides of Mauna Kea in Hawaii. The water there must have been much colder and much lower (about 100 metres) during glacial times. All coral would have died, and any coral surfaces would have been eroded by the sea. Once conditions started to warm, and sea level was rising, the previous coral reefs provided a base for the upward growth of

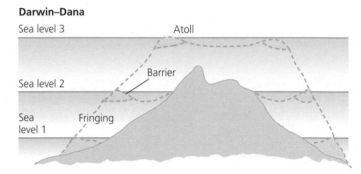

Darwin–Dana

Sea level 3   Atoll

Sea level 2   Barrier

Sea level 1   Fringing

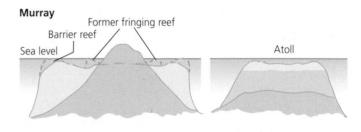

Murray

Former fringing reef
Barrier reef
Sea level   Atoll

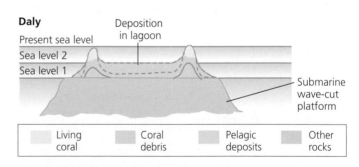

Daly   Deposition in lagoon
Present sea level
Sea level 2
Sea level 1
Submarine wave-cut platform

Living coral   Coral debris   Pelagic deposits   Other rocks

Figure 8.46 Theories of coral formation

coral. This theory helps account for the narrow, steep-sided reefs that comprise most atolls, some of which have 75° slopes.

Darwin's theory still receives considerable support. While Daly was correct in principle, it is now believed that erosion of the old reefs was much less rapid than was previously believed, and that the time available during the glacial low sea-level stages was inadequate for the formation of these bevelled platforms. Much of the erosional modification is now believed to be due to sub-aerial karstic (limestone) processes such as carbonation-solution.

☐ The value of coral

Coral reefs are among the most biologically rich ecosystems on Earth. About 4000 species of fish and 800 species of reef-building corals have been described. Coral reefs have often been described as 'the rainforests of the sea'. They resemble tropical rainforests in two ways: both thrive under nutrient-poor conditions (where nutrients are largely tied up in living matter), yet support rich communities through incredibly efficient recycling processes. Additionally, both exhibit very high levels of species diversity. However, coral reefs and other marine ecosystems contain a greater variety of life forms than do land habitats.

The NPP (net primary productivity) of coral reefs is $2500 \text{g/m}^{-2}/\text{year}$ and its biomass is 2kg/m^{-2}.

Coral reefs are not only important for their biodiversity; they are important to people too:

- **Seafood** – in LICs, coral reefs contribute about one-quarter of the total fish catch, providing food for up to a billion people in Asia alone; if properly managed, reefs can yield, on average, 15 tonnes of fish and other seafood per km² per year.
- **New medicines** – coral reef species offer particular hope because of the array of chemicals produced by many of these organisms for self-protection; corals are already being used for bone grafts, and chemicals found within several species appear useful for treating viruses, leukaemia, skin cancer and other tumours.
- **Other products** – reef ecosystems yield a host of other economic goods, ranging from corals and shells made into jewellery and tourism curios, to live fish and corals used in aquariums; and sand and limestone used by the construction industry.

Coral reefs also offer a wide range of environmental services, some of which are difficult to quantify but are of enormous importance to nearby inhabitants. These services include:

- **Recreational value** – the tourism industry is one of the fastest-growing sectors of the global economy; coral reefs are a major draw for snorkellers, scuba divers and recreational fishers (Figure 8.47).
- **Coastal protection** – coral reefs buffer adjacent shorelines through wave action and the impact of storms (Figure 8.48); the benefits of this protection are widespread, and range from maintenance of highly productive mangrove fisheries and wetlands to

Figure 8.47 Coral reef, Antigua

Figure 8.48 Storm waves batter a coral platform, Antigua

supporting local economies that are built around ports and harbours, which in the tropics are often sheltered by nearby reefs.

Coral reefs are therefore of major biological and economic importance. Countries such as Barbados, the Seychelles and the Maldives rely on tourism. Tourists attracted to Florida's reefs bring in US$1.6 billion annually, and the global value of coral reefs in terms of fisheries, tourism and coastal protection is estimated to be US$375 billion!

> ### Section 8.3 Activities
>
> Outline the main value of coral reefs.

☐ Threats to coral reefs

Global warming, sea-level rise, overfishing, destruction of the coastal habitat and pollution from industry, farms and households are endangering coral reefs. There are natural threats, too. Dust storms from the Sahara have introduced bacteria into Caribbean coral. Many areas of coral in the Indian Ocean were destroyed by the 2004 tsunami.

According to the World Resources Institute, 58 per cent of the world's coral reefs are at high or medium risk of degradation, with more than 80 per cent of South East Asia's extensive reef systems under threat.

Coral bleaching

Reef-building corals need warm, clear water. Unfortunately, pollution, sedimentation, global climate change and several other natural and anthropogenic pressures threaten this fundamental, biological need, effectively halting photosynthesis of the zooxanthellae and resulting in the death of the living part of the coral reef.

As noted above (page 248), coral lives in a symbiotic relationship with algae called zooxanthellae. Zooxanthellae live within the coral animal tissue and carry out photosynthesis, providing energy not only for themselves but for the coral too. Zooxanthellae give the coral its colour. However, when environmental conditions become stressful, zooxanthellae may leave the coral, leaving it in an energy deficit and without colour – a process that is referred to as **coral bleaching**. If the coral is re-colonised by zooxanthellae within a certain time, the coral may recover, but if not the coral will die. Coral bleaching can be caused by increases in water temperatures of as little as 1–2 °C above the average annual maxima. The shallower the water, the greater the potential for bleaching. As well as being caused by unusually warm waters – particularly if the water temperature exceeds 29 °C – bleaching may also be the result of changes in salinity, excessive exposure to ultraviolet radiation and climate change.

1998 coral bleaching

The 1998 episode of coral bleaching and mortality was the largest ever recorded on coral reefs globally, with major effects in the Arabian/Persian Gulf, East Africa, throughout the Indian Ocean, in South East Asia, parts of the western Pacific and the Caribbean and Atlantic region. Overall, it was estimated that 16 per cent of the world's area of coral reefs was severely damaged.

In 1998, there was extensive and intensive bleaching that affected the majority of coral reefs around Puerto Rico and the northern Caribbean. In the south-west region, a large number of coral colonies bleached completely (100 per cent of the living surface area) down to 40 metres deep. Maximum temperatures measured during 1998 in several reef localities ranged from 30.15 °C (20 metres deep) to 31.78 °C at the surface.

Like other parts of the wider Caribbean region, there was moderate to severe coral bleaching in 1998, but generally there were low levels of mortality. At one site in Barbados, approximately 20 per cent of bleached corals did not survive, but most reefs are showing signs of recovery from hurricanes, and sediment and bleaching damage from the previous 10 years.

Global assessment 2004

A 2004 report estimated that about one-fifth of the world's coral reefs are so damaged they are beyond repair. While the percentage of reefs recovering from past damage has risen, 70 per cent of the world's reefs are threatened or have already been destroyed, which is an increase from 59 per cent in 2000. Almost half of the reefs severely damaged by coral bleaching in 1998 are recovering, but other reefs are so badly damaged that they are unrecognisable as coral reefs.

The destruction of reefs is cause for economic, as well as ecological, concern, especially for the communities that depend on coral reefs for the fish they provide and the revenue they draw as tourist attractions.

The report states that the main causes of reef decline are climate change, which causes bleaching; poor land-management practices, which damage the reefs with sediments, nutrients and other pollutants; overfishing and destructive fishing practices; and coastal development. Other threats loom, especially climate change. Increased water temperatures have already been blamed for the single most destructive event for corals, the 1998 bleaching (see above).

The most damaged reefs are in the Persian Gulf where 65 per cent have been destroyed, followed by reefs in South and South East Asia where 45 and 38 per cent, respectively, are considered destroyed. There are also more recent reports that many reefs in the wider Caribbean have lost 80 per cent of their corals.

The report also provides some good news: the percentage of recovering reefs has increased when compared with the last global assessment. Most of the recovered reefs are in the Indian Ocean, part of the Great Barrier Reef off the coast of Australia and in the western Pacific, especially in Palau. In 2004, Australia increased protection of the Great Barrier Reef from 4 per cent to 33 per cent, and 34 per cent of Ningaloo Reef Marine Park was made off-limits from fishing.

Initiative in Climate Change and Coral Reefs, 2010

According to the Global Coral Reef Monitoring Network and the International Coral Reef *Initiative in Climate Change and Coral Reefs* published in 2010, the world's coral reefs were probably the first ecosystem to show major damage as a result of climate change. Reefs will suffer catastrophic collapse from climate change within the next few decades unless there are major and immediate reductions in greenhouse gas emissions. Global climate change will cause irreparable damage to coral reefs in our lifetime for several reasons:

- Increasing sea surface temperatures will cause more coral bleaching and mortality during summer. The abundance of many coral species will be reduced and some species may become extinct.

- Ocean temperatures will increase beyond the current maximum of natural variability, making bleaching a frequent, or eventually an annual, event.
- Increasing ocean acidification will reduce calcification in corals and other calcifying organisms, resulting in slower growth, weaker skeletons and eventual dissolution.
- A predicted increase in severe tropical storms will result in the destruction of corals and the erosion of coastlines.

Already, 19 per cent of the world's coral reefs have effectively been lost, and 35 per cent more are seriously threatened with destruction, mostly due to direct human threats. Climate change will cause even more dramatic losses, not just to tropical but also to coldwater corals.

Assessment of coral reefs, 2011

The assessment of coral reefs published in 2011 found that the majority of them were threatened by human activity:

- Over 60 per cent were under threat from local sources such as fishing, pollution and coastal development.
- When thermal stress (global warming) was added, the figure rose to over 75 per cent.

Changes in climate and ocean chemistry still represented significant and growing threats. As discussed above, rising temperatures have led to increased sea-surface temperatures, which have led to coral bleaching. Ocean acidification is another growing threat to coral. By 2050, over 90 per cent of coral will be affected by local and global threats.

Reefs remain highly valuable to people around the world and provide a number of benefits – food, disease prevention, tourism and shoreline protection. Degradation and loss of reef will result in significant social and economic impacts.

Climate change, coral and people

About 500 million people depend on coral reefs for some food, coastal protection, building materials and income from tourism. Among these, about 30 million people are dependent on coral reefs to provide their livelihoods, build up their land and support their cultures. Global climate change threatens these predominantly poor people, with many living in 80 small developing countries. Human well-being will be reduced for many people in rapidly growing tropical countries; 50 per cent of the world's population were estimated to live on coasts in 2015. This growth is putting unsustainable pressures on coastal resources. In 2009, the United Nations Environment Programme estimated that the coral reef area of 284 000 km^2 provides the world with more than US$100 billion per annum in goods and services. Even moderate climate change will seriously deplete that value.

Evidence of climate-change damage to coral reefs

Mass coral bleaching was unknown in the long oral history of many countries such as the Maldives and Palau, before their reefs were devastated in 1998. About 16 per cent of the world's corals bleached and died in 1998. In that year, 500 to 1000-year-old corals died in Vietnam, in the Indian Ocean and in the western Pacific.

Coral bleaching was only recorded as minor local incidents before the first large-scale bleaching was observed in 1983. The hottest years on record in the tropical oceans were in 1997/98, 2003, 2004 and 2005; the major bleaching years for Caribbean corals were in 1998 and 2005. Records for hurricanes in the wider Caribbean were broken in 2005. The bottom cover of corals on Caribbean reefs has dropped by more than 80 per cent since 1977, with much of this decline due to disease, coral bleaching or coral disease following bleaching. Any coral recovery was often reversed by other human pressures or by more bleaching and disease.

Between 50 and 90 per cent of corals died from bleaching on many reefs in the Indian Ocean in 1998. This caused major losses in tourism incomes and reduction in fish habitats. The first major bleaching event on the Great Barrier Reef was in 1998; since then, there has been major bleaching in 2002 and 2006. The growth rate of some coral species has declined by 14 per cent on the Great Barrier Reef since 1990, either due to temperature stress or ocean acidification or both. Other reef areas report similar results.

Ocean temperatures have risen in all oceans in the last 40 years, as seen from satellite images and other measures over 135 years from the National Oceanic and Atmospheric Administration of the USA.

Coral bleaching and death has reduced tourism incomes in countries like the Maldives, Palau and throughout the Caribbean. Sea-level rise has already threatened some coral island countries, such as Kiribati, Tuvalu, the Marshall Islands and the Maldives, with inundation, erosion, loss of agriculture and displacement of people and cultures.

☐ Management strategies to protect coral reefs

Global climate change seriously threatens the future of coral reefs. Current scientific thought is that coral reefs may become one of the first ecosystem casualties of climate change and could become functionally extinct if carbon dioxide levels rise to above 450 ppm – which could happen by 2030. It could affect the livelihoods of up to 500 million people whose lives depend on coral, and reduce the $100 billion that coral provides to the human economy. Global climate change and other human activities have affected about 19 per cent of the world's coral. Global temperatures are expected to rise by at least 2 °C, leading to widespread coral bleaching, extinction of coral species, more fragile skeletons and greater risk of storm damage.

This will make low-lying coastal communities more vulnerable to coastal hazards. This damage and the probable sea-level rise of 0.8–1.2 metres will seriously affect communities on Kiribati, the Marshall Islands, the Maldives and Tuvalu, and many communities will cease to exist. More than 3000 scientists at the 2008 International Coral Reef Symposium urged that greenhouse gases be reduced by 2018 in order to preserve coral reefs.

To avoid permanent damage and support people in the tropics, it is recommended that:

- the world community reduce the emissions of greenhouse gases and develop plans to sequester carbon dioxide
- damaging human activities (sedimentation, overfishing, blasting coral) be limited, to allow coral to recover from climate change threats
- assistance be provided to LICs
- alternative livelihoods be developed that reduce the pressure on coral reefs
- local coastal management practices be introduced
- strategies be developed to cope with climate-change damage
- the management, monitoring and enforcement of regulations be improved
- more coral reefs be designated as Marine Protected Areas (MPAs) to act as reservoirs of biodiversity, including many remote and uninhabited reefs that are still in good condition.

Section 8.3 Activities

1 Outline the main human impacts on coral reefs.
2 Examine the effects of sea-level change on coral reefs and describe the consequences.
3 To what extent is it possible to manage coral reefs?

8.4 Sustainable management of coasts

Human pressures on coastal environments create the need for a variety of coastal management strategies. These may be long term or short term, sustainable or non-sustainable. Successful management strategies require a detailed knowledge of coastal processes. Rising sea levels, more frequent storm activity and continuing coastal development are likely to increase the need for coastal management.

☐ Shoreline management plans (SMPs)

SMPs are plans in England and Wales designed to develop sustainable **coastal defence** schemes. Sections of the coast are divided up into littoral cells and plans are drawn up for the use and protection of each zone.

Defence options include:

- do nothing
- maintain existing levels of coastal defence
- improve the coastal defence
- allow retreat of the coast in selected areas.

Coastal management involves a wide range of issues:

- planning
- coastal protection
- cliff stabilisation and ground movement studies
- coastal infrastructure, including seawalls, esplanades, car parks, paths
- control of beaches and public safety
- recreational activities and sport
- beach cleaning
- pollution and oil spills
- offshore dredging
- management of coastal land and property.

☐ Coastal defence

Coastal defence covers protection against coastal erosion (coast protection) and flooding by the sea.

The coastal zone is a dynamic system that extends seawards and landwards from the shoreline. Its limits are defined by the extent of natural processes and human activities. Coastal zone management is concerned with the whole range of activities that take place in the coastal zone and promotes integrated planning to manage them. Conflicting activities in coastal areas include housing, recreation, fishing, industry, mineral extraction, waste disposal and farming.

Hard engineering structures

The effectiveness of seawalls depends on their cost and their performance. Their function is to prevent erosion and flooding (Figure 8.49), but much depends on:

- whether they are sloping or vertical
- whether they are permeable or impermeable
- whether they are rough or smooth
- what material they are made from (clay, steel or rock, for example).

Figure 8.49 A seawall and esplanade, Norfolk, UK

In general, flatter, permeable, rougher walls perform better than vertical, impermeable smooth walls.

Cross-shore structures such as **groynes**, breakwaters, piers and strongpoints have been used for many decades. Their main function is to stop the drifting of material. Traditionally, groynes were constructed from timber, brushwood and wattle. However, modern cross-shore structures are often made from rock (Figure 8.50). They are part of a more complex form of management that includes beach nourishment and offshore structures.

Managed natural retreat

Managed retreat allows nature to take its course – erosion in some areas, deposition in others. Benefits include less money spent, and the creation of natural environments. In parts of East Anglia in the UK, **hard engineering** structures are being replaced by bush defences, and some farmland is being sacrificed to erosion and being allowed to develop into saltmarsh.

Issues of coastal management

Tables 8.6, 8.7 and 8.8 present some of the issues involved in coastal management: different methods that can be put in place (hard and **soft engineering**), their costs, advantages and disadvantages. Figure 8.51 shows some of these methods of protection.

Figure 8.50 A rock strongpoint, Norfolk, UK

Table 8.6 Cost-benefit analysis of coastal defence

Costs	Benefits
• Cost of building • Maintenance/repair • Increased erosion downdrift due to beach starvation or reduced longshore drift • Reduced access to beach during works • Reduced recreational value • Reduced accessibility • Smaller beach due to scour • Disruption of ecosystems and habitats • Visually unattractive • Works disrupt natural processes	• Protected buildings, roads and infrastructure (gas, water, sewerage, electricity services) • Land prices rise • Peace of mind for residents • Employment on coastal defence works

Table 8.7 Conflicts and management strategies – relationships between human activities and coastal zone problems

Human activity	Agents/consequences	Coastal zone problems
Urbanisation and transport	Land-use changes (e.g. for ports, airports); road, rail and air congestion; dredging and disposal of harbour sediments; water abstraction; wastewater and waste disposal	Loss of habitats and species diversity; visual intrusion; lowering of groundwater table; saltwater intrusion; water pollution; human health risks; eutrophication; introduction of alien species
Agriculture	Land reclamation; fertiliser and pesticide use; livestock densities; water abstraction	Loss of habitats and species diversity; water pollution; eutrophication; river channelisation
Tourism, recreation and hunting	Development and land-use changes (e.g. golf courses); road, rail and air congestion; ports and marinas; water abstraction; wastewater and waste disposal	Loss of habitats and species diversity; disturbance; visual intrusion; lowering of groundwater table; saltwater intrusion in aquifers; water pollution; eutrophication; human health risks
Fisheries and aquaculture	Port construction; fish-processing facilities; fishing gear; fish-farm effluents	Overfishing; impacts on non-target species; litter and oil on beaches; water pollution; eutrophication; introduction of alien species; habitat damage and change in marine communities
Industry (including energy production)	Land-use changes; power stations; extraction of natural resources; process effluents; cooling water; windmills; river impoundment; tidal barrages	Loss of habitats and species diversity; water pollution; eutrophication; thermal pollution; visual intrusion; decreased input of fresh water and sediment to coastal zones; coastal erosion

Table 8.8 Coastal management

Type of management	Aims/methods	Strengths	Weaknesses
HARD ENGINEERING	To control natural processes		
Cliff-base management	To stop cliff or beach erosion		
Seawalls	Large-scale concrete curved walls designed to reflect wave energy	Easily made; good in areas of high density	Expensive; lifespan about 30–40 years; foundations may be undermined
Revetments	Porous design to absorb wave energy	Easily made; cheaper than seawalls	Lifespan limited
Gabions	Rocks held in wire cages, absorb wave energy	Cheaper than seawalls and revetments	Small scale
Groynes	To prevent longshore drift	Relatively low cost; easily repaired	Cause erosion on downdrift side; interrupt sediment flow
Rock armour	Large rocks at base of cliff to absorb wave energy	Cheap	Unattractive; small-scale; may be removed in heavy storms
Offshore breakwaters	Reduce wave power offshore	Cheap to build	Disrupt local ecology
Rock strongpoints	To reduce longshore drift	Relatively low cost; easily repaired	Disrupt longshore drift; erosion downdrift
Cliff-face strategies	To reduce the impacts of sub-aerial processes		
Cliff drainage	Removal of water from rocks in the cliff	Cost-effective	Drains may become new lines of weakness; dry cliffs may produce rockfalls
Cliff grading	Lowering of slope angle to make cliff safer	Useful on clay (most other measures are not)	Uses large amounts of land – impractical in heavily populated areas
SOFT ENGINEERING	Working with nature		
Offshore reefs	Waste materials, e.g. old tyres weighted down, to reduce speed of incoming wave	Low technology and relatively cost-effective	Long-term impacts unknown
Beach nourishment	Sand pumped from seabed to replace eroded sand	Looks natural	Expensive; short-term solution
Managed retreat	Coastline allowed to retreat in certain places	Cost-effective; maintains a natural coastline	Unpopular; political implications
'Do nothing'	Accept that nature will win	Cost-effective!	Unpopular; political implications
Red-lining	Planning permission withdrawn; new line of defences set back from existing coastline	Cost-effective	Unpopular; political implications

a Gabions (Brunei)

b Groynes (UK)

c Gabion mattress (UK)

d Revetment (UK)

e Cliff drains (UK)

Figure 8.51 Some methods of coastal protection

Section 8.4 Activities

1 Distinguish between hard and soft engineering; cliff-face and cliff-base management; managed retreat and 'hold the line'.
2 With the use of an annotated diagram, explain how seawalls and groynes work.

Along many parts of the USA's eastern seaboard, seawalls have protected buildings but not beaches. Many beaches along the east coast have disappeared in the last 100 years or so, such as Marshfield, Massachusetts and Monmouth Beach, New Jersey. As sea level rises, the beaches and barrier islands (barrier beaches) that line the coasts of the Atlantic Ocean and the Gulf of Mexico from New York to the Mexican border, are in retreat. This natural retreat does not destroy the beaches or barrier islands; it just moves them inland.

The problem is that much of the shore cannot retreat naturally because there are industries and properties worth billions of dollars on them. Many important cities and tourist centres, such as Miami, Atlantic City and Galveston (Texas), are sited on barrier islands. Consequently, many shoreline communities have built seawalls and other protective structures to defend them from the power of destructive waves. Such fortifications, which can cost millions of dollars for a single kilometre, protect structures, at least for the short term, but they accelerate erosion elsewhere. The first great seawall was built at Galveston after a hurricane in 1900 devastated the city and killed more than 6000 people. The city survived a later hurricane, but lost its beach. Now, the rising sea level is making protection by the seawalls less effective. Much of the city is less than 3 metres above sea level.

Three factors put the east coast of the USA at particularly high risk from changing sea levels. First, the flat topography of the coastal plains from New Jersey southward means that a small rise in sea level can make the ocean advance a long way inland. A rise of just a few millimetres each year in sea level could push the ocean a metre inland, while a rise of a few metres could threaten large areas such as southern Florida. Miami, in particular, faces severe problems as it is the lowest-lying US city facing the open ocean. Few places in metropolitan Miami are more than 3 metres above sea level.

Second, much of the North American coast is sinking relative to the ocean, so local sea levels are rising faster than global averages. The level of tides along the coasts shows that subsidence varies between 0.5 and 19.5 millimetres a year. By contrast, the west coast, in particular Alaska, is rising (Figure 8.52).

Skagway, Alaska (+19.5)

Juneau, Alaska (+13.8)

San Diego (+0.4)   Los Angeles (+0.4)

0

New York (−1.5)   Boston (−1.0)   Miami Beach (−1.1)
              Atlantic City (−2.9)
              Galveston, Texas (−5.1)

         Grand Isle, Louisiana (−8.9)

         Sabine Pass, Texas (−12.0)

Source: *New Scientist*

Figure 8.52 Relative sea-level change in the USA

Third, extensive coastal development has accelerated erosion. While sea level rises, apartment blocks, resorts and second homes have developed rapidly along the shoreline. By 1990, 75 per cent of Americans lived within 100 kilometres of a coast (including the Great Lakes shores).

Until the late 1970s, most Americans assumed they could successfully protect their coastline against the rising sea. Now they are considering an alternative: strategic retreat. The term 'retreat' does not mean abandoning the shore, but moving back from it. Instead of protecting the coast with seawalls, buildings are moved away from the rising sea, and new buildings are not allowed to be too close to the sea. Engineers have stopped challenging nature and have begun to work with natural coastal processes.

In the long term, this makes the most economic sense. While it is impossible and impractical to abandon coastal cities such as Boston and New York (Figure 8.53), state and federal governments are discouraging some new coastal development, especially in areas presently undeveloped.

Figure 8.53 Coastal protection in New York

The nature of erosion further complicates the issue. It is far from a uniform process. Most erosion occurs during coastal storms, especially at high tide. Superstorm Sandy, which affected New York in October 2012, is an excellent example. Wind-driven waves create storm surges that flood low-lying areas, causing severe damage. In addition, annual storm intensities are very variable, so coastal geologists try to plan for '100-year' storms; that is, an intensity likely to be experienced only once every century. This can make their plans seem excessively cautious to coastal residents, especially in areas that have not experienced a severe storm for many years.

Erosion is a dynamic process that varies from storm to storm and from point to point along the shore. Nevertheless, in many places there are observable cycles of erosion and deposition. During calm conditions, moderate currents often redeposit large quantities of sediment removed during a severe storm. This natural compensation reduces total erosion, but it can also disguise the real hazards of storms.

For example, storm damage at Chatham, a town on Cape Cod in Massachusetts, illustrates the dynamics of erosion. In 1987, winter storms broke through a barrier beach separating Chatham Harbour from the Atlantic. This dramatically increased erosion in areas exposed to the full strength of ocean waves. By 1988, more than 20 metres had been eroded from the shore.

The events at Chatham are part of a 140-year cycle of erosion and deposition. The ocean moves sand along the eastern shore of Cape Cod, forming and then eroding a barrier island that protects Chatham Harbour. Once the ocean breaks through, currents deposit sand on the north side of the inlet, building the island southward. The inlet moves south, as the ocean erodes the southern island. Eventually, the northern island builds far enough south that the ocean again breaks through during a storm.

Some ocean currents passing through the inlet erode the mainland. As the inlet moves south, currents in the harbour deposit sand in areas eroded earlier. Many homes now threatened were built over 50 years ago on sand dunes that were deposited during the last cycle of erosion and beach building.

In other parts of Cape Cod, and elsewhere along the American coast, erosion is changing the shoreline permanently. The sea is eroding about a metre a year from the glacial banks (which now form sandy cliffs) on the east side of Cape Cod. These cliffs and Cape Cod were formed 15 000 years ago at the end of the Ice Age. The Cape is less than 2 kilometres wide at its narrowest point: if nature takes its course, the northern end will be left an island within a few thousand years. Human interference, such as building seawalls, could accelerate this.

Further south, erosion has isolated the Cape Hatteras lighthouse, located on a barrier island off the coast of North Carolina. The lighthouse was built in 1870 about 460 metres from the ocean, but by the 1930s the ocean had eroded all but about 30 metres. Except for a small promontory around the lighthouse itself, the shore has receded nearly 500 metres since 1870.

Erosion is evident at many other places along the coast of the Atlantic and the Gulf of Mexico. Major resorts such as Miami Beach and Atlantic City have pumped in dredged sand to replenish eroded beaches. Erosion threatens islands to the north and south of Cape Canaveral, although the cape itself appears safe. Resorts built on barrier beaches in Virginia, Maryland and New Jersey have also suffered major erosion.

Overall losses are not well known. Massachusetts loses about 26 hectares a year to rising seas. Nearly 10 per cent of that loss is from the island of Nantucket, south of Cape Cod. However, these losses pale into insignificance when compared with Louisiana, which is losing 40 hectares of wetlands a day – about 15 500 hectares a year.

Florida's extreme measures to combat erosion are well known. Intense development of Miami Beach in the 1920s started the widespread exploitation of coastal areas exposed to major storms and erosion. At the same time, coastal towns in New Jersey, such as Sea Bright and Monmouth Beach, began building seawalls and groynes to prevent erosion. Since 1945, there have been many developments in coastal areas near large cities, especially for holiday homes and retirement communities.

Hard defences can cost millions of dollars a kilometre, and they require maintenance. Despite their cost, seawalls have failed at several places, including in Texas, South Carolina and California. This is usually due to flaws in construction or poor maintenance.

Many US coastal geologists believe that the best compromise between building defences and leaving the shore to be eroded is pumping sand from other locations, usually offshore, to replace eroded sand. The main limitations include the cost and the possible loss of the new sand. For example, between 1976 and 1980, the US Army Corps of Engineers spent $64 million on beach replenishment and flood prevention at Miami Beach. Erosion quickly removed 30 metres of the new sand, but then the beach stabilised at 60 metres wide. Other coastal resorts, including Atlantic City and Virginia Beach, Virginia, have chosen to add sand rather than build structures to keep out the sea.

Elsewhere, land-use management has been introduced. Regulations vary widely. North Carolina, Maine and Massachusetts are in the forefront of restricting development. In Massachusetts, for example, there are restrictions on new developments of natural areas, although it is neither practical nor possible to abandon Boston's city centre or the international airport, both of which are built on low-lying land facing the harbour. The Massachusetts Wetlands Protection Act limits building on coastal land. The regulations ban seawalls or permanent structures to control erosion on coastal dunes, as these are dynamic areas that supply the sand to beaches. Similarly, North Carolina was one of the first coastal states to legislate that land be left between the shore and new buildings to allow for erosion. Since 1979, small new buildings must be located inland of a line that marks 30 times the annual rate of erosion from the shore. In 1983, the state doubled the distance from the sea for large buildings. In addition, in 1986 the state banned the construction of hard defences, such as rock strongpoints and groynes. Although this was a controversial decision, people and developers have adjusted, in part because the state's beaches are a major economic asset. Moreover, few people want huge seawalls and tiny beaches.

Other states, such as South Carolina and Texas, impose few limitations, or even encourage coastal development. For example, developers have built high-rise condominiums close to the shore line at Myrtle Beach, South Carolina. Similarly, at Galveston, Texas, a new beach-front apartment block was built at the west end of the Galveston seawall, where the rate of erosion is 5 metres a year.

Rising sea levels and retreating coasts could pose continuing tough economic and environmental issues for Americans in the future. Some forms of protective hard engineering are certainly justified for major coastal cities such as New York, but it might not be justified for less developed areas such as Carolina Beach, North Carolina.

Section 8.4 Activities

1 Suggest why the US eastern seaboard needs coastal protection.

2 Identify the forms of coastal management that have been used. To what extent have they been effective?

A proposal by Kaipara Excavators is to mine designated areas of the continental shelf close to Auckland for 2 million m³ of sand at depths of 25–60 metres, over an area of 500 km², in the course of 35 years. If such sand were taken from the entire area, it would amount to no more than a fraction of a millimetre per year. But it is feared that the operation will affect nearby beaches. From an environmental viewpoint, disturbing such a large area, even once in 30 years, would disturb the bottom communities that together form the sea soil. Before 1970, sand was mined from the Omaha Beach, off the Whangateau Harbour. This has been blamed for the Omaha sand spit changing shape and eroding badly in the early 1970s, before remedial groynes were built.

Sand has become a very important mineral for the expansion of society. It is used in the making of glass and concrete, for roads, reclamation, on building sites and for the replenishment of beaches. Clean sand is a rare commodity on land, but common in sand dunes and beaches. On average, people 'use' over 200 kilograms of sand per person each year. This sand is taken from what are essentially non-renewable resources. In New Zealand, the cost of 1 m³ of replenishment sand is about NZ$40 (US$30).

Central to the need for using any resource is the question of how long one can do so before it runs out – its 'sustainability'. Although sand is one of the world's most plentiful resources (perhaps as much as 20 per cent of the Earth's crust is sand), clean sand is becoming rare, particularly since muddy deposits from soil erosion worldwide are now filling the coastal shelves and basins.

In 1994, the Minister of Conservation granted commercial sand extractors five resource consents (coastal permits) to dredge sand from the nearshore seabed at Mangawhai and Pakiri (Figure 8.54). The permits allowed a total of up to 165 000 m³ of sand to be excavated annually for 10 years. The permits ended in 2004.

A working party, chaired by the Auckland Regional Council (ARC), was set up to oversee the study. This was required to investigate:

- the overall extent and volume of the sand
- the long-term sustainable level of near-shore extraction (less than 25 metres deep)
- adverse effects on the environment.

Their specific objectives were to:

- establish a sediment budget and quantify sediment transport
- determine the long-term shoreline trend and short-term fluctuations

- determine the broad sediment characteristics and composition of the sand resource
- determine the relationship (if any) between extraction and the long-term shoreline trend.

The study concluded that there were very large amounts of 'modern' sand in dunes, on the beach, near shore and offshore. There were also extremely large amounts of Pleistocene sands underlying the modern sand. However,

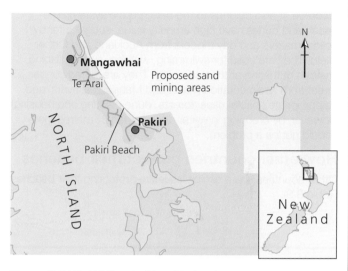

Figure 8.54 Pakiri Beach, New Zealand

the amount of sand is static, with extraction exceeding inputs. Since the end of the last Ice Age, the shoreline has widened by 150–200 metres, although since 1920 the shoreline has varied by 40 metres but without a long-term trend. Moreover, where extraction occurred, the retreat of shoreline could not be related to extraction. It was found that in the embayment no effects due to extraction could be proven.

The study offered the following options:

1 Sand extraction should be continued at its present rate.
2 Sand extraction in the near-shore area should be phased out since it will eventually cause the gradual retreat of the embayment shoreline.

The Pakiri–Mangawhai sand system is bounded by rocky promontories in the north and south. Deep waves arrive from the north-east through the gap between Great Barrier Island and the Hen and Chicken Islands. Sea winds are predominantly easterly (westerly winds are land winds). ⇨

The Great Barrier Reef supports 1500 species of fish, 400 species of coral and 4000 species of molluscs. It is a major feeding ground for many endangered species and is a nesting ground for many species of turtle. It was placed on the World Heritage List in 1981.

The reef is now carefully managed, but previously it suffered from the effects of tourism, agriculture and recreational and commercial fishing. Each year, 77 000 tonnes of nitrogen, 11 000 tonnes of phosphorus and 15 million tonnes of sediment are washed into the coastal waters from Queensland.

The Great Barrier Reef Marine Park Authority is responsible for the management and development of the reef. It follows the Agenda 21 philosophy; namely, that resources must be used and managed in such a way that they are not destroyed or devalued for future generations.

The main type of management is that of land-use zoning (Table 8.10). This means that some areas can be used for some things, such as recreation or fishing; while other areas are used for other activities, such as scientific research and conservation. The main aims of zoning are to:

- ensure permanent conservation of the area
- provide protection for selected species and ecosystems
- separate conflicting activities
- preserve some untouched areas
- allow human use of the reef as well as protecting the reef.

The Great Barrier Reef and Australia's coal industry

A 2012 study showed that over half the coral had died during the previous 27 years, due to cyclones, bleaching and the coral-eating crown-of-thorns starfish. At the same time, agricultural runoff, global warming and the development of new gas and coal ports along the Queensland coastline have emerged as additional threats to the fragmented coral ecosystem.

The development of coal resources is dividing many communities. Environmentalists claim that the government is showing a lack of commitment to tackling climate change and reducing the country's dependence on fossil fuels. Australia is the world's second largest exporter of coal (behind Indonesia) and shipped over $30 billion worth of coal in 2014. The sector employs 50 000 people directly and a further 150 000 in related industries. However, around one-third of Australia's coal mines operate at a loss. Critics argue that it is pointless to invest in a declining industry. The Galilee Basin in Queensland (Figure 8.57) contains some of the world's largest coal deposits. If developed, protestors claim that Australia would emit 700 million tonnes of CO_2 into the atmosphere for over 50 years. The resulting global warming would kill off much of the reefs. In addition, dredging and dumping of coal spill are already damaging the reefs, and this situation would in all likelihood escalate with the mining of the Galilee Basin coal deposits.

Table 8.10 Land-use zoning along the Barrier Reef

	Bait netting and gathering	Camping	Collecting (recreational – not coral)	Collecting (commercial)	Commercial netting (see also bait netting)	Crabbing and oyster gathering	Diving, boating, photography	Line fishing (bottom fishing, trolling, etc.)	Research (non-manipulative)	Research (manipulative)	Spear fishing	Tourist and education facilities and programme	Traditional hunting, fishing and gathering	Trawling
General Use 'A'	Yes	Permit	Limited	Permit	Yes	Yes	Yes	Yes	Yes	Permit	Yes	Permit	Permit	Yes
General Use 'B'	Yes	Permit	Limited	Permit	Yes	Yes	Yes	Yes	Yes	Permit	Yes	Permit	Permit	No
Marine National Park 'A'	Yes	Permit	No	No	No	Limited	Yes	Limited	Permit	Permit	Yes	Permit	Permit	Yes
Marine National Park 'B'	No	Permit	No	No	No	No	Yes	No	Permit	Permit	No	Permit	No	No
Scientific Research	No	No	No	No	No	No	No	No	Permit	Permit	No	No	No	No
Preservation Zone	No	No	No	No	No	No	No	No	Permit	Permit	No	No	No	No

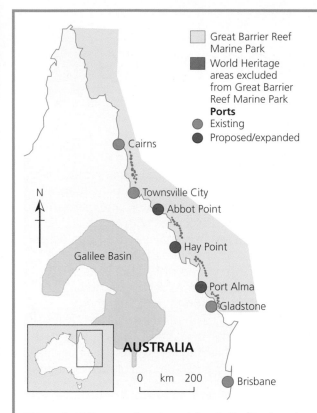

Great Barrier Reef Marine Park

World Heritage areas excluded from Great Barrier Reef Marine Park

Ports
- ● Existing
- ● Proposed/expanded

Cairns

N

Townsville City

Abbot Point

Galilee Basin

Hay Point

Port Alma

Gladstone

AUSTRALIA

0   km   200

Brisbane

Figure 8.57 Proposed ports and the Galilee Basin – threats to the Great Barrier Reef from the coal industry

Those in favour of the coal mines point out that development would create 15 000 construction jobs in Queensland and provide cheap energy to millions of people across Asia. In 2014, the Great Barrier Reef Marine Park Authority (an independent but government-funded body that manages the reef) approved plans to dredge the port area and dump 3 million tonnes of spoil (waste) within the marine park. Protestors claim that dredging destroys feeding grounds for dugongs (a type of manatee) and other animals.

In 2015, the Australian government published *Reef 2050*, allocating $80 million for reef protection and targets for reducing sediment, nitrogen and pesticide loads. However, there is no plan to tackle climate change. There is little sign that Australia is prepared to lead on action to tackle global warming. It remains one of the world's biggest emitters of CO_2 per person. Prospects of the government slowing the development of the mining industry also appear slim.

Section 8.4 Activities

1 Using examples, describe the variety of pressures that affect coral reefs.
2 Why is the Great Barrier Reef a World Heritage Site?
3 Study Table 8.10.
   a What is meant by the term *land-use zoning*?
   b How do the types of activities that are allowed in the Preservation Zone compare with the activities that are allowed in General Use 'A'?
4 a Suggest why the Australian government is keen to develop the Galilee Basin coal deposits.
   b Suggest the likely impact of the development of the coal industry on the Great Barrier Reef.

9 Hazardous environments

9.1 Hazards resulting from tectonic processes

☐ Global distribution of tectonic hazards

Distribution of earthquakes

Tectonic hazards include seismic activity (earthquakes), volcanoes and tsunamis. Most of the world's earthquakes occur in clearly defined linear patterns (Figure 9.1). These linear chains generally follow plate boundaries. For example, there is a clear line of earthquakes along the centre of the Atlantic Ocean in association with the Mid-Atlantic Ridge (a constructive plate boundary). Similarly, there are distinct lines of earthquakes around the Pacific Ocean. In some cases, these linear chains are quite broad, for example the line of earthquakes along the west coast of South America and around the eastern Pacific associated with the subduction of the Nazca Plate beneath the South American Plate – a destructive plate

boundary. Broad belts of earthquakes are associated with **subduction zones** (where a dense ocean plate plunges beneath a less dense continental plate), whereas narrower belts of earthquakes are associated with constructive plate margins, where new material is formed and plates are moving apart. Collision boundaries, such as in the Himalayas, are also associated with broad belts of earthquakes, whereas conservative plate boundaries, such as California's San Andreas fault line, give a relatively narrow belt of earthquakes (this can still be over 100 kilometres wide). In addition, there appear to be isolated occurrences of earthquakes. These may be due to human activities, or to isolated plumes of rising **magma**, known as 'hotspots'.

Distribution of volcanoes

Most volcanoes are found at plate boundaries (Figure 9.1) although there are some exceptions, such as the volcanoes of Hawaii, which occur over hotspots. About three-quarters of the Earth's 550 historically active volcanoes lie along the Pacific Ring of Fire. This includes many of the world's most recent volcanoes, such as Mt Pinatubo

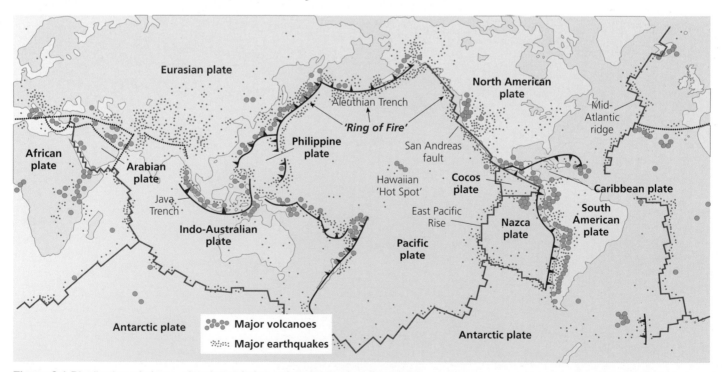

Figure 9.1 Distribution of plates, plate boundaries, volcanoes and earthquakes

in the Philippines, Mt Unzen (Japan), Mt Agung (Java), Mt Chichon (Mexico), Mt St Helens (USA) and Nevado del Ruiz (Colombia). Other areas of active vulcanicity include Iceland, Montserrat in the Caribbean and Mt Nyiragongo in Democratic Republic of Congo. Most volcanoes that are studied are above land, but some submarine volcanoes, such as Kick 'em Jenny off Grenada in the Caribbean, are also monitored closely.

Volcanoes are found along the boundaries of the Earth's major plates. Although the deeper levels of the Earth are much hotter than the surface, the rocks are usually not molten because the pressure is so high. However, along the plate boundaries there is molten rock – magma – which supplies the volcanoes.

Most of the world's volcanoes are found in the Pacific Rim or Ring of Fire (Figure 9.1). These are related to the subduction beneath either **oceanic** or **continental crust**. Subduction in the oceans provides chains of volcanic islands known as 'island arcs', such as the Aleutian Islands formed by the Pacific Plate subducting beneath the North American Plate. Where the subduction of an oceanic crust occurs beneath the continental crust, young fold mountains are formed. The Andes, for example, have been formed where the Nazca Plate subducts beneath the South American Plate.

Not all volcanoes are formed at plate boundaries. Those in Hawaii, for example, are found in the middle of the ocean (Figure 9.2). The Hawaiian Islands are a line of increasingly older volcanic islands that stretch north-west across the Pacific Ocean. These volcanoes can be related to the movement of plates above a hot part of the fluid mantle. A mantle **plume** or **hotspot** – a jet of hot material rising from deep within the mantle – is responsible for the volcanoes. Hotspots can also be found beneath continents, as in the case of the East African Rift Valley, and can produce isolated volcanoes. These hotspots can play a part in the break-up of continents and the formation of new oceans.

At subduction zones, volcanoes produce more viscous **lava**, tend to erupt explosively and produce much ash. By contrast, volcanoes that are found at **mid-ocean ridges** or hotspots tend to produce relatively fluid basaltic lava, as in the case of Iceland and Hawaii. At mid-ocean ridges, hot fluid rocks from deep in the mantle rise up due to convection currents. The upper parts of the mantle begin to melt and basaltic lava erupts, forming new oceanic crust. By contrast, at subduction zones a slab of cold ocean floor slides down the subduction zone, warming up slowly. Volatile compounds such as water and carbon dioxide leave the slab and move upwards into the mantle so that it melts. The hot magma is then able to rise.

Huge explosions occur wherever water meets hot rock. Water vaporises, increasing the pressure until the rock explodes. Gases from within the molten rock can also build up high pressures. However, the likelihood of a big, explosive eruption depends largely on the viscosity of the magma and hence its composition. Gases dissolve quite easily in molten rock deep underground due to the very high pressures there. As magma rises to the surface, the pressure drops and some of the gas may become insoluble and form bubbles. In relatively fluid magma, the bubbles rise to the surface. By contrast, viscous magma can trap gas so that it builds up enough pressure to create a volcanic eruption.

The style of eruption is greatly influenced by the processes operating at different plate boundaries, which produce magma of different, but predictable, composition. Some minerals melt before others in a process called **partial melting**. This alters the composition of molten rock produced. Partial melting of the Earth's mantle produces basalt. At subduction zones, the older and deeper slabs experience greater partial melting and this produces a silica-rich magma.

Source: *Advanced Geography: Concepts & Cases* by P. Guinness & G. Nagle (Hodder Education, 1999), p.339

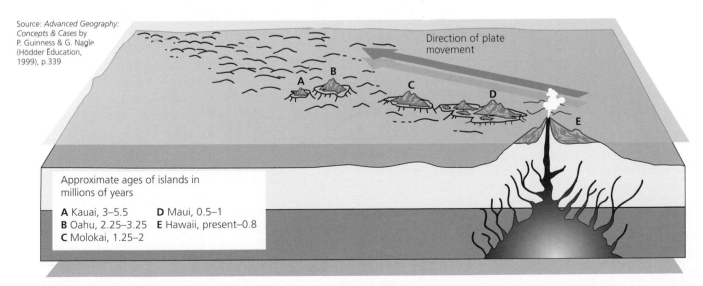

Figure 9.2 Hotspots and the evolution of Hawaii

Tsunamis

Up to 90 per cent of the world's tsunamis occur in the Pacific Ocean. This is because they are associated with subduction zones and, as Figure 9.1 shows, most subduction zones are found in the Pacific Ocean.

☐ Earthquakes and resultant hazards

An earthquake is a series of vibrations or seismic (shock) waves that originate from the focus – the point at which the plates release their tension or compression suddenly (Figure 9.3). The **epicentre** marks the point on the surface of the Earth immediately above the focus of the earthquake. A large earthquake can be preceded by smaller tremors known as **foreshocks** and followed by numerous **aftershocks**. Aftershocks can be particularly devastating because they damage buildings that have already been damaged by the first main shock. Seismic waves are able to travel along the surface of the Earth and also through the body of the Earth.

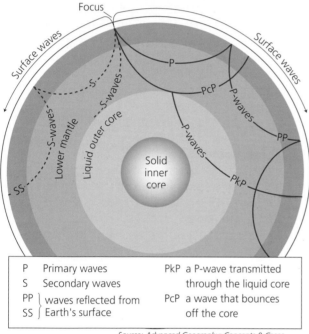

P	Primary waves	PkP	a P-wave transmitted
S	Secondary waves		through the liquid core
PP	waves reflected from	PcP	a wave that bounces
SS	Earth's surface		off the core

Source: *Advanced Geography: Concepts & Cases*
by P. Guinness & G. Nagle (Hodder Education, 1999), p.334

Figure 9.3 Seismic waves

Following an earthquake, two types of **body waves** (waves within the Earth's interior) occur. The first are P-waves (primary waves or pressure waves) and the second are the transverse S-waves. These are a series of oscillations at right-angles to the direction of movement.

P-waves travel by compression and expansion, and are able to pass through rocks, gases and liquids. S-waves travel with a side-to-side motion, and are able to pass through solids but not liquids and gases, since they have no rigidity to support sideways motion. In 1909, Andrija Mohorovičić, a Yugoslavian geophysicist who was studying earthquakes in Croatia, detected four kinds of seismic wave, two of them pressure waves and two of them shear waves. Seismographs close to the earthquake epicentre showed slow-travelling P-waves and S-waves. By contrast, those further away from the shock showed faster-moving S-waves and P-waves. These shock waves are reflected or refracted when they meet rock with different densities. If the shock waves pass through denser rocks, they speed up. If they pass through less dense rocks, they slow down. Mohorovičić deduced that the slower waves had travelled from the focus of the earthquake through the upper layer of the crust. By contrast, the faster waves must have passed through the denser material in the Earth's core; this denser material speeded up the waves and deflected them. He suggested that a change in density from 2.9g/cm^3 to 3.3g/cm^3 marks the boundary between the Earth's crust and the mantle below. This boundary is known as the 'Mohorovičić Discontinuity' or quite simply the 'Moho'.

Later geologists found a shadow zone, an area between 105° and 142° from the source of the earthquake, within which they could not detect shock waves. The explanation was that the shock waves had passed from a solid to a liquid. Thus S-waves would stop and P-waves would be refracted. The geologists concluded that there was a change in density from 5.5g/cm^3 at 2900 kilometres to a density of 10g/cm^3. This was effectively the boundary between the mantle and the core. Within the Earth, there is an inner core of very dense solid material – the density of the inner core goes up to as much as 13.6g/cm^3 at the centre of the Earth.

When P- and S-waves reach the surface, some of them become surface waves. Love waves cause the earth to move sideways whereas Rayleigh waves cause the earth to move up and down. Surface waves often do the most damage in an earthquake.

The nature of rock and sediment beneath the ground influences the pattern of shocks and vibrations during an earthquake. Unconsolidated sediments such as sand shake in a less predictable way than solid rock. Hence the damage is far greater to foundations of buildings. P-waves from earthquakes can turn solid sediments into fluids like quicksand by disrupting sub-surface water conditions. This is known as **liquefaction** or **fluidisation** and can wreck foundations of large buildings and other structures.

Resultant hazards of earthquakes

Most earthquakes occur with little, if any, advance warning. Some places, such as California and Tokyo, which have considerable experience of earthquakes, have developed 'earthquake action plans' and information programmes to increase public awareness about what to do in an earthquake.

Most problems are associated with damage to buildings, structures and transport systems (Table 9.1). The collapse of building structures is the direct cause of many injuries and deaths, but it also reduces the effect of the emergency services. In some cases, more damage is caused by the aftershocks that follow the main earthquake, as they shake the already weakened structures. Aftershocks are more subdued but longer lasting and more frequent than the main tremor. Buildings partly damaged during the earthquake may be completely destroyed by the aftershocks.

Table 9.1 Earthquake hazards and impacts

Primary hazard	Impacts
• Ground shaking • Surface faulting	• Loss of life • Loss of livelihood • Total or partial destruction of building structure
Secondary hazard	• Interruption of water supplies
• Ground failure and soil liquefaction • Landslides and rockfalls • Debris flow and mudflow • Tsunamis	• Breakage of sewage disposal systems • Loss of public utilities such as electricity and gas • Floods from collapsed dams • Release of hazardous material • Fires • Spread of chronic illness

Some earthquakes involve surface displacement, generally along fault lines. This may lead to the fracture of gas pipes, as well as causing damage to lines of communication. The cost of repairing such fractures is considerable.

Earthquakes may cause other geomorphological hazards such as **landslides**, liquefaction (the conversion of unconsolidated sediments into materials that act like liquids) and tsunamis. For example, the Good Friday earthquake (magnitude 8.5), which shook Anchorage (Alaska) in March 1964, released twice as much energy as the 1906 San Francisco earthquake, and was felt over an area of nearly 1.3 million km². More than 130 people were killed, and over $500 million of damage was caused. It triggered large **avalanches** and landslides that caused much damage. It also caused a series of tsunamis through the Pacific as far as California, Hawaii and Japan.

The relative importance of factors affecting earthquakes varies a great deal. For example, the Kobe earthquake of January 1995 had a magnitude 7.2 and caused over 5000 deaths. By contrast, the Northridge earthquake that affected parts of Los Angeles in January 1994 was 6.6 on the **Richter Scale** but caused only 57 deaths. On the other hand, an earthquake of force 6.6 at Maharashtra in India in September 1993 killed over 22 000 people.

So why did these three earthquakes have such differing effects? Kobe and Los Angeles are on known earthquake zones and buildings are built to withstand earthquakes. In addition, local people have been prepared for earthquake events. By contrast, Maharashtra has little experience of earthquakes. Houses were unstable and quickly destroyed, and people had little idea of how to manage the situation.

Another earthquake in an area not noted for seismic activity shows that damage is often most serious where buildings are not designed to withstand shaking or ground movement. In the 1992 Cairo earthquake, many poor people in villages and the inner-city slums of Cairo were killed or injured when their old, mud-walled homes collapsed. At the same time, many wealthy people were killed or injured when modern high-rise concrete blocks collapsed – some of these had actually been built without planning permission.

Earthquakes and plate boundaries

The movement of oceanic crust into the subduction zone creates some of the deepest earthquakes recorded, from 700 kilometres below the ground. When the oceanic crust slides into the hotter fluid mantle, it takes time to warm up. As the slab descends, it distorts and cracks and eventually creates earthquakes. However, subduction is relatively fast so by the time the crust has cracked it has slid several hundred kilometres down into the mantle.

In areas of active earthquake activity, the chances of an earthquake increase with increasing time since the last earthquake. Plates move at a rate of between 1.5 and 7.5 centimetres a year (the rate at which fingernails grow). However, a large earthquake can involve a movement of a few metres, which could occur every couple of hundred years rather than movements of a few centimetres each year. Many earthquakes are caused by the pressure created by moving plates. This increases the stress on rocks; the rocks deform and eventually give way and snap. The snapping is the release of energy; namely, the earthquake. The size of the earthquake depends upon the thickness of the descending slab and the rate of movement. Along mid-ocean ridges, earthquakes are small because the crust is very hot, and brittle faults cannot extend more than a few kilometres. The strength of an earthquake is measured by the Richter Scale and the **Mercalli Scale**.

The Richter and Mercalli Scales

In 1935, Charles Richter of the California Institute of Technology developed the Richter Scale to measure the magnitude of earthquakes. The scale is logarithmic, so an earthquake of 5.0 on the Richter Scale is 10 times more powerful than one of 4.0 and 100 times more powerful than one of 3.0. Scientists are increasingly using the **Moment Magnitude Scale M**, which measures the amount of energy released and produces figures that are similar to the Richter Scale. For every increase on the scale of 0.1, the amount of energy released increases by over 30. Every increase of 0.2 represents a doubling of the energy released.

By contrast, the Modified Mercalli Intensity Scale relates ground movement to commonplace observations around light bulbs and bookcases (Table 9.2). It has the advantage that it allows ordinary eyewitnesses to provide information on how strong the earthquake was. It is important to remember that these scales are only used to measure the 'strength' of an earthquake, not to predict earthquakes. Table 9.3 gives some idea of the number and magnitude of earthquakes experienced around the world each year.

Table 9.2 The Modified Mercalli Scale

1	Rarely felt.
2	Felt by people who were not moving, especially on upper floors of buildings; hanging objects may swing.
3	The effects are notable indoors, especially upstairs. The vibration is like that experienced when a truck passes.
4	Many people feel it indoors, a few outside. Some are awakened at night. Crockery and doors are disturbed and standing cars rock.
5	Felt by nearly everyone; most people are awakened. Some windows are broken, plaster becomes cracked and unstable objects topple. Trees may sway and pendulum clocks stop.
6	Felt by everyone; many are frightened. Some heavy furniture moves, plaster falls. Structural damage is usually quite slight.
7	Everyone runs outdoors. Noticed by people driving cars. Poorly designed buildings are appreciably damaged.
8	Considerable amount of damage to ordinary buildings; many collapse. Well-designed ones survive but with slight damage. Heavy furniture is overturned and chimneys fall. Some sand is fluidised.
9	Considerable damage occurs even to buildings that have been well designed. Many are moved from their foundations. Ground cracks and pipes break.
10	Most masonry structures are destroyed, sub-wooden ones survive. Railway tracks bend and water slops over river banks. Landslides and sand movements occur.
11	No masonry structure remains standing, bridges are destroyed. Broad fissures occur in the ground.
12	Total damage. Waves are seen on the surface of the ground, objects are thrown into the air.

Table 9.3 Annual frequency of occurrence of earthquakes of different magnitude based on observations since 1900

Descriptor	Magnitude (Richter Scale)	Annual average	Hazard potential
Great	≥ 8	1	Total destruction, high loss of life
Major	7–7.9	18	Serious building damage, major loss of life
Strong	6–6.9	120	Large losses, especially in urban areas
Moderate	5–5.9	800	Significant losses in populated areas
Light	4–4.9	6200	Usually felt, some structural damage
Minor	3–3.9	49000	Typically felt but usually little damage
Very minor	≤ 3	9000 per day	Not felt, but recorded

Factors affecting earthquake damage

The extent of earthquake damage is influenced by a variety of factors:

- **Strength and depth of earthquake and number of aftershocks** – The stronger the earthquake, the more damage it can do, for example an earthquake of 6.0 on the Richter Scale is 100 times more powerful than one of 4.0. The more aftershocks there are, the greater the damage that is done. Earthquakes that occur close to the surface (shallow-focus earthquakes) potentially should do more damage than earthquakes deep underground (deep-focus earthquakes) as more of the energy of the latter is absorbed by overlying rocks.
- **Population density** – An earthquake that hits an area of high population density, such as the Tokyo region of Japan, could inflict far more damage than one that hits an area of low population and building density.
- **The type of buildings** – HICs generally have better-quality buildings, more emergency services and the funds to recover from disasters. People in HICs are more likely to have insurance cover than those in LICs.
- **The time of day** – An earthquake during a busy time, such as rush hour, may cause more deaths than one at a quiet time. Industrial and commercial areas have fewer people in them on Sundays; homes have more people in them at night.

- **The distance from the centre (epicentre) of the earthquake** – The closer a place is to the centre (epicentre) of the earthquake, the greater the damage that is done.
- **The type of rocks and sediments** – Loose materials may act like liquid when shaken, a process known as 'liquefaction' ('fluidisation'); solid rock is much safer and buildings should be built on flat areas formed of solid rock.
- **Secondary hazards** – An earthquake may cause mudslides, tsunamis (high sea waves) and fires; also contaminated water, disease, hunger and hypothermia.
- **Economic development** – This affects the level of preparedness and effectiveness of emergency response services, access to technology and quality of health services.

Deaths following an earthquake can be substantial, as Table 9.4 shows quite clearly.

Table 9.4 The world's worst earthquakes by death toll in the twenty-first century

Country	Year	Death toll (est.)	Richter Scale
Haiti	2010	300000	7.0
South East Asia	2004	248000	9.1
Kashmir, Pakistan	2005	86000	7.6
Chengdu, China	2008	78000	7.9
Bam, Iran	2003	30000	6.6
Tohuku, Japan	2011	15891	9.0
Gorkha, Nepal	2015	9000	7.8

Case Study: Earthquake in Haiti – 12 January 2010, 16:53 local time, 7.0 magnitude

The country of Haiti occupies the western part of Hispaniola, a Caribbean island that it shares with the Dominican Republic. Haiti is characterised by poverty, environmental degradation, corruption and violence. On 12 January 2010, an earthquake recorded as 7.0 on the Richter Scale occurred 25 kilometres south-west of Port-au-Prince at a depth of just 13 kilometres (Figure 9.4). Aftershocks were as strong as 5.9, occurring just 9 kilometres below the surface and 56 kilometres south-west of the city. A third of the population were affected. About 300 000 people died as a result of the earthquake, 250 000 more were injured and some 1 million made homeless.

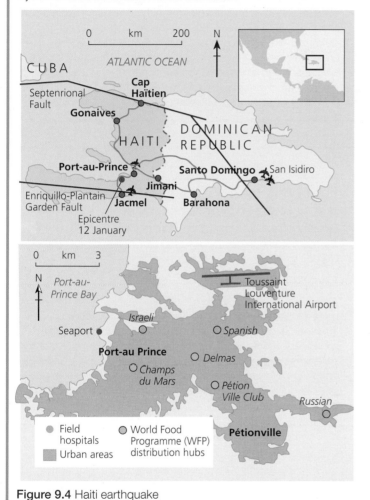

Figure 9.4 Haiti earthquake

Hispaniola sits on the Gonave microplate, a small strip of the Earth's crust squeezed between the North American and Caribbean tectonic plates. This makes it vulnerable to rare but violent earthquakes. The Dominican Republic suffered a serious 'quake in 1946, but the Enriquillo-Plantain Garden fault that separates the plates on the Haitian side of the border had been accumulating stress during more than a century of inactivity. Two things magnified its destructive power: its epicentre was just 25 kilometres south-west of Port-au-Prince and its focus was only 13 kilometres below ground.

The region is hopelessly ill-suited to withstand a shaking. Most of Port-au-Prince's 2 million residents live in tin-roofed shacks perched on unstable, steep ravines. After a school collapsed in the suburb of Pétionville in 2008, the capital's mayor said that 60 per cent of its buildings were shoddily constructed and unsafe even under normal conditions.

The Red Cross estimated that 3 million people – a third of Haiti's population – might need emergency aid. Seven days after the earthquake, the UN had managed to get food to only 200 000 people. Help – including doctors, trained sniffer dogs, and tents, blankets and food – was pledged from other countries, including Mexico, Venezuela, China, UK, France, Germany, Canada and Cuba.

Financial assistance also poured in. The UN released $10 million from its emergency fund, and European countries pledged $13.7 million. Haiti's institutions were weak even before the disaster. Because the 'quake devastated the capital, both the government and the UN, which has been trying to build a state in Haiti since 2004, were seriously affected, losing buildings and essential staff.

Following the Haiti earthquake, plans were discussed for the rescue, rehabilitation and reconstruction of the country. Reconstructing Haiti is a challenge to an international community that has failed over decades to lift the island state out of poverty, corruption and violence. Since 2000, more than $4 billion has been given to Haiti to rebuild communities and infrastructure devastated by tropical storms, floods and landslides, but mismanagement, a lack of coordination and attempts by global institutions to use Haiti as an economic test-bed are believed to have frustrated all efforts. A foreign debt of $1.5 billion has weighed down the economy.

Case Study: Tōhoku, Japan, earthquake and tsunami, 2011

The earthquake that occurred off the east coast of Japan in 2011 was magnitude 9.0 M. The epicentre was approximately 70 kilometres east of Tōhoku at a depth of about 30 kilometres. It was the most powerful earthquake ever to hit Japan, and the fourth most powerful since 1900. The earthquake caused a tsunami that generated some waves in excess of 12 metres, which killed thousands and damaged a large part of the Sendai area.

There were nearly 16 000 deaths, more than 2500 people missing and over 225 000 people forced either to live in temporary housing or relocate permanently. More than 125 000 buildings totally collapsed and a further 1 million buildings were damaged. The earthquake and tsunami caused widespread and severe structural damage to roads and railways. Some 4.4 million households in north-eastern Japan were left without electricity and 1.5 million without water. More than 1.5 million households were reported to have lost access to water supplies. The tsunami caused accidents at a number of nuclear power stations, in particular Fukushima Daiichi. ⇒

Estimates suggested insured losses from the earthquake alone at US$14.5–34.6 billion. The World Bank estimated that the economic cost was US$235 billion, making it the costliest natural disaster ever.

One minute before the earthquake was felt in Tokyo, the Earthquake Early Warning System sent out warnings to millions of people. It is believed that this may have saved many lives.

The tsunami began to hit the coastline just 10 to 30 minutes after the main earthquake. The damage from the tsunami was far greater than from the earthquake. Many of the waves were higher than the protective sea walls. It is likely that many people thought the sea walls would protect them, but these had been built on the experience of smaller tsunamis in the past. Of the casualties, over 90 per cent died by drowning. Victims aged 60 or older accounted for over 65 per cent of the deaths. A number of children were orphaned as a result of the tsunami.

Japan has invested the equivalent of billions of dollars on anti-tsunami seawalls along at least 40 per cent of its 35 000 kilometre coastline; the tsunami simply washed over the top of some seawalls, collapsing some in the process. About 10 per cent of Japan's fishing ports were damaged in the disaster.

Eleven reactors were automatically shut down following the earthquake. However, at Fukushima Daiichi, tsunami waves overtopped seawalls and destroyed backup power systems, leading to three large explosions and radioactive leakage. Over 200 000 people were evacuated from the area.

Japan declared a state of emergency following the failure of the cooling system at Fukushima Daiichi. Radiation levels inside the plant were up to 1000 times normal levels; outside the plant they were up to 8 times normal levels.

The earthquake and tsunami created a major humanitarian crisis and an economic one. Over 340 000 people were displaced in the Tōhoku region, and there were widespread shortages of food, water, shelter, medicine and fuel for survivors. Aid organisations donated around $1 billion in emergency relief. The short-term economic impact has been the suspension of industrial production in many factories, and the long-term issue has been the cost of rebuilding, which has been estimated at US$122 billion.

Case Study: Nepal earthquake, 2015

The Gorkha (Nepal) earthquake in April 2015 killed over 9000 people and injured more than 23 000. It had a magnitude of 7.8 M, and occurred about 80 kilometres north-west of the capital, Kathmandu. It was a shallow earthquake, with the focus approximately 8 kilometres beneath the surface. It was the worst natural disaster to affect Nepal since 1934.

The earthquake triggered a number of avalanches, killing at least 19 people on Mt Everest and over 250 in Langtang Valley.

Hundreds of thousands of people were made homeless. According to UNESCO, more than 30 monuments in the Kathmandu Valley collapsed in the quakes.

In addition, there was a major aftershock of 7.3 M in May 2015. Over 200 people were killed and more than 2500 were injured by this aftershock. The earthquakes were caused by a release of built-up stress along a fault line where the Indian Plate is colliding against the Eurasian plate.

Economic loss

The US Geological Survey estimated economic losses from the earthquake of about 35 per cent of GDP. Rebuilding the economy could exceed US$5 billion, or about 20 per cent of Nepal's GDP.

Rescue and relief

About 90 per cent of the soldiers from the Nepalese army helped with the rescue operation. However, rainfall and aftershocks complicated the rescue efforts, with potential secondary effects like additional landslides and further building collapses being of concern. Impassable roads and a damaged communications infrastructure posed substantial challenges to rescue efforts. Survivors were found up to a week after the earthquake.

Earthquakes and human activity

Human activities can trigger earthquakes, or alter the magnitude and frequency of earthquakes, in three main ways:

- through underground disposal of liquid wastes
- by underground nuclear testing and explosions
- by mining and fracking
- by increasing crustal loading.

Disposal of liquid waste

In the Rocky Mountain Arsenal in Denver, Colorado, wastewater was injected into underlying rocks during the 1960s (Figure 9.5). Water was contaminated by chemical warfare agents, and the toxic wastes were too costly to transport off-site for disposal. Thus it was decided to dispose of it down a well over 3500 metres deep. Disposal began in March 1962 and was followed soon afterwards by a series of minor earthquakes, in an area previously free of earthquake activity. None of the earthquakes caused any real damage, but they did cause alarm. Between 1962 and 1965, over 700 minor earthquakes were monitored in the area.

The injection of the liquid waste into the bedrock lubricated and reactivated a series of deep underground faults that had been inactive for a long time. The more wastewater was put down the well, the larger the number of minor earthquakes. When the link was established, disposal stopped. In 1966, the well was filled in and the number of minor earthquake events detected in the area fell sharply.

a Earthquake frequency

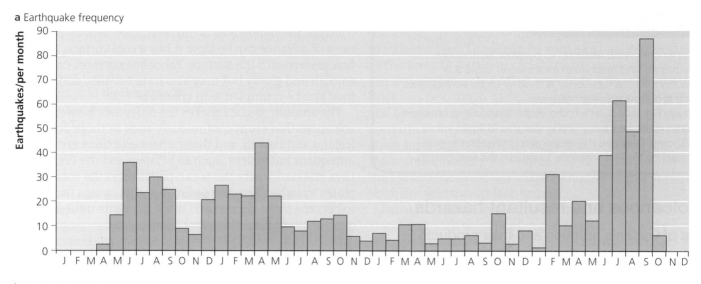

b Contaminated waste injected

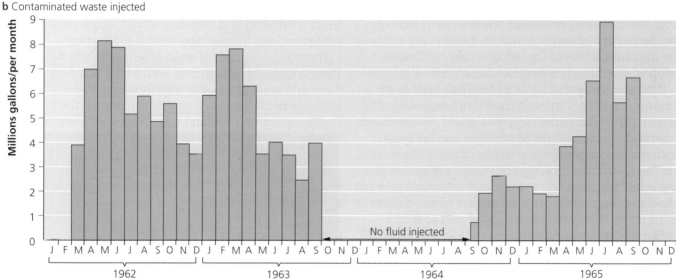

Source: *Advanced Geography: Concepts & Cases* by P. Guinness & G. Nagle (Hodder Education, 1999), p.338

Figure 9.5 Increasing earthquake frequency associated with underground liquid-waste disposal, Rocky Mountain Arsenal, Colorado, USA

Underground nuclear testing

Underground nuclear testing has triggered earthquakes in a number of places. In 1968, testing of a series of 1200 tonne bombs in Nevada set off over 30 minor earthquakes in the area over the following three days. Since 1966, the Polynesian island of Moruroa has been the site of over 80 underground nuclear explosion tests by France. More than 120 000 people live on the island. In 1966, a 120 000 tonne nuclear device was detonated, producing radioactive fallout that was measured over 3000 kilometres downwind.

Fracking

It is believed that fracking (hydraulic fracturing) of shale rocks for shale gas can trigger earthquakes. The use of high-powered water to break up shale rocks is thought to have triggered two earthquakes in Lancashire, UK, in 2011.

It is one reason that Chinese engineers have not tried to develop the Sichuan province for shale gas, as the area is known to be tectonically active, having experienced a major earthquake there in 2008.

Increased crustal loading

Earthquakes can be caused by adding increased loads on previously stable land surfaces. For example, the weight of water behind large reservoirs can trigger earthquakes. In 1935, the Colorado River was dammed by the Hoover Dam to form Lake Mead. As the lake filled, over a period of ten years, and the underlying rocks adjusted to the new increased load of over 40 km^3 of water, long-dormant faults in the area were reactivated, causing over 6000 minor earthquakes. Over 10 000 events were recorded up to 1973, about 10 per cent of which were strong enough to be felt by residents. None caused damage.

Volcanic gases

Volcanic gases are an example of a direct or primary hazard. Cameroon lies just north of the equator in West Africa. It contains a large number of deep crater lakes, such as Lake Nyos, formed as a result of tectonic activity. Lake Nyos is nearly 2 kilometres wide and over 200 metres deep. In August 1986, a huge volume of gas escaped from the lake and swept down into neighbouring valleys for a distance of up to 25 kilometres (Figure 9.9). The ground-hugging clouds of gas were up to 50 metres thick and travelled at speeds of over 70 kilometres per hour. Some 1700 people were suffocated, 3000 cattle died and all other animal life in the area was killed. The only people who escaped were sleeping on the upper floors of houses. Plants, however, were unaffected.

The gas was carbon dioxide. Because it is heavier and denser than oxygen, the 50 metre cloud deprived people and animals of oxygen, so they were asphyxiated. The source of carbon dioxide was a basaltic chamber of magma, deep beneath Cameroon. It had been leaking into and accumulating in Lake Nyos for some time. Due to its depth, water in the lake became stratified into layers of warmer water near the surface and colder denser water near the bottom of the lake. The cold dense water absorbed the carbon dioxide, which was then held down by the weight of the overlying waters.

The disaster occurred after the water at the bottom of the lake was disturbed. The cause of the disturbance is unclear. It could have been a deep volcanic eruption, an earthquake, a change in water temperature or a climatic event. Whatever the cause, the effect was like an erupting champagne bottle. Once the overlying pressure was reduced, carbon dioxide escaped into the surrounding area, causing rapid death among people and animals.

It is likely that such a tragedy will happen again. It is believed that only about 66 per cent of the carbon dioxide escaped from the lake, and that it has begun to build up again. It may take several decades for the gas cloud to occur again, or maybe even centuries, but the potential for a disaster is there. The authorities are trying to drain the lake of carbon dioxide with pumps.

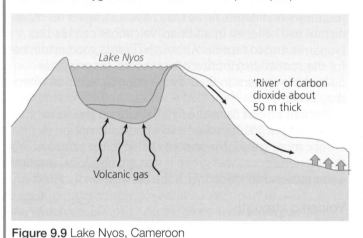

Figure 9.9 Lake Nyos, Cameroon

Lake Nyos

'River' of carbon dioxide about 50 m thick

Volcanic gas

Section 9.1 Activities

1 Explain why the disaster at Lake Nyos affected animals but not plants.
2 Cameroon is not close to a tectonic boundary. How do you explain the tectonic hazard in an area that is not close to a known boundary?

☐ Secondary hazards of tectonic events

Lahars and mudflows

One hazard that is closely associated with volcanic activity is the lahar, or mudflow:

■ Rain brings soot and ash back to ground and this becomes a heavily saturated mudflow.
■ Heat from volcanoes melts snow and ice – the resulting flow picks up sediment and turns it into a destructive lahar.

Nevado del Ruiz is a volcano in Colombia that rises to an altitude of 5400 metres and is covered with an icecap 30 metres thick, covering an area of about 20 km². In 1984, small-scale volcanic activity resumed, and large-scale activity returned in November 1985. Scientists monitoring the mountain recorded earthquakes, and soon after a volcanic eruption threw hot, pyroclastic material onto the icecap, causing it to melt.

Condensing volcanic steam, ice-melt and pyroclastic flows combined to form lahars that moved down the mountain, engulfing the village of Chinchina, killing over 1800 people and destroying the village (Figure 9.10).

Conditions worsened as further eruptions melted more ice, creating larger lahars that were capable of travelling further down the mountain into the floodplain of the Rio Magdalena. Within an hour, it had reached the city of Armero, 45 kilometres away. Most of Armero, including 22 000 of its 28 000 residents, were crushed and suffocated beneath lahars up to 8 metres thick. Those who were saved were those who just happened to be further up the slope. Images of people trapped in the mud were relayed across the world.

The volcanic eruption was relatively small but the presence of the icecap made the area especially hazardous.

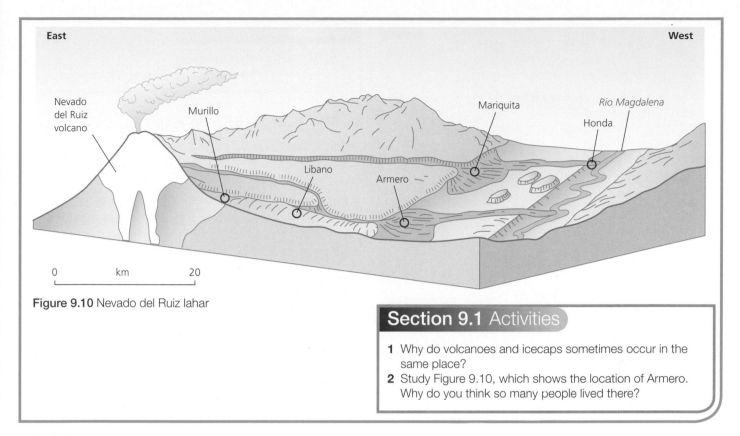

East

West

Nevado del Ruiz volcano

Murillo

Libano

Armero

Mariquita

Honda

Rio Magdalena

0   km   20

Figure 9.10 Nevado del Ruiz lahar

Section 9.1 Activities

1 Why do volcanoes and icecaps sometimes occur in the same place?
2 Study Figure 9.10, which shows the location of Armero. Why do you think so many people lived there?

☐ Tsunamis

The term 'tsunami' is the Japanese for 'harbour wave'. Ninety per cent of tsunamis occur in the Pacific basin. They are generally caused by earthquakes (usually in subduction zones) but can be caused by volcanoes (for example, Krakatoa in 1883) and landslides (for example, Alaska in 1964).

Tsunamis have the potential to cause widespread disaster, as in the case of the South Asian tsunami on 26 December 2004 (Figure 9.11). Owing to the loss of life among tourists, it came to be seen as a global disaster, killing people from nearly 30 countries. Between 180 000 and 280 000 people were killed in this tsunami.

Figure 9.11 Tsunami damage in Phuket, Thailand

The cause of the tsunami was a giant earthquake and landslide caused by the sinking of the Indian Plate under the Eurasian Plate. Pressure had built up over many years and was released in the earthquake that reached 9.0 on the Richter Scale.

The main impact of the Boxing Day tsunami, as it came to be known, was on the Indonesian island of Sumatra, the closest inhabited area to the epicentre of the earthquake. More than 70 per cent of the inhabitants of some coastal villages died. Apart from Indonesia, Sri Lanka suffered more from the tsunami than anywhere else. At least 31 000 people are known to have died there, when the southern and eastern coastlines were devastated.

Potential waves due to earthquakes and landslides

A lake formed by a landslide in northern Pakistan could have burst its banks at any time, possibly triggering a giant wave that could sweep down the Himalayan valley and swamp dozens of villages (Figure 9.12). The level of the Attabad lake, which was formed by a landslide in early January 2010, rose alarmingly fast to within a few metres of its limit.

Pakistani authorities were concerned that immense water pressure could cause the lake wall to collapse suddenly, sending a wave up to 60 metres high into the valley below and affecting up to 25 000 people.

The Attabad lake started to form after a landslide blocked the Karakoram highway, which links Pakistan and

Figure 9.12 Area of potential lake bursts in the Himalayas

China. The water level rose rapidly, swelled by meltwater from nearby glaciers, swamping 120 houses and displacing about 1300 people. Another 12000 people were evacuated from the potential flood zone downstream.

The world's largest landslide dam was formed in 1911 on the Murghab River in Tajikistan. The 550 metre dam has never breached because lake outflows are greater than inflows.

Geomorphologists estimate that 35 natural dams have formed over 500 years in the Pakistani section of the Himalayas. The latest was the Hattian dam, formed by the 2005 earthquake. Lakes formed by landslides developed in Nepal during 2013 and 2014, triggering fears that villages downstream could be destroyed by lake bursts.

Case Study: Peruvian tsunami, 2010

Tsunamis can be caused by forces that are not tectonic. For example, in 2010 a massive ice block, measuring 500 metres by 200 metres, broke from a glacier and crashed into a lake in the Peruvian Andes, causing a 23 metre tsunami and sending muddy torrents through nearby towns, killing at least one person.

The chunk of ice detached from the Hualcan glacier about 320 kilometres north of the capital, Lima. It plunged into a lagoon known as lake 513, triggering a tsunami that breached 23 metre-high levees and damaged Carhuaz and other villages.

Around 50 homes and a water-processing plant serving 60000 residents were wrecked. Due to global warming, there has been an increase in the number of glaciers melting, breaking and falling on overflowing lakes.

Section 9.1 Activities

1 Outline the causes of tsunamis.
2 Outline the short-term and long-term impacts of tsunamis.

☐ The perception of risk

At an individual level, there are three important influences upon an individual's response to any hazardous event:

1 **Experience** – the more experience a person has of environmental hazards, the greater the adjustment to the hazard.
2 **Material well-being** – those who are better off have more choices.
3 **Personality** – is the person a leader or a follower, a risk-taker or risk-minimiser?

Ultimately, there are just three choices:

1 Do nothing and accept the hazard.
2 Adjust to the situation of living in a hazardous environment.
3 Leave the area.

It is the adjustment to the hazard that we are interested in. The level of adjustment will depend, in part, upon the risks caused by the hazard. This includes:

- identification of the hazards
- estimation of the risk (probability) of the environmental hazard
- evaluation of the cost (loss) caused by the environmental hazard.

☐ Hazard mapping, risk assessment and preparedness

Hazard mapping includes a body of theory that includes risk, prediction, prevention, event and recovery. **Vulnerability** refers to the geographic conditions that increase the susceptibility of a community to a hazard or to the impacts of a hazard event. **Risk** is the probability of a hazard event causing harmful consequences (expected losses in terms of death, injuries, property damage, economy and environment).

A **hazard** is a threat (whether natural or human) that has the potential to cause loss of life, injury, property damage, socio-economic disruption or environmental degradation. In contrast, a **disaster** is a major hazard event that causes widespread disruption to a community or region, with significant demographic, economic and/or environmental losses.

A number of stages can be observed in the build-up to a disaster and in its aftermath (Table 9.7).

Rehabilitation refers to people being able to make safe their homes and be able to live in them again. This can be a very long drawn-out process, taking up to a decade for major construction projects.

As well as dealing with the aftermath of a disaster, governments try to plan to reduce impacts of future events. This was seen after the South Asian tsunami of 2004. Before the event, a tsunami early-warning system was not in place in the Indian Ocean. Following the event, as well as emergency rescue, rehabilitation and reconstruction, governments and aid agencies in the region developed a system to reduce the impacts of future tsunamis. It is just part of the process needed to reduce the impact of hazards and to improve safety in the region.

Managing the earthquake hazard

People deal with earthquakes in a number of ways. These include:

- doing nothing and accepting the hazard
- adjusting to living in a hazardous environment, for example strengthening their home
- leaving the area.

The main ways of preparing for earthquakes include:

- better forecasting and warning
- improved building design and building location
- establishing emergency procedures.

There are a number of ways of predicting and monitoring earthquakes, which involve the measurement of:

- small-scale ground surface changes
- small-scale uplift or subsidence
- ground tilt
- changes in rock stress
- micro-earthquake activity (clusters of small 'quakes)
- anomalies in the Earth's magnetic field
- changes in radon gas concentration
- changes in electrical resistivity of rocks.

One particularly intensively studied site is Parkfield in California, on the San Andreas fault. Parkfield, with a population of fewer than 50 people, claims to be the earthquake capital of the world. It is heavily monitored by instruments:

- Strain meters measure deformation at a single point.
- Two-colour laser geodimeters measure the slightest movement between tectonic plates.
- Magnetometers detect alterations in the Earth's magnetic field, caused by stress changes in the crust.

Table 9.7 Aspects of the temporal sequences or phases of disasters, with reported durations and selected features of each phase

Stage	Duration	Features
I Preconditions		
Phase 1	Everyday life (years, decades, centuries)	'Lifestyle' risks, routine safety measures, social construction of vulnerability, planned developments and emergency preparedness.
Phase 2	Premonitory developments (weeks, months, years)	'Incubation period' – erosion of safety measures, heightened vulnerability, signs and problems misread or ignored.
II The disaster		
Phase 3	Triggering event or threshold (seconds, hours, days)	Beginning of crisis; 'threat' period: impending or arriving flood, fire, explosion; danger seen clearly; may allow warnings, flight or evacuation and other pre-impact measures. May merge with …
Phase 4	Impact and collapse (instant, seconds, days, months)	… the disaster proper: concentrated death, injury, devastation; impaired or destroyed security arrangements; individual and small groups cope as isolated survivors. Followed by or merging with …
Phase 5	Secondary and tertiary damages (days, weeks)	… exposure of survivors, post-impact hazards, delayed deaths.
Phase 6	Outside emergency aid (weeks, months)	Rescue, relief, evacuation, shelter provision, clearing dangerous wreckage, 'organised response'; national and international humanitarian efforts.
III Recovery and reconstruction		
Phase 7	Clean-up and 'emergency communities' (weeks, years)	Relief camps, emergency housing; residents and outsiders clear wreckage, salvage items; blame and reconstruction debates begin; disaster reports, evaluations, commissions of enquiry.
Phase 8	Reconstruction and restoration (months, years)	Reintegration of damaged community with larger society; re-establishment of 'everyday life', possibly similar to, possibly different from, pre-disaster; continuing private and recurring communal grief; disaster-related development and hazard-reducing measures.

Nevertheless, the 1994 Northridge earthquake was not predicted and it occurred on a fault that scientists did not know existed. Technology helps, but not all of the time.

Learning to live with earthquakes

Most places with a history of earthquakes have developed plans that enable people to deal with them. The aim is to reduce the effect of the earthquakes and thus save lives, buildings and money. The ways of reducing earthquake impact include earthquake prediction, building design, flood prevention and public information.

Preparation

Earthquakes killed about 1.5 million people in the twentieth century, and the number of earthquakes appears to be rising. Most of the deaths were caused by the collapse of unsuitable and poorly designed buildings. More than a third of the world's largest and fastest-growing cities are located in regions of high earthquake risk, so the problems are likely to intensify.

It is difficult to stop an earthquake from happening, so prevention normally involves minimising the prospect of death, injury or damage by controlling building in high-risk areas, and using aseismic designs (Figure 9.13). In addition, warning systems can be used to warn people of an imminent earthquake and inform them of what to do when it does happen. Insurance schemes are another form of preparation, by sharing the costs between a wide group of people.

The seismic gap theory states that over a prolonged period of time all parts of a plate boundary must move by almost the same amount. Thus if one part of the plate boundary has not moved and others have, then the part that has not moved is most likely to move next. This theory has been used successfully to suggest that an earthquake was likely in the Loma Prieta segment of the San Andreas fault. The Loma Prieta earthquake occurred in 1989. Following the 2004 South Asian tsunami, geologists identified a seismic gap in the Central Kuril segment of the Kuril-Kamchatka trench. Two earthquakes measuring 8.3 and 8.2 on the Richter Scale occurred in November 2006 and January 2007 within the Central Kuril segment.

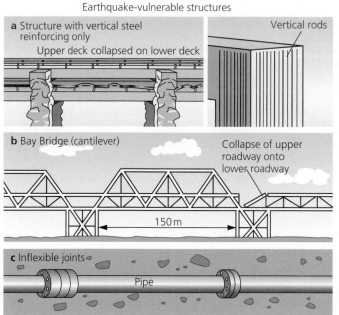

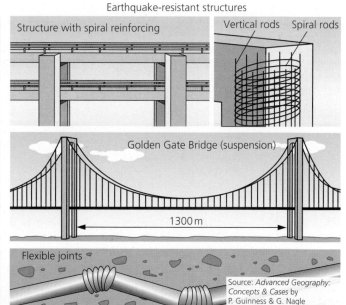

Source: *Advanced Geography: Concepts & Cases* by P. Guinness & G. Nagle (Hodder Education, 1999), p.347

Figure 9.13 Aseismic design

Building design

Increasingly, as the availability of building land is reduced, more and more people are living in seismic areas. This increases the potential impact of an earthquake. However, buildings can be designed to withstand the ground-shaking that occurs in an earthquake (Figure 9.14). Single-storey buildings are more suitable than multi-storey structures, because this reduces the number of people at risk, and the threat of collapse over roads and evacuation routes. Some tall buildings are built with a 'soft storey' at the bottom, such as a car park raised on pillars. This collapses in an earthquake, so that the upper floors sink down onto it and this cushions the impact. Basement isolation – mounting the foundations of a building on rubber mounts that allow the ground to move under the building – is widely used. This isolates the building from the tremors.

Building reinforcement strategies include building on foundations built deep into underlying bedrock, and the use of steel-constructed frames that can withstand shaking. Land-use planning is another important way of reducing earthquake risk (Figure 9.15).

Safe houses

The earthquake in Haiti was a reminder that billions of people live in houses that cannot withstand shaking. Yet safer ones can be built cheaply – using straw, adobe and old tyres, for example – by applying a few general principles (Figure 9.16).

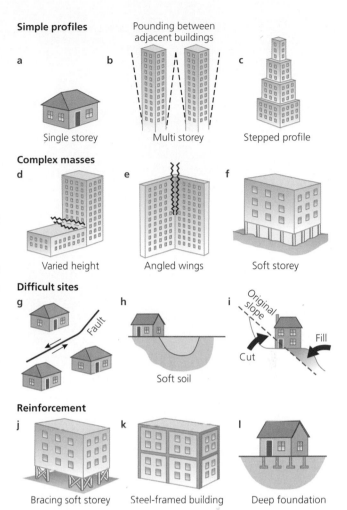

Figure 9.14 Building design

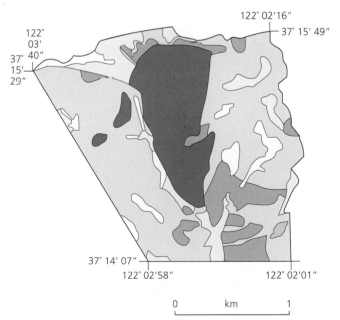

Relative stability	Map area	Geologic conditions	Recommended land use		
			Houses	Roads	
				Public	Private
Most stable		Flat to gentle slopes; subject to local shallow sliding, soil creep and settlement	Yes	Yes	Yes
		Gentle to moderately steep slopes in older stabilised landslide debris; subject to settlement, soil creep, and shallow and deep landsliding	Yes	Yes	Yes
		Steep to very steep slopes; subject to mass-wasting by soil creep, slumping and rock fall	Yes	Yes	Yes
		Gentle to very steep slopes in unstable material subject to sliding, slumping and soil creep	No	No	No
		Moving shallow (>3 m) landslide	No	No	No
Least stable		Moving, deep landslide, subject to rapid failure	No	No	No

Source: *Advanced Geography: Concepts & Cases* by P. Guinness & G. Nagle (Hodder Education, 1999), p.348

Figure 9.15 Land-use planning, San Francisco–San José California

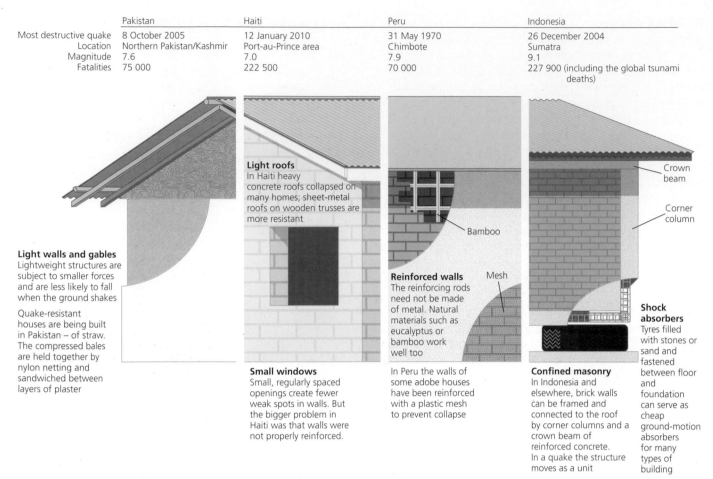

	Pakistan	Haiti	Peru	Indonesia
Most destructive quake	8 October 2005	12 January 2010	31 May 1970	26 December 2004
Location	Northern Pakistan/Kashmir	Port-au-Prince area	Chimbote	Sumatra
Magnitude	7.6	7.0	7.9	9.1
Fatalities	75 000	222 500	70 000	227 900 (including the global tsunami deaths)

Light walls and gables
Lightweight structures are subject to smaller forces and are less likely to fall when the ground shakes

Quake-resistant houses are being built in Pakistan – of straw. The compressed bales are held together by nylon netting and sandwiched between layers of plaster

Light roofs
In Haiti heavy concrete roofs collapsed on many homes; sheet-metal roofs on wooden trusses are more resistant

Small windows
Small, regularly spaced openings create fewer weak spots in walls. But the bigger problem in Haiti was that walls were not properly reinforced.

Bamboo

Reinforced walls
The reinforcing rods need not be made of metal. Natural materials such as eucalyptus or bamboo work well too

Mesh

In Peru the walls of some adobe houses have been reinforced with a plastic mesh to prevent collapse

Crown beam

Corner column

Shock absorbers
Tyres filled with stones or sand and fastened between floor and foundation can serve as cheap ground-motion absorbers for many types of building

Confined masonry
In Indonesia and elsewhere, brick walls can be framed and connected to the roof by corner columns and a crown beam of reinforced concrete. In a quake the structure moves as a unit

Figure 9.16 A safe house

In wealthy cities in fault zones, the added expense of making buildings earthquake-resistant has become a fact of life. Concrete walls are reinforced with steel, for instance, and a few buildings even rest on elaborate shock absorbers. Strict building codes were credited with saving thousands of lives when a magnitude 8.8 earthquake hit Chile in February 2010. But in less developed countries, like Haiti, conventional earthquake engineering is often unaffordable. However, cheap solutions do exist.

In Peru in 1970, an earthquake killed 70 000 or more people, many of whom died when their houses crumbled around them. Heavy, brittle walls of traditional adobe – cheap, sun-dried brick – cracked instantly when the ground started buckling. Subsequent shakes brought roofs thundering down. Existing adobe walls can be reinforced with a strong plastic mesh installed under plaster; in a 'quake, these walls crack but do not collapse, allowing occupants to escape. Plastic mesh could also work as a reinforcement for concrete walls in Haiti and elsewhere.

Other engineers are working on methods that use local materials. Researchers in India have successfully tested a concrete house reinforced with bamboo. A model house for Indonesia rests on ground-motion dampers – old tyres filled with bags of sand. Such a house might be only a third as strong as one built on more sophisticated shock absorbers, but it would also cost much less – and so be more likely to be adopted in Indonesia. In northern Pakistan, straw is available. Traditional houses are built of stone and mud, but straw is far more resilient, and warmer in winter. However, cheap ideas aren't always cheap enough.

Since 2007, some 5000 houses in Peru have been strengthened with plastic mesh or other reinforcements.

Controlling earthquakes

In theory, by altering the fluid pressure deep underground at the point of greatest stress in the fault line, a series of small and less damaging earthquake events may be triggered. This could release the energy that would otherwise build up to create a major event. Additionally, a series of controlled underground nuclear explosions might relieve stress before it reached critical levels.

Prediction and risk assessment

There are a number of methods of detecting earthquakes – distortion of fences, roads and buildings are some examples; changing levels of water in boreholes is another. As strain can change the water-holding capacity or porosity of rocks by closing and opening their tiny cracks, then water levels in boreholes will fluctuate with increased earthquake activity. Satellites can also be used to measure the position of points on the surface of the Earth to within a few centimetres. However, predicting earthquakes is not simple. Some earthquakes are very irregular in time and may only occur less than once every 100 years. By contrast, other parts of the Earth's surface may continually slip and produce a large number of very small earthquakes. In addition, different parts of a fault line may behave differently. Areas that do not move are referred to as 'seismic gaps'; areas that move and have lots of mini earthquakes may be far less hazardous.

Earthquake prediction is only partly successful, although it offers a potentially valuable way of reducing the impact of earthquakes. Some aspects are relatively easy to understand. For example, the location of earthquakes is closely linked with the distribution of fault lines. However, the timing of earthquakes is difficult to predict. Previous patterns and frequencies of earthquake events offer some clues as to what is likely to happen in the future, but the size of an earthquake event is difficult to predict.

The most reliable predictions focus on:

- measurement of small-scale ground surface changes
- small-scale uplift or subsidence
- ground tilt
- changes in rock stress
- micro-earthquake activity (clusters of small 'quakes)
- anomalies in the Earth's magnetic field
- changes in radon gas concentration
- changes in electrical resistivity of rocks.

Measurements of these are made using a variety of instruments (Table 9.8).

Table 9.8 Monitoring for earthquake prediction

Instrument	Purpose
Seismometer	To record micro-earthquakes
Magnetometer	To record changes in the Earth's magnetic field
Near-surface seismometer	To record larger shocks
Vibreosis truck	To create shear waves to probe the earthquake zone
Strain meter	To monitor surface deformation
Sensors in wells	To monitor changes in groundwater levels
Satellite relays	To relay data to the US Geological Survey
Laser survey equipment	To measure surface movement

Source: C. Park, The Environment, Routledge 1997

One particularly intensively studied site is Parkfield in California, on the San Andreas fault – see page 277.

Predicting volcanoes

Scientists are increasingly successful in predicting volcanoes. Since 1980, they have correctly predicted 19 of Mt St Helens' 22 eruptions, and Alaska's Redoubt volcano in 1989. There have been false alarms: in 1976, 72 000 residents of Guadeloupe were forced to leave their homes, and in 1980 Mammoth Lake in California suffered from a reduction in tourist numbers owing to mounting concern regarding volcanic activity.

Volcanoes are easier to predict than earthquakes since there are certain signs. The main ways of predicting volcanoes include monitoring using:

- **seismometers** to record swarms of tiny earthquakes that occur as the magma rises (Figure 9.17)
- chemical sensors to measure increased sulphur levels
- lasers to detect the physical swelling of the volcano
- ultrasound to monitor low-frequency waves in the magma, resulting from the surge of gas and molten rock, as happened at Pinatubo, El Chichón and Mt St Helens
- observations, such as of Gunung Agung (Java, Indonesia).

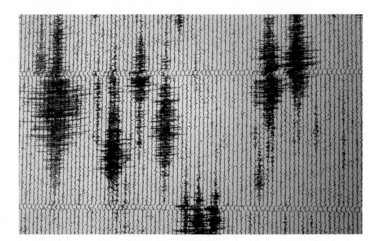

Figure 9.17 A seismograph reading (Montserrat)

However, it is not always possible to state exactly when a volcanic eruption will happen. The US Geological Survey predicted the eruption of Mt Pinatubo in 1991, and successfully evacuated the area. However, it was unsuccessful in predicting a volcanic eruption at Mammoth Mountain Ski Area in California, USA – the false prediction reduced visitor numbers to the resort and caused economic distress to local business people.

In Montserrat in the Caribbean, volcanic activity has made over 60 per cent of the southern and central parts of the island uninhabitable. Plymouth was evacuated three times in 1995 and 1996. The volcano was responsible for 19 deaths – all of them farmers – caught out by an eruption during their return to the Exclusion Zone. Volcanic dust is another hazard, as it is a potential cause of silicosis and aggravates asthma. There are many hazards around Plymouth (Figure 9.18).

Figure 9.18 Hazard sign in Plymouth, Montserrat

Volcanic management includes monitoring and prediction (Figure 9.19). GPS is used to monitor changes in the surface of the volcano – volcanoes typically bulge and swell before an eruption. The development of 'risk maps' can be used to good effect, as in the case of Montserrat. There are risks on other Caribbean islands too. St Vincent and St Kitts are high-risk islands, whereas St Lucia, Grenada and Nevis are lower risk.

Figure 9.19 Montserrat Volcanic Observatory

Living with a volcano

People often choose to live in volcanic areas because they are useful in a variety of ways:

- Some countries, such as Iceland and the Philippines, were created by volcanic activity.
- Some volcanic soils are rich, deep and fertile, and allow intensive agriculture, for example in Java. However, in other areas, for example Sumatra and Iceland, the soils are poor. In Iceland, this is because the climate is too cool to allow chemical weathering of the lava flows, while in Sumatra the soils are highly leached.
- Volcanic areas are important for tourism, for example St Lucia and Iceland.

- Some volcanoes are culturally symbolic and are part of the national identity, such as Mt Fuji in Japan.

Managing tsunamis: tsunami warning systems

At present, it is impossible to predict precisely where and when a tsunami will happen. In most cases, it is only possible to raise the alarm once a tsunami has started. In the cases of submarine volcanoes, it is possible to monitor these to predict the risk of tsunami. For example, Kick 'em Jenny, north of Grenada, has erupted ten times since the late 1970s and grown by 50 metres. Volcanologists believe it could cause a tsunami and threaten Venezuela.

The first effective tsunami warning system was developed in 1948 in the Pacific, following the 1946 tsunami. The system consisted of over 50 tidal stations and 31 seismographic stations, spread between Alaska, Hong Kong and Cape Horn. Following an earthquake, tidal gauges in the region establish whether a tsunami has formed. The earthquake epicentre is also plotted and magnitude investigated. The warning system has been improved by the use of satellites, and it is now operated by the US National Oceanic and Atmospheric Administration (NOAA).

In theory, there is time to issue warnings. A tsunami off the coast of Ecuador will take 12 hours to reach Hawaii, 20 hours to reach Japan. A tsunami from the Aleutians will take 5 hours to reach Hawaii. However, the impacts will vary with shoreline morphology.

Other tsunami early warning systems include those in Japan and Kamchatka (Russia). However, many LICs lack early warning systems, as was so tragically exposed in the 2004 Boxing Day tsunami. Following the 2010 Chile earthquake, a tsunami warning was issued. Fortunately, there was little evidence of any particularly large waves affecting areas other than part of the Chilean coast.

During the 2010 Indonesian tsunami, in which over 400 people died, the tsunami early warning system that had been put in place failed to work. The system had been vandalised in the Mentawai Islands, which were worst affected by the tsunami. In the 2011 Tōhuko tsunami, although Japan has seawalls along its coastline, the tsunami was higher than 12 metres and so the seawalls did not protect the people against the hazard.

Section 9.1 Activities

1 Examine the ways in which it is possible to predict a volcanic eruption.
2 Comment on the methods to predict earthquake activity.
3 Suggest how housing and other buildings can be made safer in the event of an earthquake.
4 To what extent is it possible to manage the impacts of tsunamis?

9.2 Hazards resulting from mass movements

☐ The nature and causes of mass movements and resultant hazards

Mass movements can be classified in a number of ways. The main ones include speed of movement and the amount of water present. In addition, it is possible to distinguish between different types of movement, such as falls, flows, slides and slumps. (See Topic 3, Section 3.3.)

The likelihood of a slope failing can be expressed by its safety factor. This is the relative strength or resistance of the slope, compared with the force that is trying to move it. The most important factors that determine movement are gravity, slope angle and pore pressure (Figures 9.20 and 9.21).

Gravity has two effects. First, it acts to move the material downslope (a slide component). Second, it acts to stick the particle to the slope (a stick component). The downslope movement is proportional to the weight of the particle and slope angle. Water lubricates particles and in some cases fills the spaces between the particles. This forces them apart under pressure. Pore pressure will greatly increase the ability of the material to move. This factor is of particular importance in movements of wet material on low-angle slopes (Table 9.9).

Table 9.9 Increasing stress and decreasing resistance

Factor	Examples
Factors contributing to increased shear stress	
Removal of lateral support through undercutting or slope steepening	Erosion by rivers and glaciers, wave action, faulting, previous rockfalls or slides
Removal of underlying support	Undercutting by rivers and waves, subsurface solution, loss of strength by exposure of sediments
Loading of slope	Weight of water, vegetation, accumulation of debris
Lateral pressure	Water in cracks, freezing in cracks, swelling, pressure release
Transient stresses	Earthquakes, movement of trees in wind
Factors contributing to reduced shear strength	
Weathering effects	Disintegration of granular rocks, hydration of clay minerals, solution of cementing minerals in rock or soil
Changes in pore-water	Saturation, softening of material pressure
Changes of structure	Creation of fissures in clays, remoulding of sands and clays
Organic effects	Burrowing of animals, decay of roots

Landslides are a common natural event in unstable, steep areas (Figure 9.22). Landslides may lead to loss of life; disruption of transport and communications; and damage to property and infrastructure. The annual repair cost for roads in the Caribbean is estimated to be US$15 million.

Figure 9.20 Landslide near Zermatt, Switzerland

Figure 9.21 Landslip on a slope in Oxford, UK

Figure 9.22 Mam Tor landslide, Derbyshire, UK

Tropical storm activity may trigger landslides. In Jamaica in 2001, tropical storm Michelle triggered a number of debris flows, many 2–3 kilometres in length. Similarly, tropical storm Mitch (1998) caused a mudflow 20 kilometres long and 2–3 kilometres wide, which killed more than 1500 people in the town of Posoltega in Nicaragua and surrounding villages.

The two main forces that trigger landslides in the Caribbean are:

- seismic activity
- heavy rainfall.

Jamaica is subject to frequent landslides. In the Blue Mountains, over 80 per cent of the slopes are greater than 2°. The area is also geologically young and heavily fractured, and the bedrock is deeply weathered, making it unstable. The largest historic landslide in the region occurred on Judgement Cliff, eastern Jamaica, where an estimated 80 million m³ of material was moved.

Human activities can increase the risk of landslides, for example by:

- **increasing the slope angle**, for instance cutting through high ground – slope instability increases with increased slope angle
- **placing extra weight on a slope**, for instance new buildings – this adds to the stress on a slope
- **removing vegetation** – roots may bind the soil together and interception by leaves may reduce rainfall compaction

☐ Impacts on lives and property

Landslides

- **exposing rock joints and bedding planes**, which may increase the speed of weathering.

There have been various attempts to manage the landslide risk. A number of landslide hazard maps have been produced for the region. Methods to combat the landslide hazard are largely labour intensive and include:

- **building restraining structures** such as walls, piles, buttresses and **gabions** – these may hold back minor landslides
- **excavating and filling steep slopes** to produce gentler slopes – this can reduce the impact of gravity on a slope
- **draining slopes** to reduce the build-up of water – this decreases water pressure in the soil
- **watershed management**, for example afforestation and agroforestry ('farming the forest') – this increases interception and reduces overland flow.

However, many settlements are located on unsuitable land because no-one else wants that land. Relocation following a disaster can also occur. For example, at Mayeyes near Ponce in Puerto Rico, the site was cleared following a landslide. Similarly, the Preston Lands landslide in 1986 in Jamaica resulted in the local community being relocated.

Section 9.2 Activities

1 Suggest why hazards due to mass movement are common throughout many parts of the Caribbean.
2 How can human activity increase the risk of landslides?

Case Study: Landslides in Puerto Rico

Approximately 70–80 per cent of Puerto Rico is hilly or mountainous (Figure 9.23). Average annual precipitation in Puerto Rico ranges from less than 1000 millimetres along the southern coast to more than 4000 millimetres in the rainforest of the Sierra de Luquillo in the north-eastern part of the island. Rain in Puerto Rico falls throughout the year, but about twice as much rain falls each month from May to October – the tropical storm season – as falls from November to April. In October 1985, a tropical wave, which later developed into tropical storm Isabel, struck the south-central coast of Puerto Rico, and produced extreme rainfall.

Figure 9.23 Puerto Rico – relief

Puerto Rico can be divided into three distinct physiographic provinces: Upland, Northern Karst and Coastal Plains. The Upland province includes three major mountain ranges and is covered by dense tropical vegetation. Slopes as steep as 45° are common. The Northern Karst province includes most of north-central and north-western Puerto Rico north of the Upland province. The Coastal Plains province is a discontinuous, gently sloping area. Puerto Rico's major cities are built primarily in the Coastal Plain province, although population growth has pushed development onto adjacent slopes of the Upland and Northern Karst provinces. Some 60 per cent of the 3.35 million population lives in the four largest cities – San Juan, Ponce, Mayaguez, and Arecibo – which are located primarily on flat or gently sloping coastal areas. However, continuing growth of these urban centres is pushing development onto surrounding steep slopes.

All major types of landslide occur in Puerto Rico. Most of the Upland province and the Northern Karst province, on account of their high relief, steep slopes and abundant rainfall, have continuing landslide problems. The drier south-western part normally experiences landslides only during exceptionally heavy rainfall.

Debris slides and debris flows – rapid downslope sliding or flowing of disrupted surface rock and soil – are particularly hazardous because they happen with little or no warning. Rock falls are common on very steep natural slopes and especially on the numerous steep road cuttings on the island.

A major tropical storm in October 1985 triggered thousands of debris flows as well as a disastrous rock slide that destroyed the Mameyes district of Ponce, killing at least 129 people. The Mameyes landslide was the worst ever landslide experienced in Puerto Rico. More than 100 homes were destroyed, and about as many were later condemned and removed because of continuing risk from landslides.

The greatest cost to public works in Puerto Rico is road maintenance. The frequency of serious storms suggests that a long-term average of perhaps five fatalities per year could occur, tens of houses be destroyed or made unfit to live in and hundreds be damaged by landslides each year.

Section 9.2 Activities

1 Suggest why Puerto Rico is so vulnerable to landslides.
2 How could the threat of landslides be reduced?

Case Study: China's landslide, 2010

China experienced its deadliest landslide in decades in 2010. At least 700 people died in north-western Gansu province when an avalanche of mud and rock engulfed the small town of Zhouqu. Zhouqu town is in a valley. Heavy rain quickly ran off the steep, barren hills, triggering mudslides and swelling the river. Landslides levelled an area about 5 kilometres long and 500 metres wide, and more than 300 houses collapsed.

Officials have warned for years that heavy tree-felling and rapid hydro development were making the mountain area around Zhouqu vulnerable to landslips. One government report in 2009 called the Bailong River a 'high-occurrence disaster zone for landslides'.

The landslide created a loose earth dam. Water levels behind the barrier fell slightly after controlled explosions created a channel to funnel off some of the water.

The landslide was the worst to hit China in 60 years, and was the most deadly single incident in a year of heavy flooding that killed nearly 1500 people.

Mudslides

Case Study: Human causes – the Italian mudslides, 1998

In May 1998, mudslides swept through towns and villages in Campania, killing nearly 300 people. Hardest hit was Sarno, a town of 35 000 people (Figure 9.24). In the two weeks before the mudslide, up to a year's rainfall had fallen. Geologically, the area is unstable – it has active volcanoes, such as Etna and Vesuvius, many mountains and scores of fast-flowing rivers. Following the

mudslide, a state of emergency was declared in the Campania region, and up to £18 million was allocated for repairing the damage. Campania is one of Italy's most vulnerable regions – since 1892, scientists have recorded at least 1173 serious landslides in Campania and Calabria. Since 1945, landslides and floods have caused an average of seven deaths every month (Table 9.10).

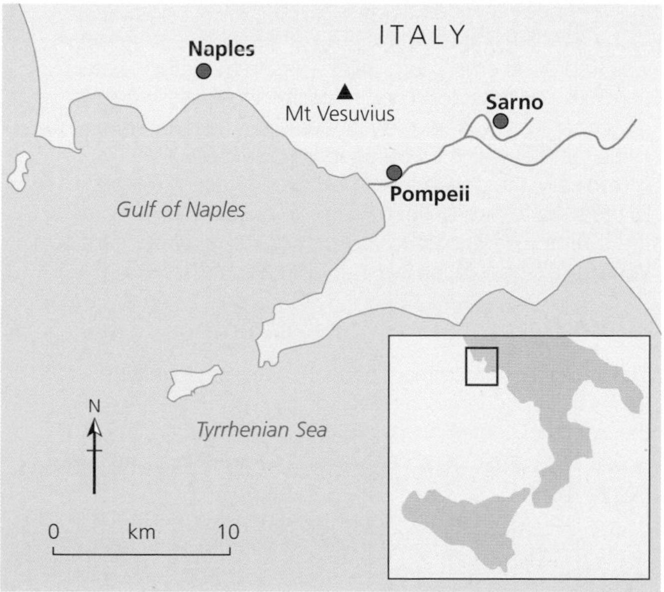

Table 9.10 Floods and landslides in Italy since 1950

Year	Region	Event	Deaths
1951	Calabria	Floods	100
1951	Polesine, Veneto	Floods	89
1954	Salerno, Campania	Floods	297
1963	Longarone, Veneto	Landslide, floods	1800
1966	Florence, Tuscany	Floods	35
1985	Val di Stava, Trentino	Landslide, floods	269
1987	Valtelina, Lombardy	Floods, landslide	53
1994	Alessandria, Piedmont	Floods	68
1996	Versilia, Tuscany	Floods, landslide	14
1998	Sarno and Siano, Campania	Mudslide, floods	285

Figure 9.24 Sarno, Italy

However, the disaster was only partially natural – much of it was down to human error. The River Sarno had dwindled to a trickle of water and part of the river bed had been cemented over. The clay soils of the surrounding mountains had been rendered dangerously loose by forest fires and deforestation. Houses had been built up hillsides identified as landslide zones, while Italy's sudden entry into the industrial age in the 1960s led to the uncontrolled building of houses and roads, and deforestation. Nowhere was this more evident than in ⇨

Campania. Over 20 per cent of the houses in Sarno were built without permission. Most are shoddily built over a 2 metre-thick layer of lava formed by the eruption of Vesuvius in 79 CE. Heavy rain can make this lava liquid, and up to 900 million tonnes of material are washed down in this way every year. Much of the region's fragility is, therefore, due to mass construction, poor infrastructure and poor planning.

It is likely that similar landslides will be experienced in Spain, Portugal, Greece and Turkey as these countries are developed. All across southern Europe, the natural means by which excess rainfall can be absorbed harmlessly are being destroyed. First, the land is cleared for development (even land that may have been designated as green-belt land). The easiest way to clear the vegetation is to set it on fire. The growing incidence of forest fires around the Mediterranean is not coincidental. Many are started deliberately by developers to ensure that the area loses its natural beauty. One of the side-effects of fire is to loosen the underlying soil.

Throughout southern Europe, the easiest way for an individual to add an extension or build a house is not to submit plans for approval but just to go ahead. In Sicily, up to 20 000 holiday homes have been built on beaches, cliffs and wetlands, in defiance of planning regulations. In Italy, 217 000 houses have been built without permission, and without proper drainage or foundations. Many stand close to an apparently dry river bed that can become a torrent during a storm. One Campanian town, Villaggio Coppola di Castelvolturno, with a population of 15 000 inhabitants, was created entirely without authorisation.

Section 9.2 Activities

1 What are the natural reasons why Italy is at risk from mudslides?
2 What human factors have increased the risk of mudslides in the region?
3 Why is the threat of mudslides increasing throughout the Mediterranean region?

Case Study: The Venezuelan mudslides

The Venezuelan mudslides of 1999 were the worst disaster to hit the country for almost 200 years (Figure 9.25). The first two weeks of December saw an unusually high amount of rainfall in Venezuela. Precipitation was 40–50 per cent above normal in most of the eastern Caribbean during 1999. On 15 and 16 December, the slopes of the 2000 metre Mt Avila began to pour forth avalanches of rock and mud, burying large parts of a 300 kilometre stretch of the central coast. The rains triggered a series of mudslides, landslides and flash floods that claimed the lives of between 10 000 and 50 000 people in the narrow strip of land between the mountains and the Caribbean Sea. Over 150 000 people were left homeless by landslides and floods in the states of Vargas and Miranda.

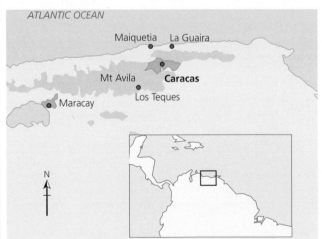

Figure 9.25 Venezuela

Hardest hit was the state of Vargas. Countless mountainside slum dwellings were either buried in the mudslide or swept out to sea. Most of the dead were buried in mudslides that were 8–10 metres deep. The true number of casualties may never be known. The mudslides also destroyed roads, bridges and factories, buried crops in the fields, destroyed telecommunications and also ruined Venezuela's tourist industry for the immediate future. The international airport of Caracas was temporarily closed and the coastal highway was destroyed or closed in many places. Flash floods damaged hundreds of containers at the seaport in Maiquetía. Hazardous materials in some containers were leaked into the ground and into the sea. Flash-flood damage halted operations at the Maiquetía seaport and hampered efforts to bring in emergency supplies immediately after the disaster. Economic damage was estimated at over US$3 billion.

The disaster was not just related to heavy rainfall. The government blamed corrupt politicians from previous governments and planners who had allowed shanty towns to grow up in steep valleys surrounding the coast and the capital, Caracas.

The immediate response was a search-and-rescue operation to find any survivors in the mudflows, landslides and buildings that had been damaged or destroyed. Few survivors were found after the first few days. The other short-term response was to provide emergency relief – accommodation, water purification tablets, food and medicines to those in need. The relief operation was severely hindered by the poor state of the infrastructure, which made operations difficult.

Ironically, the government had already been planning to redistribute part of Venezuela's population away from the overcrowded coast to the interior. Up to 70 per cent of Venezuela's population live in this small area.

Government plans for rebuilding

The Venezuela government announced a plan to restore Venezuela's northern coastal region by rebuilding thousands of homes there, expanding the country's main airport and constructing canals that can direct rivers away from communities.

The plan includes building 40 000 new homes in the hard-hit state of Vargas. The resort towns of Macuto and Camuri Chico were restored as tourist destinations and $100 million will be spent to expand Venezuela's main international airport. The country's main seaport, also in Vargas, was 'modernised'.

The towns that were utterly devastated by the disaster, where most structures were swept out to sea, were not rebuilt. Instead, these towns, including the coastal community of Carmen de Uria (Figure 9.26), were turned into parks, bathing resorts and other outdoor facilities.

Figure 9.26 Landslide at Carmen de Uria

In 2005, floods and mudslides brought on by heavy rains in the northern and central coast of Venezuela caused 14 deaths. Some 18 000 people were affected, while 2840 houses were damaged and a further 363 destroyed. In many cases, those that were affected in the 1999 mudslides were also affected in 2005.

Section 9.2 Activities

1 What were the causes of the Venezuelan mudslides?
2 Why were the impacts so great?

Table 9.11 Examples of hazards in mountainous areas

Hazards	Disaster event
Rockslides	Elm, Swiss Alps, 1881 Vaiont Dam, Italian Alps, 1963
Mud and debris flows	European Alps, 1987 Huanuco Province, Peru, 1989
Debris torrents	Coast Range, British Colombia 1983–84 Rio Colorado, Chile, 1987
Avalanches	Hakkari, Turkey, 1989 Western Iran, 1990
Earthquake-triggered mass movements	Campagna, Italy, 1980 Mt Ontake, Japan, 1984
Vulcanism-triggered mass movements	Mt St Helens, USA, 1980 Nevado del Ruiz, Colombia, 1985
Weather-triggered mass movements from volcanoes	Mt Kelut, Indonesia, 1966 Mt Semeru, Java, 1981
Natural dams and dam-break floods	
Landslide dams	Indus Gorge, Western Himalayas, 1841 Ecuadorean Andes, 1987 Sichuan earthquake, 2008
Glacier dams	'Ape Lake', British Colombia, 1984
Moraine dams	Khumbu, Nepal, Himalaya,1985
Avalanche dams	Santa River, Peruvian Andes, 1962
Vegetation dams	New Guinea Highlands, 1970
Artificial dam failures	Buffalo Creek, Appalachians, USA, 1972 Shanxi Province, China, 1989

Table 9.11 summarises some of the hazards that are experienced in mountainous areas around the world.

Avalanches

Avalanches are mass movements of snow and ice. Newly fallen snow may fall off older snow, especially in winter, while in spring partially thawed snow moves, often triggered by skiing. Avalanches occur frequently on steep slopes over 22°, especially on north-facing slopes where the lack of Sun inhibits the stabilisation of the snow. They are also very fast. Average speeds in an avalanche are 40–60 kilometres per hour, but speeds of up to 200 kilometres per hour have been recorded in Japan.

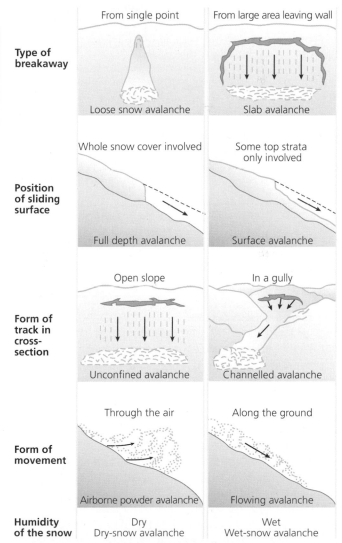

Figure 9.27 A classification of avalanches

Avalanches are classified in a number of ways (Figure 9.27). At first, a distinction was made between airborne powder-snow avalanches and ground-hugging avalanches. Later classifications have included:

- **the type of breakaway** – from a point formed with loose snow, or from an area formed of a slab
- **position of the sliding surface** – the whole snow cover or just the surface
- **water content** – *dry* or *wet avalanches*
- **the form of the avalanche** – whether it is channelled in cross-section or open.

Although avalanches cannot be prevented, it is possible to reduce their impact (Figures 9.28 and 9.29). So why do avalanches occur? The underlying processes in an avalanche are similar to those in a landslide. Snow gets its strength from the interlocking of snow crystals and cohesion caused by electrostatic bonding of snow crystals. The snow remains in place as long as its strength is greater than the stress exerted by its weight and the slope angle.

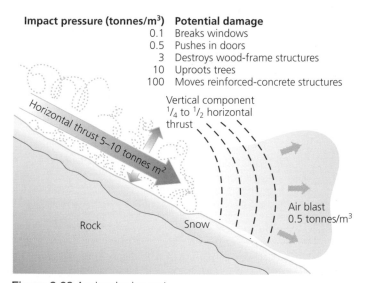

Figure 9.28 Avalanche impact

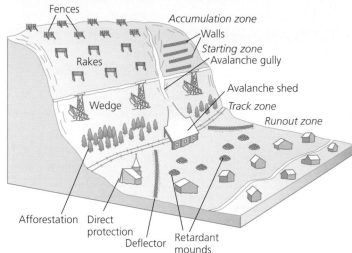

Figure 9.29 Measures to reduce the impact of avalanches

The process is complicated by the way in which snow crystals constantly change. Changes in overlying pressure, compaction by freshly fallen snow, temperature changes and the movement of meltwater through the snow cause the crystal structure of the snow to change. It may become unstable and move downslope as an avalanche.

Loose avalanches, comprising fresh snow, usually occur soon after a snowfall. By contrast, slab avalanches occur at a later date, when the snow has developed some cohesion. The latter are usually much larger than loose avalanches and cause more destruction. They are often started by a sudden rise in temperature that causes melting. The meltwater lubricates the slab, and makes it unstable. Many of the avalanches occur in spring (Table 9.12) when the snowpack is large and temperatures are rising. There is also a relationship between the number of avalanches and altitude (Table 9.13).

Table 9.12 Occurrence of avalanches in the French Alps

December	10%
January	22%
February	32%
March	23%
April	13%

Table 9.13 Avalanches and altitude in the Swiss Alps

Altitude (m)	No. of avalanches	% of total
Above 3000	326	3
2500–3000	2210	24
2000–2499	3806	41
1500–1999	2632	28
Below 1500	394	4

1 Suggest reasons why avalanches are clustered in the months January to March. Give details on at least **two** reasons.

2 Table 9.13 shows the distribution of avalanches with altitude in Switzerland. The tree-line is at about 1500 metres and the snow line is at 3000 metres. Describe the distribution of avalanches with altitude. How do you explain this pattern?

Case Study: The European avalanches of 1999

The avalanches that killed 75 people in the Alps in February 1999 were the worst in the area for nearly 100 years. Moreover, they occurred in an area that was thought to be fairly safe. In addition, precautionary measures had been taken, such as an enormous avalanche wall to defend the village of Taconnaz, and a second wall to stop the Taconnaz glacier advancing onto the motorway that runs into the mouth of the Mt Blanc tunnel. However, the villages of Montroc and Le Tour, located at the head of the Chamonix Valley, had no such defences.

The avalanche that swept through the Chamonix Valley killed 11 people and destroyed 18 chalets (Figure 9.30). Rescue work was hampered by the low temperatures (−7 °C), which caused the snow to compact and made digging almost impossible. The avalanche was about 150 metres wide, 6 metres high and travelled at a speed of up to 90 kilometres per hour. It crossed a stream and even travelled uphill for some 40 metres. Residents were shocked, since they had not experienced an avalanche so powerful, so low in the mountains and certainly not one capable of moving uphill.

Nothing could have been done to prevent the avalanche. Avalanche warnings had been given the day before, as the region had experienced up to 2 metres of snow in just three days. However, buildings in Montroc were not considered to be at risk. In fact, they were classified as being in the 'white zone', almost completely free of danger. By contrast, in the avalanche danger zones no new buildings have been developed for many decades. Avalanche monitoring is so well established and elaborate that it had caused villagers and tourists in the 'safe' zone to think that they really were safe. In Montroc, the experience was the equivalent of the eruption of an extinct volcano – the last time the snow above Montroc had caused an avalanche was in 1908.

Meteorologists have suggested that disruption of weather patterns resulting from global warming will lead to increased snowfalls in the Alps that are heavier and later in the season. This would mean that the conventional wisdom regarding avalanche 'safe' zones would need to be re-evaluated.

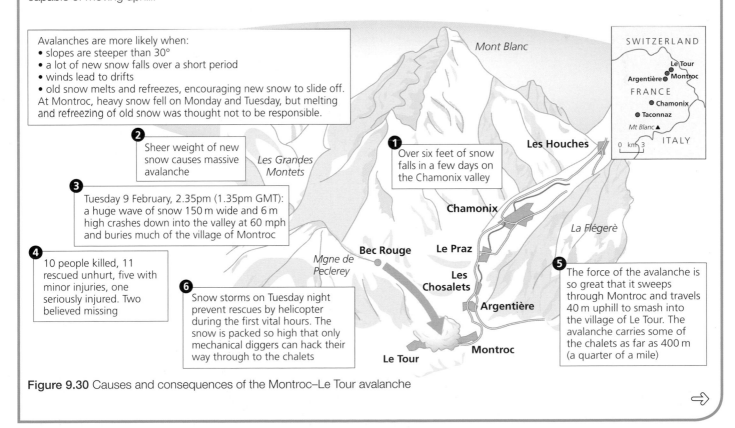

Avalanches are more likely when:
• slopes are steeper than 30°
• a lot of new snow falls over a short period
• winds lead to drifts
• old snow melts and refreezes, encouraging new snow to slide off. At Montroc, heavy snow fell on Monday and Tuesday, but melting and refreezing of old snow was thought not to be responsible.

2 Sheer weight of new snow causes massive avalanche

3 Tuesday 9 February, 2.35pm (1.35pm GMT): a huge wave of snow 150 m wide and 6 m high crashes down into the valley at 60 mph and buries much of the village of Montroc

4 10 people killed, 11 rescued unhurt, five with minor injuries, one seriously injured. Two believed missing

6 Snow storms on Tuesday night prevent rescues by helicopter during the first vital hours. The snow is packed so high that only mechanical diggers can hack their way through to the chalets

1 Over six feet of snow falls in a few days on the Chamonix valley

5 The force of the avalanche is so great that it sweeps through Montroc and travels 40 m uphill to smash into the village of Le Tour. The avalanche carries some of the chalets as far as 400 m (a quarter of a mile)

SWITZERLAND
Le Tour
Argentière ● Montroc
FRANCE
● Chamonix
● Taconnaz
Mt Blanc ▲
ITALY
0 km 3

Mont Blanc

Les Houches

Les Grandes Montets

Chamonix

La Flégerè

Le Praz

Bec Rouge

Mgne de Peclerey

Les Chosalets

Argentière

Montroc

Le Tour

Figure 9.30 Causes and consequences of the Montroc–Le Tour avalanche

Snowslides 2009–10

In December 2009 and January 2010, dozens of people were caught in the path of avalanches. The increase in snowslide activity sent ominous rumblings through the communities of Europe's Alpine resorts. Residents live in fear of seeing a repeat of early 1999 (see above, when 75 people were killed over a period of three weeks), or even of 1950–51, when more than 265 people died in three months.

Heavy snowfall combined with rain and an easing of the extreme cold prompted Météo France, the national meteorological service, to raise the avalanche warning to level 4 (out of 5), meaning 'high risk'.

In 2009, scientists in London warned that global warming, in the form of rising temperatures and melting permafrost, could make avalanches more frequent.

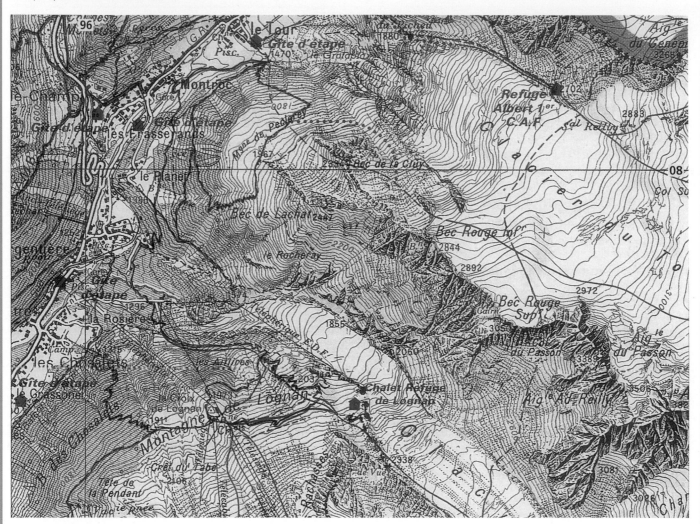

Figure 9.31 Survey map of the Alps – area affected by 1999 avalanches

Section 9.2 Activities

1 What is an *avalanche*?
2 What are the factors that increase the risk of an avalanche?
3 What were the conditions in Europe in February 1999 that led to widespread avalanches?
4 How and why may the threat of avalanches change in the next decades?

5 Study Figure 9.30.
  a Describe the site of Montroc and Le Tour.
  b What are the attractions for tourists shown on Figure 9.31? Use the grid provided to give grid references.
  c What is the map evidence to suggest that the area is at risk of hazardous events?

Prediction and hazard mapping

Landslides and other forms of mass movement are widespread and cause extensive damage and loss of life each year. With careful analysis and planning, together with appropriate stabilisation techniques, the impacts of mass movement can be reduced or eliminated.

Assessment of the hazards posed by potential mass movement events are based partly on past events, to evaluate their magnitude and frequency. In addition, mapping and testing of soil and rock properties determines their susceptibility to destabilising processes. Maps showing areas that could be affected by mass movement processes are important tools for land-use planners.

For example, valleys in the Cascade Range of Washington and Oregon, USA, have experienced extensive mudflows from volcanic activity over the last 10000 years. Hazard maps prepared before the eruption of Mt St Helens and Mt Pinatubo proved extremely useful, as the mudflows generated by these eruptions had very similar distributions to those produced in earlier times (Figure 9.32).

In addition to assessment, prediction and early warning, some engineering schemes can be applied to reduce the damage of mass wasting (Figure 9.33).

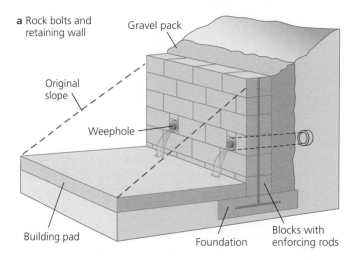

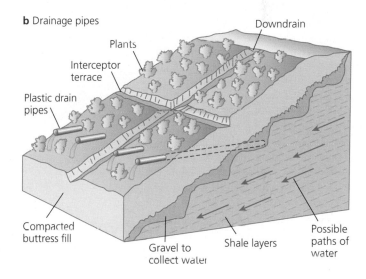

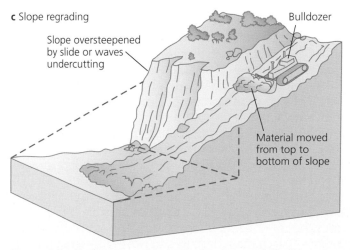

Figure 9.33 Engineering techniques for slope stabilisation

Figure 9.32 Hazard map of Mt Pinatubo

These include retaining devices, drainage pipes, grading of slope and diversion walls (Figure 9.34). Concrete blocks or gabions may be used to strengthen slopes. Slopes subject to creep can be stabilised by draining or pumping water from saturated sediment. Oversteepened slopes can be made gentler by **regrading**. However, not all communities can afford such measures and so may opt for low-cost sustainable forms of management.

Eliminating or restricting human activities in areas where slides are likely may be the best way to reduce damage and loss of life. Land that is susceptible to mild failures may be suitable for some forms of development (for example, recreation or parkland) but not others (for example, residential or industrial). Early-warning systems can provide forecasts of intense rain. High-risk areas can then be monitored and remedial action taken.

Figure 9.34 Engineering techniques, Brunei

9.3 Hazards resulting from atmospheric disturbances

☐ Large-scale tropical disturbances – tropical storms (cyclones)

Tropical cyclones are known as 'hurricanes' in the Atlantic, Caribbean and north-west Pacific; they are known as 'typhoons' in the north-western Pacific; and they are called 'tropical cyclones' in the Indian Ocean and south Pacific.

Tropical storms bring intense rainfall and very high winds, which may in turn cause storm surges and coastal flooding, and other hazards such as (inland) flooding and mudslides. Tropical storms are also characterised by enormous quantities of water. This is due to their origin over tropical seas. High-intensity rainfall, as well as large totals – up to 500 millimetres in 24 hours – invariably cause flooding. Their path is erratic, so it is not always possible to give more than 12 hours' notice of their position. This is insufficient for proper evacuation measures. In North America and the Caribbean, tropical storms are referred to as hurricanes.

Tropical storms develop as intense low-pressure systems over tropical oceans. Winds spiral rapidly around a calm central area known as the 'eye'. The diameter of the whole tropical storm may be as much as 800 kilometres, although the very strong winds that cause most of the damage are found in a narrower belt up to 300 kilometres wide. In a mature tropical storm, pressure may fall to as low as 880 millibars (mb). This, and the strong contrast in pressure between the eye and outer part of the tropical storm, leads to very strong winds.

Tropical storms move excess heat from low latitudes to higher latitudes. They normally develop in the westward-flowing air just north of the equator (known as an 'easterly wave'). They begin life as a small-scale tropical depression, a localised area of low pressure that causes warm air to rise. This causes thunderstorms that persist for at least 24 hours, and may develop into tropical storms, which have greater wind speeds of up to 117 kilometres per hour. However, only about 10 per cent of tropical disturbances ever become tropical storms, with wind speeds above 118 kilometres per hour.

For tropical storms to form, a number of conditions are needed (Figure 9.35):

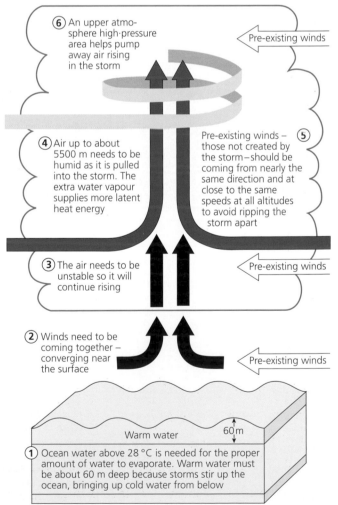

Figure 9.35 Formation of a tropical storm

- Sea temperatures must be over 27°C to a depth of 60 metres (warm water gives off large quantities of heat when it is condensed – this is the latent heat that drives the tropical storm).
- The low-pressure area has to be far enough away from the equator so that the Coriolis force (the force caused by the rotation of the Earth) creates rotation in the rising air mass – if it is too close to the equator, there is insufficient rotation and a tropical storm would not develop.
- Conditions must be unstable – some tropical low-pressure systems develop into tropical storms (not all of them), but scientists are unsure why some do and others do not.

Tropical storms are the most violent, damaging and frequent hazard to affect many tropical regions (Figure 9.36). They are measured on the Saffir–Simpson Scale, which is a 1–5 rating based on the tropical storm's intensity (Table 9.14). It is used to give an estimate of the potential property damage and flooding expected along the coast from a tropical storm landfall. Wind speed is the determining factor in the scale, as storm surge values are highly dependent on the slope of the continental shelf and the shape of the coastline in the landfall region. Tropical storms can also cause considerable loss of life. Hurricane Georges (1998) killed more than 460 people, mainly in Dominican Republic and Haiti.

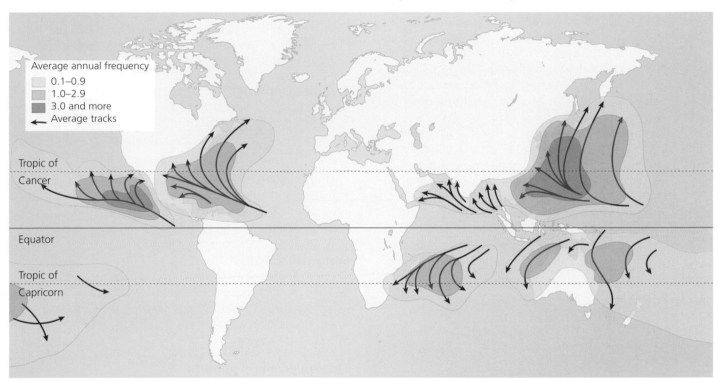

Figure 9.36 Distribution of tropical storms

Table 9.14 Saffir–Simpson Scale of tropical storm strength

Category 1	Winds 119–153 km/hour; storm surge generally 1.2–1.5 m above normal	No real damage to building structures. Damage primarily to unanchored mobile homes. Also, some coastal road flooding and minor pier damage.
Category 2	Winds 154–177 km/hour; storm surge generally 1.8–2.4 m above normal	Some damage to roofing materials, doors and windows. Considerable damage to vegetation, mobile homes and piers. Coastal and low-lying escape routes flood 2–4 hours before arrival of the tropical storm eye. Small craft in unprotected anchorages break moorings.
Category 3	Winds 178–209 km/hour; storm surge generally 2.7–3.6 m above normal	Some structural damage to small residences and utility buildings. Mobile homes are destroyed. Flooding near the coast destroys smaller structures, with larger structures damaged by floating debris. Land below 1.5 m above mean sea level may be flooded inland 13 km or more. Evacuation of low-lying residences close to the shoreline may be necessary.
Category 4	Winds 210–249 km/hour; storm surge generally 3.9–5.5 m above normal	Some complete roof structure failures on small residences. Complete destruction of mobile homes. Extensive damage to doors and windows. Land below 3 m above sea level may be flooded, requiring massive evacuation of residential areas as far inland as 10 km.
Category 5	Winds greater than 249 km/hour; storm surge generally greater than 5.5 m above normal	Complete roof failure on many residences and industrial buildings. Some complete building failures, with small utility buildings blown over or blown away. Complete destruction of mobile homes. Severe and extensive window and door damage. Low-lying escape routes are cut by rising water 3–5 hours before arrival of the centre of the tropical storm. Major damage to lower floors of all structures located less than 4.5 m above sea level and within 500 m of the shoreline. Massive evacuation of residential areas on low ground within 8–16 km of the shoreline may be required.

There are a number of significant factors that affect the impact of tropical storms:

- Tropical storm paths are unpredictable, which makes effective management of the threat difficult. It was fortunate for Jamaica that Hurricane Ivan (2004) (Figure 9.37) suddenly changed course away from the most densely populated parts of the island where it had been expected to hit. In contrast, it was unfortunate for Florida's Punta Gorda when Hurricane Charley (2004) moved away from its predicted path.
- The strongest storms do not always cause the greatest damage. Only six lives were lost to Hurricane Frances in 2004, but 2000 were taken by Jeanne when it was still categorised as just a 'tropical storm' and had not yet reached full hurricane strength.
- The distribution of the population throughout the Caribbean islands increases the risk associated with tropical storms. Much of the population lives in coastal settlements and is exposed to increased sea levels and the risk of flooding.
- Hazard mitigation depends upon the effectiveness of the human response to natural events. This includes urban planning laws, emergency planning, evacuation measures and relief operations such as re-housing schemes and the distribution of food aid and clean water.
- LICs continue to lose more lives to natural hazards, due to inadequate planning and preparation. By way of contrast, insurance costs continue to be greatest in American states such as Florida, where multi-million-pound waterfront homes proliferate.

Figure 9.37 Damage in Grenada after Hurricane Ivan

Tropical-storm management

Information regarding tropical storms is received from a number of sources including:

- satellite images
- aircraft that fly into the eye of the tropical storm to record weather information

- weather stations at ground levels
- radars that monitor areas of intense rainfall.

Preparing for tropical storms
Housing is particularly vulnerable to tropical storms. Hurricane Luis (1995) caused damage to 90 per cent of Antigua's houses, while Hurricane Gilbert (1988) made 800 000 people temporarily homeless in Jamaica. To limit damage to houses, owners are now encouraged to fix tropical storm straps to roofs and put storm shutters over windows. Houses built on stilts allow flood waters to pass away safely.

There are a number of ways in which national governments and agencies can help prepare for a tropical storm. These include **risk assessment**, land-use zoning, floodplain management and reducing the vulnerability of structures and organisations.

Risk assessment
The evaluation of risks of tropical cyclones can be shown in a hazard map. Particular information may be used to estimate the probability of cyclones striking a country:

- analysis of climatological records to determine how often cyclones have struck, their intensity and locations
- history of winds speeds, frequencies of flooding, height, location and storm surges over a period of about 50–100 years.

Land-use zoning
The aim is to control land use so that the most important facilities are placed in the least vulnerable areas. Policies regarding future development may regulate land use and enforce building codes for areas vulnerable to the effects of tropical cyclones.

Floodplain management
A plan for floodplain management should be developed to protect critical assets from flash, riverine and coastal flooding.

Reducing vulnerability of structures and infrastructures

- New buildings should be designed to be wind and water resistant. Design standards are usually incorporated into building codes.
- Communication and utility lines should be located away from the coastal area or installed underground.
- Areas of building should be improved by raising the ground level to protect against flood and storm surges.
- Protective river embankments, levees and coastal dikes should be regularly inspected for breaches due to erosion and opportunities should be taken to plant mangrove trees to reduce breaking wave energy.
- Vegetation cover should be increased to help reduce the impact of soil erosion and landslides, and facilitate the absorption of rainfall to reduce flooding.

Figure 9.38 What to do before, during and after a tropical storm

There are many other things that individuals can do to prepare for a tropical storm, and to learn how to act during and after a storm (Figure 9.38).

A tropical storm watch is issued when there is a threat of tropical storm conditions within 24–36 hours. A tropical storm warning is issued when tropical storm conditions (winds of 120 kilometres per hour or greater, or dangerously high water and rough seas) are expected in 24 hours or less. A tropical storm warning is issued when there are risks of tropical storm winds within 24 hours. A tropical storm watch is issued when tropical storm winds are expected within 36 hours.

The emergency relief offered after a tropical storm can take many forms – food supplies, clean water, blankets and medicines. Much of this is provided in tropical storm shelters. In some communities, emergency electricity generators may be needed. The community normally becomes involved in the clean-up operation, and electricity and phone companies work to restore power lines and communications.

Long-term redevelopment may include construction of new buildings in areas away from the coastline and on high ground. Long-term reconstruction in Grenada following Hurricane Ivan concentrated on housing and community projects, water supply and sanitation, transport and communications, agriculture, fisheries and small businesses, schools and government expenses.

Section 9.3 Activities

1 In what ways is it possible to prepare for tropical storms?
2 How can governments help prepare for tropical storms?
3 What are the main actions that should be taken during a tropical storm?

Case Study: Cyclone Nargis, 2 May 2008

Cyclone Nargis was a strong tropical cyclone (Figure 9.39). It formed on 27 April 2008, made landfall by 2 May and died out by 3 May. It involved winds of up to 165 kilometres per hour (sustained for 3 minutes) and winds of over 215 kilometres per hour (sustained for over 1 minute). At its peak, air pressure dropped to 962 millibars. Around 146 000 people were killed and it caused damage estimated at $10 million. As well as Burma (Myanmar), parts of Bangladesh, India and Sri Lanka were affected. However, it was the Burmese government's actions – or rather their lack of them – that caused widespread anger and disbelief.

The Burmese government identified 15 townships in the Irrawaddy delta that had suffered the worst. Seven of them had lost 90–95 per cent of housing, with 70 per cent of their population dead or missing. The land in the Irrawaddy delta is very low-lying. It is home to an estimated 7 million of Burma's 53 million people. Nearly 2 million of the densely packed area's inhabitants live on land that is less than 5 metres above sea level, leaving them extremely vulnerable. As well as the cost in lives and homes, there is the agricultural loss to the fertile delta, which is seen as Burma's 'rice bowl'.

It was the worst ever natural disaster in Burma. There were over 80 000 deaths in Labutta and a further 10 000 in Bogale. The UN estimated that 1.5 million people were severely affected by Cyclone Nargis. Thousands of buildings were destroyed; 75 per cent of the buildings in the town of Labutta collapsed and a further 20 per cent had their roofs ripped off. Up to 95 per cent of buildings in the Irrawaddy delta were destroyed. ⇨

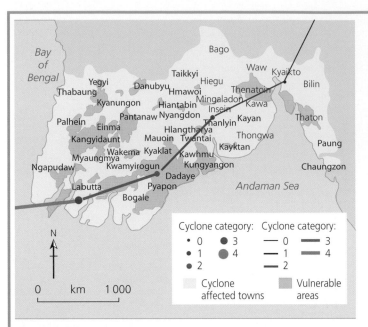
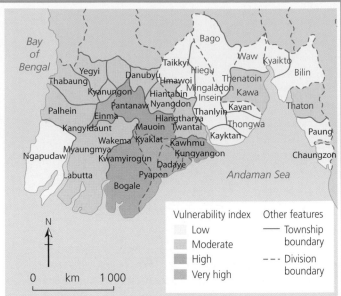

Figure 9.39 Cyclone Nargis

According to aid agencies trying to get into Burma, up to 1 million people could have died from the cyclone due to lack of relief. Relief efforts were delayed for political reasons. Burma's political leaders declined international aid; the World Food Programme said the delays were 'unprecedented in modern humanitarian relief efforts'. Within two weeks, an earthquake in China had deflected aid and sympathy away from Burma.

On 6 May, the Burmese junta (military government) finally asked the UN for aid, but accepted it only from India. Many nations and organisations hoping to deliver relief were unable to do so – the Burmese government refused to issue visas to many of them. On 9 May, the junta officially declared that its acceptance of international aid would be limited to food, medicines and some other specified supplies as well as financial aid, but would allow additional foreign aid workers to operate in the country.

India is one of the few countries to maintain close relations with Burma. It launched Operation Sahayata, under which it supplied two ships and two aircraft. However, the Burmese government denied Indian search and rescue teams and media access to critical cyclone-hit areas. On 16 May, India's offer to send a team of 50 medical personnel was accepted. Cyclone survivors needed everything – emergency shelter to keep them dry, all basic food and medicines.

Many Burmese people were displeased with their government, which had provided no warning of the cyclone. According to some reports, Indian meteorologists had warned Burma of Cyclone Nargis 48 hours before it hit the country's coast. People also believed the mayhem caused by the cyclone and associated flooding was further exacerbated by the government's unco-operative response.

The delays attracted international condemnation. More than a week after the disaster, only 1 out of 10 people who were homeless, injured or threatened by disease and hunger had received any kind of aid. More than two weeks later, relief had only reached 25 per cent of people in need. Some news stories stated that foreign aid provided to disaster victims was modified to make it look as if it came from the military regime, and state-run television continuously ran images of General Than Shwe ceremonially handing out disaster relief.

Uninterrupted referendum

Despite objections raised by the Burmese opposition parties and foreign nations in the wake of the natural disaster, the junta proceeded with a previously scheduled constitutional referendum. However, voting was postponed from 10 to 24 May in Yangon and other areas hardest hit by the storm.

☐ Small-scale tropical disturbances – tornadoes

Tornadoes are small and short-lived but highly destructive storms. Because of their severe nature and small size, comparatively little is known about them. Measurement and observation within them are difficult. A few low-lying, armoured probes called 'turtles' have been placed successfully in tornadoes. Tornadoes consist of elongated funnels of cloud that descend from the base of a well-developed cumulonimbus cloud, eventually making contact with the ground beneath. In order for a **vortex** to be classified as a tornado, it must be in contact with the ground *and* the cloud base. Within tornadoes are rotating violent winds, perhaps exceeding 100 metres per second. Pressure gradients in a tornado can reach an estimated 25 millibars per 100 metres (this compares with the most extreme pressure gradients of about 20 millibars per 100 kilometres in a larger-scale cyclone).

How tornadoes form

Moisture, instability, lift and wind shear are the four key ingredients in tornado formation (Figure 9.40). Most tornadoes, but not all, rotate *cyclonically*; that is, anticlockwise in the northern hemisphere and clockwise south of the equator. The standard explanation is that warm moist air meets cold dry air to form a tornado. Many thunderstorms form under these conditions (near warm fronts and cold fronts), which never even come close to producing tornadoes. Even when the large-scale environment is extremely favourable for tornado-type thunderstorms, not every thunderstorm spawns a tornado. The most destructive and deadly tornadoes develop from **supercells**, which are rotating thunderstorms with a well-defined low-pressure system called a **mesocyclone**.

Tornadoes can last from several seconds to more than an hour. The convectional activity that creates the source cloud is itself highly variable, and a single cloud can spawn a number of different tornado vortices, either simultaneously or in sequence, beneath different areas of the cloud, as parts of it develop and decay. Movement is generally with the parent cloud, perhaps with the funnel twisting sinuously across the ground beneath. Once contact with the ground is made, the track of a tornado at ground level may frequently extend for only a few kilometres, though there are examples of sustained tracks extending over hundreds of kilometres. The diameter of the funnel is rarely more than 200 metres; track length and width are therefore limited.

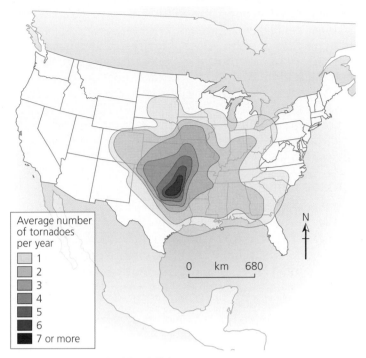

Figure 9.41 Tornado Alley, USA

Average number of tornadoes per year

- 1
- 2
- 3
- 4
- 5
- 6
- 7 or more

Tornadoes, being associated with extreme atmospheric instability, show both seasonal and locational preference in their incidence. 'Favoured' areas are temperate continental interiors in spring and early summer, when insolation is strong and the air may be unstable, although many parts of the world can be affected by tornado outbreaks at some time or another. The Great Plains of the USA, including Oklahoma, Texas and Kansas, have a high global frequency (Figure 9.41), and tornadoes are particularly likely to be experienced here at times when cool, dry air from the Rockies overlies warm, moist 'Gulf' air. Some areas of the USA experience tornadoes from a specific direction, such as north-west in Minnesota or south-east in coastal south Texas. This is because of an increased frequency of certain tornado-producing weather patterns, for example tropical storms in south Texas, or north-west-flow weather systems in the upper Midwest.

Some tropical storms in the USA fail to produce any tornadoes, while others cause major outbreaks. The same tropical storm may produce none for a while, and then erupt with tornadoes – or vice versa. Hurricane Andrew (1992), for example, spawned several tornadoes across the Deep South after crossing the Gulf, but produced none during its rampage across southern Florida. Katrina (2005) spawned numerous tornadoes after its devastating landfall.

The size and strength of tropical cyclones is not related to the birth of tornadoes. Relatively weak tropical storms like Danny (1985) have spawned significant supercell tornadoes well inland, as have larger, more intense storms like Beulah (1967) and Ivan (2004). In general,

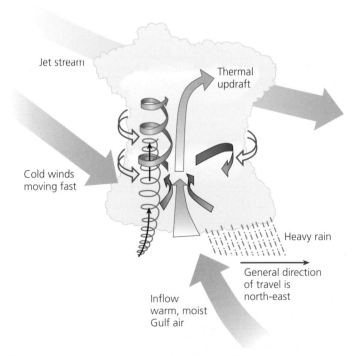

Figure 9.40 Formation of tornadoes in the USA

Jet stream

Thermal updraft

Cold winds moving fast

Heavy rain

General direction of travel is north-east

Inflow warm, moist Gulf air

the bigger and stronger the wind fields with a tropical cyclone, the bigger the area of favourable wind shear for supercells and tornadoes. But supercell tornadoes (whether or not in tropical cyclones) also depend on instability, lift and moisture. Surface moisture isn't lacking in a tropical cyclone, but sometimes instability and lift are too weak. This is why tropical systems tend to produce more tornadoes in the daytime, and near any fronts that may become involved in the cyclone circulation.

Tornado damage

About a thousand tornadoes hit the USA each year. On average, tornadoes kill about 60 people per year – most from flying or falling (crushing) debris. A tornado's impact as a hazard is extreme. They bring intense precipitation (heavy rainfall or hail), very high wind speeds and gusts of wind, and large-scale changes in pressure gradients (pressure imbalances).

There are three damaging factors at work. First, the winds are often so strong that objects in the tornado's path are simply removed or very severely damaged. Second, strong rotational movement tends to twist objects from their fixings, and strong uplift can carry some debris upwards into the cloud. Third, the very low atmospheric pressure near the vortex centre is a major source of damage. When a tornado approaches a building, external pressure is rapidly reduced, and unless there is a nearly simultaneous and equivalent decrease in internal pressure, the walls and roof may explode outwards in the process of equalising the pressure differences.

Most tornado damage is due to multiple-vortex tornadoes or very small, intense single-vortex tornadoes. The winds in most multiple-vortex tornadoes may only be strong enough to do minor damage to a particular house. But one of the smaller subvortices, perhaps only a few metres across, may strike the house next door with winds over 300 kilometres per hour, causing complete destruction. Also, there are great differences in construction from one building to the next, so that even in the same wind speed, one may be flattened while the other is barely touched.

Although winds in the strongest tornadoes may far exceed those in the strongest tropical storms, tropical storms typically cause much more damage individually and over a season, and over far bigger areas. Economically, tornadoes cause about a tenth as much damage per year, on average, as tropical storms. Tropical storms tend to cause much more overall destruction than tornadoes because of their much larger size, longer duration and the variety of ways they damage property. The destructive core in tropical storms can be tens or hundreds of kilometres across, last many hours and damage structures through storm surge and flooding caused by heavy rain, as well as from wind. Tornadoes,

in contrast, tend to be a few hundred metres in diameter, last for minutes and primarily cause damage from their extreme winds.

Tornado damage scale

Dr T. Theodore Fujita developed a damage scale for winds, including tornadoes, which is supposed to relate the degree of damage to the intensity of the wind. Work on a new **Enhanced F-Scale** started in 2006 (Table 9.15). The Enhanced F-Scale is a much more precise way to rank tornado damage than the original, because it classifies damage F0–F5 calibrated by engineers across more than 20 different types of buildings. A team of meteorologists and engineers has worked on this for several years. The idea is that a 'one size fits all' approach does not work in rating tornado damage, and a tornado scale needs to take into account the typical strengths and weaknesses of different types of construction. This is because the same wind does different things to different kinds of buildings. In the Enhanced F-Scale, there are different, customised standards for assigning any given F rating to a well-built, well-anchored wood-frame house compared with a garage, school, skyscraper, unanchored house, barn, factory, utility pole or other type of structure. In a real-life tornado track, these ratings can be mapped together more smoothly to produce an accurate damage analysis.

Table 9.15 Enhanced Fujita Scale

EF number	3-second gust (mph)
0	65–85
1	86–110
2	111–135
3	136–165
4	166–200
5	200+

Note: this scale was created in the USA, so measurements are imperial.

Managing tornadoes

The main problem with anything that could realistically stand a chance of affecting a tornado (for example an atomic bomb) is that it would be even more deadly and destructive than the tornado itself. Lesser things (like huge piles of dry ice) would be too hard to deploy in the right place fast enough, and would probably not have a significant effect on the tornado.

Nor is there any proof that seeding can or cannot change tornado potential in a thunderstorm. This is because there is no way of knowing that the things a thunderstorm does after it has been seeded would not have happened *anyway*. This includes any presence or lack of rain, hail, wind gusts or tornadoes. Because the effects of seeding are impossible to prove or disprove, there is a great deal of controversy among meteorologists about whether it works and, if so, under what conditions and to what extent.

Case Study: Tornadoes in Indiana

Indiana is in what is considered to be 'Tornado Alley' (see Figure 9.41 and Table 9.16), a swathe of states extending from the south-east USA to the interior plains. Although the state lacks the high frequency of tornadoes seen in places like Kansas and Oklahoma, it makes up for it in the intensity of its tornadoes.

Table 9.16 Indiana tornado disasters

Date	Place	Damage
13 April 1852	New Harmony	16 killed
14 May 1886	Anderson	43 killed
23 March 1913	Terre Haute	21 killed
11 March 1917	New Castle	21 killed
23 March 1917	New Albany	45 killed
28 March 1920	Allen through Wayne counties	39 killed by three tornadoes
17 April 1922	Warren through Delaware counties	14 killed
18 March 1925	'Tri-State Tornado': Posey, Gibson and Pike counties	74 killed
26 March 1948	Coatesville destroyed	20 killed
21 May 1949	Sullivan and Clay counties	14 killed
11 April 1965	'Palm Sunday Outbreak': 11 tornadoes, 20 counties	137 killed
3 April 1974	'Super Outbreak': 21 tornadoes hit 39 counties	47 killed
2 June 1990	37 tornadoes hit 31 counties	8 killed

Tornadoes can occur in any month, but March–June is considered tornado season in Indiana. Historically, the most destructive tornadoes strike in March and April. June holds the record for the most tornadoes in Indiana on any given day (37), and for the most in a single month (44). Both records were set in 1990, which is also the year when the state experienced the most tornadoes (49).

From 1950 to November 2001, 1024 tornadoes caused more than $1.7 billion in damage in Indiana, and killed 223 people.

Indiana was one of three mid-western states in the path of the deadliest tornado in American history. On 18 March 1925, the Tri-State Tornado travelled a record 352 kilometres on the ground from Missouri through Illinois and into Indiana, where it struck Posey, Gibson and Pike counties. The town of Griffin lost 150 homes, and 85 farms near Griffin and Princeton were devastated. About half of Princeton was destroyed, with losses totalling nearly $2 million. The funnel finally dissipated just outside Princeton, 3½ hours after it had begun. Nearly 700 people died, 74 of them in Indiana. Murphysboro in Illinois lost 234 people, a record for a single community.

In April 1965, 11 tornadoes struck 20 counties in central and northern Indiana, killing 137 people. More than 1700 people were injured and property damage exceeded $30 million. It was Indiana's worst tornado disaster. The tornadoes that devastated Indiana were part of an outbreak in which nearly 50 tornadoes struck the Great Lakes region on 11–12 April, causing 271 deaths and more than 3400 injuries.

The most destructive tornado outbreak of the twentieth century was the 'Super Outbreak' of 3–4 April 1974. During a 16-hour period, 148 tornadoes hit 13 states, including Indiana. The path of destruction stretched over 4000 kilometres. More than 300 people died and more than 5000 were injured. The most notable tornado in this group destroyed much of Xenia, Ohio. In Indiana, 21 tornadoes struck 39 counties, killing 47 people. Seven produced damage rated F5, the maximum possible, and 23 more were rated F4. This was one of only two outbreaks with over 100 confirmed tornadoes, the other being during tropical storm Beulah in 1967 (115 tornadoes).

Section 9.3 Activities

1 Briefly explain how tornadoes are formed.
2 Using examples, outline the factors that affect tornado damage.
3 To what extent is it possible to manage the risk of tornado damage?

9.4 Sustainable management of hazardous environments

Case Study: The use of geo-materials for erosion and sediment control

In Malaysia, early research on bio-engineering involved studies on plant selection for the re-vegetation of cut slopes along highways. Research in 2000–01 involved gully erosion control and vegetation establishment on degraded slopes. These techniques have incorporated the coppicing abilities of cut stems and the soil-binding properties of roots into civil designs, to strengthen the ground and to control erosion. Bio-engineering designs have great potential and application in Malaysia because in deforested upland sites landslides are common, particularly during the wetter months between November and January. Post-landslide restoration works involving conventional civil designs are costly and sometimes not practical at remote sites. Due to cost constraints, the remoteness of the sites and low risk to lives and property, bio-engineering was the option taken for erosion control, slope stabilisation and vegetation establishment.

⇨

The study took place at Fraser's Hill, in the state of Pahang, Malaysia. The area receives 20–410 millimetres of rainfall each month. The temperature is moderate, ranging from 18 to 22 °C annually, with high humidity, ranging from 85 to 95 per cent every month. The surrounding vegetation is lower montane forest.

Two study plots were chosen, and a control plot. Initial works involved soil nailing, using 300 live stakes of *angsana* tree branches and 200 cut stems of *ubi kayu*. Subsequently, major groundworks involved the installation of **geo-structures** (structures constructed from **geo-materials** such as bamboo and brush bundles, coir rolls and straw wattles). The volume of sediment trapped by the geo-structures was measured every two weeks, while plant species that were established on the retained sediments and on geo-materials were identified. The number of live stakes that produced shoots and roots was also recorded. Ten 1 metre-tall saplings of *Toona sinensis*, a fast-growing tree species, were planted at the toe of the slope for long-term stability.

The first slope failure was caused by seepage of drainage water into the cut slope of the access road. The total area affected by the landslide was about 0.25 hectare. Two large trees, 4–5 metres tall, were uprooted and ground vegetation and debris were washed downhill, preventing road access. The second and more extensive failure was located uphill and was a rotational failure. It covered an area of about 0.75 hectare. The landslide was probably triggered by seepage of water from a badly damaged toe drain beside the road.

Bio-engineering design: After six months

The bio-engineering designs involved the installation of 11 bamboo bundles ('faschines') and 16 brush bundles along rills and gullies. At suitable sites along contours, 10 coir rolls and 5 straw wattles were installed, using live stakes and steel wiring. Lighter geo-materials such as straw wattles and brush faschines were positioned on the upper slope face, while heavier geo-materials such as coir rolls were positioned lower down.

At the end of six months, the situation at each study site was assessed (Table 9.17).

Table 9.17 Selected geo-materials and total volume of sediment retained over six months at the two study sites

Geo-materials	Total sediment retained m³	Total number of migrant species
Bamboo faschine	1.7	14
Brush faschine	1.0	17
Coir roll	2.2	20
Straw wattle	0.2	26
Total sediment retained by different geo-materials	**5.1**	–
Total number of migrant species	–	**77**

Live stakes and *Toona sinensis* saplings

At the end of six months, the live stakes had become living trees. A high percentage of *angsana* stakes (93 per cent) sprouted shoots and roots after a month, and 75 per cent of *ubi kayu* stems sprouted leaves within a week. Thus, live stakes were effective in stabilising unstable slopes.

Vegetation cover on slopes helped reduce soil erosion because shoots helped reduce the intensity of raindrops falling on the exposed soil. Furthermore, root-reinforced soils functioned like micro-soil nails to increase the shear strength of surface soils.

Slope stability

The indicator poles at both study sites moved less than 8°, unlike the indicator poles from the control plot, which moved about 20°. Without erosion-control measures, there was aggressive soil erosion during heavy downpours, which caused scouring of the steep slope below the tarred road and resulted in an overhang of the road shoulder.

Trapped sediments and vegetation establishment

A total of 57 geo-structures retained 5.1 m³ of sediment after six months. The retained sediments and decomposing geo-materials also trapped moisture and provided ideal conditions for the germination of incoming seeds. After six months, it was found that 77 plant species were established.

After one year

A year after the study was first implemented, about 75 per cent of one study site was covered by vegetation, while 90 per cent of the second plot was re-vegetated. There was no more incidence of landslide at these two plots. However, at the control plot there was further soil erosion, which resulted in further undercutting of the slope face.

At the control plot, after one year, only seven plant species were present. These were weeds. The poor vegetation cover was probably due to unstable soil conditions caused by frequent soil erosion and minor landslides. It is believed that vegetation cover can provide a layer of roots beneath the soil layer and this contributes additional shear strength to the soil and slope stability.

The geo-structures were installed at a cost of about US$3078, which was cheaper than restoration works using conventional civil structures such as rock gabions, which cost about US$20 000. As the site is quite remote, higher transportation and labour costs would have contributed to the higher cost of constructing a rock gabion at this site. On the other hand, the geo-materials that are abundantly available locally are relatively cheap to make or purchase, and this contributed to the low project cost. The geo-structures were non-polluting, required minimal post-installation maintenance, were visually attractive and could support greater biodiversity within the restored habitats. The geo-materials used in this project, such as faschines, coir rolls and straw wattles, biodegrade after about a year and become organic fertilisers for the newly established vegetation.

After 18 months, the restored cut slopes were almost covered by vegetation, and there was no further incident of landslides. The geo-structures installed on site were cost-effective and visually attractive. The restored cut slopes were more stable and supported higher biological diversity.

Assessment of costs

The geo-structures cost approximately $3000 to install. In contrast, a rock gabion would have cost about $20 000 to install (as the area is remote, transport costs would increase, and there would be increased emissions of greenhouse gases). Moreover, the geo-structures were visually attractive, could support biodiversity, were locally available, and took just two weeks to install. In terms of a cost-benefit analysis, therefore, the geo-structure has a great deal to offer.

☐ The sustainable livelihoods approach for volcano-related opportunities

In an article entitled 'Living with Volcanoes: The sustainable livelihoods approach for volcano-related opportunities', Ilan Kelman and Tamsin Mather outlined ways in which people could have a sustainable livelihood in volcanic areas.

The destructive forces of volcanoes are well known, for example Mt Pelée in Martinique killed approximately 30 000 people in St Pierre, while in 1985 lahars from Nevado del Ruiz, Colombia, killed approximately 25 000 people. National/regional impacts are represented by the 1783–84 eruptions of Laki on Iceland, which killed 24 per cent of Iceland's population and caused thousands of deaths elsewhere in Europe. Global volcano-related impacts have been noticeable through weather alteration, as was the case following the 1991 Mt Pinatubo eruption in the Philippines.

However, human fatalities linked to volcanoes have been relatively few. The death toll attributed to volcanoes since 1 CE is approximately 275 000. As with many disasters, volcano-related disasters also have psychological impacts.

Literature dealing with the volcanic risk perception tends to focus on threats and dangers from volcanoes, along with possible preparation measures, whereas information regarding perceptions of volcano-related benefits or opportunities are more limited.

The contributions of volcanoes to society are widespread. For example, the Mt Etna region represents just under 7 per cent of the land area of Sicily, yet is home to over 20 per cent of the population. Reasons for this intense human activity on the lower slopes of the volcano are not difficult to find, including fertile soils and a reliable freshwater supply. The Soufrière volcano on St Vincent brings agricultural, mining, quarrying and tourism benefits to St Vincent and the Grenadines. There are also geothermal resources, and the use of volcanic materials for making items such as basalt hammers and pumice, along with the archaeological and artistic gains from volcanism.

Dealing with environmental hazards

As exemplified by Mt Etna in Italy and Mt Mayon in the Philippines, people have good reasons for living near or on volcanoes, including good farmland and reliable water supplies. This sometimes yields dangers, despite the rewards. To balance the dangers or potential dangers with the gains or potential gains from environmental hazards, including volcanoes, a four-option framework has been developed (Table 9.18).

Table 9.18 Options and consequences for dealing with environmental hazards

Option for dealing with environmental hazards	Main implications
1 Do nothing	Disasters occur.
2 Protect society from hazards	Not always feasible and leads to risk transference, which augments vulnerability.
3 Avoid hazards	Not always feasible and can exacerbate other problems, augmenting vulnerability.
4 Live with the hazards and risks	Livelihoods are integrated with environmental threats and opportunities.

The first option is to do nothing, accepting that volcanic disasters will happen. Depending on the volcano, this option might be more viable or less viable. Mt Etna in Italy frequently erupts, so doing nothing could lead to a disaster, depending on the extent and characteristics of an eruption. In contrast, Mt Jefferson in the USA has not erupted in several centuries and doing nothing could be an option there.

The second option is to try to protect society from volcanic hazards, such as by strengthening roofs against tephra fall, building structural defences against lahars, pumping sea water onto lava (Heimaey, Iceland 1973), diverting lava (Mt Etna) or slowly degassing (Lake Nyos, Cameroon). However, this protection option is not always feasible. For example, not all gas releases could be averted through degassing. Large pyroclastic flows and lava flows are challenging to stop or even to redirect, although structures could be designed to afford some level of protection against these hazards. Moreover, reliance on protective measures could lead to a false sense of security, without tackling the root causes of vulnerability.

The third option is to avoid volcanic hazards, but that is not always feasible. Volcanic impacts are often not local and are sometimes even global, so all places on Earth have the potential for being severely affected by volcanic activity. Additionally, with global population increasing, constraints on land and resources frequently leave little option other than to inhabit areas that are potentially affected by volcanic hazards.

Moreover, avoiding volcanic hazards could cause further problems. Volcanic activity can yield advantages that might outweigh the problems. Moving away from volcanoes could yield other concerns, perhaps exposure to other environmental hazards or perhaps social challenges. After Montserrat's volcano started erupting in 1995, some families moved to England, only to be disappointed at the low standard of education in English schools. Many Montserratians were shocked, too, at the level of crime risk to which they were exposed on neighbouring Caribbean islands.

The fourth option – living with risk – means accepting that environmental hazards are a part of life and of a productive livelihood. A component of living with risk is localising disaster risk reduction. Disaster risk reduction, including pre-disaster activities such as preparedness and mitigation and post-disaster activities such as response and recovery, is best achieved at the local level with community involvement. The most successful outcomes are seen with broad support and action from local residents, rather than relying on external specialists or interventions. Although the long dormancy periods of volcanoes and significant uncertainties about eruptive pathways might make community interest in disaster risk reduction wane, few communities are vulnerable only to volcanic hazards.

☐ The sustainable livelihoods approach

Sustainable livelihoods can be defined as creating and maintaining means of individual and community living that are flexible, safe and healthy from one generation to the next. The sustainable livelihoods approach is important in its application to volcanic scenarios in four ways:

1 Understanding, communicating and managing vulnerability and risk and local perceptions of vulnerability and risk beyond the immediate threats to life.
2 Maximising the benefits to communities of their volcanic environment, especially during quiescent periods, without increasing vulnerability.
3 Managing crises.
4 Managing reconstruction and resettlement after a crisis.

Applying the sustainable livelihoods approach

Managing vulnerability and risk

The first application of the sustainable livelihoods approach to volcanoes is understanding, communicating and managing vulnerability and risk along with local perceptions of vulnerability and risk beyond immediate threats to life.

Thinking ahead of the event ensures that:

■ local livelihoods are preserved, meaning that the population has an easier post-disaster recovery except for cases of extreme destruction
■ the affected population is confident that their livelihoods will remain, so they will be more willing to shelter and evacuate without putting their lives at risk for the sake of livelihoods.

Examples include attempts to prevent lava blocking Heimaey's harbour and balancing ski access to Ruapehu during active episodes, especially in light of the continuing lahar threat. In these instances, it was decided that saving only lives without considering livelihoods was unacceptable. Risk and vulnerability have been managed to achieve a balance between lives and livelihoods: living with volcanic risk.

Maximising community benefits sustainably

The second application is maximising the benefits to communities of their volcanic environment, especially during quiescent periods, while decreasing vulnerability. The livelihood benefits of volcanoes can be placed into three main categories: physical resources (for example, mining), energy resources (for example, heat) and social resources (for example, tourism).

Volcanoes play an important role in the formation of precious metal ores. However, if the volcano's activity increases, the mining resources, equipment and expected income could be jeopardised. The 2006 eruption of a 'mud volcano' in eastern Java, which was highly destructive to local livelihoods, resulted from borehole drilling.

Managing crises

The third application is managing crises. Emergency response and humanitarian relief are adopting the sustainable livelihoods approach, such as for the sectors of transitional settlement and shelter and food security.

Managing reconstruction and resettlement

The fourth application is managing reconstruction and/or resettlement after a volcanic crisis. Montserrat provides a good example. Resettlement in the island's north, away from the most dangerous zones due to volcanic activity, included housing construction that was completed without sufficient attention to local culture, other hazards or livelihoods. The resettlement saved lives, but did not adopt a local approach to living with risk. Long-term problems emerged that the sustainable livelihoods approach might have prevented.

Disadvantages

Volcano-related evacuations have sometimes forced people to choose between staying in poorly managed shelters with no livelihood prospects and returning home to their livelihoods despite a high risk of injury or death from the volcano. This issue was witnessed in Montserrat, exacerbated by economic structures that encouraged farming in the exclusion zone (Figure 9.42).

Figure 9.42 Cattle in the exclusion zone, Montserrat

Figure 9.43 It is not always possible to see volcanic impacts as positive

☐ Towards reducing volcanic impacts

Considering livelihoods is important in successful volcanic disaster risk reduction because they contribute to living with volcanic risk based on a localised approach. Living with volcanoes at the local level requires changes of perception and action, resulting in advantages for volcanic disaster risk reduction, although there can also be potential disadvantages (Figure 9.43). With the local population involved in monitoring, understanding, communicating, making decisions and taking responsibility for aspects of volcanic disaster risk reduction – with external guidance and assistance where requested – disadvantages can be minimised.

Three points emerge from applying the sustainable livelihoods approach to localised living with volcanic risk:

- First, not all livelihoods near volcanoes are volcano-related. Productive agriculture could be due to floodwaters rather than volcanic deposits.
- Second, not all volcanic activity necessarily yields livelihoods, or livelihoods that should be encouraged. Tourism and research activities in active craters (Figure 9.44), for example, tend to be discouraged

in vulcanology. That level of risk-taking could also make the livelihood vulnerable. For example, if tourists were killed by a volcano, the area's tourism could suffer.

- Third, resource availability does not always imply resource use. Mining could be deemed too externally dependent or too environmentally and socially destructive to be worthwhile pursuing.

Volcanic risk perception and communication studies show that not everyone living by a volcano understands or accepts the actual or potential implications of the volcano. Risk and disasters emerge from volcanoes, but livelihood opportunities emerge from volcanoes too. Those opportunities form an integral part of volcanic disaster risk reduction.

Despite volcanic benefits, living with volcanic risk is not always feasible and volcanoes should not be relied on for livelihoods without careful consideration of potential drawbacks. Other approaches – do nothing, protect and avoid – should be considered, as well as appropriate combinations of the approaches for different combinations of volcanic risks, volcanic benefits and societal desires.

Figure 9.44 The world's only drive-in active volcanic crater, St Lucia

Case Study: Montserrat

The Soufrière volcano on Montserrat is a well-used example of the effects of a volcano in a LIC. It is over 15 years since the main eruption in 1997 in which 19 people died. The capital city, Plymouth, was abandoned, and became a modern-day

Pompeii. Much of the southern third of the island became an exclusion zone (Figure 9.45). So how have things changed since 1997?

Figure 9.45 Plymouth and Soufrière, Montserrat

By 2002, Montserrat was experiencing something of a boom. The population, which had dropped in size from over 11 000 before the eruption to less than 4000 in 1999, had risen to over 8000. The reason was very clear. There were many jobs available on the island. There were many new buildings, including new government buildings, a renovated theatre, new primary schools and lots of new housing in the north of the island. There was even a new football pitch and stadium built at Blakes Estate (Figure 9.46). There were plans to build a new medical school and a school for hazard studies. To date, these have not been built.

Figure 9.46 Montserrat football pitch

However, by the summer of 2009 it was very clear that conditions on Montserrat had changed. The population had fallen to a little over 5200. There are two main reasons for this. The first is the relative lack of jobs. Although there was an economic boom in the early 2000s, once the new buildings were built many of the jobs disappeared. There are still plans to redevelop the island – a new urban centre and a new port are being built at Little Bay but they will not be complete until 2020. The museum has been built but not much else (Figure 9.47). Thus there are some jobs available but not so many as there were previously. Second, one of the new developments on Montserrat was a new airstrip. Once this was built, the UK and US governments stopped subsidising the ferry that operated between Antigua and Montserrat. This made it more difficult to get to Montserrat, both for visitors and for people importing basic goods. Thus the number of tourists to the island fell and the price of goods on the island rose. Many Montserratians were against the airstrip and campaigned unsuccessfully for the port to be kept open. It is possible to charter a boat and sail to Montserrat but it is far more expensive than taking a ferry.

Thus with fewer jobs in construction, a declining tourist sector and rising prices, many Montserratians left the island for a second time. Many went to Antigua and others went to locations such as Canada, the USA and the UK. Much of the aid that was given to Montserrat following the eruptions of 1997 has dried up. The UK provided over $120 million of aid but announced in 2002 that it was phasing out aid to the island. Nevertheless, in 2004 it announced a £40 million aid deal over three years.

Figure 9.47 Montserrat museum

The volcano has been relatively quiet for the last few years. However, there was an event in May 2006 that was relatively unreported. The Soufrière dome collapsed, causing a tsunami that affected some coastal areas of Guadeloupe, and English Harbour and Jolly Harbour in Antigua. The Guadeloupe tsunami was 1 metre high and the one in Antigua between 20 and 30 centimetres. No-one was injured in the tsunami but flights were cancelled between Venezuela and Miami, and to and from Aruba, due to the large amount of ash in the atmosphere.

So while volcanic activity in Montserrat is currently quiet, the volcano continues to have a major impact on all those who remain on the island. The economic outlook for the island does not look good – and that is largely related to the lack of aid, the difficulty and cost of reaching Montserrat and the small size of the island and its population.

☐ Sustainable development and hurricanes

The achievement of equity, risk reduction and long-term development through local participation in recovery planning and institutional co-operation is the central issue in recovery and sustainable development.

The recovery phase is the least investigated and least understood of the four phases of a disaster (Figure 9.48) – pre-disaster mitigation, emergency preparedness, emergency response and recovery.

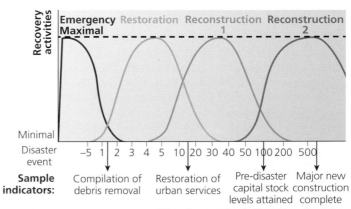

Figure 9.48 The four-stage model

Various LICs are trying to integrate recovery with sustainable development initiatives. Jamaica, for example, has shifted disaster recovery responsibilities from its national emergency management agencies to government agencies charged with environmental protection and long-term economic development and to community-based private voluntary organisations active in development initiatives such as housing, healthcare, watershed management and agriculture.

The four-stage model has four clear aspects:

1 Take emergency measures for removal of debris, provision of temporary housing and search and rescue.
2 Restore public services (electricity and water).
3 Replace or reconstruct capital stock to pre-disaster levels.
4 Initiate reconstruction that involves economic growth and development.

However, several studies suggest that the four stage model may be inaccurate. For example, stage 3 may occur in some areas while some areas are still in stage 1, and some groups may not have services restored as quickly, for example shanty-town residents, poor communities and immigrant groups. Many actual recoveries may not be so clear cut (Figure 9.49).

Successful recovery requires:

- integration of interested parties (government, non-governmental organisations, community groups)
- monitoring of programmes/enforcement of policies
- recognition of all people's rights (elderly, young, women, homeless, poor, migrants, refugees)
- leadership – ideally community-based (bottom-up) development
- resources.

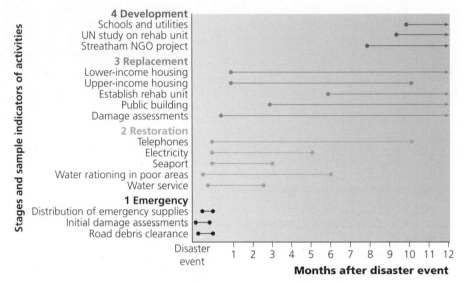

Figure 9.49 Recovery in Montserrat following Hurricane Hugo

Case study: Recovery after Hurricane Hugo

When Hurricane Hugo struck Montserrat on 17 September 1989, it was the first hurricane to hit the island in over 60 years. Eleven people were killed, 3000 people were made homeless and up to 98 per cent of buildings were damaged. All government buildings and schools were either partially or totally destroyed. Damage exceeded over US$360 million – devastating for the island.

Yet for some parts of Montserrat, the recovery was considered to be very successful. The village of Streatham near Plymouth was a small agricultural village of about 300 people. All the homes were damaged. The recovery was organised by local people – they rebuilt more than 20 homes and a new community centre, introduced new agricultural practices and improved the settlement's water supply.

Summary

Hurricanes in the Caribbean are serious. People and property are placed at risk, and government attempts to protect both are limited. Much of the construction in the Caribbean is informal, and lacks adequate building standards. Moreover, population growth suggests more people will be at risk in the future. In addition, global warming may potentially increase the frequency and magnitude of hurricanes due to increased atmospheric energy.

It might be more realistic to think of recovery as a process in which political, economic and demographic factors, as well as location, are important. Some groups are slower to recover than others. This raises questions about fairness and equity.

Top-down programmes managed by central government and international NGOs do not necessarily work well because they may be vulnerable to political manipulation.

Opportunities to relocate people and structures out of floodplains and other high-risk locations may be missed. Long-term sustainable development should include reduced environmental degradation – for example deforestation, soil erosion, habitat degradation – and improved housing and living conditions. One positive example of this was Streatham on Montserrat.

External donor organisations and charities must not just treat the symptoms of hurricanes – they must also address the causes of disasters alongside refocusing on long-term community development.

Promoting bottom-up recovery

A bottom-up community-based approach to recovery will be more effective than the traditional top-down approach, as it will respond to local people's needs and priorities. At Streatham, the disaster was used as a unique opportunity for change, and brought to the fore problems that are usually low in priority.

Strategies for long-term mitigation include:

- strengthening the housing stock
- improving land-use patterns
- environmental protection
- increased understanding of natural hazards.

Section 9.4 Activities

1 To what extent is the management of the Soufrière volcano on Montserrat an example of sustainable development? Give reasons for your answer.
2 Briefly explain the main methods of dealing with earthquakes.
3 In what ways is it possible to manage the risk of volcanoes?
4 Outline the advantages of geo-engineering over hard engineering structures for slope stabilisation.
5 Compare the main characteristics of the emergency phase following a natural disaster with that of the reconstruction/replacement phase.

Hot arid and semi-arid environments

10.1 Hot arid and semi-arid climates

☐ Global distribution and climatic characteristics

Figure 10.1 and Tables 10.1 and 10.2 show the distribution of arid environments. While Africa has the greatest proportion of these, Australia is the most arid continent with about 75 per cent of the land being classified as arid or **semi-arid**. Most arid areas are located in the tropics, associated with the subtropical high-pressure belt. However, some are located alongside cold ocean currents (such as the Namib and Atacama deserts), some are located in the lee of mountain ranges (such as the Gobi and Patagonian deserts), while others are located in continental interiors (such as the Sahara and the Australian deserts).

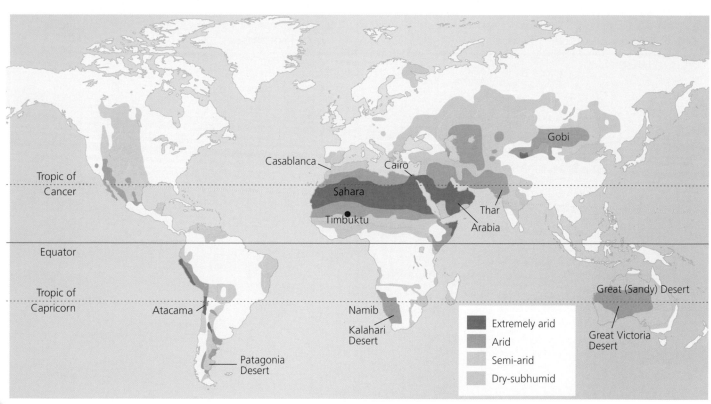

Figure 10.1 The global distribution of arid areas

Table 10.1 The extent of global arid areas (as a percentage of the global land area)

Classification	Semi-arid	Arid	Extremely arid	Total
Köppen (1931)	14.3	12.0	–	26.3
Thornthwaite (1948)	15.3	15.3	–	30.6
Meigs (1953)	15.8	16.2	4.3	36.3
Shantz (1956)	5.2	24.8	4.7	34.7
UN (1977)	13.3	13.7	5.8	32.8

Table 10.2 Distribution of arid lands by continent (as a percentage of the global total)

Continent	Percentage arid
Africa	37
Asia	34
Australasia	13
North America	8
South America	6
Europe	2

Definitions of aridity

There are many definitions of the term 'arid'. Literary definitions use such terms as 'inhospitable', 'barren', 'useless', 'unvegetated' and 'devoid of water'. Scientific definitions have been based on a number of criteria including climate, vegetation, drainage patterns and erosion processes. What they share is a consideration of moisture availability, through the relationship between precipitation and evapotranspiration.

Most modern systems for defining **aridity** are based on the concept of water balance; that is, the relationship that exists between inputs in the form of precipitation (P) and the losses arising from evaporation and transpiration (E). The actual amount of evapotranspiration that will occur depends on the amount of water available, hence geographers use the concept of potential evapotranspiration (PE), which is a measure of how much evapotranspiration would take place if there was an unlimited supply of water.

Meigs' (1953) classification is probably the most widely used today. It was produced for UNESCO and was concerned with food production. Arid areas that are too cold for food production (such as polar and mountainous regions) were omitted. Meigs based his classification scheme on Thornthwaite's (1948) indices of moisture availability (Im):

$$Im = (100 S - 60D)/PE$$

where PE is potential evapotranspiration, S is moisture surplus and D is moisture deficit, aggregated on an annual basis and taking soil moisture storage into account.

When P = PE throughout the year the index is 0.

When P = 0 throughout the year, the index is –60.

When P greatly exceeds PE throughout the year, the index is 100 (see Figure 10.2).

Meigs identified three types of arid area:

- semi-arid: –40 < Im <20
- arid: –56 < Im < –40
- hyper-arid (extremely arid): < –56 Im.

Grove (1977) attached mean annual precipitation to the first two categories: 200–500 millimetres for arid and 25–200 millimetres for semi-arid, but these are only approximate. Hyper-arid areas have no seasonal precipitation and occur where twelve consecutive months without precipitation have been recorded. According to these definitions, arid areas cover about 36 per cent of the global land area.

Aridity is a permanent water deficit, whereas drought is an unexpected short-term shortage of available moisture.

Rainfall effectiveness (P–E) is influenced by a number of factors:

- **Rate of evaporation** – this is affected by temperature and wind speed, and in hot, dry areas evaporation losses are high
- **Seasonality** – winter rainfall is more effective than summer rainfall since evaporation losses are lower
- **Rainfall intensity** – heavy intense rain produces rapid runoff with little infiltration
- **Soil type** – impermeable clay soils have little capacity to absorb water, whereas porous sandy soils may be susceptible to drought.

Another classification is based on rainfall totals. This states that semi-arid areas are commonly defined as having a rainfall of less than 500 millimetres per annum, while arid areas have less than 250 millimetres and **extremely arid areas** less than 125 millimetres per annum. In addition to low rainfall, dry areas have variable rainfall. For example, annual rainfall variability in a rainforest area might be 10 per cent. If the annual rainfall is about 2000 millimetres, this means that in any one year the rainfall would be somewhere between 1800 millimetres and 2200 millimetres. As rainfall total decreases, variability increases. For example, areas with a rainfall of 500 millimetres have an annual variability of about 33 per cent. This means that in such areas, rainfall could range between 330 millimetres and 670 millimetres. This variability has important consequences for vegetation cover, farming and the risk of flooding.

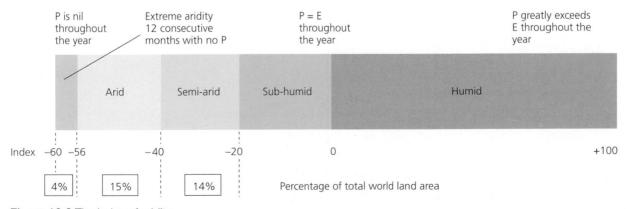

Figure 10.2 The index of aridity

All three areas are considered as part of the arid zone. This is because:

- the division between the three is arbitrary and varies depending on the classification used
- annual precipitation is highly variable and in any one year could be extremely low
- these areas share the same processes and landforms
- in the twentieth century, climate change and human activities have caused the expansion of some arid areas into semi-arid areas
- semi-arid areas are often termed 'deserts' by their inhabitants.

It is important to remember that there are other factors that influence arid areas. There are hot deserts (tropical and subtropical) and cold deserts (high latitude and high altitude). Coastal deserts, such as the Atacama and the Namib, have very different temperature and humidity characteristics from deserts of continental interiors, such as central areas of the Sahara. There are also shield deserts, as in India and Australia, which are tectonically inactive, and mountain and basin deserts, such as south-west USA, which are undergoing mountain building.

Causes of aridity

Arid conditions are caused by a number of factors. The main cause is the global atmospheric circulation. Dry, descending air associated with the **subtropical high-pressure belt** is the main cause of aridity around 20°–30°N (Figure 10.3a). Here, the stable, adiabatically warmed, subsiding body of air prevents rising air from reaching any great height. Convection currents are rarely able to reach sufficient height for condensation and precipitation. After the air has subsided, it spreads out from the centre of high pressure (Figure 10.3a). It thereby prevents the incursion of warm maritime air into the region, reinforcing its aridity. The distribution of land and sea prevents the formation of a single zone of high pressure – rather it is divided into discrete cells such as those over South America and Africa. Tropical and subtropical deserts cover about 20 per cent of the global land area. These are large arid zones composed of central arid areas surrounded by relatively small, marginal semi-arid belts. Rainfall is very unreliable and largely associated with seasonal movements of the inter-tropical convergence zone.

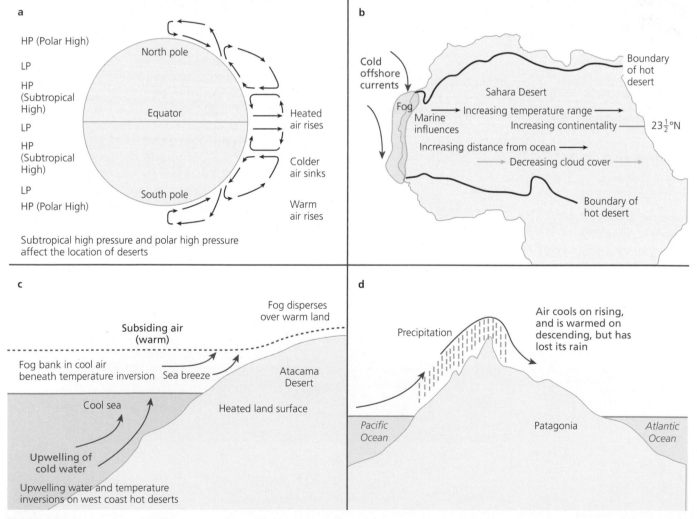

Figure 10.3 The causes of aridity

In addition, distance from sea, **continentality**, limits the amount of water carried by winds (Figure 10.3b). Precipitation and evapotranspiration are usually lower than in arid areas, resulting from subtropical high-pressure belts. Cold winters are common. These areas are characterised by a relatively small area of aridity surrounded by an extensive area of semi-aridity. The three major northern hemisphere deserts (Gobi and Turkestan in Asia and the Great Basins in North America) are mid-continental and receive little rain. The major central deserts of Australia and Africa also receive little rain as the precipitation is lost when air masses pass over the land. **Rainshadow effects** further increase the aridity of continental interiors.

In other areas, such as the Atacama and Namib deserts, **cold offshore currents** limit the amount of condensation into the overlying air (Figure 10.3c). Cold currents reinforce climatic conditions, causing low sea-surface evaporation, high atmospheric humidity, low precipitation (very low rainfall, with precipitation mainly in the form of fog and dew) and a small temperature range.

Others are caused by intense rainshadow effects, as air passes over mountains. This is certainly true of the Patagonian desert (Figure 10.3d). This can reinforce aridity that has been caused primarily by atmospheric stability or continentality. The prevailing winds in the subtropics are trade winds, which blow from the north-east in the northern hemisphere and the south-east in the southern hemisphere. Where the trade winds meet mountain barriers, such as the Andes or the Rockies, the air is forced to rise. **Orographic or relief rainfall** is formed on the windward side; but on the lee side dehydrated air descends, creating a rainshadow effect. If the mountain ranges are on the east side of the continent then the rainshadow effect creates a much larger extent of arid land. For example, in Australia the Great Dividing Range intercepts rain on the east coast, creating a rainshadow effect to the west.

A final cause, or range of causes, is human activities. Many of these have given rise to the spread of desert conditions into areas previously fit for agriculture. This is known as **desertification**, and is an increasing problem.

Section 10.1 Activities

1 Explain the term *rainfall effectiveness*.
2 Describe the location of the world's dry areas as shown on Figure 10.1.
3 Briefly explain why there are deserts on the west coast of southern Africa and the west coast of South America.
4 Explain the main causes of aridity.

☐ Key features of hot arid and semi-arid environments

Desert rainfall

The main characteristic of deserts is their very low rainfall totals. Some coastal areas have extremely low rainfall: Lima in Peru receives just 45 millimetres of rain and Swakopmund in Namibia just 15 millimetres. Very often, they may receive no rain in a year. Desert rain is also highly variable. The inter-annual variability (V) is expressed:

$$V\% = \frac{\text{mean deviation from the average}}{\text{the average rainfall}} \times 100\%$$

Variability in the Sahara is commonly 50–80 per cent, compared with just 20 per cent in temperate humid areas. Moreover, individual storms can be substantial. In Chicama, Peru, 394 millimetres fell in a single storm in 1925, compared with the annual average of just 4 millimetres! Similarly, at El Djem in Tunisia 319 millimetres of rain fell in three days in September 1969, compared with the annual average of 275 millimetres.

However, many desert areas receive low-intensity rainfall. Analysis of figures for the Jordan desert and for Death Valley in south-west USA show that most rainfall events produce 3–4 millimetres, similar to temperate areas. In coastal areas with cold offshore currents, the formation of fog can provide significant amounts of moisture. In the coastal regions of Namibia, fog can occur up to 200 times a year, and extend 100 kilometres inland. Fog provides between 35 and 45 millimetres of precipitation per annum. Similarly, in Peru fog and low cloud provide sufficient moisture to support vegetation growth.

Temperature

Deserts exhibit a wide variation in temperature. Continental interiors show extremes of temperature, both seasonally and diurnally. In contrast, coastal areas have low seasonal and diurnal ranges. The temperature in coastal areas is moderated by the presence of cold, upwelling currents. Temperature ranges are low – in Callao in Peru the average diurnal range is just 5 °C, but it has a seasonal range of 8 °C. In contrast, in the Sahara the annual range can be up to 20 °C. Mean annual temperatures are also lower in coastal areas: 17 °C in the Namib and 19 °C in the Atacama.

Continental interiors have extremes of temperature, often exceeding 50 °C. Daily (**diurnal**) ranges may exceed 20 °C. In winter, frost may occur in a high-altitude interior desert.

Wind in deserts

Hot arid and semi-arid climates are characterised by high wind-energy environments. This is partly due to the lack of vegetation, and so therefore there is a lesser degree of friction with air movement.

☐ Classification of desert climates

Semi-arid outer tropical climate (BShw)

Bordering the deserts, these areas have long dry winters dominated by subsiding air. Brief, erratic rains occur, associated with the ITCZ at its poleward limit. However, owing to the hot temperatures and rapid evaporation, this climate zone is less effective for plant growth. Years of average rainfall may be followed by many years of drought, as in the case of the Sahel region south of the Sahara.

Semi-arid: poleward of hot deserts (BShs)

Summer months are dry and very hot. During winter, occasional rain is associated with mid-latitude depressions. These areas are very variable in terms of rainfall – years of drought may be followed by storms, bringing hundreds of millimetres of rain. Winter rain generally supports coarse grass and drought-tolerant plants.

Hot desert climates (BWh)

In the subtropics, descending air affects the very dry western parts of land masses between 20° and 25° and strongly influences adjacent areas. Even if the air contains a considerable amount of water vapour, relative humidity is low. Stable, subsiding air prevents convective updraughts, which rarely reach sufficient height for cumulonimbus clouds to develop. Occasionally they may develop and result in sheetwash and flash flooding.

During the day, temperatures may reach 50–55 °C, and at night, due to the clear skies, they may fall to 20–25 °C. During winter, daytime temperatures may reach 15–20 °C whereas at night it may be cold enough to allow dew to form.

Table 10.3 Climate data for some arid and semi-arid cities

Cairo

	J	F	M	A	M	J	J	A	S	O	N	D	Yr
Temperature													
Daily max. (°C)	19	21	24	28	32	35	35	35	33	30	26	21	**28**
Daily min. (°C)	9	9	12	14	18	20	22	22	20	18	14	10	**16**
Average monthly (°C)	14	15	18	21	25	28	29	28	26	24	20	16	**22**
Rainfall													
Monthly total (mm)	4	4	3	1	2	1	0	0	1	1	3	7	**27**
Sunshine													
Daily average	6.9	8.4	8.7	9.7	10.5	11.9	11.7	11.3	10.4	9.4	8.3	6.4	**9.5**

Casablanca

	J	F	M	A	M	J	J	A	S	O	N	D	Yr
Temperature													
Daily max (°C)	17	18	20	21	22	24	26	26	26	24	20	18	**22**
Daily min (°C)	8	9	11	12	15	18	19	20	18	15	12	10	**14**
Average monthly (°C)	13	13	15	16	18	21	23	23	22	20	17	14	**18**
Rainfall													
Monthly total (mm)	78	61	54	37	20	3	0	1	6	28	58	94	**440**
Sunshine													
Daily average	5.2	6.3	7.3	9.0	9.4	9.7	10.2	9.7	9.1	7.4	5.9	5.3	**7.9**

Timbuktu, Mali

	J	F	M	A	M	J	J	A	S	O	N	D	Yr
Temperature													
Daily max (°C)	31	35	38	41	43	42	38	35	38	40	37	31	**37**
Daily min (°C)	13	16	18	22	26	27	25	24	24	23	18	14	**21**
Average monthly (°C)	22	25	28	31	34	34	32	30	31	31	28	23	**29**
Rainfall													
Monthly total (mm)	0	0	0	1	4	20	54	93	31	3	0	0	**206**
Sunshine													
Daily average	9.1	9.5	9.6	9.7	9.8	9.4	9.6	9	9.3	9.5	9.5	8.9	**9.4**

1 Explain the term *rainfall variability*.
2 Compare and contrast the seasonal and monthly temperature ranges for Casablanca and Timbuktu.
3 Using the geographical locations of Casablanca and Timbuktu, suggest reasons for the differences you have noted in their seasonal and monthly temperature ranges.
4 Compare and contrast the precipitation totals for Cairo, Casablanca and Timbuktu. Suggest reasons for the differences you have identified.

10.2 Landforms of hot arid and semi-arid environments

☐ Weathering processes

Weathering in deserts is superficial and highly selective. The traditional view was that all weathering in deserts was mechanical due to the relative absence of water. However, it is increasingly realised that **chemical weathering** is important, and that water is important for mechanical weathering, especially exfoliation. Weathering is greatest in shady sites and in areas within reach of soil moisture. Chemical weathering is enhanced in areas that experience dew or coastal fog. As rainfall increases, weathering increases; and soils tend to have more clay, less salt and more distinct horizons. Salt weathering is frequent in arid areas because desert rocks often have soluble salts, and these salts can disintegrate rocks through salt crystal growth and hydration.

Salt crystallisation causes the decomposition of rock by solutions of salt (Figure 10.4). There are two main types of **salt crystal growth**. First, in areas where temperatures fluctuate around 26–28°C, sodium sulphate (Na_2SO_4) and sodium carbonate (Na_2CO_3) expand by about 300 per cent. This creates pressure on joints, forcing them to crack. Second, when water evaporates, salt crystals may be left behind. As the temperature rises, the salts expand and exert pressure on rock. Both mechanisms are frequent in hot desert regions where low rainfall and high temperatures cause salts to accumulate just below the surface.

Experiments investigating the effectiveness of saturated salt solutions have shown a number of results:

■ The most effective salts are sodium sulphate, magnesium sulphate and calcium chloride.
■ The rate of disintegration of rocks is closely related to porosity and permeability.

Figure 10.4 Salt crystallisation

■ Surface texture and grain size control the rate of rock breakdown. This diminishes with time for fine materials and increases over time for coarse materials.
■ Salt crystallisation is more effective than insolation weathering, hydration or freeze–thaw. However, a combination of freeze–thaw and salt crystallisation produces the highest rates of breakdown.

The most effective salts are, in descending order, sodium sulphate (Na_2SO_4), magnesium sulphate ($MgSO_4$) and calcium chloride ($CaCl_3$). Sodium sulphate caused a 100 gramme block of stone to break down to about 30 grammes – a loss of 70 per cent (Figure 10.5). Similarly, magnesium sulphate reduced a 95 gramme block to just over 40 grammes, a loss of over 50 per cent. The least effective salts were common salt (NaCl) and sodium carbonate (Na_2CO_3).

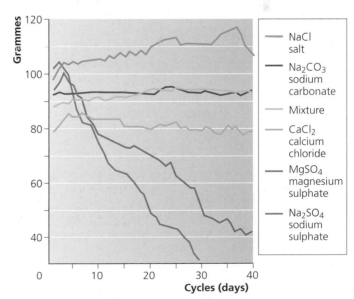

Figure 10.5 The effects of common salts on sandstone under laboratory conditions

Thermal fracturing refers to the break-up of rock as a result of repeated changes in temperature over a prolonged period of time. **Disintegration** or **insolation weathering** is found in hot desert areas where there is a large diurnal temperature range. In many desert areas, daytime temperatures exceed 40°C, whereas night-time temperatures are little above freezing. Rocks heat up by day and contract by night. As rock is a poor conductor of heat, stresses occur only in the outer layers. This causes peeling or exfoliation to occur. Griggs (1936) showed that moisture is essential for this to happen. In the absence of moisture, temperature change alone did not cause the rocks to break down. The role of salt in insolation weathering has also been studied. The expansion of many salts such as sodium, calcium, potassium and magnesium has been linked with exfoliation. However, some geographers find little evidence to support this view.

In some instances, rocks may be split in two. **Block disintegration** is most likely to result from repeated heating and cooling. Such rocks are known as *kernsprung*. A more localised effect is **granular disintegration**. This occurs due to certain grains being more prone to expansion and contraction than others – this exerts great pressure on the grains surrounding them and forces them to break off.

Hydration is the process whereby certain minerals absorb water, expand and change. For example, gypsum is changed to anhydrate. Although it is often classified as a type of chemical weathering, mechanical stresses occur as well. When anhydrite ($CaSO_4$) absorbs water to become gypsum ($CaSO_4.2H_2O$), it expands by about 0.5 per cent. More extreme is the increase in volume of up to 1600 per cent by shales and mudstones when clay minerals absorb water.

Freeze–thaw occurs when water in joints and cracks freezes at 0°C. It expands by about 10 per cent and exerts pressure up to a maximum of 2100 kg/cm^2 at –22°C. This greatly exceeds most rocks' resistance. However, the average pressure reached in freeze–thaw is only 14 kg/cm^2.

Freeze–thaw is most effective in environments where moisture is plentiful and there are frequent fluctuations above and below freezing point. It can occur in deserts at high altitude and in continental interiors in winter.

☐ Processes of erosion, transport and deposition by wind

The importance of wind in deserts has been hotly debated by geographers. At the end of the nineteenth century, it was considered to be a very effective agent in the formation of desert landforms. By contrast, in the twentieth century the role of wind was played down, in part because much of the research into deserts took place in high-relief, tectonically active areas such as the south-west USA. It was argued that:

■ wind-eroded landscapes were only superficially eroded
■ some features, such as playas, were formed by other processes, especially tectonic ones
■ desert surfaces were protected from the wind by crusts, salts and gravel
■ wind erosion depends on the availability of abrasive sands and only operates over a limited height range
■ water is still very active.

However, in the middle of the twentieth century the use of aerial photography and satellites showed major features aligned with prevailing wind systems, such as yardangs in the Sahara, Iran, Peru and Arabia. In addition, examination of desert playas, which are large and frequent, showed that some are tectonic but others are aeolian. Dunes on the lee side of playas suggest that the dunes were deposited by excavating winds. In the Qattara Depression in the Sahara, 3335 km^3 of material has been removed by the wind. Moreover, meteorological observations of dust storms have illustrated the importance of winds. The Great American Dust Storm of 12 November 1933, which marked the beginning of the Dust Bowl, stretched from Canada to western Ohio and the Missouri Valley, an area larger than France, Italy and Hungary! The increased frequency of dust storms in the USA was due to severe drought and poor agricultural techniques. Although the land was not desert, it took on desert characteristics. In the 1970s, there was an increase in the number of dust storms in the Sahel region of Africa. Some of these storms travelled across the Atlantic to reach the Caribbean and were also associated with an increase in asthma there, partly as a result of increased dust, and partly as a result of the transfer of bacteria by winds across the Atlantic Ocean. Finally, as already noted, during the last glaciations some areas in the tropics were wetter and some were drier. It is estimated that the rate of dust removal and deposition was 100 times greater than it is at present.

Erosion

By itself, wind can only blow away loose, unconsolidated material so gradually lowering the surface by **deflation**. At wind speeds of 40 kilometres per hour, sand grains will move by surface creep and saltation. Much transport of sand will therefore be limited to a metre or so above the surface (Figure 10.6). Most abrasion (corrasion) occurs in this zone, by sand particles hitting against rocks. Higher wind speeds will cause dust storms. Extremely rare gusts of over 150 kilometres per hour are needed to roll pebbles along the ground. Fine dust is moved easily by light winds.

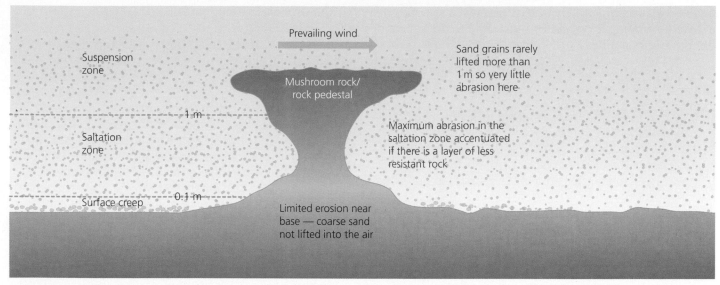

Figure 10.6 Wind transport and erosion

Transport

Sand-sized particles are well suited to transport by the wind. Sand movement occurs when wind speeds exceed 20 kilometres per hour. Grains initially begin to roll (traction), and then follow a bouncing action, known as saltation. The saltating grains are typically 0.15–0.25 millimetres in diameter. In contrast, larger grains (0.26–2 millimetres) move by surface creep and smaller grains (0.05–0.14 millimetres) move through **suspension**.

Deposition

Deposition occurs when the wind speed is reduced. The form taken by the deposited material is influenced by:

- the nature of any surface irregularity
- the amount and type of material carried by the wind, itself controlled by velocity
- the flow pattern of the dominant wind (shaping the material being deposited)
- the presence or absence of vegetation and groundwater.

Deserts occupy about 20 per cent of the world and their area is expanding. For example, the arid belt of the southern Sahara, known as the Sahel, has extended considerably into the savanna lands of Ethiopia, causing widespread famine there.

Sand drifts (temporary pockets of sand found in 'wind shadow' areas) and sand sheets (wide areas of flat or undulating sandscape) are common in desert areas. However, the formation of different types of sand dune seems to typify desert deposition.

☐ Characteristic landforms of wind action

As well as blowing away layers of unconsolidated material over wide areas, deflation can be localised to produce deflation basins. It is not fully understood why

localised deflation occurs – faulting may produce the initial depression that is enlarged by the eddying nature of the wind; differential erosion could also cause the initial trigger, as could solution weathering in a past **pluvial** period. Many such basins are found in the Sahara west of Cairo. Some basins have become so deflated that their bases reach the water table and so form an oasis, for example the Baharia and Farafra oases and the massive Qattara Depression, which lies 128 metres below sea level.

Selective deflation causes various different types of desert landscape:

- **hammada desert** – all loose material is blown away, leaving large areas of bare rock, often strewn with large, immovable weathered rocks
- **desert pavement** – pebbles are concentrated, for example by a flash flood, and packed together into a mosaic; the tops are then worn flat by wind erosion and perhaps become shiny as a coat of desert varnish develops
- **reg desert** – here, the finest material has been deflated, leaving a gravelly or stony desert
- **erg desert** – this is the classic sandy desert.

Wind erosion only takes place when the wind is loaded with loose materials, especially sand grains. Dust particles are ineffective. The wind throws the particles of sand against rock faces, creating abrasion or corrasion (a sand-blast effect). Large rock fragments, too heavy to be transported by the wind, are worn down on the windward side – these worn fragments are called ventifacts.

In areas of homogeneous rock, the wind will smooth and polish the surface. However, if the rocks are heterogeneous, for example weakened by joints or faults, some dramatic landforms will result from wind erosion, with rock faces etched, grooved, fluted and honeycombed, forming towers, pinnacles and natural arches. Undercutting (abrasion occurring at about 1 metre above the ground) is common, and produces distinctive landforms, including:

- **gours** – mushroom pinnacles where the base has been undercut, and bands of hard and soft rock have been differentially sand-blasted
- **zeugens** – develop where differing rock strata lie horizontally; after being eroded by the wind, the rocks form small plateau-like blocks that are isolated residuals of the original plateau, called 'mesas' (if quite large) and 'buttes' (if relatively small) in parts of Colorado, USA
- **yardangs** – occur where hard and soft rocks lie side by side; the softer rocks are worn down to form troughs, while the harder rocks stand up as wind-worn ridges or yardangs.

Wind-borne material is in constant motion and consequently attrition of this material occurs, the particles becoming rounder and smaller. Wind rounds material more effectively than running water because:

- wind speeds are greater
- distances over which the attrition takes place are often much greater
- the grains are not protected by a film of water.

Section 10.2 Activities

Outline the ways in which wind can erode desert surfaces.

Sand dunes

Only about 25–33 per cent of the world's deserts are covered by dunes, and in North America only 1–2 per cent of the deserts are ergs (sandy deserts) (Table 10.4). Large ergs are found in the Sahara and Arabia. The sand that forms the deserts comes from a variety of sources: alluvial plains, lake shores, sea shores and from weathered sandstone and granite.

The geometry of dunes is varied and depends on the supply of sand, the wind regime, vegetation cover and the shape of the ground surface.

Some dunes are formed in the lee of an obstacle. A **nebkha** is a small dune formed behind a tree or shrub, whereas a **lunette dune** is formed in the lee of a depression (Figure 10.7). Lunettes may reach a height of about 10 metres. They are asymmetric in cross-section, with the steeper side facing the wind. However, most dunes do not require an obstacle for their formation.

Barchan dunes are crescent-shaped and are found in areas where sand is limited but there is a constant wind supply. They have a gentle windward slope and a steep leeward slope up to 33°. Variations include barchan ridges and tranverse ridges, the latter forming where sand is abundant, and where the wind flow is checked by a topographic barrier, or increased vegetation cover. Barchans can be as wide as 30 metres.

Parabolic dunes have the opposite shape from barchans – they are also crescent-shaped but point downwind. They occur in areas of limited vegetation or soil moisture.

Linear dunes or **seifs** are commonly 5–30 metres high and occur as ridges 200–500 metres apart. They may extend for tens, if not hundreds, of kilometres. They are found in areas where there is a seasonal change in wind direction. It is believed that some regularity of turbulence is responsible for their formation (Figure 10.8).

Where the winds come from many directions, star dunes may be formed. Limbs may extend from a central peak. Star dunes can be up to 150 metres high and 2 kilometres wide.

Dune types can merge. Crescent barchans can be transformed into longitudinal seif dunes, depending on the wind regime. The overemphasis on barchans and seif dunes is somewhat misleading. Less than 1 per cent of sand dunes are of these types. Dunes are not necessarily longitudinal or transverse – many are oblique. Grain size is also important – coarse sand is associated with rounded dunes, of subdued size and long wavelength. Fine sand produces stronger relief with smaller wavelengths.

Section 10.2 Activities

Comment on the regional distribution, and relative importance, of linear dunes, as shown in Table 10.4.

Table 10.4 The relative importance (percentage figures) of major dune types in the world's deserts

Desert	Thar	Takla Makan	Namib	Kalahari	Saudi Arabia	Ala Shan	South Sahara	North Sahara	North–east Sahara	West Sahara	Average
Linear dunes	13.96	22.12	32.84	85.85	49.81	1.44	24.08	22.84	17.01	35.49	**30.54**
Crescent dunes	54.29	36.91	11.80	0.59	14.91	27.01	28.37	33.34	14.53	19.17	**24.09**
Star dunes	–	–	9.92	–	5.34	2.87	–	7.92	23.92	–	**5.00**
Dome dunes	–	7.40	–	–	–	0.86	–	–	0.8	–	**0.90**
Sheets and streaks	31.75	33.56	45.44	13.56	23.24	67.82	47.54	35.92	39.25	45.34	**38.34**
Undifferentiated	–	–	–	–	6.71	–	–	–	4.50	–	**1.12**

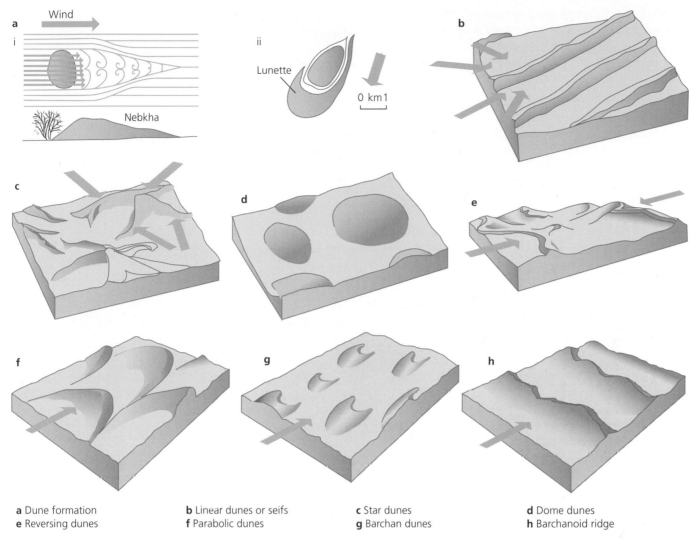

a Dune formation   **b** Linear dunes or seifs   **c** Star dunes   **d** Dome dunes
e Reversing dunes   **f** Parabolic dunes   **g** Barchan dunes   **h** Barchanoid ridge

Figure 10.7 Some sand dune types

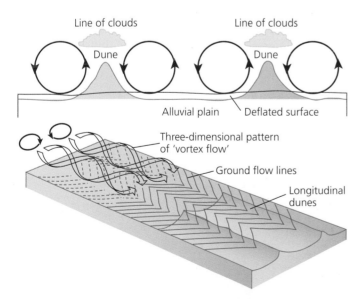

Figure 10.8 Turbulence and the formation of seif dunes

Water action and its characteristic landforms

Despite the low rainfall in arid regions, rivers play an important part in the development of many arid landforms. The **hydrological regime** (annual and seasonal pattern) is very irregular and can be unpredictable. Rainfall may be irregular and **episodic** (sporadic) but some desert areas experience occasional heavy downpours. These downpours may generate sudden **flash floods** and **sheet floods** (where the water is not confined in an identifiable channel).

Rivers in arid lands can be divided into three types:

- **Exogenous** rivers have their origin in humid areas – they are exotic rivers. The Nile flows through the Sahara but rises in the monsoonal Ethiopian Highlands and in equatorial Lake Victoria.
- **Endoreic** rivers flow into inland lakes. The Jordan River flows into the Dead Sea and the Bear River flows into the Great Salt Lake.

- **Ephemeral** rivers flow only after rainstorms. They can generate high amounts of discharge because torrential downpours exceed the infiltration capacity of the soils. Most ephemeral streams consist of many braided channels separated by islands of sediment.

Even areas of low-intensity rain can generate much overland flow. This is because of the lack of vegetation and the limited soil development. The presence of duricrusts also reduces the ability of water to infiltrate the soil.

Surface runoff is typically in the form of sheet flow, where water flows evenly over the land. The runoff may become concentrated into deep, steep-sided valleys known as **wadis** or **arroyos**.

Stream flow in dry areas is seasonal, and in some cases erratic. This increases the potential for flooding due to a combination of:

- high velocities
- variable sediment concentrations
- rapid changes in the location of channels.

According to some geographers, erosion is most effective in dry areas. This is because of the relative lack of vegetation. When it rains, a large proportion of rain will hit bare ground, compact it and lead to high rates of overland runoff. By contrast, in much wetter areas such as rainforests, the vegetation intercepts much of the rainfall and reduces the impact of rainsplash. At the other extreme, areas that are completely dry do not receive enough rain to produce much runoff. Hence it is the areas that have variable rainfall (and a variable vegetation cover) that experience the highest rates of erosion and runoff (Figure 10.9). Moreover, as the type of agriculture changes, the rate of erosion and overland runoff change (Figure 10.10). Under intense conditions this creates gullies.

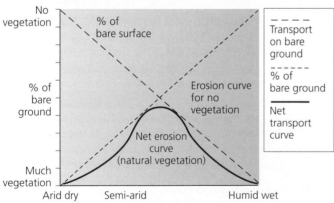

Rates of transport (and therefore denudation) tend to be highest in semi-arid areas and especially in more humid areas where vegetation is removed

Figure 10.9 Rainfall, vegetation cover and soil erosion

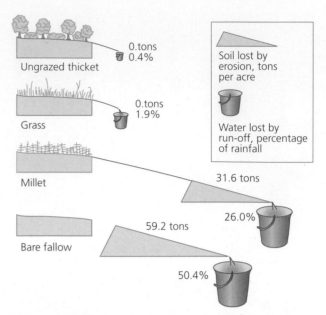

Figure 10.10 The impact of vegetation type on runoff and soil erosion

There is a paradox that in deserts most runoff occurs on low-angle slopes. This is due to particle size. Coarse debris makes up the steeper slopes, while fine material makes up the low-angle slopes. Coarse debris allows more infiltration, so there is less overland flow on steeper slopes.

High concentrations of sediment in runoff from desert uplands illustrate (a) the erodibility of unvegetated areas and (b) the contemporary nature of the work of water in deserts. Desert streams are cloudy and muddy – up to 75 per cent of the flow may be solid matter. This solid matter is important for the formation of alluvial fans (and for silt building up behind dams). An **alluvial fan** is a cone of sediment occurring between a mountain and a lowland plain; that is, the **piedmont zone** (literally the foot of the mountain) (Figure 10.11). They can be up to 20 kilometres wide and up to 300 metres at the apex. They generally form when a heavily sediment-laden river emerges from a canyon. The river, no longer confined to the narrow canyon, spreads out laterally, losing height, energy and velocity, so that deposition occurs; larger particles are deposited first and finer materials are carried further away from the mountain. If a number of alluvial fans merge, the feature is known as a **bajada (bahada)**.

Figure 10.11 An alluvial fan

On a larger scale, **pediments** are gently sloping areas (< 7°) of bare rock where there is a distinct break with the mountain region (Figure 10.12). One idea is that they are the result of lateral planation. Another hypothesis involves sub-surface weathering. This is likely to be accentuated at the junction of the mountain and the plain because of the concentration of water there through percolation. The weathering will produce fine-grained material that can be removed, in the absence of vegetation cover, by sheetfloods, wind and other processes.

Salt lakes (chottes/playas) are found in the lowest part of the desert surface, where ephemeral streams flow into inland depressions, for example the Chott el Djerid of Tunisia. After flowing into the depression, water evaporates, leaving behind a thick crust. Sodium chloride is the most common salt found in such locations, but there could also be gypsum (calcium sulphate), sodium sulphate, magnesium sulphate and potassium and magnesium chlorides.

In some semi-arid areas, water action creates a landscape known as badlands. These are areas where soft and relatively impermeable rocks are moulded by rapid runoff that results from heavy but irregular rainfall. There is insufficient vegetation to hold the regolith and bedrock together, and rainfall and runoff are powerful enough to create dramatic landforms. Badlands generally have the following features:

- wadis of various sizes with debris-covered bottoms
- gullies that erode headwards, leading to their collapse
- slope failure and slumping
- alluvial fans at the base of slopes
- natural arches formed by the erosion of a cave over time.

An excellent example of badland topography is in southern Tunisia around Matmata.

Wadis are river channels that vary in size from a few metres in length to over 100 kilometres. They are generally steep-sided and flat-bottomed. They may be formed by intermittent flash floods or they may have been formed during wetter pluvial periods in the Pleistocene. The relative infrequency of flash floods in some areas where wadis are found could suggest that they were formed at a time when storms were more frequent and more intense. In contrast, **arroyos** are channels that have enlarged by repeated flooding. They are common in semi-arid areas on alluvium and solid rock.

Mesas are plateau-like features with steep sides at their edges. **Buttes** are similar but much smaller. Water has eroded most of the rock, leaving a thin pillar. **Inselbergs** (see Topic 7, Section 7.2) may be the result of deep chemical weathering during wetter pluvial periods. Overlying sediments were subsequently removed by river activity. They are isolated steep-sided hills. A good example is Uluru (Ayers Rock) in Australia.

High runoff and sediment yields cause much dissection and high drainage densities (total length of water channel per km²). In the Badlands of the USA, drainage densities can be as high as 350 km/km², whereas in a typical temperate region it is 2–8 km/km². In contrast, in sandy deserts (high infiltration), drainage densities can be as low as 0–1 km/km².

Section 10.2 Activities

1 Study Figure 10.9, which shows the relationship between rainfall, vegetation cover and erosion.
  a Why is there limited erosion in areas where rainfall is very low?
  b Why is there limited erosion in areas where rainfall is very high?
  c Why are there high rates of erosion in areas with about 600 millimetres of rain?
2 Figure 10.10 shows the effects of crop type on runoff and erosion. Describe what happens when scrubland (ungrazed thicket) is used for either pastoral agriculture (grass) or arable agriculture (millet). Describe and explain the effects of the removal of vegetation on runoff rates and erosion rates.

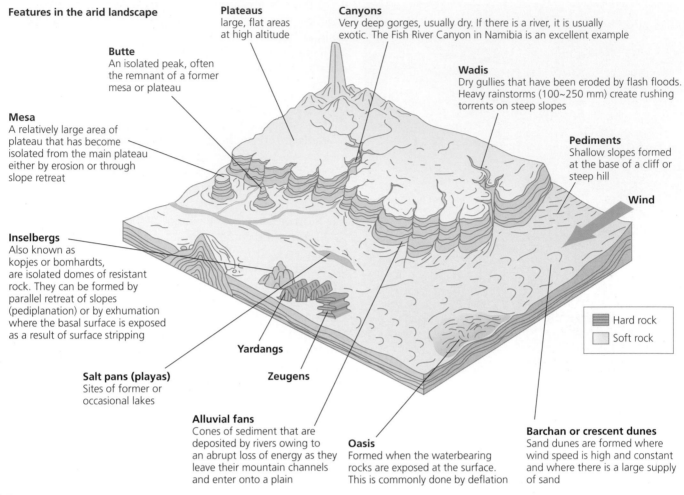

Features in the arid landscape

Plateaus
large, flat areas at high altitude

Canyons
Very deep gorges, usually dry. If there is a river, it is usually exotic. The Fish River Canyon in Namibia is an excellent example

Butte
An isolated peak, often the remnant of a former mesa or plateau

Wadis
Dry gullies that have been eroded by flash floods. Heavy rainstorms (100~250 mm) create rushing torrents on steep slopes

Mesa
A relatively large area of plateau that has become isolated from the main plateau either by erosion or through slope retreat

Pediments
Shallow slopes formed at the base of a cliff or steep hill

Wind

Inselbergs
Also known as kopjes or bomhardts, are isolated domes of resistant rock. They can be formed by parallel retreat of slopes (pediplanation) or by exhumation where the basal surface is exposed as a result of surface stripping

Hard rock
Soft rock

Salt pans (playas)
Sites of former or occasional lakes

Yardangs

Zeugens

Alluvial fans
Cones of sediment that are deposited by rivers owing to an abrupt loss of energy as they leave their mountain channels and enter onto a plain

Oasis
Formed when the waterbearing rocks are exposed at the surface. This is commonly done by deflation

Barchan or crescent dunes
Sand dunes are formed where wind speed is high and constant and where there is a large supply of sand

Figure 10.12 Desert landforms

☐ Relative roles of aeolian and river processes

Evidence of past climate change in deserts

During the Pleistocene Ice Age, high latitudes contained more ice (30 per cent of the world surface) than today (10 per cent of the world surface), while low-latitude areas experienced increased rainfall – episodes known as 'pluvials'. Some deserts, however, received less rainfall – these dry phases are known as 'interpluvials'.

There is widespread evidence for pluvial periods in deserts (Figure 10.13):

- shorelines marking higher lake levels around dry, salty basins
- fossil soils of more humid types, including horizons containing laterite
- spring deposits of lime, called 'tufa', indicating higher groundwater levels
- river systems now blocked by sand dunes

- animal and plant remains in areas that are now too arid to support such species
- evidence of human habitation, including cave paintings.

Wetter conditions existed in the tropics, causing lakes to reach much higher levels and rivers to flow into areas that are now dry. On the margins of the Sahara, Lake Chad may have been 120 metres deeper than it currently is, and may have extended hundreds of kilometres north of its present position.

The evidence for drier conditions includes sand dune systems in areas that are now too wet for sand movement to occur. Dunes cannot develop to any great degree in continental interiors unless the vegetation cover is sparse enough to allow sand movement. If the rainfall is much over 150 millimetres, this is generally not possible. Satellite imagery and aerial photographs have shown that some areas of forest and savanna, with 750–1500 millimetres of rain, contain areas of ancient degraded dunes. Today, about 10 per cent of the land area between 30 °N and 30 °S is covered by active sand deserts, but about 18 000 years ago this area was about 30 per cent sand desert.

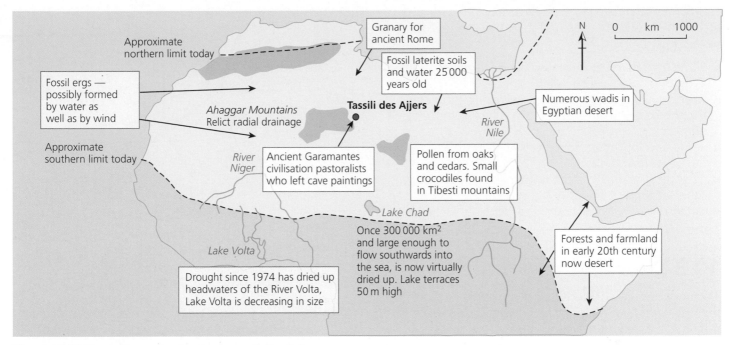

Figure 10.13 The evidence for climate change in the Sahara

The following labels appear on the map:

- Approximate northern limit today
- Fossil ergs — possibly formed by water as well as by wind
- Ahaggar Mountains Relict radial drainage
- Approximate southern limit today
- Granary for ancient Rome
- Fossil laterite soils and water 25 000 years old
- **Tassili des Ajjers**
- Numerous wadis in Egyptian desert
- *River Nile*
- Ancient Garamantes civilisation pastoralists who left cave paintings
- *River Niger*
- Pollen from oaks and cedars. Small crocodiles found in Tibesti mountains
- *Lake Chad*
- Once 300 000 km² and large enough to flow southwards into the sea, is now virtually dried up. Lake terraces 50 m high
- Forests and farmland in early 20th century now desert
- *Lake Volta*
- Drought since 1974 has dried up headwaters of the River Volta, Lake Volta is decreasing in size
- N 0 km 1000

Case Study: Climate change in Australia

Glacial periods triggered decreased rainfall and increased windiness. At least eight episodes of dune building have occurred over the last 370 000 years. The largest sand dune system in the world is the Simpson desert, which was formed only 18 000 years ago. The Simpson desert covers 159 000 km² and consists of linear dunes 10–35 metres high and up to 200 kilometres long (Figure 10.14). They run parallel to each other, with an average spacing over 510 metres. The dunes are fixed (vegetated) except for their crests, which are mobile. The Simpson desert dunes form part of a continental anticlockwise swirl that relates to the dominant winds of the subtropical anticyclone system.

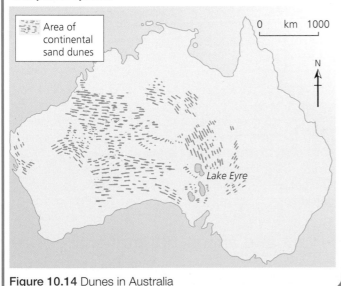

Area of continental sand dunes

0 km 1000

N

Lake Eyre

Figure 10.14 Dunes in Australia

Section 10.2 Activities

Examine the evidence to suggest that some deserts in the past were **a** wetter and **b** drier.

☐ Equifinality: different processes, same end product

A question frequently asked is whether desert landforms are the result of wind action or water action. This is a simplification because there are other processes than wind and water acting in desert regions. For example, stone pavements are surfaces of coarse debris lying above finer material. They could be caused by:

- deflation of fine material, leaving coarse material behind
- removal of fine material by rainsplash or sheetwash
- vertical sorting by frost action or hydration.

It may even be a combination of processes.
  Similarly, depressions can be caused by a variety of processes:

- deflation removing finer material, as in the Qattara Depression, Sahara
- tectonic, for example block faulting in the basin and range region of the USA
- solution of limestone during a pluvial period as, for example, in Morocco
- animal activity as, for example, in Zimbabwe where herds are concentrated near water holes and accentuate the initial depression.

Gully development is normally associated with river activity. However, there are different reasons for their development:

- Some are related to increased discharge due to climate change.
- Others are caused by the removal of vegetation by people exposing the surface to accelerated rates of erosion.
- Some develop where concrete structures have been built to improve runoff.
- Tectonic disturbances can initiate gully development, especially uplift.
- Catastrophic flooding may be responsible for some gully development.

10.3 Soils and vegetation

An ecosystem is the interrelationship between plants and animals and their living and non-living environments. A biome is a global ecosystem, such as the tropical rainforest, savanna or hot desert ecosystem.

Deserts have low rates of **biomass productivity** – on average, net primary productivity of $90\,g/m^2/year$. This is due to the limited amount of organic matter caused by extremes of heat and lack of moisture. Productivity can generally be positively correlated with water availability. Despite the diversity of life forms found in deserts, desert flora and fauna are relatively species poor. At the continental scale, species diversity of lizards and rodents has been correlated with increasing precipitation. Desert vegetation is simple in that its structure is poorly developed and its cover becomes increasingly open and discontinuous with increasing aridity. Given the extreme conditions, it is not surprising that the **biodiversity** is limited.

Energy flow in deserts is controlled by water, which can be very irregular. The impact of herbivores in deserts is similar to that in other ecosystems, with about 2–10 per cent of the primary production being directly consumed. Some studies have indicated that 90 per cent or more of seed production may be eaten by seed-eaters such as ants and rodents. Energy flow in hot deserts is less than in semi-arid areas due to the lack of rainfall, and less biomass.

Owing to the low and irregular rainfall, inputs to the nutrient cycle (dissolved in rain and as a result of chemical weathering) are low (Figure 10.15). Most of the nutrients are stored in the soil, and there are very limited stores in the biomass and litter. This is due to the limited amount of biomass and litter in the desert environment. In some deserts, nutrient deficiency (especially of nitrogen and/or phosphorus) may become critical. The rapid growth of annuals following a rain event rapidly depletes the store of available nutrients, while their return in decomposition is relatively slow. Desert ecosystems are characterised by smaller stores of nutrients in the soil (due to low rates of chemical weathering), and low amounts of nutrients in

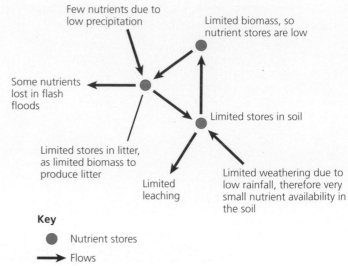

Figure 10.15 Nutrient cycle in a desert

the biomass, due to the dry conditions. In semi-arid areas the amount of nutrients available increases with rainfall and chemical weathering.

Decomposition, like growth, is slow. Microbial decomposers are limited. Two important processes are involved in nutrient cycling:

- the fragmentation, erosion and transport of dead organic matter (DOM) by wind and runoff
- consumption of DOM by detritivores such as termites, ants and mites, which are relatively abundant in deserts.

In the absence of leaching, nutrients may accumulate in the upper layers of the soil. A large proportion of the nutrients is held either in young tissue or in the fertile islands surrounding large plants where, as a result of slightly lower temperatures and higher humidity, decomposition is lower and DOM accumulates.

Desert ecosystems are sometimes considered to be fragile, due to the extreme climatic conditions and the relative lack of biodiversity. Nevertheless, despite the extreme short-term variability of the desert environment, the desert ecosystem is considered, in the long term, to be both stable and resilient. This is due to the adaptations of desert organisms to survive water stress – in some cases for years. The hogweed (*Boerhavia repens*) plant in the Sahara takes just 8–10 days from seed germination to seed production. It therefore produces seeds before the water runs out, and flowers at a time when insect pollinators are abundant. Desert ecosystems may appear fragile, but in fact they are very resilient.

☐ Plant and animal adaptations

Desert vegetation is generally ephemeral (it appears or flowers after rain). Some desert vegetation has a very short life cycle, some less than eight weeks. Vegetation is generally shallow-rooted, small in size and with small leaves. In contrast, vegetation in semi-arid areas is succulent (able to store water), and more vegetation is located near to water sources.

Desert plants and animals have acquired similar morphological, physiological and behavioural strategies that, although not unique to desert organisms, are often more highly developed and effectively utilised than their moist counterparts.

The two main strategies are avoidance and tolerance of heat and water stress (Table 10.5). The *evaders* comprise the majority of the flora of most deserts. They can survive periods of stress in an inactive state or by living permanently or temporarily in cooler and/or moister environments, such as below shrubs or stone, in rock fissures or below ground. Of desert animals, about 75 per cent are subterranean, nocturnal or active when the surface is wet. In such ways, plants and animals can control their temperature and water loss.

Section 10.3 Activities

1 Describe the typical nutrient cycle of a desert as shown in Figure 10.15.
2 Explain why deserts have low values for NPP (net primary productivity).

Table 10.5 Adaptations of plants and animals to hot desert environments

Strategy	Plants	Animals
Stress-evading strategies		
	Inactivity of whole plants Cryptobiosis* of whole plant Dormancy of seeds	Dormancy in time (diurnal and seasonal) and space (take refuge in burrows) Cryptobiosis of mature animals (aestivation of snails, hibernation) Cryptobiosis of eggs, shelled embryos, larvae: permanent habitation or temporary use of stress-protected microhabitats
Structural and physiological stress-controlling strategies		
Strategies reducing water expenditure	Small surface : volume ratio Regulation of water loss by stomatal movements Xeromorphic features Postural adjustments	Small surface : volume ratio Regulation and restriction of water loss by concentrated urine, dry faeces, reduction of urine flow rate Structures reducing water Postural adjustment
Strategies to prevent death by overheating	Transpiration cooling High heat tolerance Mechanisms decreasing and/or dissipating heat load	Evaporative cooling High heat tolerance Mechanisms decreasing and/or dissipating heat load
Strategies optimising water uptake	Direct uptake of dew, condensed fog and water vapour Fast formation of water roots after first rain Halophytes: uptake of saline water, high salt tolerance, salt-excreting glands	Direct and indirect uptake of dew, condensed fog and water vapour (e.g. arthropods, water enrichment of stored food) Fast drinking of large quantities of water (large mammals), uptake of water from wet soil (e.g. snails) Uptake of highly saline water, high salt tolerance, salt-excreting glands
Strategies to control reproduction in relation to environmental conditions	'Water clocks' of seed dispersal and germination Suppression of flowering and sprouting in extreme years	Sexual maturity, mating and birth synchronised with favourable conditions Sterility in extreme drought years

* Cryptobiosis: an ametabolic state of life in response to adverse environmental stress; when the environment becomes hospitable again, the organism returns to its metabolic state.

Temperature adaptations

Desert plants and animals are able to function at higher temperatures than their mesis (moist environment) counterparts. Some cacti such as the prickly pear can survive up to 65 °C, while crustose lichens can survive up to 70 °C. The upper lethal levels for animals are lower, although arthropods, particularly beetles and scorpions, can tolerate 50 °C. Plants and animals are able to modify the heat of the desert environment in a number of ways:

- Changing the orientation of the whole body enables the organism to minimise the areas and/or time they are exposed to maximum heat – many gazelle, for example, are long and thin.
- Light colours maximise reflection of solar radiation.
- Surface growth (spines and hairs) can absorb or reflect heat, which **a** keeps the undersurface cooler and **b** creates an effective boundary layer of air, which insulates the underlying surface.
- Body size is especially important in controlling the amount of heat loss – evaporation and metabolism

are proportional to the surface area of the plant or animal. The smaller the organism, the larger the surface area to volume ratio and the greater the heat loss.

- Large desert animals such as the camel and the oryx can control heating by means of evaporative cooling. Cooling by transpiration is also thought to be most effective in cacti and small-leaved desert shrubs because of their surface-area-to-volume ratio.

Water loss

Physical droughts refer to water shortages over an extended period of time. **Physiological drought** occurs when drought conditions are experienced by plants despite there being sufficient soil moisture. In hot arid areas, this is associated with high rates of evapotranspiration. To reduce water loss, desert plants and animals have many adaptations. Again, a small surface-area-to-volume ratio is an advantage (Figure 10.16b). Water regulation by plants can be controlled by diurnal closure of stomata, and xerophytic plants have a mix of thick, waxy cuticles, sunken stomata and leaf hairs (Figure 10.16e). The most drought-resistant plants are the **succulents**, including cacti, which possess well-developed storage tissues (Figure 10.16a); small surface-to-volume ratios and rapid stomatal closure especially during the daytime; deep tap roots (Figure 10.16b and 10.16c); and very small leaves (Figure 10.16d).

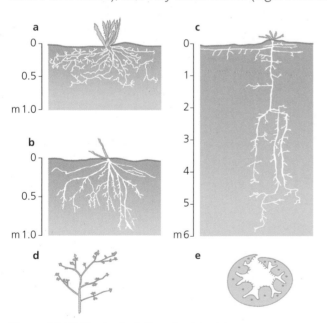

Figure 10.16 Plant adaptations to drought

Some plants and many arthropods are drought-resistant. The creosote bush can survive up to a year without rain. Rapid uptake of water is also a characteristic of many desert organisms, including lichen, algae and camels. Animals can rapidly imbibe, and salt-tolerant plants have a high cell osmotic pressure that allows the efficient uptake of alkaline water. The roots of many desert plants can exert a greater suction pressure so they can extract water from fine water-retentive soils.

Reproduction

Desert survival is also dependent on an organism's ability to reproduce itself. Desert fauna and flora have developed several different strategies. High seed production and efficient dispersal are more essential than in humid environments. Some seeds have built-in 'water clocks' and will not germinate until a certain amount of water becomes available. Some arid-zone shrubs, such as the ironwood and the smokewood, have seeds with coats so tough that germination can only take place after severe mechanical abrasion during torrential flash floods. In both plants and animals, reproduction is suppressed during periods of drought.

☐ Desert soils

Desert soils, called **aridisols**, have a low organic content and are only affected by limited amounts of leaching. Soluble salts tend to accumulate in the soil either near the water table or around the depth of moisture percolation. As precipitation declines, this horizon occurs nearer to the surface. Desert soils also have a limited clay content. In semi-arid areas, there is a deeper soil, more chemical weathering and more biomass in the soil, due to the higher rainfall.

The accumulation of salt in desert soils is important. Salt concentrations may be toxic to plants. This is more likely in areas where there is a high water table or in the vicinity of salt lakes. Soils with a saline horizon of NaCl (sodium chloride) are called **solonchaks** and those with a horizon of Na_2CO_3 (sodium carbonate) are termed **solonetz**. Solonchaks are white alkali soils, whereas solonetz are black alkali soils. A high concentration of salt can cause the breakdown of soil structure, increase water stress and affect the health of plants.

Sometimes the concentration of salts becomes so great that crusts are formed on the surface or sub-surface. There are different types of hard crust (duricrusts). Calcrete or caliche is formed of calcium carbonate and is the most common crust in warm desert environments. It can be up to 40 metres, comprising boulders, gravels, silt and calcareous materials. It predominates in areas of between 200 and 500 millimetres.

Silcrete is a crust cemented by silica. It may produce an impermeable hard pan in a soil. Silcretes are found in areas that have more than 50 millimetres but less than 200 millimetres of rain, such as southern Africa and Australia.

Section 10.3 Activities

1 Describe the ways in which plants have adapted to drought, as shown in Figure 10.16.
2 Describe and explain the main characteristics of desert soils.

Case Study: Salinisation in Pakistan

Irrigation has been practised in Pakistan since at least the eighth century CE. Much of the irrigation takes place along the Indus and Punjab rivers. The irrigation system here is among one of the most complex in the world, and provides Pakistan with most of its food and commercial crops, such as wheat, cotton, rice, oil seed, sugar cane and tobacco. Hence the health of the irrigated area is essential to the health of the national economy.

Many of the drainage canals are in a poor state. Many are unlined and seepage is a major problem. Consequently, there has been a steady rise in the water table, which has caused widespread waterlogging and salinisation. Up to 40 000 hectares of irrigated land are lost annually.

There have been attempts to rectify the problem. Two main methods are used: pumping water from aquifers (to reduce the water table) and vertical and horizontal drainage of saline water. These have met with some success. In parts of the lower Indus plain, water tables have been reduced by as much as 7 metres, and up to 45 per cent of saline soils have been reclaimed. However, the use of reclaimed land for agriculture only results in salinisation again.

☐ Desertification

Desertification is defined as land degradation in humid and semi-arid areas; that is, not including non-desert (arid) areas. It involves the loss of biological and economic productivity and it occurs where climatic variability (especially rainfall) coincides with unsustainable human activities (Figure 10.17). For example, if the surface cover is removed and the surface colour becomes lighter, its reflectivity (albedo) changes. It reflects more heat, absorbs less, and so there will be less convectional heating, less rain and possibly more drought. Desertification occurs in discontinuous and isolated patches – it is not the general extension of deserts as a consequence of natural events like prolonged droughts, as in China in the 2000s.

Desertification leads to a reduction in vegetation cover and accelerated soil erosion by wind and water, lowering the carrying capacity of the area affected. Desertification is one of the major environmental issues in the world today. At present, 25 per cent of the global land territory and nearly 16 per cent of the world's population are threatened by desertification.

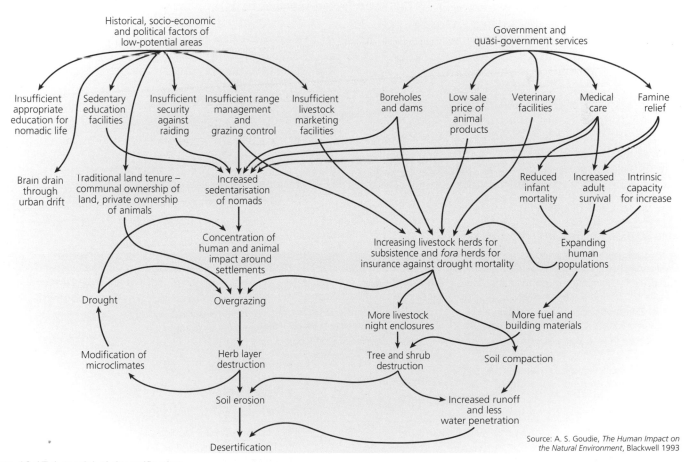

Source: A. S. Goudie, *The Human Impact on the Natural Environment*, Blackwell 1993

Figure 10.17 A model of desertification

Causes of desertification

Desertification can be a natural process intensified by human activities. All the areas affected by desertification are marginal and characterised by highly variable rainfall. An exception to this are the parts of the rainforest desertified following inappropriate farming techniques.

Natural causes of desertification include temporary drought periods of high magnitude and long-term climate change towards aridity. Many people believe that it is a combination of increasing animal and human population numbers, which causes the effects of drought to become more severe. Desertification occurs when already fragile land in arid and semi-arid areas is over-exploited. This overuse can be caused by overgrazing, when pastoralists allow too many animals to graze on a fixed area of land; overcultivation, where the growing of crops exhausts soil nutrients; and deforestation, when too few trees are left standing after use as firewood, to act as windbreaks or to prevent soil erosion.

■ **Overgrazing** is the major cause of desertification worldwide. Vegetation is lost both in the grazing itself and in being trampled by large numbers of livestock. Overgrazed lands then become more vulnerable to erosion as compaction of the soils reduces infiltration, leading to greater runoff, while trampling increases wind erosion. Fencing, which confines animals to specific locations, and the provision of water points and wells, have led to severe localised overgrazing. Boreholes and wells also lower the water table, causing soil salinisation.
■ **Overcultivation** leads to diminishing returns, where the yield decreases season by season, requiring an expansion of the areas to be cultivated simply to maintain the same return on the agricultural investment. Reducing fallow periods and introducing irrigation are also used to maintain output, but all these contribute to further **soil degradation** and erosion by lowering soil fertility and promoting salinisation.
■ **Deforestation** is most obvious where land has been cleared to extend the area under cultivation and in the surrounds of urban areas where trees are stripped for firewood. The loss of vegetation cover increases rainsplash erosion, and the absence of root systems allows easy removal of the soil by wind and water.

Other factors are involved, including the following:

■ The mobility of some people has been limited by governments, especially where their migratory routes crossed international boundaries. Attempts to provide permanent settlements have led to the concentration of population and animals, with undesirable consequences.
■ Weak or non-existent laws have failed to provide environmental protection for marginal land by preventing or controlling its use.
■ Irrational use of water resources has caused water shortages or salinisation of soil.
■ International trade has promoted short-term exploitation of land by encouraging cash crops for export. This has disrupted local markets and created a shortage of staple foods.
■ Civil strife and war diverts resources away from environmental issues.
■ Ignorance of the consequences of some human actions, and the use of inappropriate techniques and equipment, have contributed to the problem.

Consequences of desertification

There are some serious consequences of desertification (Table 10.6).

Table 10.6 Consequences of desertification

Environmental	Economic	Social and cultural
• Loss of soil nutrients through wind and water erosion	• Reduced income from traditional economy (pastoralism and cultivation of food crops)	• Loss of traditional knowledge and skills
• Changes in composition of vegetation and loss of biodiversity as vegetation is removed	• Decreased availability of fuelwood, necessitating purchase of oil/kerosene	• Forced migration due to food scarcity
• Reduction in land available for cropping and pasture	• Increased dependence on food aid	• Social tensions in reception areas for migrants
• Increased sedimentation of streams because of soil erosion and sediment accumulations in reservoirs	• Increased rural poverty	
• Expansion of area under sand dunes		

Combating desertification

There are many ways of combating desertification, which depend on the perceived causes (Table 10.7).

Table 10.7 The strategies for preventing desertification, and their disadvantages

Cause of desertification	Strategies for prevention	Problems and drawbacks
Overgrazing	• Improved stock quality: through vaccination programmes and the introduction of better breeds, yields of meat, wool and milk can be increased without increasing the herd size. • Better management: reducing herd sizes and grazing over wider areas would both reduce soil damage.	• Vaccination programmes improve survival rates, leading to bigger herds. • Population pressure often prevents these measures.
Overcultivation	• Use of fertilisers: these can double yields of grain crops, reducing the need to open up new land for farming. • New or improved crops: many new crops or new varieties of traditional crops with high-yielding and drought-resistant qualities could be introduced. • Improved farming methods: use of crop rotation, irrigation and grain storage can all increase and reduce pressure on land.	• Cost to farmers. • Artificial fertilisers may damage the soil. • Some crops need expensive fertiliser. • Risk of crop failure. • Some methods require expensive technology and special skills.
Deforestation	• Agroforestry: combines agriculture with forestry, allowing the farmer to continue cropping while using trees for fodder, fuel and building timber. Trees protect, shade and fertilise the soil. • Social forestry: village-based tree-planting schemes involve all members of a community. • Alternative fuels: oil, gas and kerosene can be substituted for wood as sources of fuel.	• Long growth time before benefits of trees are realised. • Expensive irrigation and maintenance may be needed. • Expensive. Special equipment may be needed.

Section 10.3 Activities

1 Suggest a definition for the term *desertification*.
2 Outline the main natural causes of desertification.
3 Briefly explain two examples of desertification caused by people.

4 Comment on the effects of desertification.
5 To what extent is it possible to manage desertification?

Case Study: Desertification in China

Parts of China are among the most seriously desertified areas in the world. More than 27 per cent, or 2.5 million km², of the country comprises desert (whereas just 7 per cent of Chinese land feeds about a quarter of the world's population). China's phenomenal economic growth over the last ten years has been at a serious environmental cost. According to the China State Forestry Administration, the desert areas are still expanding by between 2460 and 10 400 km² per year. Up to 400 million people are at risk of desertification in China – the affected area could cover as much as 3.317 million km² or 34.6 per cent of the total land area. Much of it is happening on the edge of the settled area – which suggests that human activities are largely to blame.

Causes of desertification

Desertification is widely distributed in the arid, semi-arid and dry sub-humid areas of north-west and northern China and western parts of the north-east of the country (Figure 10.18). Much of the country is affected by a semi-permanent high-pressure belt, which causes aridity. In addition, continental areas experience intensive thunderstorms, which can cause accelerated soil erosion. Drought plagues large parts of northern China. In addition to dry weather, human activities such as livestock overgrazing, cultivation of steep slopes, rampant logging and excessive cutting of branches for firewood are at the root of the crisis.

Figure 10.18 Desertification in China

- In the Inner Mongolia Autonomous Region, over 133 000 hectares of rangeland has been seriously degraded by overgrazing. The density of animals now exceeds the carrying capacity of the land.
- On the loess plateau, cultivation of steep slopes is the main cause of desertification. On slopes of less than 5°, the loss of topsoil per annum is about 15 tonnes/hectare. In contrast, on slopes of over 25° it rises to 120–150 tonnes/hectare. However, there is very little loss of soil on terraced slopes.
- Illegal collection of fuelwood and herbal medicines has removed more than one-third of vegetation in the Qaibam basin since the 1980s.

Rates and types of desertification

In China, the main types of desertification include sandy desertification caused by wind erosion; land degradation by water erosion; soil salinisation; and other land degradation caused by engineering construction of residential areas and communications, and industrial activities such as coal mining and oil extraction (Tables 10.8 and 10.9).

Table 10.8 Land in China desertified by different processes

Types of desertification	Area (km²)	% of total
Wind erosion	379 600	44.1
Water erosion	394 000	45.7
Salinisation	69 000	8.3
Engineering construction	19 000	1.9
Total	**861 600**	**100.0**

Table 10.9 Soil degradation in China (million hectares)

		Negligible	Light	Moderate	Strong	Extreme
Water erosion	Loss of topsoil	15.8	105.9	44.9	3.8	0.2
	Terrain deformation	0.5	7.9	5.9	24.0	–
	Off–site effects	0.2	0.2	0.2	–	–
Wind erosion	Loss of topsoil	1.7	65.9	2.5	+	+
	Terrain deformation	+	7.2	5.5	57.9	–
	Off–site effects	+	2.0	6.5	0.2	–
Chemical deterioration	Fertility decline	32.4	31.7	4.8	–	–
	Salinisation	0.5	6.8	2.6	–	–
	Desertification	–	+	–	–	–
Physical deterioration	Aridification	–	23.7	–	–	–
	Compaction and crusting	–	0.5	–	–	–
	Waterlogging	3.8	–	–	–	–
Total degradation	All types	55.0	251.9	72.9	86.0	0.25

Note: (–) no significant occurrence; (+) less than 0.1 but more than 0.01 million hectares; for calculation of the totals (+) is equivalent to 0.05 million hectares.

Sandy desertification through wind erosion

In northern China, the main land degradation is sandy desertification caused by wind erosion, an area that covers about 379 600 km² and is mainly distributed in the arid and semi-arid zones where the annual rainfall is below 500 millimetres.

Sandy desertification in northern China has been caused mainly by irrational human economic activities, and the growth rate of desertified land increased from 1560 km²/year during the 1950s to 2100 km²/year between the mid-1970s and the late 1980s, and since the late 1980s has increased to 2460 km²/year.

Desertification through water erosion

Soil loss through water erosion is the most serious land degradation in China. By a rough estimate, annual soil loss caused by water erosion has reached about 5 billion tonnes, of which about two-fifths pours into the seas.

Salinisation

About 69 000 km² of China's farmland has been salinised, mainly in the arid and semi-arid regions of north-west China and the sub-humid regions of the North China Plain.

Desertification caused by engineering construction

A new type of desertification has spread very quickly, with some large-scale projects such as the development of oilfields and mining; construction of residential areas; and communications. These developments have led to increased wind and water erosion.

Some consequences of desertification

Desertification brings many adverse impacts. It causes a decrease in farmland availability, declining crop productivity, falling incomes, disruptions to communications and may eventually cause out-migration. Desertification also causes an increase in sandstorms, silting of rivers and reservoirs and increased soil erosion.

- Desertification causes annual direct economic losses valued at US$6.5 billion.
- In the north-west, where the problems are the biggest, desertification has escalated rapidly (see above).
- Each year, another 180 000 hectares of farmland in China is salinised, causing productivity to fall by 25–75 per cent, and about 200 000 hectares turns into desert; about 2 million hectares of pasturage are degraded each year.
- Erosion claims about 5 billion tonnes of China's topsoil each year, washing away nutrients equivalent to 54 million tonnes of chemical fertiliser – twice the amount China produces in a year.
- In the 1950s, dust storms affected Beijing once every seven or eight years, and only every two or three years in the 1970s. By the early 1990s, dust storms were an annual problem.

- Desertification has led to a heavy loss of land in pastoral, dry farming areas in northern China and in hilly areas in southern China.
- The government fears that encroaching sands will reach Beijing by the year 2035 as any serious drought turns farmlands into dunes in northern parts of the country, just 100 kilometres away. These dunes are advancing at a rate of 3.5 kilometres a year.

Possible solutions

The China National Research and Development Centre on Combating Desertification (RDCCD) was established to assist the government in implementing the UN Convention to Combat Desertification, which was established to enable China and other countries to combat desertification by developing profitable techniques and environmentally improved practices, and to meet the needs of poverty alleviation.

There are a number of effective measures that can be taken to combat desertification (Table 10.10).

Table 10.10 Methods of combating desertification

Method	Description
Fixing sand by planting	Effective; economic – source of additional fuel and forage
Fixing sand by engineering	Cover sand with straw, clay, pebbles, branches – used successfully along railways, motorways and near cities
Fixing sand using chemical methods	Create a protective layer over the sand – used in areas of high economic value such as airports and railways
Water-saving techniques	Spray irrigation, drip irrigation, prevention of seepage in channels, water transport in pipes
Integrated water management	Terraces, check dams, silt arresters

In 1978, China implemented a forest shelterbelt development programme in its northern, north-western and north-eastern regions, where 16 million hectares of plantations were established, which increased the forest cover from 5 per cent to 9 per cent. It also brought 10 per cent of the desertified land under control and protected 11 million hectares of farmland. In 1991, this was extended as a nationwide campaign.

To protect Beijing, the government issued a ban on the foraging and distribution of three wild plants – facai, liquorice and ephydra – grown in the country's dry western regions. It also plans to build a second green belt of forest around Beijing to achieve a forest coverage of 49.5 per cent by 2020.

Major problems in combating desertification

China is a developing country, and as economic growth exerts great pressure on its funds, the state input to combat desertification is limited. In addition:

- Public awareness needs to be raised, and education regarding desertification improved.
- Legislation is incomplete and the legal enforcement system is imperfect.
- The speed at which desertification is being tackled lags behind the rate of development.
- There is a shortage of funds to combat desertification.

Conclusions

China suffers as a result of desertification. This is the result of a combination of natural reasons and human ones. Economic growth and population pressure are placing a great strain on China's environment. Nevertheless, there have been a number of strategies to tackle desertification, and some of these have had impressive results. However, despite these successes, the rate of desertification appears to be exceeding the rate of environmental restoration. Unless China can tackle its desertification problem, there will be an increase in problems related to its overall development and standards of living.

Section 10.3 Activities

1 Outline **a** the causes and **b** the impacts of desertification in China.
2 To what extent is it possible to manage desertification in China?

☐ Soil degradation

Soil degradation is the decline in quantity and quality of soil. It includes:

- erosion by wind and water
- biological degradation – for example, the loss of humus and plant/animal life
- physical degradation – loss of structure, changes in permeability
- chemical degradation – acidification, declining fertility, changes in pH
- salinisation
- chemical toxicity.

Causes of degradation

The universal soil loss equation (USLE) A = RKLSCP is an attempt to predict the amount of erosion that will take place in an area on the basis of certain factors that increase susceptibility to erosion (Table 10.11).

The complexity of soil degradation means that it is hard to make a single statement about its underlying causes. Soil degradation encompasses several issues at various spatial and time scales (Figure 10.19):

- **Water erosion** – water erosion accounts for about 60 per cent of soil degradation; there are many types of erosion, including surface, gully, rill and tunnel erosion.

Table 10.11 Factors relating to the universal soil loss equation

Factor	Description
Ecological conditions	
Erosivity of soil R	Rainfall totals, intensity and seasonal distribution. Maximum erosivity occurs when the rain falls as high-intensity storms. If such rain is received when the land has just been ploughed or full crop cover is not yet established, erosion will be greater than when falling on a full canopy. Minimal erosion occurs when rains are gentle, and fall onto frozen soil or land with natural vegetation or a full crop cover.
Erodibility K	The susceptibility of a soil to erosion. Depends on infiltration capacity and the structural stability of soil. Soils with a high infiltration capacity and high structural stability that allow the soil to resist the impact of rainsplash have the lowest erodibility values.
Length–slope factor LS	Slope length and steepness influence the movement and speed of water down the slope, and thus its ability to transport particles. The greater the slope, the greater the erosivity; the longer the slope, the more water is received on the surface.
Land-use type	
Crop management C	Most control can be exerted over the cover and management of the soil, and this factor relates to the type of crop and cultivation practices. Established grass and forest provide the best protection against erosion and, of agricultural crops, those with the greatest foliage and thus greatest ground cover are optimal. Fallow land or crops that expose the soil for long periods after planting or harvesting offer little protection.
Soil conservation P	Soil conservation measures can reduce erosion or slow the runoff of water, such as contour ploughing and use of bunds, strips and terraces.

Source: adapted from Hugget et al, Physical Geography – a Human Perspective, Arnold 2004

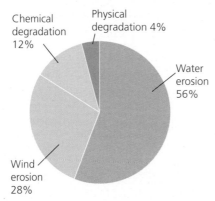

Figure 10.19 Types of soil degradation

- **Wind erosion** involves removal of material, especially of fine-grained loess and silt-sized materials or smaller.
- **Acidification** is the change in the chemical composition of the soil, which may trigger the circulation of toxic metals.
- **Eutrophication** (nutrient enrichment) may degrade the quality of groundwater; over-abstraction of groundwater may lead to dry soils.
- **Salt-affected soils** are typically found in marine-derived sediments, coastal locations and hot arid areas where capillary action brings salts to the upper part of the soil; soil salinity has been a major problem in Australia following the removal of vegetation for dryland farming.
- **Atmospheric deposition** of heavy metals and persistent organic pollutants may change soils so that they become less suitable to sustain the original land cover and land use.
- **Climate change** will probably intensify the problem; it is likely to affect hydrology and hence land use.

Climate change, higher average temperature and changing precipitation patterns may have three direct impacts on soil conditions. The higher temperatures cause higher decomposition rates of organic matter. Organic matter in soil is important as a source of nutrients and it improves moisture storage. More floods will cause more water erosion, while more droughts will cause more wind erosion.

Besides these direct effects, climate change may:

- create a need for more agricultural land to compensate for the loss of degraded land
- lead to higher yields for the major European grain crops due to the carbon dioxide fertilisation – the increase in carbon dioxide in the atmosphere leads to increased plant growth by allowing increased levels of photosynthesis.

These two indirect effects appear to balance out.

Human activities

Human activities have often led to degradation of the world's land resources (Figure 10.20 and Table 10.12). A global assessment of human-induced soil degradation has shown that damage has occurred on 15 per cent of the world's total land area (13 per cent light and moderate; 2 per cent severe and very severe). These impacts frequently lead to a reduction in yields. Land conservation and rehabilitation are essential parts of sustainable agricultural development. While severely degraded soil is found in most regions of the world, the negative economic impact of degraded soil may be most severe in countries that are most dependent on agriculture for their income.

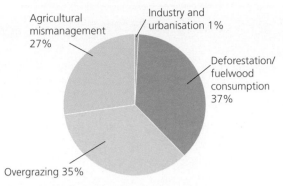

Figure 10.20 Causes of soil degradation

Table 10.12 Human activities and their impact on soil erosion

Action	Effect
Removal of woodland or ploughing established pasture	The vegetation cover is removed, roots binding the soil die and the soil is exposed to wind and water. Particularly susceptible to erosion if on slopes.
Cultivation	Exposure of bare soil surface before planting and after harvesting. Cultivation on slopes can generate large amounts of runoff and create rills and gullies.
Grazing	Overgrazing can severely reduce the vegetation cover and leave the surface vulnerable to erosion. Grouping of animals can lead to overtrampling and creation of bare patches. Dry regions are particularly susceptible to wind erosion.
Roads or tracks	They collect water due to reduced infiltration that can cause rills and gullies to form.
Mining	Exposure of the bare soil.

Managing soil degradation

Abatement strategies, such as afforestation, for combating accelerated soil erosion are lacking in many areas. To reduce the risk of soil erosion, farmers are encouraged towards more extensive management practices such as organic farming, afforestation, pasture extension and benign crop production. Nevertheless, there is a need for policy-makers and the public to intensify efforts to combat the pressures on and risks to the soil resource.

Methods to reduce or prevent erosion can be mechanical, for example physical barriers such as embankments and windbreaks, or they may focus on vegetation cover and soil husbandry. Overland flow of water can be reduced by increasing infiltration.

Mechanical methods include building bunds, terracing, contour ploughing and planting shelterbelts (trees or hedgerows). The key is to prevent or slow the movement of rainwater downslope. Contour ploughing takes advantage of the ridges formed at right-angles to the slope, which act to prevent or slow the downward accretion of soil and water. On steep slopes and those with heavy rainfall, such as areas in South East Asia that experience the monsoon,

contour ploughing is insufficient and so terracing is practised. The slope is broken up into a series of level steps, with bunds (raised levees) at the edge. The use of terracing allows areas to be cultivated that would not otherwise be suitable. In areas where wind erosion is a problem, shelterbelts of trees or hedgerows are used. The trees act as a barrier to the wind and disturb its flow. Wind speed is reduced, which therefore reduces its ability to disturb the topsoil and erode particles.

Preventing erosion by different cropping techniques largely focuses on:

■ maintaining a crop cover for as long as possible
■ keeping in place the stubble and root structure of the crop after harvesting
■ planting a grass crop.

A grass crop maintains the action of the roots in binding the soil, minimising the action of wind and rain on a bare soil surface. Increased organic content allows the soil to hold more water, thus preventing aerial erosion and stabilising the soil structure. In addition, care is taken over the use of heavy machinery on wet soils and ploughing on soils sensitive to erosion, to prevent damage to the soil structure.

There are three main approaches in the management of salt- and chemical-affected soils:

■ flush the soil and leach the salt away
■ apply chemicals, for example gypsum (calcium sulphate), to replace the sodium ions on the clay and colloids with calcium ones
■ reduce evaporation losses in order to limit the upward movement of water in the soil.

Soil degradation is a complex issue. It is caused by the interaction of physical forces and human activities. Its impact is increasing and it is having a negative effect on food production. Some areas are more badly affected than others, but in a globalised world the impacts are felt worldwide. The methods of dealing with soil degradation depend on the cause of the problem, but also on the resources available to the host country. Degradation is a problem that is not going to go away and is likely to increase over the next decades as population continues to grow, and people use increasingly marginal areas.

Section 10.3 Activities

1 Explain the meaning of the term *soil degradation*.
2 Outline the natural causes of soil degradation.
3 a Comment on the human causes of soil degradation.
  b To what extent is it possible to manage soil degradation?
4 Study Table 10.13, which shows annual soil losses from a small catchment in the Lake Victoria basin.
  a Describe how the range of soil loss varies with the type of land cover.
  b Suggest reasons for the patterns you have identified in **a**.

Table 10.13 Annual soil losses from a small catchment in the Lake Victoria basin

Land use	Land cover (%)	Range of soil loss (tonnes/ha/year)
Annual cropland	6	65–93
Rangeland	15	42–68
Bananas/Coffee	63	36–47
Bananas	6	22–32
Forest	1	0
Papyrus marsh	9	0

10.4 Sustainable management of hot arid and semi-arid environments

Sustainable management is management that meets the needs of the present generation without compromising the ability of future generations to meet their own needs. It is a process by which human potential (quality of life) is improved and the environment (resource base) is used and managed to supply humanity on a long-term basis.

Nearly three-quarters of the world's drylands are degraded and it has been estimated that desertification costs an estimated $42 million each year. Are there sustainable options for the world's drylands?

☐ Changing land-use trends

Case Study: Game farming in the Eastern Cape province of South Africa

A shift from pastoralism to **game farming** has been identified in the Eastern Cape province of South Africa since the 1980s (Figure 10.21). Examples include Bushbuck Ridge Game Farm, AddoAfrique Estate and Kichaka Lodge. In some cases, this change has been made by private landowners to diversify their operations. In other cases, private landowners have removed all stock and replaced it with game. In a survey of the Eastern Cape region of South Africa, it was found that 2.5 per cent of the study area had converted entirely from stock to game farming. A total of 41 game species was recorded on the 63 game farms surveyed. Most farmers expressed a positive attitude towards game farming and are trying to implement conservation measures. The main activity for which game is utilised is hunting – both recreational and trophy hunting. The foreign ecotourist and the hunting market have been strong driving forces behind the introduction of **extra-limital** (non-native) species to the region.

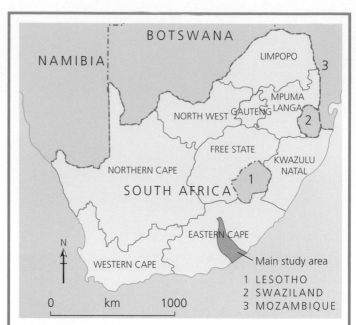

Figure 10.21 Location of the Eastern Cape province of South Africa

This change in land use has drawn the attention of scientists worldwide, and specifically with reference to desertification of rangelands. Desertification currently affects about one-sixth of the world's population and 70 per cent of all drylands, which amounts to 3.6 billion hectares. Widespread poverty is one of the key impacts of desertification.

In South Africa, the thicket vegetation of the Eastern Cape has been recognised as being particularly vulnerable to degradation, due mainly to years of overgrazing. Over 95 per cent of this vegetation is under threat from overgrazing by domestic stock; bush clearing for agriculture and urban development; coastal resort development; and invasion by alien vegetation.

The average game farm size is 4496 hectares. Most of the game farms are concentrated in the south and central regions of the Eastern Cape. Land-use changes first started to occur in the 1970s, and were characterised by two basic trends that included either the landowners themselves changing from being stock farmers to game farmers, or investors purchasing stock farms and financing their conversion to game farms.

Utilising game has provided an important secondary income to most mixed farmers. The impetus behind the growing game industry can be attributed to a number of socio-political, economic and ecological motivations. For example:

- Recently changed labour legislation stipulates increased wages for workers on farms. This has made landowners regard game farming as an alternative to stock farming, as it is considered to be potentially less labour-intensive than traditional stock farming.
- Increased stock theft, especially of small domestic stock, has rendered stock farming economically less viable.
- Vermin such as the jackal and caracal sometimes come from adjacent game farms' statutory reserves and this has resulted in increased stock losses.
- Decades of overgrazing has led to rangeland degradation, thereby reducing livestock production. By (re)introducing (indigenous) game species that are better adapted to their

natural environment, periodic droughts can be survived both economically and demographically.

- Game is considered to contribute, in the long term, to **veld** restoration (rather than its degradation).
- There is good potential for foreign exchange earnings from trophy hunting and tourism.

A total of 41 game species were recorded on the 63 farms surveyed. The high diversity that was recorded was not, however, found on any single farm. Rather, 11–15 species occurred on a third of the game farms, with only five game farmers maintaining more than 20 species.

Game farming has been described as a potential ecologically sustainable form of land use, but the introduction of extra-limital species may threaten this state. In order to guarantee tourist satisfaction, farmers have found it necessary to erect game-proof fences around their farms with the purpose of introducing 'hunting' or 'tourist' species, whether indigenous or extra-limital. Kudu and bushbuck, both indigenous to thicket vegetation, are among the most desired hunting species in the Eastern Cape. Promotion of these animals as hunting species may promote ecologically sound farming practices, without the introduction of extra-limital species.

There is also the ecological risk of allowing certain species to hybridise by keeping such species in the same fenced area. Some farmers in the survey had both blue and black wildebeest species on their property, and some had both Blesbok and Bontebok antelope; both pairs of species have the ability to hybridise.

Section 10.4 Activities

1 Define *sustainable development*.
2 To what extent is game farming a form of sustainable development? Justify your views.

Case Study: The establishment of drought-resistant fodder in the Eastern Cape

Pastureland in the Eastern Cape is especially fragile due to drought and overgrazing (Figures 10.22 and 10.23). In the former homelands Ciskei and Transkei, there are additional problems of population pressure and, sometimes, the absence of secure land-ownership policies. During periods of prolonged drought, levels of cattle, sheep and goats decrease significantly. However, trying to decrease herd size has proved unpopular and unsuccessful. An alternative is to produce drought-resistant **fodder crops** such as the American aloe and prickly pear, saltbrush and the indigenous gwanish.

Figure 10.22 Gully erosion due to overgrazing

Figure 10.23 Concentration of sheep at a waterhole – note the irrigation scheme in the background

The American aloe (Figure 10.24) has traditionally been used for fencing, for kraals (animal compounds) and for soil conservation, but has also been used as a fodder in times of drought. It has a number of advantages:

- It requires little moisture (annual rainfall in this region is around 450 mm).
- It is not attacked by any insects.
- Although low in protein, it raises milk production in cows.
- It can be used for soil conservation.
- After 10 years, it produces a pole that can be used for fencing or building.
- It can act as a windbreak.
- The juice of the aloe is used in the production of tequila.

Figure 10.24 American aloe

Saltbrush provides protein-rich fodder that is eaten by sheep and goats. Goats, in particular, thrive on saltbrush. It requires less than half the water need by other crops such as lucerne, and once established it requires no irrigation. It remains green throughout the year and therefore can provide all-year fodder. However, it is difficult to propagate and needs high-quality management.   ⇨

The spineless cactus or prickly pear (Figure 10.25) features prominently in the agriculture of many countries, such as Mexico, Peru and Tunisia, where it is used as fodder and as a fruit crop for 2–3 months each year. This plant is becoming more widespread in the Eastern Cape. Two varieties are common: one, insect-resistant, is used as fodder in times of drought, while the other, which needs to be sprayed to reduce insect damage, yields high-quality fruit. The fruit is sold at prices comparable with apples and oranges. Pruning is needed annually. This provides up to 100 tonnes of fodder per hectare per year.

Figure 10.25 Prickly pear

In the former Ciskei region of the Eastern Cape, drought in the 1980s prompted the government to embark on a series of trials with prickly pear, saltbrush and American aloe in order to create more fodder. One of the main advantages of the prickly pear is its low water requirements. This makes it very suitable to the region where rainfall is low and unreliable. Although there are intensive irrigation schemes in the region, such as at Keiskammahoek, these are expensive and are inappropriate to the area and to the local people.

Although prickly pear is mainly used as a fodder and fruit crop, it is also used for the production of carminic acid for the cochineal dye industry and as a means of soil conservation. Nevertheless, prickly pear has been described by some development planners as a 'weed, the plant of the poor, a flag of misery … inconsistent with progress'.

Section 10.4 Activities

1 Outline the advantages of the American aloe plant.
2 Comment on the advantages and disadvantages of using the prickly pear.

Case Study: Essential oils in the Eastern Cape

About 65 per cent of the world's production of essential oils is from LICs such as India, China, Brazil, Indonesia, Mexico, Egypt and Morocco. However, the USA is also a major producer of essential oils such as peppermint and other mints. The South African essential oils industry has only recently emerged in this area. Currently, the South African essential oils industry exports mainly to HICs in Europe (49 per cent), the USA (24 per cent) and Japan (4.5 per cent). The most significant essential oils produced by South Africa are eucalyptus, citrus, geranium and buchu.

Globally, the essential oils industry – valued at around $10 billion – is enjoying huge expansion. Opportunities include increasing production of existing products and extending the range of crops grown. Developing the essential oils industry in South Africa would achieve much-needed agricultural and agri-processing diversification in the province.

Currently, the South African essential oils industry comprises about 100 small commercial producers, of which less than 20 per cent are regular producers.

Several factors make South Africa an attractive essential oils market:

■ Much of the demand is in the northern hemisphere and seasonal effects make southern hemisphere suppliers globally attractive.
■ South Africa traditionally has strong trade links with Europe, as a major importer of fragrance materials.
■ South Africa is being established as a world-class agricultural producer in a wide range of products.

The Eastern Cape is set to become one of the main contributors to South Africa's burgeoning essential oils

industry, with 10 government-sponsored trial sites currently in development throughout the province. Six of these form part of the Essential Oil Project of Hogsback, where approximately 8 hectares of communal land are being used. A project at Keiskammahoek has been operational since 2006. These trials form part of a strategy to develop a number of essential oil clusters in the Eastern Cape.

The production of essential oils holds considerable potential as a form of sustainable agricultural development in the former Ciskei region of the Eastern Cape. Not only are the raw materials already here, but it is a labour-intensive industry and would employ a large number of currently unemployed and underemployed people.

The essential oils industry has a number of advantages:

■ It is a new or additional source of income for many people.
■ It is labour-intensive and local in nature.
■ Many plants are already known and used by local people as medicines, and they are therefore culturally acceptable (Figure 10.26).
■ In their natural state, the plants are not very palatable nor of great value and will not therefore be stolen.
■ Many species are looked upon as weeds. Removing these regularly improves grazing potential as well as supplying raw materials for the essential oils industry.

Some species such as geranium, peppermint and sage require too much land, labour and water to be very successful. Wild als (*Artemesia afra*) is an indigenous mountain shrub, used for the treatment of colds. Its oil has a strong medicinal fragrance and is used in deodorants and soaps. Double

Figure 10.26 A herbalist's preparation table

cropping in summer when the plant is still growing and in autumn at the end of the growing season yields the best results. Demand for *Artemisia* has not outstripped the supply of naturally growing material but it is increasingly being cultivated as a second crop. It requires minimal input in terms of planting, tillage and pest control, and it is relatively easy to establish and manage. Moreover, it can stabilise many of the maize fields and slopes where soil erosion is a problem. The local people are very enthusiastic about growing it, especially when they are given appropriate economic incentives.

Khakibush or *Tagetes* is an aromatic. In the former Ciskei area, it is a common weed in most maize fields. Oil of tagetes is an established essential oil, although its market is limited. Local people are again quite enthusiastic about collecting khakibush if the incentives are there. Harvesting takes place over a period of up to three months and provides a great deal of extra employment, as well as eradicating a weed. At present, the supply of khakibush and those in the maize fields is sufficient to meet demand. An increase in demand might lead to the establishment of *Tagetes* as secondary crop in maize fields – not just as a 'weed'.

Section 10.4 Activities

1 Suggest why the essential oils industry has developed in the Eastern Cape province.
2 To what extent could the essential oils industry be considered a form of sustainable development?

Case Study: Developing sustainable farming in Egypt

The Nile provides Egypt with almost all of its water, 85 per cent of which goes to agriculture – but population growth and increased demands for water is putting a strain on water resources. Up to 95 per cent of Egypt's population lives in the Nile Valley and Delta, increasing the pressure on land resources. The same area accounts for the bulk of Egyptian food production. Although one-third of Egypt's annual share of the Nile is used for irrigation, it contains pollutants and pesticides from upstream countries and from Egypt itself. Since chemical pesticides were first introduced to Egypt in the early 1950s, a million tonnes have been released into the environment. To compound the matter, Ethiopia is building the Grand Ethiopian Renaissance Dam on the Blue Nile, which is likely to cut supplies of fresh water to Egypt.

However, Egypt is developing forms of sustainable agriculture. One of the leading individuals is Faris Farrag, who has developed aquaponics at his farm outside Cairo called 'Bustan' (Arabic for orchard). Aquaponics is an integrated form of farming that originated in Central America. It enables farmers to increase yields by growing plants and farming fish in the same closed freshwater system.

Bustan is the first commercial aquaponics farm in Egypt. Water circulates from tanks containing fish through hydroponic trays that grow vegetables including cucumber, basil, lettuce, kale, peppers and tomatoes. Each tank contains about a thousand tilapia fish, which are native to Egypt and are known for resisting slight water pH and temperature variations. Water from the pond is then used to water the olive trees that produce a high-quality olive oil. This organic and closed system mimics natural processes and enables waste to be efficiently reused. The fish tanks provide 90 per cent of the nutrients plants need to grow. The ammonia that results from the fish breathing is naturally transformed into nitrogen and absorbed by the plants before being sent back to the fish tanks, ammonia-free and healthy.

Just outside the fish tanks lies a large pond covered with a slimy layer of duckweed, a highly nutritious floating plant that is regularly scraped, dried and fed to the fish as vegetable protein.

Bustan uses 90 per cent less water than traditional farming methods in Egypt. It produces 6–8 tonnes of fish per year and can potentially yield 45 000 heads of lettuce if it were to grow just a single type of vegetable. Hydroponics can make lettuce grow 20 per cent faster than average.

Bustan is a labour-intensive farm and uses sustainable biological pest-control methods, such as ladybirds to kill aphids, in order to avoid chemical inputs. Farrag intends to establish a permaculture system by introducing chickens that would feed on compost and produce natural fertilisers for the soil.

This method of farming could serve as a means of income generation for unemployed women, as well as a means of education for sustainable farming. However, it is quite costly, especially for those on a low income. Inside Bustan, the pumps used to filter water require a source of energy, mainly oil. Farrag has invested more than $43 500 to develop this scheme.

Small hydroponic units could be established for rooftops, balconies and kitchens. Vertical and rooftop farming, in light of the country's serious water and food crisis, is also an effective way to grow organic food while cutting transportation costs, emissions and waste.

Section 10.4 Activities

Research Bustan online and watch a video clip. Find out about fish farms in the desert and rooftop farms in Cairo.

11 Production, location and change

11.1 Agricultural systems and food production

⟳ □ Factors affecting agricultural land use and practices

A wide range of factors combine to influence agricultural land use and practices on farms. These can be placed under the general headings of physical, economic, political and social/cultural factors.

Physical factors

North America, for example, has many different physical environments. This allows a wide variety of crops to be grown and livestock kept. New technology and high levels of investment have steadily extended farming into more difficult environments. Irrigation has enabled farming to flourish in the arid west, while new varieties of wheat have pushed production northward in Canada. However, the physical environment remains a big influence on farming. There are certain factors that technology and investment can do little to alter. So relief, climate and soils set broad limits as to what can be produced. This leaves the farmer with some choices, even in difficult environments. The farmer's decisions are then influenced by economic, political and social/cultural factors.

Figure 11.1 shows the **agricultural regions** of the USA. Look at relief and climate maps of the USA in an atlas and see how the agricultural regions vary according to different physical conditions. Temperature is a critical factor in crop growth as each type of crop requires a minimum growing temperature and a minimum length of growing season. Latitude, altitude and distance from the

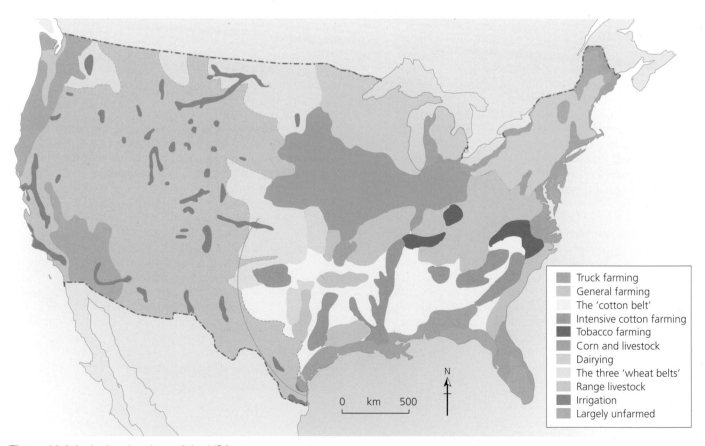

Truck farming
General farming
The 'cotton belt'
Intensive cotton farming
Tobacco farming
Corn and livestock
Dairying
The three 'wheat belts'
Range livestock
Irrigation
Largely unfarmed

0   km   500

N

Figure 11.1 Agricultural regions of the USA

sea are the major influences on temperature. Precipitation is equally important – not just the annual total, but also the way it is distributed throughout the year. Long, steady periods of rainfall to infiltrate into the soil are best, making water available for crop growth. In contrast, short heavy downpours can result in rapid surface runoff, leaving less water available for crop growth and soil erosion.

Soil type and fertility have a huge impact on **agricultural productivity**. Often, areas that have never been cleared for farming were ignored because soil fertility was poor or was perceived to be poor. In some regions, wind can have a serious impact on farming, for example causing bush fires in some states such as California. Locally, aspect and the angle of slope may also be important factors in deciding how the land is used.

Cotton, for example, needs a frost-free period of at least 200 days. Rainfall should be over 625 millimetres a year, with not more than 250 millimetres in the autumn harvest season. Cotton production is now highly mechanised. Irrigation has allowed cotton to flourish in the drier western states of California, New Mexico and Texas. In contrast, the area under cotton has fallen considerably in the south. A crop pest called the cotton boll weevil, which caused great destruction to cotton crops in the past, has been a big factor in the diversification of agriculture in the southern states.

In contrast, corn is grown further north than cotton. Corn needs a growing season of at least 130 days. For the crop to ripen properly, summer temperatures of 21 °C are needed, with warm nights. Precipitation should be over 500 millimetres, with at least 200 millimetres in the three summer months.

In Canada, the USA's northern neighbour, farming is severely restricted by climate. Less than 8 per cent of the total area of the country is farmed; 70 per cent of Canada lies north of the **thermal limit for crop growth** and most farms are within 500 kilometres of the main border (apart from Alaska) with the USA. Other high-latitude countries such as Russia also suffer considerable climatic restrictions on agriculture.

Water is vital for agriculture. **Irrigation** is an important factor in farming, not just in North America but in many other parts of the world as well. Figure 11.2 shows the divide by world region between **rainfed water** for crop use and irrigation water. The figures in the circles refer to the total amount of rainfed water used. Here, the highest totals are for East Asia, South Asia and Sub-Saharan Africa. The highest proportion of irrigation water use is in the Middle East and North Africa, and South Asia. Irrigated farming accounts for 70 per cent of global annual water consumption. This rises to over 90 per cent in some countries such as India. Table 11.1 compares the main types of irrigation (Figures 11.3 and 11.4). This is an example of the **ladder of agricultural technology**, with surface irrigation being the most traditional method and sub-surface (drip) irrigation the most advanced technique.

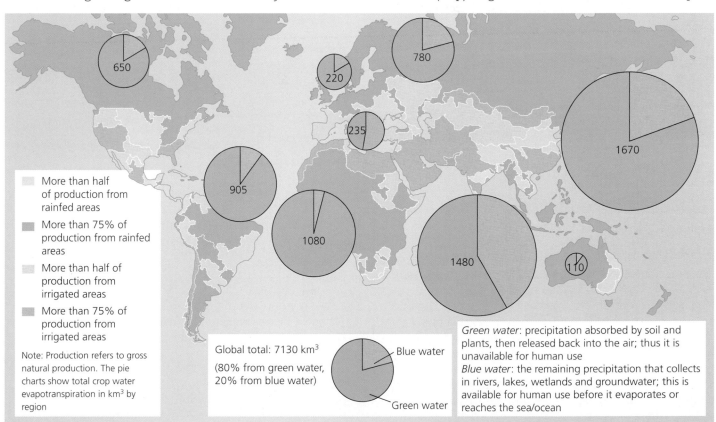

Figure 11.2 World distribution of rainfed and irrigation water for crop use

Table 11.1 Main types of irrigation

Method of irrigation	Efficiency (%)
Surface – used in over 80 % of irrigated fields worldwide	
Furrow: Traditional method; cheap to install; labour-intensive; high water losses; susceptible to erosion and salinisation **Basin:** Cheap to install and run; needs a lot of water; susceptible to salinisation and waterlogging	20–60 50–75
Aerial (using sprinklers) – used in 10–15 % of irrigation worldwide	
Costly to install and run; low-pressure sprinklers preferable	60–80
Sub-surface ('drip') – used in 1 % of irrigation worldwide	
High capital costs; sophisticated monitoring; very efficient	75–95

Source: 'The Water Crisis: A Matter of Life and Death',
Understanding Global Issues, p7

Figure 11.4 A sprinkler system irrigating crops in northern Spain

Figure 11.5 Goats feeding from a bowl because the ground is frozen, in cold central Asia

Figure 11.3 An irrigation canal in northern Spain

Economic factors and agricultural technology

Economic factors include transport, markets, capital and technology. The role of government is a factor here too, but this is considered in the next section, 'Political factors'.

The costs of growing different crops and keeping different livestock vary. The market prices for agricultural products also vary and can change from year to year. The necessary investment in buildings and machinery can mean that some changes in farming activities are very expensive. These will be more difficult to achieve than other, cheaper changes. Thus it is not always easy for farmers to react quickly to changes in consumer demand.

In most countries, there has been a trend towards fewer but larger farms. Large farms allow **economies of scale** to operate, which reduce the unit costs of production. As more large farms are created, small farms find it increasingly difficult to compete and make a profit. Selling to a larger neighbouring farm may be the only economic solution. The EU is an example of a region where average farm size varies significantly. Those countries with a large average farm size have more efficient agricultural sectors than countries with a small average farm size.

Distance from markets has always been an important influence on agricultural practices. Heinrich von Thünen published a major theoretical work on this topic in 1826. He was mainly concerned with the relationships between three variables:

Figure 11.6 A food market in Morocco

- the distance of farms from the market
- the price received by farmers for their products
- the economic rent (the profit from a unit of land).

Von Thünen argued that the return a farmer obtained for a unit of his product was equal to its price at market less the cost of transporting it to the market. Thus the nearer a farmer was to the market, the greater his returns from the sale of his produce (Figure 11.7). The logic of this is that land closest to the market would be the most intensively farmed land, with farming intensity decreasing with increasing distance from the market. At a certain distance from the market, transport costs would be so high that they would equal the profit from farming and therefore make cultivation illogical (Figure 11.7). Farmers setting out to maximise their profits would choose that activity or combination of farming activities that would give the best economic rent (profit). Although this theory is almost two centuries old, it still holds a basic logic.

Agricultural technology is the application of techniques to control the growth and harvesting of animal and vegetable products. The development and application of

agricultural technology requires investment and thus it is an economic factor. Advances in agricultural technology can be traced back to the Neolithic Revolution. Table 11.2 shows the last two sections of a timeline of agricultural advance published in Wikipedia (the table only shows major and selected advances and thus omits a whole range of smaller improvements).

Table 11.2 Timeline of agricultural technology

Year	Event
Agricultural Revolution	
1700	Agricultural Revolution begins in the UK
1809	French confectioner Nicolas Appert invents canning
1837	John Deere invents steel plough
1863	International 'Corn Show' in Paris, with corn varieties from different countries
1866	George Mendel publishes his paper describing Mendelian inheritance
1871	Louis Pasteur invents pasteurisation
1895	Refrigeration introduced in the USA for domestic food preservation, and in the UK for commercial food preservation
1930	First use of aerial photos in earth sciences and agriculture
Green Revolution	
1944	Green Revolution begins in Mexico
2000	Genetically modified plants cultivated around the world
2005	Lasers used to replace stickers by writing on food to 'track and trace' and identify individual pieces of fresh fruit

Source: Wikipedia

The status of a country's agricultural technology is vital for its **food security** and other aspects of quality of life. An important form of aid is the transfer of agricultural technology from more advanced to less advanced nations (Figure 11.8). China is now playing a major role in this process. Eighty per cent of the population in rural Sub-Saharan Africa is reliant on agriculture as a source of income and employment. Yet agricultural productivity has stagnated. The agricultural sector is mostly made up of small-scale farms. But small farmers face serious barriers to their development:

- They have limited access to new technologies, such as new crop varieties and better methods of storage.
- They have difficulty accessing finance and suffer from a lack of investment in areas such as roads, agricultural equipment and silos.
- They lack support from areas such as market boards and advisory services.

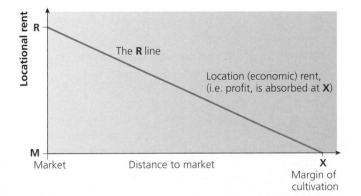

Figure 11.7 The relationship between economic rent and distance from the market

Figure 11.8 China agrees to help improve African food production

- They contend with market constraints such as an inability to produce the right amount or quality for customers, price variations and inadequate storage systems.

Small farmers in other parts of the world such as India face similar problems (Figure 11.9).

TECA is an FAO (Food and Agriculture Organization of the United Nations) initiative that aims at improving access to information and knowledge-sharing about proven technologies in order to enhance their adoption in agriculture, livestock, fisheries and forestry, thus addressing food security, climate change, poverty alleviation and sustainable development.

Agro-industrialisation, or 'industrial agriculture', is the form of modern farming that refers to the industrialised production of livestock, poultry, fish and crops. This type of large-scale, **capital-intensive farming** originally developed in Europe and North America and then spread to other HICs. It has been spreading rapidly in many MICs and LICs since the beginning of the **Green Revolution**. Industrial agriculture is heavily dependent on oil for every stage of its operation. The most obvious examples are fuelling farm machinery, transporting produce and producing fertilisers and other farm inputs. Table 11.3 shows the general characteristics of agro-industrialisation. Not all farms and regions involved in agro-industrialisation will display all these characteristics. For example, intensive market gardening units may be relatively small but the capital inputs are extremely high.

Table 11.3 The characteristics of agro-industrialisation

Very large farms
Concentration on one (monoculture) or a small number of farm products
A high level of mechanisation
Low labour input per unit of production
Heavy usage of fertilisers, pesticides and herbicides
Sophisticated ICT management systems
Highly qualified managers
Often owned by large agribusiness companies
Often vertically integrated with food processing and retailing

Regions where agro-industrialisation is clearly evident on a large scale include:
- the Canadian Prairies
- the corn and wheat belts in the USA

Figure 11.9 A smallholding in northern India

Figure 11.10 Industrial agriculture on the Canadian Prairies

- the Paris basin (Figure 11.10)
- East Anglia in the UK
- the Russian steppes
- the Pampas in Argentina
- Mato Grosso in Brazil
- the Murray–Darling basin in Australia.

Agro-industrialisation is a consequence of the **globalisation of agriculture**, profit ambitions of large agribusiness companies and the push for cheaper food production. Over the last half-century, every stage in the food industry has changed in the attempt to make it more efficient (in an economic sense). Vertical integration has become an increasingly important process, with extended linkages between the different stages of the food industry (Figure 11.11).

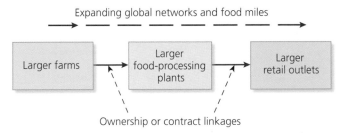

Figure 11.11 Agro-industrialisation – increasing vertical integration

Political factors

The influence of government on farming has steadily increased in many countries. For example, in the USA the main sectors of federal farm policy over the last half-century have been the following:

- **Price support loans** – loans that tide farmers over until they sell their produce; the government sets a price for each farm product it is willing to support, and if the farmer cannot sell the product for at least this price, they can keep the loan and let the government keep the crop that secured it.
- **Production controls** – these limit how much a farmer can produce of surplus crops; farmers lose price-support loans and other benefits if they don't comply.
- **Income supplements** – these are cash payments to farmers for major crops in years when market prices fail to reach certain levels.

The decisions made by individual farmers are therefore heavily influenced by government policies such as those listed above. However, in centrally planned economies the state has far more control. This was the case for many years in the former Soviet Union and China. Although much has changed in both of these countries in recent decades, the influence of government on farming still remains stronger than in most other parts of the world.

An agricultural policy can cover more than one country, as evidenced by the **EU's Common Agricultural Policy (CAP)**. The CAP is a set of rules and regulations governing agricultural activities in the EU. The need for Europe to ensure a reliable and adequate supply of food in the post-Second World War period was one of the main reasons for its introduction in 1960. It is expensive to run: each year, every EU taxpayer contributes about £80 to the CAP.

Social/cultural factors

What a particular farm and neighbouring farms have produced in the past can be a significant influence on current farming practices. There is a tendency for farmers to stay with what they know best, and often a sense of responsibility from one generation to the next to maintain a family farming tradition. Tradition matters more in some farming regions than others.

A traditional rainforest system

Shifting cultivation is a traditional farming system that developed a long time ago in tropical rainforests. An area of forest is cleared to create a small plot of land that is cultivated until the soil becomes exhausted. The plot is then abandoned and a new area cleared. Frequently, the cultivators work in a circular pattern, returning to previously used land once the natural fertility of the soil has been renewed. Shifting cultivation is also known as 'slash and burn' and by various local names such as *chitimene* in Central Africa.

In the Amazon rainforest, shifting cultivation has been practised for thousands of years by groups of Amerindians who initially had no contact with the outside world. It is likely that there are some isolated groups where this situation still exists, but for most Amerindians there are now varying degrees of contact with mainstream Brazilian society. As a result, there has been a gradual blending of modern ideas with traditional practices.

Legal rights and land tenure

Land tenure refers to the way in which land is or can be owned. In the past, inheritance laws had a huge impact on the average size of farms. In some countries, it has been the custom on the death of a farmer to divide the land equally between all his sons, but rarely between daughters. Also, dowry customs may include the giving of land with a daughter on marriage. The reduction in the size of farms by these processes often reduced them to operating at only a subsistence level.

Women face widespread discrimination around the world with regard to land and property. The agrarian reforms implemented in many countries from the 1950s and through the 1970s were 'gender blind'. They were often based on the assumption that all household members would benefit equally, when this was simply not the case. For example, many women in LICs lose their homes, inheritance and possessions, and sometimes even their children when their partners die. This may force women to adopt employment practices that increase their chances of contracting HIV.

In many societies, women have very unequal access to, and control over, rural land and associated resources. The UN's Food and Agriculture Organization has stated that 'denying large segments of rural society equitable access to land and to the benefits of land tenure regularisation creates unanticipated costs and is a major contributing factor to extreme poverty, dependence and rural migration leading to land abandonment, social instability and many other negative conditions because of the unforeseen externalities that arise.' It is now generally accepted that societies with well-recognised property rights are also the ones that thrive best economically and socially.

Section 11.1 Activities

1 List the main physical factors that can influence farming.
2 Look at Figure 11.1. Suggest why almost all of the USA's range livestock and irrigated farming is in the west.
3 Summarise the information presented in Table 11.1.
4 Describe and explain the relationship shown in Figure 11.7.
5 Discuss the characteristics of agro-industrialisation.
6 Briefly state the importance of advances in agricultural technology.
7 Give an example of how one social/cultural factor can affect farming.

☐ Agricultural systems

Individual farms and general types of farming can be seen to operate as a **system**. A farm requires a range of **inputs**, such as labour and energy, so that the **processes** (throughputs) that take place on the farm can be carried out. The aim is to produce the best possible **outputs**, such as milk, eggs, meat and crops. A profit will only be made if the income from selling the outputs is greater than expenditure on the inputs and processes. Figure 11.12 illustrates the agricultural system. It shows how physical, cultural, economic and behavioural factors form the inputs. Decision-making at different scales, from the individual farmer to governments and international organisations such as the EU, influence the processes. The nature and efficiency of the processes dictate the range, scale and quality of the outputs. Agricultural systems are dynamic human systems that change as farmers attempt to react to a range of physical and human factors.

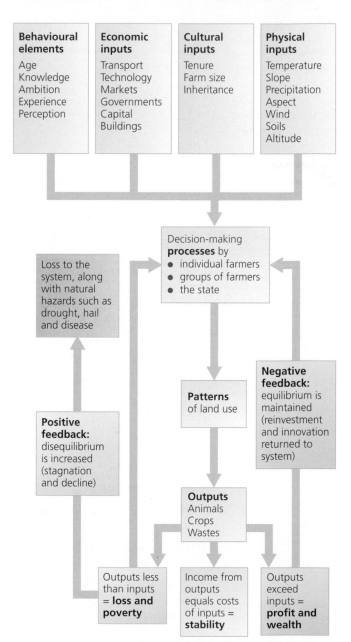

Source: *AQA AS Geography* by A. Barker, D. Redfern & M. Skinner (Philip Allan Updates, 2008), p.211

Figure 11.12 The agricultural system

Different types of agricultural system can be found within individual countries and around the world. The most basic distinctions are between:

- arable, pastoral and mixed farming
- subsistence and commercial farming
- extensive and intensive farming
- organic and non-organic farming.

Arable farms cultivate crops and are not involved with livestock. An arable farm may concentrate on one crop such as wheat or may grow a range of different crops. The crops grown on an arable farm may change over time. For example, if the market price of potatoes increases, more farmers will be attracted to grow this crop. **Pastoral farming** involves keeping livestock such as dairy cattle, beef cattle and sheep. **Mixed farming** involves cultivating crops and keeping livestock together on a farm.

Subsistence and commercial farming

Subsistence farming is the most basic form of agriculture, where the produce is consumed entirely or mainly by the family who work the land or tend the livestock. If a small surplus is produced, it may be sold or traded. Examples of subsistence farming are shifting cultivation and nomadic pastoralism. Subsistence farming is generally small-scale and labour-intensive, with little or no technological input.

In contrast, the objective of **commercial farming** is to sell everything that the farm produces. The aim is to maximise yields in order to achieve the highest profits possible. Commercial farming can vary from small-scale to very large-scale (Figure 11.13).

Figure 11.13 Much farmland in Nepal is terraced to counteract the steep slopes

Extensive and intensive farming

Extensive farming is where a relatively small amount of agricultural produce is obtained per hectare of land, so such farms tend to cover large areas. Inputs per unit of land are low. Extensive farming can be both arable and pastoral in nature. Examples include wheat farming in the Canadian Prairies and sheep farming in Australia. In contrast, **intensive farming** is characterised by high inputs per unit of land to achieve high yields per hectare. Examples of intensive farming include market gardening, dairy farming and horticulture. Intensive farms tend to be relatively small in terms of land area.

Organic farming

Organic farming has become increasingly popular in recent decades as people seek a healthier lifestyle. In 2010, 37 million hectares of land were organically farmed worldwide – three times more than in 1999.

Organic farmers do not use manufactured chemicals, and so this type of farming is practised without chemical fertilisers, pesticides, insecticides or herbicides. Instead, animal and green manures are used, along with mineral fertilisers such as fish and bone meal. Organic farming therefore requires a higher input of labour than mainstream farming. Weeding is a major task with this type of farming. Organic farming is less likely to result in soil erosion and is less harmful to the environment in general. For example, there will be no nitrate runoff into streams and much less harm to wildlife.

Organic farming tends not to produce the 'perfect' potato, tomato or carrot. However, because of the increasing popularity of organic produce it commands a substantially higher price than mainstream farm produce.

Section 11.1 Activities

1 Describe the inputs, processes and outputs for the agricultural system shown in Figure 11.12.
2 How can the increase in a country's wealth influence its demand for energy?
3 Examine the differences between **a** commercial and subsistence farming and **b** intensive and extensive farming.
4 Describe the characteristics of organic farming.

Characteristics and location

Sheep farming in Australia occupies an area of about 85 million hectares, making it one of Australia's major land uses. It is a classic example of extensive farming, which can be seen to operate clearly as a system. The main physical input is the extensive use of natural open ranges, which are often fragile in nature. Australia's sheep farms are located predominantly in inland and semi-arid areas. Human inputs are low compared with most other types of agriculture, with very low use of labour and capital per hectare. The main processes are grazing, lambing, dipping and shearing. The outputs are lambs, sheep, wool and sheep skins.

Australia is the world's leading sheep-producing country, with a total of about 120 million sheep. As well as being the largest wool producer and exporter, Australia is also the largest exporter of live sheep and a major exporter of lamb and mutton. The sheep and wool industry is an important sector of Australia's economy.

Sheep are raised throughout southern Australia in areas of moderate to high rainfall and in the drier areas of New South Wales and Queensland. Merinos, which produce very high-quality wool for clothing, make up 75 per cent of the country's sheep. Merino sheep are able to survive in harsh environments and yet produce heavy fleeces. Sixteen per cent of Australia's sheep are bred for meat production and are a mixture of breeds such as Border Leicester and Dorset. The remaining 9 per cent are a mixture of Merino and cross-bred sheep used for wool and meat production. There are about 60 000 sheep farms in Australia overall, carrying from a few hundred sheep to over 100 000 animals.

Sheep and wool production occurs in three geographical zones (Figure 11.14):

- high rainfall coastal zone
- wheat/sheep intermediate zone
- pastoral interior zone.

About a quarter of all sheep are farmed in the pastoral zone (Figure 11.15). Sheep farming in Australia in general is extensive in nature but this type of agriculture is at its most extensive in the pastoral zone, which is the arid and semi-arid inland area. Here, summer temperatures are high, rainfall is low and the area is prone to drought. Because of the lack of grass in this inhospitable environment, sheep are often left to

eat saltbush and bluebush. In the pastoral zone, the density of sheep per hectare is extremely low due to the poor quality of forage. The overall farming input in terms of labour, capital, energy and other inputs is also very low – it is in fact the lowest input per hectare of farmland in the country. Not surprisingly, farms can be extremely large.

Figure 11.15 Sheep farm in Australia's pastoral zone

In the coastal and intermediate zones, the best land is reserved for arable farming, dairy and beef cattle and market gardening. Sheep are frequently kept on the more marginal areas, for example on higher and colder land in the New South Wales highlands where more profitable types of farming are not viable.

About two-thirds of Australia's sheep are on farms that support more than 2000 animals. The smallest sheep farms are generally those on the better-quality land, where it is possible to keep many more animals per hectare than in the pastoral zone.

Farming issues

The main issues in Australian sheep farming areas are:

- weed infestation, which is difficult to control on very large extensive farms that yield relatively small profits per hectare
- destruction of wildlife habitats due to sheep grazing, particularly in marginal areas
- the occurrence of periodic droughts that make farming even more difficult in low-rainfall areas
- soil loss from wind erosion and loss of soil structure – in some areas, this is transforming traditional 'mainstream' farming areas into marginal lands
- animal welfare, particularly in the most inhospitable environments where the low human input means that individual animals may not be seen for long periods
- increasing concern about the shortage of experienced sheep shearers.

Regarding the last point, many shearers have left the industry because of poor working conditions and the attraction of better-paid jobs in the mining industry and elsewhere. The number of experienced shearers fell by about a quarter

Source: *IGCSE Geography* by P. Guinness & G. Nagle (Hodder Education, 2009), p.112

Figure 11.14 Australia's three 'sheep' geographical zones

N

Pastoral zone

Wheat and sheep zone

High rainfall zone

between 2003 and 2006. A good shearer can shear up to 200 sheep in one day.

Sheep farming in Australia is a major user of land resources in a generally fragile landscape. Changes in farming systems are required in some locations to address the issues facing the industry. Failure to do so will result in the progressive decline in utility of the resource base for the sheep and wool industry.

Section 11.1 Activities

1 Why is sheep farming in Australia considered to be 'extensive farming'?
2 Describe the three geographical zones in which sheep are kept.
3 Briefly discuss the main issues affecting sheep farming in Australia today.

Case Study: An arable system – intensive rice production in the Lower Ganges valley

Location

An important area of intensive subsistence rice cultivation is the Lower Ganges valley (Figures 11.16 and 11.17) in India and Bangladesh. The Ganges basin is India's most extensive and productive agricultural area and its most densely populated. The delta region of the Ganges occupies a large part of Bangladesh, one of the most densely populated countries in the world. Rice contributes over 75 per cent of the diet in many parts of the region. The physical conditions in the Lower Ganges valley and delta are very suitable for rice cultivation:

- Temperatures of 21 °C and over throughout the year allow two crops to be grown annually. Rice needs a growing season of only 100 days.
- Monsoon rainfall over 2000 millimetres provides sufficient water for the fields to flood, which is necessary for wet rice cultivation.
- Rich alluvial soils have built up through regular flooding over a long time period during the monsoon season.
- There is a seasonal dry period, which is important for harvesting the rice.

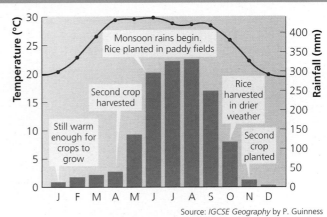

Source: *IGCSE Geography* by P. Guinness & G. Nagle (Hodder Education, 2009), p.114

Figure 11.17 Climate graph for Kolkata

A water-intensive staple crop

Rice is the staple or main food crop in many parts of Asia. This is not surprising considering its high nutritional value. Current rice production systems are extremely water-intensive; 90 per cent of agricultural water in Asia is used for rice production. The International Rice Research Institute estimates that it takes 5000 litres of water to produce 1 kilogram of rice. Much of Asia's rice production can be classed as intensive subsistence cultivation where the crop is grown on very small plots of land using a very high input of labour. Rice cultivation by small farmers is sometimes referred to as 'pre-modern intensive farming' because of the traditional techniques used, in contrast to intensive farming systems in HICs such as market gardening, which are very capital intensive.

'Wet' rice is grown in the fertile silt and flooded areas of the lowlands, while 'dry' rice is cultivated on terraces on the hillsides. A **terrace** is a levelled section of a hilly cultivated area. Terracing is a method of soil conservation. It also prevents the rapid runoff of irrigated water. Dry rice is easier to grow but provides lower yields than wet rice. ⇒

Source: *IGCSE Geography* by P. Guinness & G. Nagle (Hodder Education, 2009), p.114

Figure 11.16 The Lower Ganges valley

The farming system

Paddy fields (flooded parcels of land) characterise lowland rice production (Figure 11.18). Water for irrigation is provided either when the Ganges floods or by means of gravity canals. At first, rice is grown in nurseries. It is then transplanted when the monsoon rains flood the paddy fields. The flooded paddy fields may be stocked with fish for an additional source of food. The main rice crop is harvested when the drier season begins in late October. The rice crop gives high yields per hectare. A second rice crop can then be planted in November, but water supply can be a problem in some areas for the second crop.

Water buffalo are used for work. This is the only draft animal adapted for life in wetlands. The water buffalo provide an important source of manure in the fields. However, the manure is also used as domestic fuel. The labour-intensive nature of rice cultivation provides work for large numbers of people. This is important in areas of very dense population where there are

Figure 11.18 Rice paddy field, Lower Ganges valley

limited alternative employment opportunities. The low incomes and lack of capital of these subsistence farmers mean that hand labour still dominates in the region. It takes an average of 2000 hours a year to farm 1 hectare of land. A high labour input is needed to:

- build the embankments (bunds) that surround the fields – these are stabilised by tree crops such as coconut and banana
- construct irrigation canals where they are required for adequate water supply to the fields
- plant nursery rice, plough the paddy field, transplant the rice from the nursery to the paddy field, weed and harvest the mature rice crop
- cultivate other crops in the dry season and possibly tend a few chickens or other livestock.

Rice seeds are stored from one year to provide the next year's crop. During the dry season when there may be insufficient water for rice cultivation, other crops such as cereals and vegetables are grown. Farms are generally small, often no more than 1 hectare. Many farmers are tenants and pay for use of the land by giving a share of their crop to the landlord.

Section 11.1 Activities

1 Describe the location of the Lower Ganges valley.
2 Why is rice cultivation in the area considered to be an intensive form of agriculture?
3 Explain why the physical environment provides good conditions for rice cultivation.
4 Describe the inputs, processes and outputs of this type of agriculture.

☐ Issues in the intensification of agriculture and the extension of cultivation

Agricultural production can be achieved in two ways, by:

- increasing the land under cultivation through, for example, irrigation, or extending farming onto marginal land
- increasing the yield per hectare when scientific advance allows such changes to occur.

The **intensification of agriculture** has occurred through the use of high-yielding crop varieties, fertilisers, herbicides and pesticides and irrigation. The result has been a substantial increase in global food production over the last 60 years. However, increasing agricultural production has not just been achieved by the more intensive farming of long-established agricultural land, but also by the **extension of cultivation** into previously unfarmed areas. This has occurred with varying degrees of success.

The industrialised farmlands of today all too frequently lack the wildflowers, birds and insects that lived there in the past. These sterilised landscapes provide relatively cheap food, but at high environmental cost. These costs are typically borne by the citizens of the countries concerned rather than by the producers. Land conversion and intensification can alter ecosystems to such an extent that serious local, regional and global consequences result:

- **local** – increased soil erosion, lower soil fertility, reduced biodiversity
- **regional** – pollution of groundwater, eutrophication of rivers and lakes
- **global** – impacts on global atmospheric conditions.

The intensification of agriculture can result in **soil degradation**. Soil degradation is a global process. It involves both the physical loss (erosion) and the reduction in quality of topsoil associated with nutrient decline and contamination. It has a significant impact on agriculture and also has implications for the urban environment, pollution and flooding. The loss of the upper soil horizons containing organic matter and nutrients and the thinning of **soil profiles** reduces crop yields on degraded soils. Soil degradation can cancel out gains from improved crop yields. The statistics on soil degradation make worrying reading:

- Globally, it is estimated that 2 billion hectares of soil resources have been degraded. This is equivalent to about 15 per cent of the Earth's land area. Such a scale of soil degradation has resulted in the loss of 15 per cent of world agricultural supply in the last 50 years.
- For three centuries ending in 2000, topsoil had been lost at the rate of 300 million tonnes a year. Between 1950 and 2000, topsoil was lost at the much higher rate of 760 million tonnes a year.
- During the last 40 years, nearly one-third of the world's cropland has been abandoned because of soil erosion and degradation.
- In Sub-Saharan Africa, nearly 2.6 million km² of cropland has shown a 'consistent significant decline' according to a March 2008 report by a consortium of agricultural institutions. Some scientists consider this to be a 'slow-motion disaster'.
- In the UK, 2.2 million tonnes of topsoil is eroded annually and over 17 per cent of arable land shows signs of erosion.
- It takes natural processes about 500 years to replace 25 millimetres of topsoil lost to erosion. The minimum soil depth for agricultural production is 150 millimetres. From this perspective, therefore, productive fertile soil can be considered a non-renewable, endangered ecosystem.

The Global Assessment of Human-induced Soil Degradation (GLASOD) is the only global survey of soil degradation to have been undertaken. Figure 11.19 is a generalised map of the findings of this survey. It shows that substantial parts of all continents have been affected by various types of soil degradation. The GLASOD calculation is that damage has occurred on 15 per cent of the world's total land area – 13 per cent light and moderate, with 2 per cent severe and very severe (Figure 11.20).

The International Forum of Soils, Society and Global Change in September 2007 referred to 'the massive degradation of land and soil around the world which is contributing to climate change and threatening food security'. The Forum noted that:

- At least a quarter of the excess carbon dioxide in the atmosphere has come from changes in land use, such as deforestation, in the last century.
- Without the cover of vegetation, land becomes more reflective. It also loses fertility and the capacity to support vegetation and agricultural crops.
- The Intergovernmental Panel on Climate Change should develop a special report on the link between land

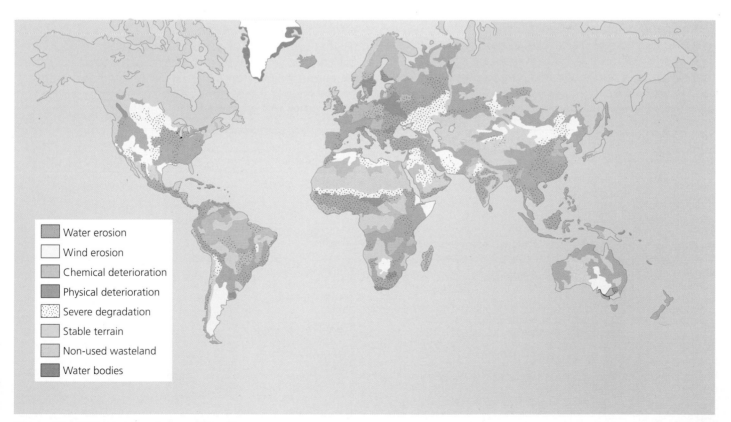

Figure 11.19 Worldwide soil degradation types

Legend:
- Water erosion
- Wind erosion
- Chemical deterioration
- Physical deterioration
- Severe degradation
- Stable terrain
- Non-used wasteland
- Water bodies

Figure 11.20 Infertile saline soil in the south of France

degradation and climate change. By addressing soils and protecting the land cover and vegetation, it is possible to obtain high value in terms of mitigating climate change.
- A better understanding of the capacity for carbon sequestration in soil is needed.
- Degradation of soil and land in already marginally productive land is a significant issue for many LICs, particularly in northern Africa, the Sahara region and

parts of Asia, including China. Many of these regions have fragile ecosystems where any human interventions can lead to serious degradation.

Research has shown that the heavy and sustained use of artificial fertiliser can result in serious soil degradation. In Figure 11.21, soil profile **a** illustrates the problems that can result. In contrast, soil profile **b** shows a much healthier soil treated with organic fertiliser. In the artificially fertilised soil, the ability of the soil to infiltrate water has been compromised by the breakdown of **soil aggregates** to fine particles that have sealed the surface. Pore spaces have been filled up by the fine soil material from the broken crumbs. This can result in ponding in surface depressions, followed by soil erosion.

It has been estimated that food production and consumption accounts for up to twice as many greenhouse emissions as driving vehicles. Figure 11.22 shows US data published in the *New Scientist*. The average US household's footprint for food consumption is 8.1 tonnes of carbon dioxide equivalent, compared with 4.4 tonnes from driving.

The environmental impact of the Green Revolution

Much of the global increase in food production in the last 50 years can be attributed to the Green Revolution, which took agro-industrialisation to LICs on a large scale. India was one of the first countries to benefit when a high-yielding variety seed programme (HVP) commenced in 1966–67. In terms of production, it was a turning point for Indian agriculture, which had virtually reached stagnation. The HVP introduced new hybrid varieties of five cereals: wheat, rice, maize, sorghum and millet. All were drought-resistant with the exception of rice, were

a Soil treated with artificial fertilisers and pesticides

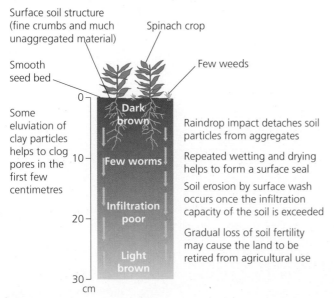

b Soil treated with organic fertiliser

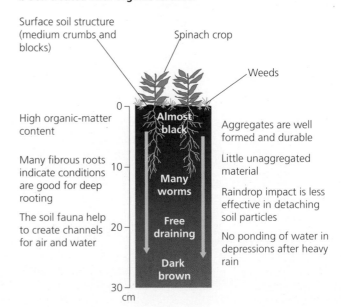

Figure 11.21 Two soil profiles

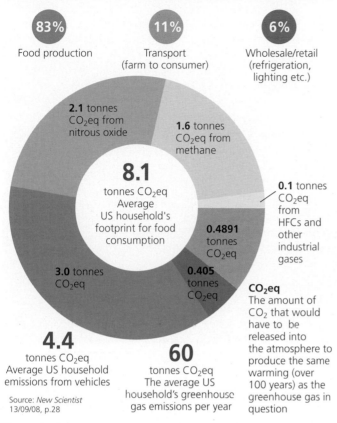

Household greenhouse gas emissions from food account for almost twice those produced by vehicle use. Most of this comes from the food production process itself, rather than food·miles, as is often believed

83%
Food production

11%
Transport
(farm to consumer)

6%
Wholesale/retail
(refrigeration, lighting etc.)

2.1 tonnes CO₂eq from nitrous oxide

1.6 tonnes CO₂eq from methane

8.1
tonnes CO₂eq
Average
US household's
footprint for food
consumption

0.1 tonnes CO₂eq from HFCs and other industrial gases

0.4891 tonnes CO₂eq

3.0 tonnes CO₂eq

0.405 tonnes CO₂eq

CO₂eq
The amount of CO₂ that would have to be released into the atmosphere to produce the same warming (over 100 years) as the greenhouse gas in question

4.4
tonnes CO₂eq
Average US household emissions from vehicles

60
tonnes CO₂eq
The average US household's greenhouse gas emissions per year

Source: *New Scientist* 13/09/08, p.28

Figure 11.22 Comparison of household greenhouse-gas emissions from food and vehicle use

very responsive to the application of fertilisers and had a shorter growing season than the traditional varieties they replaced. Although the benefits of the Green Revolution are clear, serious criticisms have also been made, many linked to the impact on the environment:

- High inputs of fertiliser and pesticide have been required to optimise production – this is costly in both economic and environmental terms.
- The problems of salinisation and waterlogged soils have increased, along with the expansion of the irrigated area, leading to the abandonment of significant areas of land.
- High chemical inputs have had a considerable negative effect on biodiversity.
- People have suffered ill-health due to contaminated water and other forms of agricultural pollution.

In the early 1990s, nutritionists noticed that even in countries where average food intake had risen, incapacitating diseases associated with mineral and vitamin deficiencies remained commonplace and in some instances had actually increased. The problem is that the high-yielding varieties introduced during the Green Revolution are usually low in minerals and vitamins.

Because the new crops have displaced the local fruits, vegetables and legumes that traditionally supplied important vitamins and minerals, the diet of many people in LICs is now extremely low in zinc, iron, vitamin A and some other micronutrients.

In India's Punjab, yield growth has flattened since the mid-1990s. Over-irrigation has resulted in a steep fall in the water table, now tapped by 1.3 million tube wells. Since the beginning of the Green Revolution in Asia, the amount of land under irrigation has tripled.

The Green Revolution has been a major factor in enabling global food supply to keep pace with population growth, but with growing concerns about a new food crisis, new technological advances may well be required to improve the global food-security situation.

Section 11.1 Activities

1 Describe the distribution of soil degradation types shown in Figure 11.19. Refer to all elements of the key and make reference to all continental areas.
2 Describe and explain the differences shown in the soil profiles illustrated in Figure 11.21.
3 Summarise the data presented in Figure 11.22.
4 Discuss the environmental impact of the Green Revolution.

11.2 The management of agricultural change

Agriculture remains vital to the lives of many individual people and communities, and to the economies of many countries, particularly in LICs. Jamaica is an example of a country where the management of agricultural change can be observed at both the level of the individual farm and the country a whole (Figure 11.23).

☐ Physical background

Jamaica has considerable variety of topography and geology (Figure 11.24). Approximately half of the island lies above 1000 metres, which has a significant influence on its various microclimates. The country has a highland interior formed by a series of mountain ranges along the major west-north-west to east-south-east axis of the island. The central mountain ranges form the main watershed for rivers, which drain either to the north or the south, except for the Plantain Garden River, which drains to the east. Flat coastal plains surround the central mountain ridge. The climate of Jamaica is mainly subtropical or tropical maritime.

☐ The importance of agriculture

Agriculture in Jamaica is dominated by the production of **traditional crops** such as sugar, bananas, coffee, cocoa

Figure 11.23 Large plantation of bananas in Jamaica

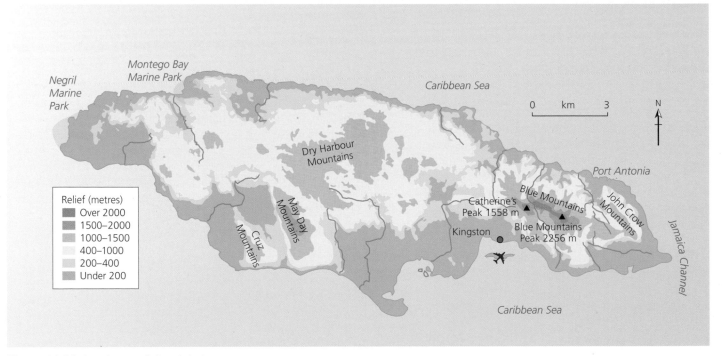

Figure 11.24 Jamaica – relief and drainage

and spices. In addition, a number of **non-traditional crops** including sweet potatoes, yams and hot peppers are cultivated for both domestic and international markets. In terms of livestock, Jamaica has well-developed beef, dairy and poultry sectors. The products of these, together with those of the pork and small ruminant industries, are mainly for domestic consumption. The maritime and inland fish sectors serve both domestic and export markets. Over the last two decades, the major export earner, sugar, has experienced a considerable decline. Both sugar and bananas in particular have had to contend with price and market insecurity as a result of preference erosion in the EU market. Additional problems for Jamaican agriculture have arisen in relation to non-traditional products such as milk, food aid and the dumping of surpluses on the local markets.

Although it has faced significant challenges, the agricultural sector continues to play an important role in terms of:

- food security
- employment
- income
- rural livelihoods (Figure 11.25).

Agriculture contributes 7 per cent to Jamaica's GDP and employs about 20 per cent of the workforce.

Figure 11.25 Jamaican farming scene – a smallholding

☐ Recent changes in Jamaican farming

Table 11.4 shows that the total amount of land in farming fell by almost 23 per cent between 1996 and 2007 as significant areas of **marginal land** were abandoned. Land devoted to crops declined by 13 per cent during this period, while land given over to pasture fell by a massive 49.6 per cent. The difficulties of making a living on marginal land were the main reason, as people sought other means of employment, particularly in urban areas.

A significant problem has been the removal of **preferential treatment** for bananas on the European market, which is creating greater competition and lowering prices. Some farmers no longer consider bananas a profitable venture and have stopped farming. This is particularly true of small farmers who are unable to achieve the economies of scale of their larger competitors. However, farmers have had to face other problems such as 'praedial larceny', by which farmers are robbed of their produce, in some cases even before the crops are mature. For small farmers, such theft can turn a modest profit into a loss, with resultant rural–urban migration. Disease is another problem. For example, Moko disease, which

Table 11.4 Area in farming in Jamaica, 1996 and 2007

Items	2007		1996		Change 1996–2007	
	Area in ha	% of total	Area in ha	% of total	Absolute change	% change
Total land in farming	**325 810**	**100.0**	**421 550**	**100.0**	**–95 740**	**–22.7**
Active farmland	202 727	62.2	273 229	64.8	–70 502	–25.8
Crops	154 524	47.4	177 580	42.1	–23 056	–13.0
Pasture	48 203	13.8	95 649	22.7	–47 446	–49.6
Inactive farmland	114 048	35.0	134 204	31.8	–20 157	–15.0
Ruinate and fallow	80 560	24.7	87 300	20.7	–6 740	–7.7
Woodland and other land on farm	33 488	10.3	46 905	11.1	–13 417	–28.6
Land identified to be in farming but no information reported	9 035	2.8	14 116	3.2		

Source: www.statinja.gov.jm

affects bananas and similar species such as plantains, has infected some farms and resulted in losses to farmers.

Climatic hazards often have a substantial impact on farming in Jamaica. In 2005, for example, agricultural GDP fell by 7.3 per cent. The reasons for this decline included:

- the long-term effects of Hurricane Ivan
- the drought that occurred between January and April 2005
- the impacts of Hurricanes Dennis and Emily and tropical storm Wilma in 2005, which caused combined losses of $994 million.

In addition, there have been economic and political difficulties. The European Commission proposed a 36 per cent cut in the price paid for raw sugar exports from African, Caribbean and Pacific countries, starting in 2006. In the 1950s, Jamaica had 20 working sugar factories, but by 2005 this number had fallen to eight.

☐ Policy responses

In response, the Jamaican government announced a new policy for a sustainable local sugar industry. The main elements of the policy were:

- to centre the industry around three products – raw sugar for export and domestic markets; molasses for the manufacture of rum; and ethanol for fuel
- to set a production target of 200 000 tonnes of raw sugar per year.

Commodity-specific policies for bananas included rationalisation of areas under production; provision of technical support for irrigation and extension; and the restructuring of the banana insurance scheme. For the cocoa industry, expansion in production, increased efficiencies and identification of more lucrative markets were the main strategies.

Jamaica has also produced a New Agricultural Development Plan, which aims to transform the farming sector by 2020. The main objectives of the Plan are to halt the decline of the agricultural sector, to restore productivity to agricultural resources and to ensure that farming communities provide meaningful livelihoods and living environments for those who depend on the agricultural sector. The New Agricultural Development Plan aims to increase production in eight key areas, through:

- The Small Ruminant Industry Development Project
- The National Organic Agriculture Project
- Protected Cultivation (Hydroponics)
- The Beekeeping Enhancement Project
- Marketing (Agribusiness Enhancement Project)
- The Fruit Tree Crop Development Project
- Ornamental Horticulture
- The Fisheries Development Project.

As exports of some traditional farm products have declined, the Jamaican government has tried to encourage **agricultural diversification**. This is exemplified by Table 11.5, which divides exports into 'traditional' and 'non-traditional'. Look at the agricultural products under these headings. Also take note of the manufacture of agricultural products.

Table 11.5 Exports of traditional and non-traditional commodities, January–December 2009

Commodities	Jan–Dec 2009
Total traditional exports	**55 026 594**
Agriculture	**3 440 439**
Banana	559
Citrus	149 080
Coffee	2 978 540
Cocoa	157 750
Pimento	154 509
Mining and quarrying	**40 645 392**
Bauxite	8 326 228
Alumina	32 302 855
Gypsum	16 309
Manufacture	**10 940 763**
Sugar	6 405 019
Rum	4 296 721
Citrus products	46 208
Coffee products	133 061
Cocoa products	59 753

Commodities	Jan–Dec 2009
Total non-traditional exports	**55 465 717**
Food	**10 538 829**
Pumpkins	31 319
Other vegetables and preparations thereof	210 663
Dasheen	122 669
Yams	1 650 806
Papayas	253 032
Ackee	1 199 038
Other fruits and fruit preparations	553 278
Meat and meat preparations	249 329
Dairy products and birds' eggs	568 153
Fish, crustaceans and molluscs	418 816
Baked products	951 342
Juices excluding citrus	604 530
Animal feed	459 184
Sauces	935 614
Malt extract and preparations thereof	333 266
Other food exports	1 776 465
Beverages and tobacco (excl. rum)	**4 672 930**
Non-alcoholic beverages	755 157
Alcoholic beverages (excl. rum)	3 912 774
Tobacco	4 999
Crude materials	**1 476 079**
Limestone	95 439
Waste and scrap metals	1 115 360
Other	265 280
Other	**38 777 879**
Mineral fuels, etc.	18 887 511
Animal and vegetable oils & fats	16 432
Chemicals (excl. ethanol)	2 546 690
Ethanol	15 151 290
Manufactured goods	1 401 815
Machinery and transport equipment	155 332
Wearing apparel	129 250
Furniture	63 388
Other domestic exports	421 261

Source: Statistical Institute of Jamaica

Poultry is an example of a farming sector in which significant benefits have accrued through:

- internal structural changes
- reorganisation of the production system
- the introduction of higher levels of technology.

The spice industry is another example. This traditionally operated at the cottage level, but with government encouragement the industry has been restructured and modernised to increase its share of international markets. The exploitation of **niche markets** has been a major aspect of the modernisation of Jamaican agriculture.

In March 2010, Jamaica's Minister of Agriculture, the Honourable Dr Christopher Tufton, announced that the government was working to prepare a new agriculture land-use policy for the island, with the aim of getting fallow fields back into production. Dr Tufton made this statement at a meeting held to discuss the Arable Lands Irrigated and Growing for the Nation (ALIGN) project. The ALIGN programme is intended to boost agricultural production in Jamaica by providing irrigation on arable land.

☐ Land degradation

Jamaica is having to address the issue of land degradation. A report published in the early 2000s stated: 'While land degradation in Jamaica is not as serious as in some parts of Africa or even like that in its Caribbean neighbour Haiti, it is a problem that must be confronted.' Some of the most seriously degraded areas of the island are in the southern coastal sections of the parishes of Clarendon, Manchester and St Catherine and particularly on the southern coastal border areas of Manchester and St Elizabeth.

☐ ICT and agriculture

The government has recognised the contribution ICT can make to enhancing the sector's efficiency and productivity. Current initiatives include:

- **Agri-Business Information System (ABIS)** – Recently developed, this computer-based information system collects and disseminates to interested parties information on crops, marketing, agricultural stakeholders and agricultural production.
- **Geographical Information Systems (GIS)** – The Forest Department and the Rural Physical Planning Division are currently using GIS as a tool in the mapping and management of Jamaica's forest and land resources. The private sector, in turn, is using GIS for the purpose of advertising and marketing agricultural products via the internet.

☐ Evaluation

The range of policies introduced by the government in recent years has undoubtedly helped to bring about beneficial changes in Jamaican agriculture. More efficient management and new agricultural technology have been introduced into both the traditional and non-traditional farming sectors. The product range has been broadened and more attention has been placed on marketing. ICT systems have played an increasing role in this push for modernisation. However, limited funding has meant that progress has not always been as rapid as hoped and climatic hazards have at times proved costly. The erosion of preferential treatment in EU markets has been a significant setback, although Jamaica did have advance warning this was going to happen. There is little that small countries like Jamaica can do in terms of world trade agreements and fluctuating international demand and prices. Nevertheless, this should not deter governments from making the best policy provisions they can.

> ### Section 11.2 Activities
>
> 1. With reference to Figure 11.24, describe the relief and drainage of Jamaica.
> 2. In what ways is the agricultural sector important to Jamaica?
> 3. Discuss the differences between Jamaica's traditional and non-traditional farm products.
> 4. How has the government tried to improve the fortunes of the country's agricultural sector?

Case Study: Kew Park farm, Jamaica

Kew Park farm is a mixed commercial farm in the west of Jamaica. It is located high in the hills of the parish of Westmoreland, overlooking the Great River valley that forms the border between Westmoreland and St James (Figures 11.26 and 11.27). Kew Park covers an area of about 385 hectares and is run along with Copse Mountain farm (about 425 hectaresa). The two farms together form one unit: Kew Park. This is a very hilly part of Jamaica. About 30 per cent of Kew Park can only be accessed on foot and about 15 per cent of the total area is not farmed at all. Good management has been essential for the farm's survival as an economic entity because of the physical hazards and economic obstacles the farm has had to face.

Most of the farmed area is allocated to beef cattle; much of the breeding research for the Jamaican Red Poll was conducted here. At present, there are five pedigree Jamaica Red Poll herds and two commercial herds on the farm – a total of about 700 animals. The cattle are raised extensively, but are confined in grass pastures by either barbed-wire fences or dry-stone walls. Other parts of the farm support a variety of agricultural activities:

- Above about 400 metres, an area of 16 hectares is planted with arabica coffee, producing the 'Estate' brand. Much of the primary processing of the coffee is done on the farm, but it is then sent to Kingston to be graded, roasted and packed. Kew Park has a licence to export the processed coffee, although much of it is sold locally. Kew Park has worked hard for 20 years to develop the quality of its coffee and its reputation. Decisions such as this can take many years really to pay off. The high quality of Kew Park coffee has resulted in the farm being one of a relatively ⟹

small number being granted a licence to export. The farm's website states: 'Kew Park Estate Coffee has been carefully expanded over the past 20 years to fit the coffee in to the land; working always for the long term sustainability of the farm, the people who work here, and the environment on which it depends.'

- 2 hectares are given over to citrus fruits (ortaniques). However, the fruit are not of prime quality as the climate is too wet and the trees are not well maintained.
- There are 2 hectares of lychees, which is a difficult crop to grow, but there is a good local market for the fruit.
- The farm supports some 2000 free-range chickens. The eggs are washed and packed on the farm and sold locally.

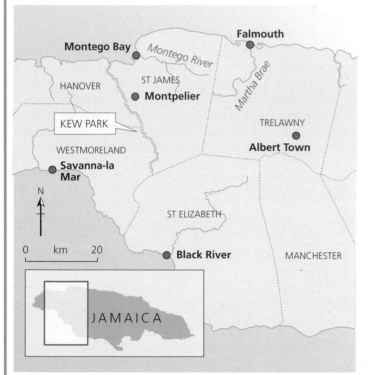

Figure 11.26 The location of Kew Park farm

Figure 11.27 Kew Park farm

Life in rural Jamaica is not easy and Kew Park provides the only full-time employment in the area. Wages are low. Some staff live in houses owned by the estate but most travel from their own homes nearby.

Farm managers have had to be constantly aware of the costs of all their inputs, such as labour, animal stock, seeds and machinery. The costs of the processing that is carried out on the farm also have to be calculated. Knowledge of local and more distant markets in terms of both access and price are important. Because Kew Park produces a number of farm products, the allocation of resources for different purposes must be done carefully. A more favourable price for one particular farm product may justify a larger share of farm resources if it is felt that the increase in price is not just a temporary upturn.

The farm also has to be aware of government agricultural policies and incentives. Often the balance between costs and revenue is marginal, which makes correct decision-making crucial.

The damage caused by a pest known as the coffee berry borer can eliminate the profit expected from a coffee crop in Jamaica. The female borer, which is a tiny beetle just 1.5 millimetres long, drills into coffee berries and lays its eggs inside. There can be up to 50 per berry. Once hatched, the young borers eat the beans from within, leaving them worthless. Combating the borer is an expensive task. Existing methods include traps, parasitic insects and insecticides. The most effective insecticide is endosulfan, which is highly toxic. Jamaica's Coffee Industry Board phased out its use in 2010.

However, a more natural solution may be at hand. Migratory warblers (birds) spend winter in Jamaica and like to feast on the coffee berry borers. Research has estimated that growers who enlist these birds to control berry borers could save as much as $96/hectare every year. This should create an economic incentive for coffee producers to manage their farms in ways that aid bird conservation. In particular, this means planting or maintaining pockets of trees instead of clear cutting, which has been the traditional method. Kew Park farm is looking at ways to attract more birds to the coffee fields, including preserving more woodland.

Figure 11.28 shows the section of the farm's website devoted to 'Kew Park Essentials'. Many of the traditional herbal remedies, spicy foods and refreshers date back to the indigenous Tainos (Arawaks); others were introduced by African slaves, indentured labourers from China and India and other migrants to the island. Even though conventional medicine is well established in Jamaica, folk medicine is still widely practised, particularly in the rural areas. Popular remedies include some of the herbs and spices grown at Kew Park. Although production of these products is not new, the marketing of them has changed considerably in recent years, reaching an international audience through the farm's website and other channels. This aspect of the farm's production has accounted for an increasing proportion of its income in recent years.

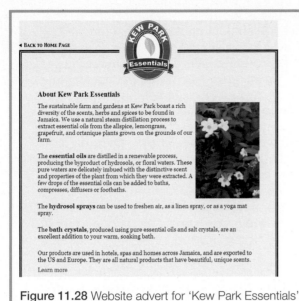

Figure 11.28 Website advert for 'Kew Park Essentials'

Evaluation

The relief and location of the farm have always presented certain challenges to Kew Park. The farm is in a hilly part of the country, some distance from Kingston, the capital city. However, careful locational choices within the farm and management of the farm's product range, along with incisive marketing, have built a good reputation for the farm. The farm has responded to international markets by extending the range of its non-traditional products in particular. Its use of ICT has been an important part of this process. However, individual farms have no influence on national and international policies and thus they must be able to react to policy changes at both these levels, as well as dealing with the challenges presented by the physical environment.

Section 11.2 Activities

1 Describe the location of Kew Park farm.
2 Discuss the farm's product range.
3 What are the main problems the farm has had to contend with?

11.3 Manufacturing and related service industry

☐ Industrial location: influential factors

Every day, decisions are made about where to locate industrial premises, ranging from small workshops to huge industrial complexes. In general, the larger the company, the greater the number of real alternative locations available. For each possible location, a wide range of factors can affect total costs and thus influence the decision-making process. The factors involved in industrial location differ from industry to industry and their relative importance is subject to change over time. These factors can be broadly subdivided into physical and human. Table 11.6 provides a brief summary and introduction to this topic.

Table 11.6 Factors of industrial location

Physical factors	Human factors
Site – the availability and cost of land: large factories in particular need flat, well-drained land on solid bedrock; an adjacent water supply may be essential **Raw materials** – industries requiring heavy and bulky raw materials that are expensive to transport generally locate as close to these raw materials as possible **Energy** – at times in the past, industry needed to be located near fast-flowing rivers or coal mines; today, electricity can be transmitted to most locations – however, energy-hungry industries, such as metal smelting, may be drawn to countries with relatively cheap hydro-electricity **Natural routeways and harbours** – these were essential in the past and are still important today as many modern roads and railways still follow natural routeways; natural harbours provide good locations for ports, and industrial complexes are often found at ports **Climate** – some industries such as the aerospace and film industries benefit directly from a sunny climate; indirect benefits, such as lower heating bills and a more favourable quality of life, may also be apparent	**Capital (money)** – business people, banks and governments are more likely to invest money in some areas than others **Labour** – increasingly, it is the quality and cost of labour rather than the quantity that are the key factors; the reputation, turnover and mobility of labour can also be important **Transport and communications** – transport costs are lower in real terms than ever before but remain important for heavy, bulky items; accessibility to airports, ports, motorways and key railway terminals may be crucial factors for some industries **Markets** – the location and size of markets is a major influence for some industries **Government influence** – government policies and decisions can have a big direct and indirect impact; governments can encourage industries to locate in certain areas and deny them planning permission in others **Quality of life** – highly skilled personnel who have a choice about where they work will favour areas where the quality of life is high

Raw materials

Industries that use raw materials directly, such as oil refining and metal smelting, are known as **processing industries**. Once the dominant type of manufacturing, processing industries are in a minority today as most industries now use components and parts made by other firms.

The processes involved in turning a raw material into a manufactured product usually result in **weight loss**, so that the transport costs incurred in bringing the raw materials to the factory will be greater than the cost of transporting the finished product to market. If weight loss is substantial, the location of the factory will be drawn towards its most costly-to-transport raw material(s). Figure 11.29 shows a simple example of weight loss, where 2 tonnes each of two raw materials are required to manufacture 1 tonne of the finished product.

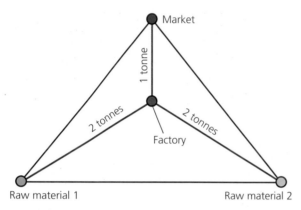

Figure 11.29 Weight-loss diagram

The clearest examples of this influence are where one raw material only is used. In the UK, sugar beet refineries are centrally located in crop-growing areas because there is a 90 per cent weight loss in manufacture (Figure 11.30). Frozen pea factories are also strategically located in the growing areas. Here, the weight loss is much less, but to achieve prime quality the peas must be processed very quickly after picking.

Figure 11.30 Sugar refinery in France

In many processing industries, technological advance has reduced the amount of raw material required per finished product and in some cases less bulky and cheaper substitutes have been found. Thus, across the industrial board as a whole, the raw material requirement per unit of finished product has been reduced.

Tidewater locations are particularly popular with industries that use significant quantities of imported raw materials, for example flour milling, food processing, chemicals and oil refining. Tidewater locations are **break-of-bulk** points where cargo is unloaded from bulk carriers and transferred to smaller units of transport for further movement. However, if raw materials are processed at the break-of-bulk point, significant savings in transport costs can be made.

Markets

Where a firm sells its products may well have a considerable influence on where the factory is located. Where the cost of distributing the finished product is a significant part of total cost and the greater part of total transport costs, a market location is logical. A small number of industries, including soft drinks and brewing, are **weight gaining** and are thus market-oriented in terms of location. The heavy weight gain for both of these industries comes from the addition of water, a ubiquitous resource. The baking industry is also cited frequently as an example of weight gain, but here it is largely a case of increase in volume rather than weight, although the impact on transport costs is similar.

However, there are other reasons for market location. Industries where fashion and taste are variable need to be able to react quickly to changes demanded by their customers. Clear examples of this can be seen on both the national and international scales. In terms of the latter, one of the reasons that the global car giants spread themselves around the world is to ensure that they can produce vehicles that customers will buy in the different world regions. Ford, for example, recognised a long time ago that Europeans prefer smaller cars than Americans do.

Energy

The **Industrial Revolution** in many countries was based on the use of coal as a fuel, which was usually much more costly to transport than the raw materials required for processing. It is therefore not surprising that outside of London most of Britain's industrial towns and cities developed on coalfields or at ports nearby. The coalfields became focal points for the developing transport networks – first canals, then rail, and finally road. The investment in both **hard** and **soft infrastructure** was massive, so even when new forms of energy were substituted for coal, many industries remained at their coalfield locations, a phenomenon known as **industrial inertia**. Apart from the advantages of the infrastructure being in place, the cost of relocating might be prohibitive.

Also, a certain number of new industries have been attracted to urban areas on coalfields because of the acquired advantages available, such as a pool of skilled labour and the existing network of linkages between firms. However, overall the coalfields have suffered considerable economic distress due to the decline of coal and the traditional industries associated with it. In many HICs, these areas are now the main 'problem' regions.

During the twentieth century, the construction of national electricity grids and gas pipeline systems made energy virtually a ubiquitous resource in HICs (Figure 11.31). As a result, most modern industry is described as **footloose**, meaning that it is not tied to certain areas because of its energy requirement or other factors. However, there are some industries that are constrained in terms of location because of an extremely high energy requirement. For example, the lure of low-cost hydro-electric power has resulted in a huge concentration of electro-metallurgical and electro-chemical industries in southern Norway.

Figure 11.31 Oil storage, Parry Sound, Ontario, Canada

Section 11.3 Activities

1 **a** What are raw materials?
   b How can raw materials influence industrial location?
2 For what reasons are companies likely to choose a market location?
3 Explain the importance of break-of-bulk points.
4 Discuss industrial inertia as a factor in industrial location.
5 What are footloose industries?

Transport

Although it was once a major locational factor, the share of industry's total costs accounted for by transportation has fallen steadily over time. For example, for most manufacturing firms in the UK, transportation now accounts for less than 4 per cent of total costs. The main reasons for this reduction are:

- major advances in all modes of transport
- great improvements in the efficiency of transport networks
- technological developments moving industry to the increasing production of higher value/lower bulk goods.

The cost of transport has two components: fixed costs and line-haul costs. **Fixed (terminal) costs** are accrued by the equipment used to handle and store goods, and the costs of providing the transport system. **Line-haul costs** refer to the cost of actually moving the goods and are largely composed of fuel costs and wages. In Figure 11.32, the costs of the main methods of freight transport are compared. While water and pipeline transport have higher fixed costs than rail (Figure 11.33) and road, their line-haul costs are significantly lower. Air transport, which suffers from both high fixed and line-haul costs, is only used for high-value freight or for goods such as flowers that are extremely perishable. Other factors affecting the cost of transport are:

- **the type of load carried** – perishable and breakable commodities that require careful handling are more costly to move than robust goods such as iron ore and coal
- **the type of journey** – those that involve transferring cargo from one mode of transport to another are more costly than those using the same mode of transport throughout
- **the degree of competition** within and between the competing modes of transport.

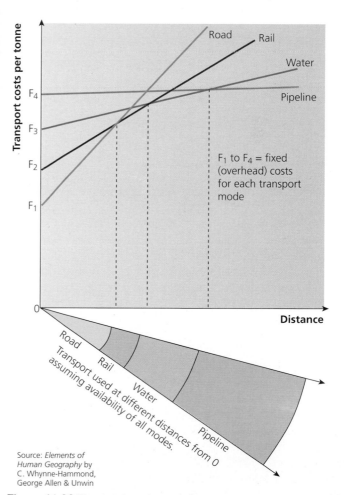

Source: *Elements of Human Geography* by C. Whynne-Hammond, George Allen & Unwin

Figure 11.32 Transport costs and distance

Figure 11.33 Timber being transported on the Trans-Siberian railway, Russia

Figure 11.34 A large capital input was required to build this container port in Indonesia

Land

The space requirements of different industries, and also of firms within the same industry, vary enormously. Technological advance has made modern industry much more space-efficient than in the past. However, modern industry is horizontally structured (on one floor) as opposed to, for example, the textile mills of the nineteenth century, which had four or five floors. In the modern factory, transportation takes up much more space than it used to – for example, consider the area required to park cars for a firm employing 300 people.

During the Industrial Revolution, entrepreneurs had a relatively free choice of where to locate in terms of planning restrictions, providing of course that they could afford to purchase the site they wanted. However, with the passage of time more and more areas have become unavailable to industry, mainly in an effort to conserve the environment. Areas such as National Parks, Country Parks and Areas of Special Scientific Interest now occupy a significant part of most countries. In urban areas, land-use zoning places a considerable restriction on where industry may locate, and green belts often prohibit location at the edge of urban areas.

Capital

Capital represents the finance invested to start up a business and to keep it in production. That part of capital invested in plant and machinery is known as 'fixed capital' as it is not mobile, compared with 'working capital' (money). Capital is obtained either from shareholders (share capital) or from banks or other lenders (loan capital). Some geographers also use the term 'social capital', which is the investment in housing, schools, hospitals and other amenities valued by the community, which may attract a firm to a particular location.

In the early days of the Industrial Revolution in present-day HICs, the availability of capital was geographically constrained by the location of the major capital-raising centres and by limited knowledge – and thus confidence – about untested locations. It was thus one of the factors that led to the clustering of industry. In the modern world, the rapid diffusion of information and the ability to raise and move capital quickly within and across international borders means that this factor has a minimal constraining influence in HICs today. However, in less developed economies the constraints of capital are usually greater, depending on the level of economic development. It is the perceived risk that is the vital factor. The political unrest that has affected so many African countries in recent decades has made it very difficult for these nations to raise the amount of capital desired.

Virtually all industries have over time substituted capital for labour in an attempt to reduce costs and improve quality. Thus in a competitive environment, capital has become a more important factor in industry. In some industries, the level of capital required to enter the market with a reasonable chance of success is so high that just a few companies monopolise the market. This has a major influence on the geography of manufacturing. The aircraft industry, for instance, has a massive barrier to entry in terms of the capital required to compete successfully.

The issue of foreign direct investment (FDI) is considered in detail in Section 14.2, pages 466–68.

Section 11.3 Activities

1 Why has the relative cost of transport declined significantly over time for most industries?
2 Distinguish between fixed transport costs and line-haul costs.
3 Discuss 'land' as a factor in industrial location.
4 How does capital influence industrial location?

Labour

The interlinked attributes of labour that influence locational decision-making are cost, quality, availability and reputation.

Although all industries have become more capital-intensive over time, labour still accounts for over 20 per cent of total costs in manufacturing industry. The cost of labour can be measured in two ways: as wage rates and as unit costs (Figure 11.35). The former is simply the hourly or weekly amount paid to employees, while the latter is a measure of productivity, relating wage rates to output. Industrialists are mainly influenced by unit costs, which explains why industry often clusters where wages are higher rather than in areas where wage rates are low. It is frequently, although not always, the high quality and productivity of labour that pushes up wages in an area. In such an area, unit costs may well be considerably lower than in an economically depressed area with poor-quality labour and lower wage rates. Certain skills sometimes become concentrated in particular areas, a phenomenon known as the **sectoral spatial division of labour**. As the reputation of a region for a particular skill or set of skills grows, more firms in that particular economic sector will be attracted to the area.

Variation in wage rates can be identified at different scales. By far the greatest disparity is at the global scale. The low wages of LICs with reasonable enough levels of skill to interest foreign companies has been a major reason for transnational investment in regions such as South East Asia and Latin America. A filter-down of industry to lower and lower wage economies can be recognised in particular in Asia –this topic is examined in more detail in Section 14.2, pages 470–71.

Recent analyses of labour costs in manufacturing have highlighted the wide variations in non-wage labour costs, which include employer social security contributions, payroll taxes, holiday pay, sick leave and other benefits.

The availability of labour as measured by high rates of unemployment is not an important location factor for most industries. The regions of the UK that have struggled most to attract new industry are the traditional industrial areas, which have consistently recorded the highest unemployment rates in the country. In such regions, although there are many people available for work, they frequently lack the skills required by modern industry. The physical dereliction and the social problems generated by unemployment also act to deter new investment. Where availability really has an impact is in sparsely populated areas because large prospective employers know that they will struggle to assemble enough workers with the skills demanded. Such regions are therefore often ruled out at the beginning of the locational search.

Figure 11.36 Factory producing tapestries in Vietnam – this is a labour-intensive process

That there have always been considerable regional differences in unemployment in the UK, a relatively small country, indicates that the **geographical mobility of labour** is limited. A major factor impeding the movement of labour from region to region is the huge differential in the cost of housing between the South East and the traditional industrial areas. In general, the degree of geographical mobility increases with skill levels and qualifications. It is the most able and financially secure that can best overcome the obstacles to mobility.

People can of course move from one type of job to another within the same town or region. Such

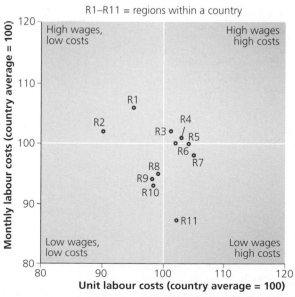

R1–R11 = regions within a country

Figure 11.35 Regional variations in labour costs

Source: *Advanced Geography: Concepts & Cases* by P. Guinness & G. Nagle (Hodder Education, 1999), p.147

movement is referred to as **occupational mobility**. However, like geographical mobility, it is limited in extent. People who have been employed in heavy industry in particular often find it very difficult to adjust to a working environment that is less physically demanding but requires much more in terms of concentration.

The reputation of a region's labour force can influence inward investment. Regions with militant trade unions and a record of work stoppages are frequently avoided in the locational search. In the USA, manufacturing firms often avoid states in the north-east where trade union membership is high, favouring instead the south and the west where union influence is minimal. Trade union membership in most countries has weakened in recent decades for two main reasons:

- Many governments have passed legislation to restrict the power of unions.
- The decline in employment in manufacturing, the historic nucleus of trade unionism, has had severe implications for membership; unions are particularly unpopular in Asia.

Economies of scale: internal and external

Both internal and external economies of scale can be recognised. Internal economies of scale occur when an increase in production results in a lowering of unit costs. This is a major reason why firms want to increase in size. The reduced costs of production can be passed on to customers and in this way a company can increase its market share. Alternatively, it can increase its profits. Economists recognise five types of **internal economies of scale**:

- **Bulk-buying economies** – as businesses grow, their bargaining power with suppliers increases.
- **Technical economies** – businesses with large-scale production can use more advanced machinery or use existing machinery more efficiently. A larger firm can also afford to invest more in research and development (R&D) and in ICT.
- **Financial economies** – larger firms find it easier to find potential lenders and to raise money at lower interest rates.
- **Marketing economies** – as a business gets larger, it is able to spread the cost of marketing over a wider range of products and sales, thus cutting the average marketing cost per unit.
- **Managerial economies** – as a company grows, there is more potential for managers to specialise in particular tasks and thus become more efficient.

However, it is possible that an increase in production at some stage might lead to rising unit costs. If this happens, **diseconomies of scale** are said to exist. In Figure 11.37, the average cost of production at output Q is C_2. Increasing output beyond this point reduces unit costs and thus economies of scale are achieved. This continues until output Q_2 is reached, when the lowest unit costs of production are achieved (C). Beyond this point, unit costs rise and diseconomies of scale are occurring.

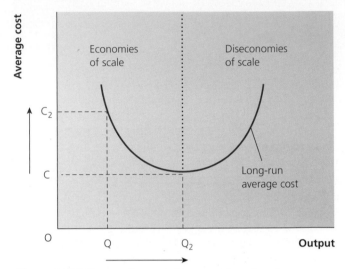

Figure 11.37 Economies and diseconomies of scale

External economies of scale (agglomeration economies) are the benefits that accrue to a firm by locating in an established industrial area. External economies of scale can be subdivided into:

- **urbanisation economies**, which are the cost savings resulting from urban location due to factors such as the range of producer services available and the investment in infrastructure already in place, and
- **localisation economies**, which occur when a firm locates close to suppliers (backward linkages) or to firms that it supplies (forward linkages) – this reduces transport costs, allows for faster delivery and facilitates a high level of personal communication between firms.

However, when an urban-industrial area reaches a certain size, urbanisation diseconomies may come into play. High levels of traffic congestion may push up transport costs. Intense competition for land will increase land prices and rents. If the demand for labour exceeds the supply, wages will rise. So locating in such a region may no longer be advantageous, with fewer new firms arriving and some existing firms relocating elsewhere. In the USA, such a process has occurred in the Santa Clara valley (Silicon Valley), with entrepreneurs looking in particular at the less crowded mountain states such as Arizona and Colorado (Figure 11.38).

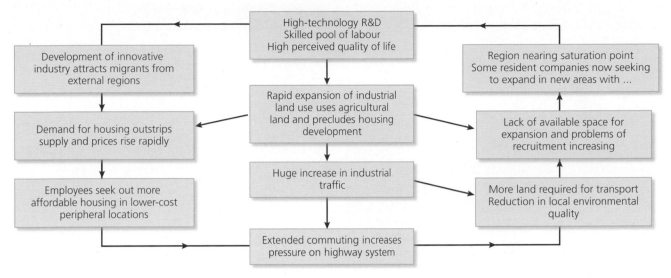

Figure 11.38 Model illustrating the problems of rapid growth in the Santa Clara valley (Silicon Valley)

Government policies

In the old-style centrally planned economies of the communist countries or former communist countries, the influence of government on industry was absolute. In other countries, the significance of government intervention has depended on:

- the degree of public ownership
- the strength of regional policy in terms of restrictions and incentives.

Governments influence industrial location for economic, social and political reasons. Regional economic policy largely developed after the Second World War, although examples of legislation with a regional element can be found before this time. There is a high level of competition both between countries and between regions in the same country to attract FDI.

Technology

Technology can influence industrial location in two main ways:

- The level of technological development in a country or region in terms of infrastructure and human skills has a major impact on the type of industry that can be attracted.
- Technological advance may induce a company or industry to move to alternative locations that have now become more suitable due to the new developments in technology.

Figure 11.39 is a useful summary of the different stages of technological development. Advances in technology can stimulate new industrial clusters, as has happened with biotechnology in a number of countries.

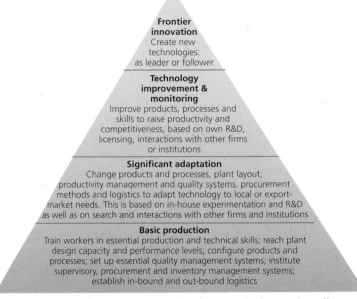

Figure 11.39 Stages of technology development by innovation effort

Industrial inertia

The importance of the factors responsible for the location of a particular industry may diminish over time. What was once a profitable location may become less so. However, many factories remain in locations that are no longer the best places to be because of the cost of moving and for other reasons such as tradition. This phenomenon is known as industrial inertia. Examples of industrial inertia are often found in industrial towns that developed on coalfields that no longer produce coal. Although coal as a source of energy was often the initial reason for factories locating in a coalfield town, the investment in hard and soft infrastructure over a long period of time has brought new advantages to such settlements. These might include improved accessibility, purpose-built industrial estates and modern hospitals and schools.

However, some industries are affected by changes of such magnitude that they have no option but to close down (Figure 11.40).

Figure 11.40 Abandoned whaling station, island of South Georgia, South Atlantic Ocean

Section 11.3 Activities

1 Distinguish between wage rates and unit costs.
2 Define the sectoral spatial division of labour.
3 With regard to labour, briefly discuss the impact on industrial location of cost, quality, availability and reputation.
4 Explain internal economies of scale.
5 How can diseconomies of scale occur?
5 What are external economies of scale?
6 Briefly review the impact of **a** government policies and **b** technology on industrial location.

Case Study: Slovakia – the changing location of EU car manufacturing

Car manufacturing is one of the world's largest industries. Within the EU, investment in car manufacturing has shifted from western to eastern Europe in recent years as countries like Slovakia have joined the EU (Figure 11.41). This is because eastern EU countries like Slovakia can manufacture cars at a lower cost than western EU countries.

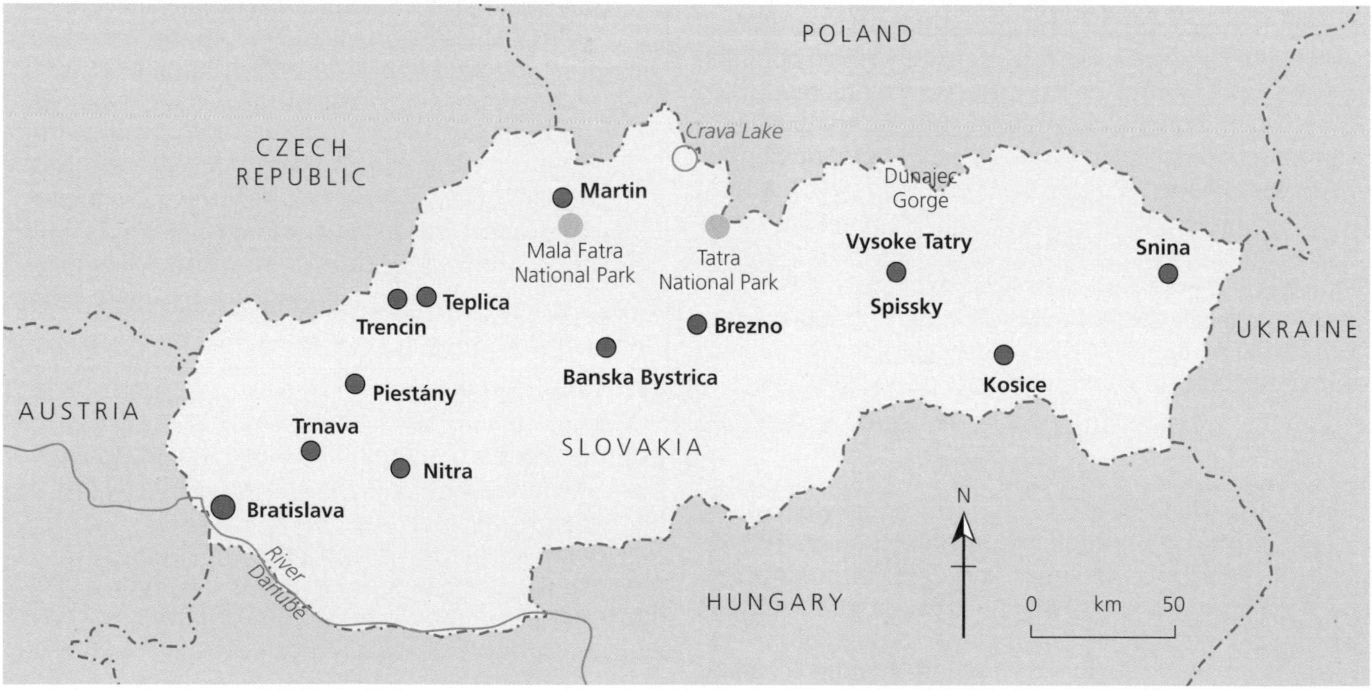

Figure 11.41 Slovakia

The location factors that have attracted the car industry to Slovakia are:

- relatively low labour costs
- low company taxation rates
- a highly skilled workforce, particularly in areas that were once important for heavy industry
- a strong work ethic, resulting in high levels of productivity
- low transport costs because of proximity to western European markets
- very low political risk because of the stable nature of the country
- attractive government incentives due to competition between Slovakia and other potential receiving countries
- good infrastructure in and around Bratislava and other selected locations
- an expanding regional market for cars as per person incomes increase.

Volkswagen expands

Prior to EU membership, Slovakia already boasted a Volkswagen (VW) plant with an output of 250 000 cars a year. The Bratislava plant is one of the top three Volkswagen factories in the world, producing the Polo, the Touareg and the SEAT Ibiza.

In addition to its car manufacturing plant in Bratislava, which was founded in 1991, VW also has a plant manufacturing components in Martin, which opened in 2000. Between the two plants, VW employs 8700 workers. A number of companies supplying parts to VW have also opened up in Slovakia.

Other recent investment

In 2006, Hyundai–Kia opened a major car factory in Slovakia. The location of the factory is near Žilina, 200 kilometres north-east of Bratislava. As with other large car plants, it is attracting some of its main suppliers to locate nearby. With its seven suppliers, the total investment is estimated to be $1.4 billion.

In 2006, Peugeot opened a large new car plant in Trnava, 50 kilometres from Bratislava. When it reaches maximum production, this state-of-the-art plant will export 300 000 cars a year to western Europe and to other parts of the world. Production reached 248 000 in 2013.

Slovakia produced a record 980 000 cars in 2013. This accounted for just over 40 per cent of Slovakia's industrial output. The car industry in Slovakia now employs more than 60 000 people.

Section 11.3 Activities

1 Why have Slovakia and other eastern EU countries been so keen to attract foreign car manufacturers?
2 Discuss the reasons for such a high level of investment in car manufacturing in Slovakia by foreign TNCs.

☐ Industrial agglomeration and functional (industrial) linkages

Industrial agglomeration is the clustering together and association of economic activities in close proximity to one another. Agglomeration can result in companies enjoying the benefits of external economies of scale (Figure 11.42). This means the lowering of a firm's costs due to external factors. For example, the grouping together of a number of companies may encourage local government to upgrade the transport infrastructure and attract bus companies to run new services. Companies may be able to share certain costs, such as security and catering. Such benefits are greatly increased if they actually do business together.

Alfred Weber published his *Theory of the Location of Industries* in 1909. At that turn of the century, transportation was a much greater element of total industrial costs than it is today, and for many industries it was the major cost. Thus it is not surprising that Weber developed his theory around the cost of transportation. However, he did recognise that other elements of total cost could also vary, particularly labour and the savings associated with agglomeration. Containerisation (Figure 11.43) has been a major factor in reducing transport costs.

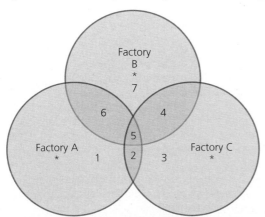

Source: *Advanced Geography: Concepts & Cases* by P. Guinness & G. Nagle (Hodder Education, 1999), p.152

Figure 11.42 Agglomeration economies

Figure 11.43 A container ship, Seattle dockside

a Vertical (or simple chain) linkages

The raw material goes through several successive processes:

b Horizontal (or simple origin) linkages

An industry relies on several other industries to provide its component parts:

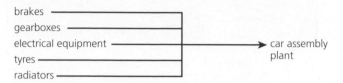

c Diagonal linkages

An industry makes a component which can be used subsequently in several industries :

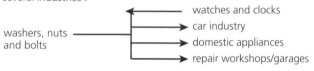

d Technological linkages

A product from one industy is used subsequently as a raw material by other industries:

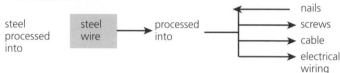

Figure 11.44 Types of industrial linkage

Weber referred to the savings that could be made when firms located together as 'agglomeration economies'. In Figure 11.42, the least transport cost location and the critical isodapanes for agglomeration for three factories are shown. Only in area 5 will all three factories benefit from agglomerating there.

The success of one company may attract other companies from the same or industry groups. Alternatively, a number of smaller firms may combine to produce components for a larger product.

Industrial (functional) linkages are the contacts and flows of information between companies that can happen more cheaply and easily when companies locate in close proximity. Three types of linkage are generally recognised:

- **backward linkages** – to firms providing raw materials, components and services needed in its production processes and activities
- **forward linkages** – to firms further processing the product or using it as a component part
- **horizontal linkages** – with other companies involved in the same processes or production, for example collaboration in research or marketing.

Figure 11.44 shows examples of the linkages that can induce companies to locate together.

Industrial estates

An **industrial estate** is an area zoned and planned for the purpose of industrial development. Industrial estates are also known as industrial parks and trading estates. A more 'lightweight' version is the business park or office park, which has offices and light industry, rather than larger-scale industry.

Industrial estates can be found in a range of locations, from inner cities to rural areas. In inner cities, they tend to be relatively small, but nevertheless important to local employment. Industrial estates are usually located close to transport infrastructure, especially where more than one transport mode meet. The logic behind industrial estates includes:

- concentrating dedicated infrastructure in a delimited area to reduce the per-business expense of that infrastructure
- attracting new business by providing an integrated infrastructure in one location
- separating industrial uses from residential areas to try to reduce the environmental and social impact of the industrial uses
- providing for localised environmental controls that are specific to the needs of an industrial area
- eligibility of industrial estates for grants and loans under regional economic development policies.

Export processing zones

There are a number of different types of **export processing zones** (EPZs), including free trade zones, special economic zones, bonded warehouses and free ports. The International Labour Organization (ILO) has defined EPZs as 'industrial zones with special incentives set up to attract foreign investors, in which imported materials undergo some degree of processing before being re-exported'. This can also include electronic data. EPZs have evolved from initial assembly and simple processing activities to include high-tech and science parks, finance zones, logistics centres and even tourist resorts. Table 11.7 summarises the different types of EPZ.

Table 11.7 Export processing zones – types of zones

	Trade	Manufacturing			Services		
	Free port	Special economic zone	Industrial free zone/EPZ	Enterprise zone	Information processing zone	Financial services zone	Commercial free zone
Physical characteristics	Entire city or jurisdiction	Entire province, region or municipality	Enclave or industrial park	Part of city or entire city	Part of city or 'zone within zone'	Entire city or 'zone within zone'	Warehouse area, often adjacent to port or airport
Economic objectives	Development of trading centre and diversified economic base	Deregulation; private sector investment in restricted area	Development of export industry	Development of SMEs in depressed areas	Development of information processing centre	Development of offshore banking, insurance, securities hub	Facilitation of trade and imports
Duty-free goods allowed	All goods for use in trade, industry, consumption	Selective basis	Capital equipment and production inputs	No	Capital equipment	Varies	All goods for storage and re-export of imports
Typical activities	Trade, service, industry, banking, etc.	All types of industry and services	Light industry and manufacturing	All	Data processing, software development, computer graphics	Financial services	Warehousing, packaging, distribution, trans-shipment
Incentives: Taxation Customs duties Labour laws Other	Simple business start-up; minimal tax and regulatory restraints; waivers with regard to termination of employment and overtime; free repatriation of capital, profits and dividends; preferential interest rates	Reduced business taxes; liberalised labour codes; reduced foreign exchange controls; no specific advantages – trade unions are discouraged within the SEZ	Profits tax abatement and regulatory relief; exemption from foreign exchange controls; free repatriation of profits; trade union freedom restricted despite EPZs being required to respect national employment regulation; 15 years' exemptions on all taxes (maximum)	Zoning relief; simplified business registration; local tax abatement; reduction of licensing requirement; trade unions are prohibited; government mandated liberal policies on hiring and firing of workers	De-monopolisation and deregulation of telecoms; access to market-priced INTELSAT services; a specific authority manages labour relations; trade union freedom restricted	Tax relief, strict confidentiality; deregulation of currency exchange and capital movements; free repatriation of profits	Exemption from import quotas; reinvested profits wholly tax-free
Domestic sales	Unrestricted within free port; outside free port, upon payment of full duty	Highly restricted	Limited to small portion of production			Limited to small portion of production	Unlimited, upon payment of full duty
Other features	Additional incentives and streamlined procedures	Developed by socialist countries	May be extended to single-factory sites				
Typical examples	Hong Kong (China), Singapore, Bahamas free port, Batam Labuan, Macao	China (southern provinces incl. Hainan and Shenzhen)	Ireland, Taiwan (China), Malaysia, Dominican Republic, Mauritius, Kenya, Hungary	Indonesia, Senegal	India (Bangalore), Caribbean	Bahrain, Dubai, Caribbean, Turkey, Cayman Islands	Jebel Ali, Colon Miami (US FTZ), Mauritius, Iran

Source: www.ilo.org

Table 11.8 shows the considerable global increase in number of EPZs, rising from 79 in 1975 to 3500 in 2006. Asia and Central America have the largest share of employment in EPZs. Apart from China, which has 40 million people working in EPZs, the rest of Asia has 15 million people employed in EPZs. In Central America, 5 million workers come into this category.

Table 11.8 The development of export processing zones

Year	1975	1986	1997	2002	2006
Number of countries with EPZs	25	47	93	116	130
Number of EPZs or similar types of zone	79	176	845	3000	3500
Employment (millions)	n.a.	n.a.	22.5	43	66
– of which China	n.a.	n.a.	18	30	40
– of which other countries with figures available	0.8	1.9	4.5	13	26

☐ The formal and informal sectors of employment

The concept of the informal sector was introduced into international usage in 1972 by the ILO. Jobs in the **formal sector** will be known to the government department responsible for taxation and to other government offices. Such jobs generally provide better pay and much greater security than jobs in the **informal sector** (Figure 11.45). Fringe benefits such as holiday and sick pay may also be available. Formal sector employment includes health and education service workers, government workers and people working in established manufacturing and retail companies.

In contrast, the informal sector is that part of the economy operating outside official government recognition. Employment is generally low-paid and often temporary and/or part-time in nature. While such employment is outside the tax system, job security will be poor with an absence of fringe benefits. About three-quarters of those working in the informal sector are employed in services. Typical jobs are shoe-shiners,

Figure 11.45 Informal sector – selling cigarettes on the street in Ulaanbaatar

street food stalls, messengers, repair shops and market traders. Informal manufacturing tends to include both the workshop sector, making for example cheap furniture, and the traditional craft sector. Many of these goods are sold in bazaars and street markets.

The government estimates that about 5 per cent of all employment in the UK is in the informal sector. This usually occurs when people insist on being paid in cash and do not declare this to the Inland Revenue. Examples may be window cleaners, part-time bar staff, cleaners and builders. In LICs, the informal sector may account for up to 40 per cent of the total economy.

A World Bank report recognises two types of informal sector activities:

- **coping strategies** (survival activities) – casual jobs, temporary jobs, unpaid jobs, subsistence agriculture, multiple job holding
- **unofficial earning strategies** (illegality in business) – unofficial business activities (tax evasion, avoidance of labour regulation and other government or institutional regulations – no registration of the company) and underground activities (crime, corruption – activities not registered by statistical offices).

The advantages of the informal sector are that it:

- provides jobs and reduces unemployment and underemployment
- alleviates poverty
- bolsters entrepreneurial activity
- facilitates community cohesion and solidarity.

Activities in the informal sector have to contend with a significant number of obstacles, for example:

- There is little access to credit for workers in the informal sector to finance their activities, although in some countries **microcredit** is being used to fill this gap.
- The World Bank estimates that the size of the informal labour market varies from the estimated 4–6 per cent in HICs to over 50 per cent in LICs. Its size and role in the economy increases during economic downturns and periods of economic adjustment and transition.
- Women in Informal Employment: Globalizing and Organizing (WIEGO) is a global research-policy network that seeks to improve the status of the working poor, especially women, in the informal economy. It does so by highlighting the size, composition, characteristics and contribution of the informal economy through improved statistics and research.

Informal sector employment can be found in all parts of urban areas but is particularly concentrated in and around the CBD where potential demand for such services is at its highest. It is also often concentrated at key tourism locations where the informal crafts sector is in clear evidence. Informal sector employment is also attracted to industrial areas offering food and other services to industrial workers.

11.4 The management of change in manufacturing industry

Case Study: India

With approximately 1.3 billion people, India has the second largest population in the world. India is the 'I' in **BRIC**, the new buzzword for the economies tipped for rapid growth: Brazil, Russia, India and China. Because of its recent rapid economic growth (Figure 11.46), India is classed as a NIC. However, unlike other Asian economies such as South Korea, Taiwan, Thailand and Malaysia, which all became NICs at an earlier date, the recent transformation of the Indian economy has been based more on the service sector than on manufacturing. This has been at least partly due to a low level of **foreign direct investment (FDI)** in manufacturing, a situation that began to change in the early 1990s with the introduction of a number of important economic reforms.

The service sector accounts for 57 per cent of India's GDP, with industry responsible for 26 per cent and agriculture for 17 per cent (Figure 11.47). The situation with regard to employment is very different. Agriculture leads with 52 per cent of the workforce. Services account for 26 per cent and industry 22 per cent.

Textiles is the largest industry in the country, employing about 45 million people and accounting for 27 per cent of India's exports. India is the second largest producer of textiles and garments in the world. The car industry has expanded significantly in recent times and is now the seventh largest in the world, with a production of 3.2 million cars in 2013. However, it is in the field of software and ICT in general that India has built a global reputation. Figure 11.48 shows the major manufacturing hubs in India for automobiles (Figure 11.49), machine tools, textiles and pharmaceuticals.

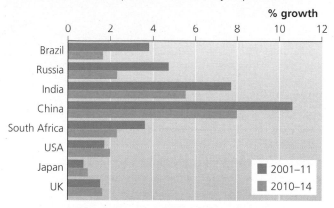

Source: Calculated from World Economic Outlook Database October 2014

Figure 11.46 Annual growth rates of the BRIC countries and three major developed economies

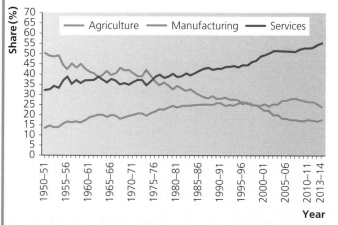

Figure 11.47 Main economic sectors' share of Indian GDP

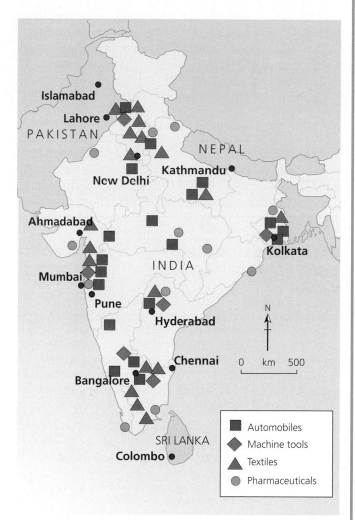

Figure 11.48 Major manufacturing hubs in India

Figure 11.49 Car manufacturing in India

Traditional industrial policy

In 1950, India was arguably the first non-communist LIC to institute a fully-fledged industrial policy. The objective of India's policy was to co-ordinate investment decisions in both the public and private sectors and to bring certain strategic industries and companies under public ownership. Following the example of the former Soviet Union, five-year plans were set up and this state-directed industrialisation model was followed from 1950 to 1980. The five-year plans were designed to bring about economic and social development within a 'socialist' framework. The main objectives of the plans were to:

- industrialise
- raise per person incomes
- achieve equity in the distribution of gains from economic development
- reduce the existing concentration of economic power
- achieve a more even regional distribution of industrial development.

The role of **heavy industry**, particularly of iron and steel, was emphasised, with the public sector playing a major role in the structural transformation of the economy from what was primarily an agricultural society. Investment in the private sector would be based not on the issue of profitability, but according to the requirements of the overall national plan. Technological

self-reliance became an important element of industrial policy. Here, the objective was to produce as much as possible inside India itself and keep imports to a minimum. This would be important for India's trade balance and also advance technical knowledge in the country.

Industrial policy measures under the five-year plans included:

- industrial licensing – a firm that wanted to manufacture a new product or sought a substantial expansion of its existing capacity had to obtain a government licence
- strict import controls
- subsidising exports
- strict controls on investment by TNCs.

The range of controls made India one of the most protected economies in the world. High tariffs made imports very expensive and thus controlled their volume. The Indian model was not just influenced by the Soviet Union but also by Fabian socialism and UK labour party thinkers like Harold Laski. In the 1980s, the model began to erode (as some liberalisation measures were introduced) and it was virtually abandoned after a serious external liquidity crisis in 1991.

The Planning Commission has taken the major role in formulating industrial policy and in guiding India's ongoing industrial revolution. It is still widely accepted by the country's mainstream political parties.

Many economists argue that India made a serious error from 1950 in taking on so many aspects of socialist central planning. The main criticisms were that there were far too many rules and regulations, which proved to be a major hindrance in the successful development of private sector industry. There was widespread corruption and massive inefficiency. Considerable aid from the West seemed to have very little impact on the industrial sector. However, opinion does differ widely on this subject.

Economic reform

The currency crisis of 1991 proved to be a major turning point, instigating bold economic reforms that resulted in rapid economic growth that is likely to double average productivity levels and living standards in India every 16 years. The economic reforms were based on:

- **liberalisation** – fewer government regulations and restrictions in the economy
- **deregulation** – changing regulatory policies and laws to increase competition among suppliers of commodities and services
- **market orientation** – more careful analysis of demand in domestic and global markets.

Tariffs on imports were significantly reduced along with other non-tariff trade barriers as a result of India's membership of the World Trade Organization (WTO). The essence was greater freedom from government control. This 'unshackling' of the economy is credited with increasing the growth rate of GDP from 3 to 3.5 per cent during the period 1950–80 and to 6–7 per cent in recent decades. The international financial institutions (IFIs) regard India as a major beneficiary of globalisation and are urging India to undertake even more reforms to open up its economy even further.

Instead of planning inputs and outputs for each company and each industry, the government adopted indicative planning. However, it maintained high tariffs by international standards and restrictions on portfolio and FDI. India has been determined to be master of its own policies and not blindly to copy the 'Western model'.

India's technological success has not been confined to the ICT industry. The country's corporations have achieved significant growth in a number of industries including, in particular, pharmaceuticals and auto components. It is one of only three countries in the world to build super-computers on its own, and one of only six countries in the world to launch satellites.

The relationship between growth of manufacturing and that of services is an issue of considerable significance for the economic development of the country.

Regional policy

Like many other countries, India adopted regional economic planning and tried to encourage a better spread of industry around the country. In the early 1970s, backward states and districts were identified and a scheme of incentives for industry to locate in these regions was introduced. This included:

- a grant of 15 per cent of fixed capital investment
- transport subsidies
- income tax concessions.

In 1977, central government decided that no more industrial licences would be granted in and around metropolitan cities and urban areas with a population of 500 000 and more. In 1980, new initiatives confirmed the government's commitment to correcting regional imbalance. The number of industrial concentrations has risen from about half a dozen in 1965 to more than 40 today, indicating a significant spatial expansion. In terms of international comparison, India has achieved a reasonable degree of success in its attempts to narrow regional imbalance.

Industrial policy and ICT

In the last two decades, India has spawned a modern, highly export-oriented ICT industry. The export intensity of Indian software is more than 70 per cent, compared with an overall export intensity of 10 per cent for the economy as a whole. The country's comparative advantage lies largely in the availability of low-cost skilled labour. The background to the success of the software industry was established in the era of traditional industrial policy, which more recent reforms have magnified. In the pre-1980 era:

- a large number of engineering colleges were established, particularly in the south of India under entrepreneurial state governments; these colleges were subsidised to a considerable degree by state and central government
- the government's philosophy in this period was to create a broad science and technology base to transform the Indian economy by stimulating domestic innovation; the benefits of this process were particularly felt by the ICT, biotechnology and pharmaceutical industries
- the government's role in the establishment of Bangalore as a hub attracting the bulk of the country's technological and scientific activity was fundamental to the development of the city as a global centre of ICT; Bangalore was favoured partly because of its distance from Pakistan and China, countries with which India has had difficult relations in the past.

NASSCOM, the Indian software association, has stated: 'The software and services industry has received immense support from the government both at the central and state level. This support in the form of tax incentives and other benefits has been instrumental in the growth of software and services exports.'

With the background set in the traditional industrial policy era, the age of reform has seen the ICT industry flourish, attracting high levels of FDI.

Future industrial policy

The direction taken by the Planning Commission in the future will have a major impact on India's economic performance (Figure 11.50). Many Indian economists stress the importance of achieving the best possible balance between the manufacturing and the service sectors. The general feeling is that India should take advantage of its strength in ICT and use it extensively in all areas of the economy to upgrade agriculture, industry and services in order to compete more effectively in the global economy. A major issue is the distribution of the gains and losses from globalisation. It seems that most gains have accrued to the Indian urban middle-class of around 100 million people, which amounts to less than 10 per cent of the country's population.

Some Japanese scholars have used a picturesque analogy to describe the gradual spread of development in Asia, with countries escaping mass poverty in a V-shaped formation that resembles a flock of flying geese.

Japan led the way after World War II, till rising wage costs in the 1960s led to the shift of low-value manufacturing to other regional economies in decadal waves that pulled millions off the farm and into the factory.

Most Asian countries that prospered used explicit industrial policies – and a rigged exchange rate – to build manufacturing prowess. Such policies went out of fashion in recent decades, but seem to have made a comeback in the entire ideological churn in the wake of the Western financial crisis.

World Bank chief economist Justin Lin says it is time to rethink development policy, with the state playing an important role even though 'the market is the basic mechanism for effective resource allocation'. There are clear signs that there has been a change in the attitude of the Indian government as well: industrial policy is making a quiet comeback in India.

The contours of the new industrial policy seem quite different from the sort of policies followed by Nehruvian India and other Asian countries in their early stages of development. 'The needs of building competitive enterprises and meeting WTO requirements need to be taken into account,' Planning Commission member Arun Maira told me during a telephonic chat. This means the new industrial policy that is emerging will not have much of the old statist and protectionist policy mix: protection through high import tariffs, preferential access to bank funds, promotion of national champions and resource allocation by a government agency.

Yet, there is a clear belief that the country needs an explicit industrial strategy. The government will choose which industries need encouragement and design suitable policies. Physical and social infrastructure will also be developed, a process that should lower transaction costs and raise the rates of return on investment.

Maira gave me three key policy parameters that will be kept in mind in designing the new economic strategy – there should be a growth in quality jobs, the Indian economy should get strategic depth in capital goods such as power and telecom equipment, and defence and security issues should be kept in mind.

There will be both technical and political economy challenges here. The technical challenge is to identify industries that need a helping hand, and one assumes that government agencies have an understanding of India's factor endowments and comparative advantages. The political economy challenge is perhaps even more complex. Comparative advantage rapidly changes in the modern economy and technology cycles are getting shorter. Policy will have to be flexible if India is not to stagnate.

A market economy has immense flexibility. Japan was a pioneer of successful industrial policy, but it lost the flexibility that it needed to fight its long economic stagnation. South Korea provides another lesson. Industrial policy there led to the formation of industrial conglomerates – the chaebol – and the gradual decline into crony capitalism.

What the Planning Commission has now set out to do is thus interesting but fraught with risks of regulatory capture and rent-seeking by favoured industrial groups. Maira says it is important that the focus of industrial policy remains on sectors rather than companies, in the attempts to forge closer collaboration between 'productive sectors and policymakers'.

By Niranjan Rajadhyaksha

Figure 11.50 India's new industrial policy

The state of India's infrastructure is also an important issue. Infrastructure in India is at a lower level than that in China and other NICs in the region. Current spending on the elements of infrastructure such as railways, roads, seaports and airports is about 6 per cent of GDP. This is about 50 per cent below what the government itself thinks needs to be spent. Because of the huge sums of money involved, the Planning Commission suggests that it can only be done by creating a partnership with the private sector.

Bangalore: India's high-tech city

Bangalore, Hyderabad and Chennai, in the south, along with the western city of Pune and the capital city Delhi, have emerged as the centres of India's high-technology industry.

Bangalore is the most important individual centre in India for high-tech industry. The city's pleasant climate, moderated by its location on the Deccan Plateau over 900 metres above sea level, is a significant attraction to foreign and domestic companies alike (Figures 11.51 and 11.52). Known as 'the Garden City', Bangalore claims to have the highest quality of life in the country. Because of its dust-free environment, large public-sector undertakings such as Hindustan Aeronautics Ltd and the Indian Space Research Organisation were established in Bangalore by the Indian government. The state government also has a long history of support for science and technology. The city prides itself on a 'culture of learning', which gives it an innovative leadership within India.

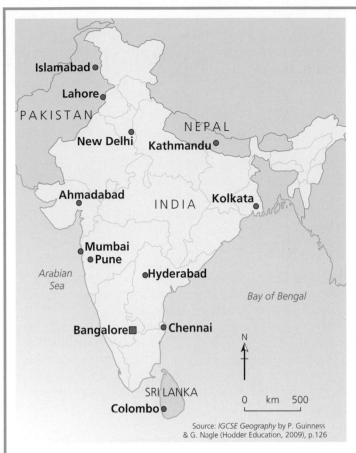

Figure 11.51 Location of Bangalore

In the 1980s, Bangalore became the location for the first large-scale foreign investment in high-technology in India when Texas Instruments selected the city above a number of other contenders. Other multinationals soon followed as the reputation of the city grew. Important backward and forward linkages were steadily established over time. Apart from ICT industries, Bangalore is also India's most important centre for aerospace and biotechnology.

India's ICT sector has benefited from the filter-down of business from HICs. Many European and North American companies that previously outsourced their ICT requirements to local companies are now using Indian companies because:

- labour costs are considerably lower
- a number of HICs have significant ICT skills shortages
- India has a large and able English-speaking workforce (there are about 80 million English-speakers in India).

Since 1981, the city's population has grown rapidly, from 2.4 million 9.6 million in 2011, while the number of vehicles has grown even faster, from fewer than 200 000 cars and scooters to over 2 million. The city's landscape has changed dramatically, with many new glass-and-steel skyscrapers and numerous cybercafés. The city has grown into a major international hub for ICT companies. Bangalore has the nickname of 'the Silicon Valley of India'.

BANGALORE

- Bangalore is the location of 925 software companies employing more than 80 000 IT workers. Bangalore accounts for nearly 40 per cent of India's software exports.
- The city has 46 integrated circuit design companies, 166 systems software companies and 108 communications software companies. Over 40 per cent of Bangalore's software exports are in these high-technology areas.
- Major companies include Tata Consulting Services (TCS), Infosys Technologies, Wipro and Kshema Technologies.
- The 170 000 m² International Tech Park was set up as a joint venture between the government of Karnataka state, the government of Singapore, and the House of TATA. The Electronic City is an industrial area with over 100 electronics companies including Infosys, Wipro, Siemens, Motorola and TI.
- The city has attracted outsourcing right across the IT spectrum from software development to IT-enabled services.
- The city boasts 21 engineering colleges.
- NASDAQ, the world's biggest stock exchange, with a turnover of over $20 trillion, opened its third international office in Bangalore in 2001.

Figure 11.52 Bangalore factfile

Section 11.4 Activities

1 Describe the trends shown in Figure 11.46.
2 Describe and explain the changing share of India's GDP by the three main economic sectors.
3 Discuss the development of traditional industrial policy in India.
4 Comment on the changes in industrial policy that began in the early 1990s.
5 How has the Indian government tried to narrow regional industrial imbalance?
6 In terms of future industrial policy, produce a 100-word summary of Figure 11.50.
7 How has government policy helped to build India's ICT industry into one of global prominence?

12 Environmental management

12.1 Sustainable energy supplies

☐ Renewable and non-renewable energy

Non-renewable sources of energy are the **fossil fuels** (Figure 12.1) and nuclear fuel. These are finite, so as they are used up the supply that remains is reduced. Eventually, these non-renewable resources could become completely exhausted. **Renewable energy** can be used over and over again. The sources of renewable energy are mainly forces of nature that are sustainable and that usually cause little or no environmental pollution. Renewable energy includes hydro-electric (HEP), biomass, wind, solar, geothermal, tidal and wave power.

Energy resources are at the core of the global economy. Energy is vital for economic growth and development. However, emissions of greenhouse gases from energy use are the main contributors to human-induced climate change. Energy production and consumption are major issues both within and between countries. Many countries that have to import a significant amount of the energy they use have become more and more concerned about energy security. There are growing concerns in some parts of the world that an inequitable availability and uneven distributions of energy sources may lead to conflict.

Figure 12.1 An oil well in Dorset, UK

Energy shortages have occurred in different parts of the world on a number of occasions in the past. Such shortages can have major economic and social consequences. But it is not just potential shortages of energy that is the concern. In poor countries, energy poverty has a major impact on people's lives and it is a major obstacle to development. The concept of fuel poverty is becoming a more and more important issue in many apparently affluent countries as people struggle to pay for rising energy bills in harsh winters.

At present, non-renewable resources dominate global energy. The environmental challenge is to transform the global **energy mix** to achieve a better balance between renewables and non-renewables. The key energy issues for individual countries are the three Ss: sustainability, security and strategy.

☐ Factors affecting the demand for and supply of energy

At the national scale, there are huge variations in energy demand and supply. Demand is primarily governed by the size of a country's population and its level of economic development. The gap between the world's richest and poorest countries in terms of energy demand is huge. Growth in energy demand is particularly rapid in middle-income countries (MICs) such as China and India. A country's energy policy can have a significant impact on demand if it focuses on efficiency and sustainability as opposed to concentrating solely on building more power stations and refining facilities. High levels of pollution due to energy consumption can be a strong stimulus to developing a cleaner energy policy.

As might be expected, global variations in energy supply occur for a number of reasons. These can be broadly subdivided into physical, economic and political factors. Table 12.1 shows examples for each of these groupings. The combination of factors operating in each country can vary significantly.

Table 12.1 Factors affecting the supply of energy

Physical	Economic	Political
• Deposits of fossil fuels are only found in a limited number of locations. • Large-scale HEP development requires high precipitation, major steep-sided valleys and impermeable rock. • Large power stations require flat land and geologically stable foundations. • Solar power needs a large number of days a year with strong sunlight. • Wind power needs high average wind speeds throughout the year. • Tidal power stations require a very large tidal range. • The availability of biomass varies widely according to climatic conditions.	• The most accessible, and lowest-cost, deposits of fossil fuels are invariably developed first. • Onshore deposits of oil and gas are usually cheaper to develop than offshore deposits. • Potential HEP sites close to major transport routes and existing electricity transmission corridors are more economical to build than those in very inaccessible locations. • In poor countries, FDI is often essential for the development of energy resources. • When energy prices rise significantly, companies increase spending on exploration and development.	• Countries wanting to develop nuclear electricity require permission from the International Atomic Energy Agency. • International agreements such as the Kyoto Protocol can have a considerable influence on the energy decisions of individual countries. • Potential HEP schemes on 'international rivers' may require the agreement of other countries that share the river. • Governments may insist on energy companies producing a certain proportion of their energy from renewable sources. • Legislation regarding emissions from power stations will favour the use of, for example, low-sulphur coal, as opposed to coal with a high sulphur content.

The key factor in supply is energy resource endowment. Some countries are relatively rich in domestic energy resources, while others are lacking in such resources and heavily reliant on imports. However, resources by themselves do not constitute supply. Capital and technology are required to exploit resources.

The use of energy in all countries has changed over time due to a number of factors:

- **Technological development** – for example: **a** nuclear electricity has only been available since 1954, **b** oil and gas can now be extracted from much deeper waters than in the past and **c** renewable energy technology is advancing steadily.
- **Increasing national wealth** – as average incomes increase living standards improve, which involves the increasing use of energy and the use of a greater variety of energy sources.
- **Changes in demand** – at one time, all of Britain's trains were powered by coal and most people also used coal for heating in their homes. Before natural gas was discovered in the North Sea, Britain's gas was produced from coal (coal gas).
- **Changes in price** – the relative prices of the different types of energy can influence demand. Electricity production in the UK has been switching from coal to gas over the last 20 years, mainly because power stations are cheaper to run on natural gas.
- **Environmental factors/public opinion** – public opinion can influence decisions made by governments. People today are much better informed about the environmental impact of energy sources than they were in the past.

Section 12.1 Activities

1 Look at Table 12.1. Select two points from each of the three categories to investigate further. Present your findings to others in your class.
2 For the UK, find out when:
  a the last steam trains (burning coal) stopped being used on Britain's general railway network
  b nuclear electricity first came online
  c North Sea gas first came online.

☐ Trends in consumption of conventional energy resources

The fossil fuels dominate the global energy situation. Their relative contributions (for 2012) are:

- **oil**: 33 per cent (Figure 12.2)
- **coal**: 30 per cent
- **natural gas**: 24 per cent.

Figure 12.2 Oil refinery, Milford Haven, UK

In contrast, HEP accounted for 6.6 per cent and nuclear energy 4.5 per cent of global energy. The main data source used in this topic is the *BP Statistical Review of World Energy*. It includes commercially traded fuels only. It excludes fuels such as wood, peat and animal waste, which, though important in many countries, are unreliably documented in terms of production and consumption statistics.

Figure 12.3 shows the regional pattern of energy consumption for 2012. Consumption by type of fuel varies widely by world region:

- **Oil** – only in Asia Pacific is the contribution of oil less than 30 per cent and it is the main source of energy in four of the six regions shown in Figure 12.3. In the Middle East, it accounts for approximately 50 per cent of consumption.
- **Coal** – only in the Asia Pacific region is coal the main source of energy. In contrast, it accounts for less than 5 per cent of consumption in South and Central America, and in the Middle East. China was responsible for 50.2 per cent of global coal consumption in 2012.
- **Natural gas** – natural gas is the main source of energy in Europe and Eurasia and it is a close second to oil in the Middle East. Its lowest share of the energy mix is 11 per cent in the Asia Pacific region.
- **Hydro-electricity** – the relative importance of hydro-electricity is greatest in South and Central America (25 per cent). Elsewhere, its contribution varies from 6 per cent in Africa to less than 1 per cent in the Middle East.
- **Nuclear energy** – nuclear energy is not presently available in the Middle East and it makes the smallest contribution of the five traditional energy sources in Asia Pacific, Africa and South and Central America. It is most important in Europe and Eurasia and North America.
- **Renewables** – consumption of renewable energy other than HEP is rising rapidly, but from a very low base. Renewables make the largest relative contribution to energy consumption in Europe and Eurasia.

In terms of usage by type of energy, some general points can be made:

The most developed countries tend to use a wide mix of energy sources, being able both to invest in domestic energy potential and to buy energy from abroad.

- The high investment required for nuclear electricity means that only a limited number of countries produce electricity this way. However, many countries that could afford the investment choose not to adopt this strategy.
- Richer nations have been able to invest more money in renewable sources of energy.
- In the poorest countries, fuelwood is an important source of energy, particularly where communities have no access to electricity.

Figure 12.4 shows how global consumption of the five major traditional sources of energy changed between 1987 and 2012. The demand for energy has grown relentlessly over this time period, with an overall increase of 60 per cent in total global energy consumption.

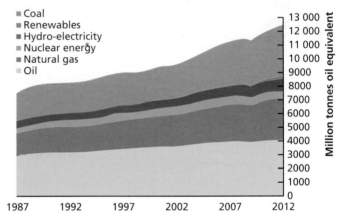

Source: *IGCSE Geography* 2nd edition, P. Guinness & G. Nagle (Hodder Education, 2014) p.207

Figure 12.4 Global consumption of major energy sources, 1987–2012

Figure 12.5 shows per person energy consumption around the world. The highest consumption countries such as the USA and Canada use more than 6 tonnes oil equivalent per person, while almost all of Africa and much of Latin America and Asia use less than 1.5 tonnes oil equivalent per person. The individual countries consuming the most energy in 2012, as a percentage of the world total, were:

- **China**: 21.9 per cent
- **USA**: 17.7 per cent
- **Russia**: 5.6 per cent
- **India**: 4.8 per cent
- **Japan**: 3.8 per cent.

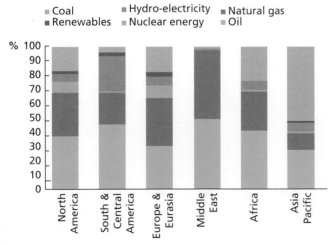

Source: *IGCSE Geography* 2nd edition, P. Guinness & G. Nagle (Hodder Education, 2014) p.208

Figure 12.3 The regional pattern of energy consumption, 2012

Source: *IGCSE Geography* 2nd edition, P. Guinness & G. Nagle (Hodder Education, 2014) p.209

Figure 12.5 World map showing energy consumption per person, 2012

Wealth is the main factor explaining the energy gap. The use of energy can improve the quality of people's lives in so many ways. That is why most people who can afford to buy cars, televisions and washing machines do so. However, there are other influencing factors, with climate at the top of the list.

Oil: global patterns and trends

Oil is the most important of the non-renewable sources of energy. Even though investment in new sources of energy is increasing rapidly, the global economy still relies on oil to a considerable extent. Oil clearly has significant advantages as a source of energy, otherwise it would not be as important as it is today. However, its disadvantages have gained increasing recognition in recent decades (Table 12.2).

Table 12.2 The advantages and disadvantages of oil

Advantages	Disadvantages
• A compact, portable source of energy; relatively easy to transport and store • Used for most forms of mechanical transportation • Flexible use – can be distilled into different fuel products • Cleaner and easier to burn than coal • Compared to most other fuel sources, it remains one of the most economical sources of energy • The oil industry has been the source of much advanced technology • Oil refining produces the world's supply of elemental sulphur as a by-product, used for many industrial applications • Has a well-established global infrastructure	• Non-renewable – takes millions of years to form • Burning oil generates CO_2, a greenhouse gas • Oil contains sulphur, which when burnt forms sulphur dioxide and sulphur trioxide; these combine with atmospheric moisture to form sulphuric acid, leading to 'acid rain' • Not as clean or efficient in use as natural gas • Serious oil spills from supertankers and pipelines • Locating additional reserves requires a very high level of investment • Political instability of some major oil producing countries and concern about the vulnerability of energy pathways • Concerns that 'peak oil' is not far away • The price of oil has varied significantly over the last decade • Some oil is now being strip-mined in the form of tar sands, creating serious environmental concerns

Figure 12.6 shows the change in daily oil consumption by world region from 1987 to 2012. From just over 60 million barrels daily in the mid-1980s, global demand rose steeply to 89.8 million barrels a day in 2012. Satisfying such a rapid rate of increase in demand requires a high level of investment and exploration, and has environmental and other consequences. The largest increase has been in the Asia Pacific region (Figure 12.7), which now accounts for 33.6 per cent of consumption. This region now uses more oil than North America, which accounts for 24.6 per cent of the world total. In contrast, Africa consumed only 4.0 per cent of global oil.

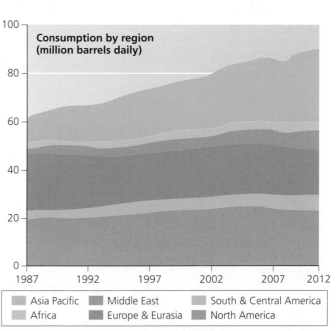

Source: *BP Statistical Review of World Energy 2013*, p.12

Figure 12.6 Oil consumption by world region, 1987–2012

Figure 12.7 Fuel station on the Mekong River, Vietnam

In 2012, the Middle East accounted for over 48 per cent of global **proven reserves**. Political instability in the Middle East is a major concern to the countries that import oil from this region. Table 12.3 shows the **reserves-to-production ratio** for the world in 2012. While the R/P ratio is over 78 years in the Middle East and even higher in South and Central America, it is only 13.6 years in Asia Pacific.

Table 12.3 Oil reserves-to-production ratio at the end of 2012

Region	Reserves/production ratio (years)
North America	38.7
South and Central America	123.0
Europe and Eurasia	22.4
Middle East	78.1
Africa	37.7
Asia Pacific	13.6
World	52.9

Source: BP Statistical Review of World Energy 2013

The price of oil increased sharply in the early years of the new century, causing major financial problems in many importing countries. It rose from $10 a barrel in 1998 to more than $130 a barrel in 2008, before falling back sharply in the global recession of 2008–09. As the global economy has slowly recovered, the price of oil has fluctuated considerably and was at a price of about $40 a barrel in late 2015.

When will global peak oil production occur?
There has been growing concern about when global oil production will peak and how fast it will decline thereafter. There are concerns that there are not enough large-scale projects underway to offset declining production in well-established oil-production areas. The rate of major new oil-field discoveries has fallen sharply in recent years. It takes 6 years on average from first discovery for a very large-scale project to start producing oil. In 2010, the International Energy Agency expected **peak oil production** somewhere between 2013 and 2037, with a fall by 3 per cent a year after the peak. The United States Geological Survey predicted that the peak was 50 years or more away.

However, in complete contrast, the Association for the Study of Peak Oil and Gas (ASPO) predicted in 2008 that the peak of global oil production could come as early as 2011, stating 'Fifty years ago the world was consuming 4 billion barrels of oil per year and the average discovery was around 30 billion. Today we consume 30 billion barrels per year and the discovery rate is now approaching 4 billion barrels of crude oil per year'. ASPO's dire warnings have not (yet) materialised. This is at least partly down to new developments, particularly the rapid growth in production of shale oil and gas in the USA, which has changed the global energy situation. The current period of slow growth in the global economy has also eased the pressure on energy resources.

Shale oil
The exploitation of **shale oil** has been a very recent development. It has been concentrated mainly in the United States. Shale oil is extracted from reserves, sometimes described as 'tight oil' reserves, held in shales and other rock formations from which it will not naturally flow freely. Shale oil has become more accessible due to advances in technology.

The rapid increase in the scale of production in the USA has fundamentally changed global energy markets as the USA has quickly regained much of its self-sufficiency in energy. Gas can also be obtained from shale, and the exploitation of shale gas has led the exploitation of shale oil by a few years. The exploitation of shale deposits has had a massive impact on US gas and oil production in recent years. Further significant production increases of shale oil and gas are forecast for the USA, and the 'shale revolution' is likely to spread to other parts of the world, albeit with a time lag. The speed of this geographical spread (diffusion) will depend on a number of factors, including the extent of opposition to this process on environmental grounds.

Although the basic technology was originally developed in the USA in the 1940s, the recent 'shale revolution' was the result of technological breakthroughs in horizontal drilling and hydraulic fracturing (fracking), which have made shale deposits economically viable. In the USA, oil production peaked in 1970 at 534 million tonnes, falling to 305 million tonnes in 2008. The subsequent rise to 499 million tonnes in 2013 has been a phenomenal turnaround. Most of this increase has been due to the rapid rise in the production of shale oil. The full extent of recoverable shale oil in the USA is still to be determined as exploration continues. A recent analysis by the EIA put the total at nearly 8 billion tonnes, a very considerable energy resource base indeed!

The geopolitical impact of changes in patterns and trends in oil

Energy security depends on resource availability – domestic and foreign – and security of supply. It can be affected by **geopolitics**, and is a key issue for many economies. Because there is little excess capacity to ease pressure on energy resources, energy insecurity is rising, particularly for non-renewable resources.

The USA, gravely concerned about the political leverage associated with imported oil, began in 1977 the construction of a **strategic petroleum reserve**. The oil was to be stored in a string of salt domes and abandoned salt mines in southern Louisiana and Texas, which could be linked up easily to pipelines and shipping routes. The initial aim was to store 1 billion barrels of oil that could be used in the event of supply discontinuation. The SPR currently holds 700 million barrels.

The Middle East is the major global focal point of oil exports. The long-running tensions that exist in the Middle East have at times caused serious concerns about the vulnerability of oilfields, pipelines and oil-tanker routes. The destruction of oil wells and pipelines during the Iraq War showed all too clearly how energy supplies can be disrupted. Middle East oil exports are vital for the functioning of the global economy. Most Middle East oil exports go by tanker through the Strait of Hormuz, a relatively narrow body of water between the Persian Gulf and the Gulf of Oman. The strait at its narrowest is 55 kilometres wide. Roughly 30 per cent of the world's oil supply passes through the Strait, making it one of the world's strategically important chokepoints. Iran has at times indicated that it could block this vital shipping route in the event of serious political tension. This could cause huge supply problems for many importing countries. Concerns about other key **energy pathways** have also arisen from time to time.

Section 12.1 Activities

1 Explain why the locations of global oil production and consumption vary so widely.
2 Define the term *reserves-to-production ratio*. Describe how this varies around the world.
3 **a** Why is the prediction of peak oil production so important?
   b Suggest why the predictions of when peak oil production will occur vary so widely.

Natural gas

Global production of natural gas increased from 2524 billion m³ in 2002 to 3364 billion m³ in 2012 (Table 12.4). All six world regions showed an increase in production. However, the largest producing world regions, Europe/Eurasia and North America, recorded the lowest percentage increases between 2002 and 2012. The highest relative change was in the Middle East.

Table 12.4 Natural gas production by world region, 2002–12

Region	2002	2012	% change
North America	763.6	896.4	17
South & Central America	107.9	177.3	64
Europe & Eurasia	966.5	1035.4	7
Middle East	247.2	548.4	122
Africa	138.2	216.2	56
Asia Pacific	300.5	490.2	63
Total world	**2523.9**	**3363.9**	**33**

Source: BP Statistical Review of World Energy 2013

On an individual country basis, natural gas production is dominated by the USA (20.4 per cent of the global total) and the Russian Federation (17.6 per cent). There is a very substantial gap between these two natural gas giants and the next largest producers, which are Iran (4.8 per cent), Qatar (4.7 per cent) and Canada (4.6 per cent).

There is a much stronger correlation between consumption and production of natural gas than for oil, due mainly to the different ways these two energy products are transported. Global consumption of natural gas in 2012 was led by Europe and Eurasia (32.6 per cent), North America (27.5 per cent) and Asia Pacific (18.8 per cent).

During the period 2002–12, proven reserves of natural gas increased substantially. The global share of proven reserves in the Middle East fell slightly, while the share held in Europe and Eurasia increased. On an individual country basis, the largest reserves in 2012 were in Iran (18 per cent), the Russian Federation (17. 6 per cent) and Qatar (13.4 per cent). In 2012, the global reserves-to-production ratio stood at 55.7 years.

Coal

Coal production is dominated by the Asia Pacific region, accounting for 67.8 per cent of the global total in 2012. Much of this coal is produced in China, which alone mines 47.5 per cent of the world total. The next largest producing countries were the USA (13.4 per cent), Australia (6.3 per cent), Indonesia (6.2 per cent), India (6.0 per cent) and the Russian Federation (4.4 per cent). Like natural gas, there is a strong relationship between the production and consumption of coal by world region. Consumption is led by Asia Pacific (69.9 per cent), Europe and Eurasia (13.9 per cent) and North America (12.6 per cent). China alone consumed 50.2 per cent of world coal in 2012.

Figure 12.8 shows the proven reserves of coal in 2002 and 2012. There is a fairly even spread between three regions: Europe/Eurasia, Asia Pacific and North America. However, total global reserves declined by 12.5 per cent over this ten-year time period. In terms of the reserves-to-production ratio (Figure 12.9), the figure for Asia Pacific at 51 years is significantly below that for other world regions. The global reserves-to-production ratio fell from 119 years in 2009 to 109 years in 2012, a significant decline in such

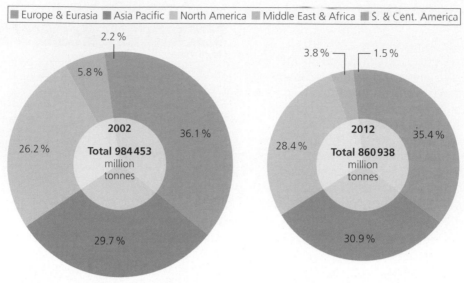

Europe & Eurasia ■ Asia Pacific ■ North America ■ Middle East & Africa ■ S. & Cent. America

2.2 %
5.8 %

2002
Total 984 453
million tonnes

36.1 %
26.2 %
29.7 %

3.8 % ─ ─ 1.5 %

2012
Total 860 938
million tonnes

35.4 %
28.4 %
30.9 %

Source: Survey of Energy Resources 2010, World Energy Council

Figure 12.8 Distribution of proven coal reserves, 2002 and 2012

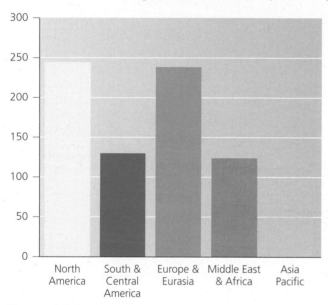

Figure 12.9 Coal: reserves-to-production ratios, 2012

Figure 12.10 A former coal mine in the Massif Central region of France, which was closed in the early 1990s – it is now a museum to the coal industry

a short time period. Coal reserves can become exhausted within a relatively short time period. In the nineteenth and early twentieth centuries, countries such as Germany, the UK and France were significant producers. Today, there are very few operational coal mines in these three countries (Figure 2.10).

Extending the 'life' of fossil fuels

There are a number of technologies that can improve the use and prolong the life of fossil fuels. These include coal gasification, clean coal technologies and the extraction of unconventional natural gas. Such techniques may be very important in buying time for more renewable energy to come online.

Coal gasification is the technology that could transform the situation. At present, electricity from coal gasification is more expensive than that from traditional power plants, but if more stringent pollution laws are passed in the future this situation could change significantly.

The coal industry in a number of areas may be on the point of a limited comeback, with the development of **clean coal technology**. This new technology has developed forms of coal that burn with greater efficiency and capture coal's pollutants before they are emitted into the atmosphere. The latest 'supercritical' coal-fired power stations, operating at higher pressures and temperatures than their predecessors, can operate at efficiency levels 20 per cent above those of coal-fired power stations constructed in the 1960s. Existing power stations can be upgraded to use clean coal technology.

Conventional natural gas, which is generally found within a few thousand metres or so of the surface of the Earth, has accounted for most of the global supply to date. However, in recent years 'unconventional' deposits have begun to contribute more to supply. The main categories of **unconventional natural gas** are:

- deep gas
- tight gas
- gas-containing shales
- coalbed methane
- geopressurised zones
- Arctic and sub-sea hydrates.

Unconventional deposits are clearly more costly to extract but as energy prices rise and technology advances, more and more of these deposits are attracting the interest of governments and energy companies.

Nuclear power: a global renaissance?

Until a few years ago, the future of nuclear power looked bleak, with a number of countries apparently 'running down' their nuclear power stations and many other nations firmly set against the idea of introducing nuclear electricity. However, heightened fears about oil supplies, energy security and climate change have brought this controversial source of power back onto the global energy agenda.

No other source of energy creates such heated discussion as nuclear power. The main concerns about nuclear power are:

- power plant accidents, which could release radiation into air, land and sea
- radioactive waste storage/disposal – no country has yet implemented a long-term solution to the nuclear-waste problem
- rogue state or terrorist use of nuclear fuel for weapons
- high construction and decommissioning costs
- the possible increase in certain types of cancer near nuclear plants.

In addition, because of the genuine risks associated with nuclear power and the level of security required, it is seen by some people as less 'democratic' than other sources of power.

By early 2015, 30 countries around the world were operating 443 nuclear reactors for electricity generation, with 66 new nuclear plants under construction. Nuclear power accounted for almost 11 per cent of the world's electricity production in 2012.

With 99 operating reactors, the USA leads the world in the use of nuclear electricity. This amounts to 32.7 per cent of the world's total , producing about 20 per cent of the USA's electricity. At one time, the rise of nuclear power looked unstoppable. However, a serious incident at the Three Mile Island nuclear power plant in Pennsylvania in 1979, and the much more serious Chernobyl disaster in Ukraine in 1986, brought any growth in the industry to a virtual halt. No new nuclear power plants have been ordered in the USA since then, although public opinion has become more favourable in recent years, as memories of the Three Mile Island and Chernobyl incidents recede into the past, and as worries about polluting fossil fuels increase.

The big advantages of nuclear power are:

- there are zero emissions of greenhouse gases – this has become increasingly important as concerns about climate change have increased
- it means reduced reliance on imported fossil fuels (which can help ease concerns about energy security)
- it is not as vulnerable to fuel price fluctuations as oil and gas – uranium, the fuel for nuclear plants, is relatively plentiful and most of the main uranium mines are in politically stable countries
- nuclear power plants have demonstrated a very high level of reliability and efficiency in recent years.

The next major consumers of nuclear energy after the USA are France (17.2 per cent of the 2012 world total), Russia (7.2 per cent) and South Korea (6.1 per cent). France obtains over 75 per cent of its electricity from nuclear. Table 12.5 shows all the countries where nuclear power accounts for more than 30 per cent of electricity production. Global nuclear power production declined slightly between 2002 and 2012, from 185.8 million tonnes oil equivalent to 183.2 million tonnes. The current decade will be crucial to the future of nuclear energy, with many countries making final decisions to extend or begin their nuclear electricity capability.

Table 12.5 The contribution of nuclear power to electricity production

Country	Contribution of nuclear power (%)
France	76.9
Slovakia	56.8
Hungary	53.6
Ukraine	49.4
Belgium	47.5
Sweden	41.5
Switzerland	37.9
Slovenia	37.2
Czech Republic	35.8
Finland	34.6
Bulgaria	31.8
Armenia	30.7
South Korea	30.4

A few countries have developed **fast breeder reactor** technology. These reactors are very efficient at manufacturing plutonium fuel from their original uranium fuel load. This greatly increases energy production – but it could prove disastrous should the plutonium get into the wrong hands, as plutonium is the key ingredient for nuclear weapons.

Section 12.1 Activities

1 Compare the changes in production of the five traditional forms of energy between 2002 and 2012.
2 Using Table 12.4, describe how the global production of natural gas changed between 2002 and 2012.
3 Outline the extent of the world's coal reserves.
4 What are the main advantages and disadvantages of nuclear power?

☐ Renewable energy resources

Table 12.6 compares renewable energy capacity at the beginning of 2004 to the end of 2013. For both years, hydro-electricity dominated renewable energy production, but most other sources of renewable energy have grown at a faster rate. Overall renewable power capacity almost doubled in the time period covered by Table 12.6. The newer sources of renewable energy making the largest contribution to global energy supply are wind power and biofuels.

Table 12.6 Capacity of renewable energy sources 2004 and 2013

		Start 2004	End 2013
Investment			
New investment (annual) in renewable power and fuels	Billion US$	39.5	214.4
Power			
Renewable power capacity (total, not including hydro)	gigawatts	85	560
Renewable power capacity (total, including hydro)	gigawatts	800	1560
Hydropower capacity (total)	gigawatts	715	1000
Bio-power capacity	gigawatts	<36	88
Bio-power generation	tera watt hours	227	405
Geothermal power capacity	gigawatts	8.9	12
Solar PV capacity (total)	gigawatts	2.6	139
Concentrating solar thermal power (total)	gigawatts	0.4	3.4
Wind power capacity (total)	gigawatts	48	318
Heat			
Solar hot water capacity (total)	gigawatts	98	326
Transport			
Ethanol production (annual)	billion litres	28.5	87.2
Biodiesel production (annual)	billion litres	2.4	26.3

Source: Renewables 2014 Global Status Report

Hydro-electric power

Of the traditional five major sources of energy, HEP is the only one that is renewable. It is by far the most important source of renewable energy. The 'big four' HEP nations of China, Brazil, Canada and the USA account for almost 53 per cent of the global total. However, most of the best HEP locations are already in use, so the scope for more large-scale development is limited. In many countries though, there is scope for small-scale HEP plants to supply local communities. However, global consumption of hydro-electricity increased from 598.5 million tonnes oil equivalent in 2002 to 831.1 million tonnes in 2012.

In 2012, the countries with the largest share of the world total were: China (23.4 per cent), Brazil (11.4 per cent) (Figure 12.11), Canada (10.4 per cent) and the USA (7.6 per cent). Consumption of hydro-electricity in China amounted to 194.8 million tonnes oil equivalent in 2012.

Figure 12.11 Inside the Itaipu hydro-electric power plant, Brazil

Although HEP is generally seen as a clean form of energy, it is not without its problems, which include:

- Large dams and power plants can have a huge negative visual impact on the environment.
- They may obstruct the river for aquatic life.
- There may be a deterioration in water quality.
- Large areas of land may need to be flooded to form the reservoir behind the dam.
- Submerging large forests without prior clearance can release significant quantities of methane, a greenhouse gas.

Newer alternative energy sources

The first major wave of interest in new alternative energy sources resulted from the **energy crisis** of the early 1970s. However, the relatively low price of oil in the 1980s, 1990s and the opening years of the present century dampened down interest in these energy sources. Then, renewed concerns about energy in recent years and corresponding price increases kick-started the alternative-energy industry again. The main drawback to the new alternative energy sources is that they invariably produce higher cost electricity than traditional sources. However, the cost gap with non-renewable energy is narrowing. Figure 12.12 shows the sharp increase in the consumption of renewable energy (other than HEP) in the last decade. In 2012, this accounted for 1.9 per cent of global primary energy consumption. The newer sources of renewable energy making the largest contribution to global energy supply are wind power and biofuels.

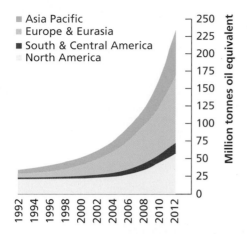

Source: *IGCSE Geography* 2nd edition, P. Guinness & G. Nagle (Hodder Education, 2014) p.212

Figure 12.12 Renewable energy consumption by world region, 1992–2012

Wind power

Wind power is arguably the most important of the new renewable sources of energy (Fig 12.13). The worldwide capacity of wind energy is approaching 400 000 megawatts, a very significant production mark (Figure 12.14). The wind industry set a new record for annual installations in 2014. Global wind energy is dominated by a relatively small number of countries. China is currently the world leader, with 31 per cent of global capacity, followed by the USA, Germany, Spain and India. Together, these five countries account for almost 72 per cent of the global total. In the last five years, for the first time ever, more new wind power capacity was installed in LICs and MICs than in the developed world.

Wind energy has reached the 'take-off' stage, both as a source of energy and a manufacturing industry. As the cost of wind energy improves further against conventional energy sources, more and more countries will expand into this sector. However, projections regarding the industry still vary considerably because of the number of variables that will impact on its future.

Figure 12.13 Wind farm in northern Spain

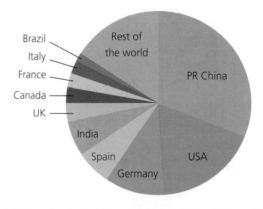

Country	Megawatts	Share (%)
PR China	114 763	31.1
USA	65 879	17.8
Germany	39 165	10.6
Spain	22 987	6.2
India	22 465	6.1
United Kingdom	12 440	3.4
Canada	9 694	2.6
France	9 285	2.5
Italy	8 663	2.3
Brazil	5 939	1.6
Rest of the world	58 275	15.8
Total Top 10	**311 279**	**84.2**
World total	**369 553**	**100**

Source: GWE

Figure 12.14 Global wind power capacity, end 2014

Costs of generating electricity from wind today are only about 10 per cent of what they were 20 years ago due mainly to advances in turbine technology. Thus, at well-chosen locations, wind power can now compete with conventional sources of energy. Wind energy operators argue that costs should fall further due to **a** further technological advances and **b** increasing economies of scale. One large turbine manufacturer has stated that it expects turbine costs to be reduced by 3.5 per cent a year for the foreseeable future. Table 12.7 summarises the advantages and disadvantages of wind power.

Table 12.7 The advantages and disadvantages of wind power

Advantages	Disadvantages
• A renewable source of energy that can produce reasonable levels of electricity with current technology • Advances in wind turbine technology over the last decade have reduced the cost per unit of energy considerably • Suitable locations with sufficient wind conditions can be found in most countries • Wind energy has reached the take-off stage both as a source of energy and as a manufacturing industry • Flexibility of location with offshore wind farms gaining in popularity • Repowering can increase the capacity of existing wind farms • Significant public support for a renewable source of power although this may be waning to an extent	• Growing concerns about the impact on landscapes as the number of turbines and wind farms increases • NIMBY (not in my back yard) protests with people concerned about the impact of local turbines adversely affecting the value of their properties • The hum of turbines can be disturbing for both people and wildlife • Debate about the number of birds killed by turbine blades • TV reception can be affected by wind farms • The development of wind energy has required significant government subsidies – some people argue that this money could have been better spent elsewhere (opportunity cost) • Many wind farms are sited in coastal locations where land is often very expensive

Public finance continues to play a strong role in the economics of the industry and this is likely to continue in the foreseeable future, particularly in the light of the current global financial situation and the fragility of commercial banks.

Apart from establishing new wind energy sites, **repowering** is also beginning to play an important role. This means replacing first generation wind turbines with modern multi-megawatt turbines that give a much better performance. The advantages are:

- more wind power from the same area of land
- fewer wind turbines
- higher efficiency, lower costs
- enhanced appearance as modern turbines rotate at a lower speed and are usually more visually pleasing due to enhanced design
- better grid integration as modern turbines use a connection method similar to conventional power plants.

As wind turbines have been erected in more areas of more countries, the opposition to this form of renewable energy has increased:

- People are concerned that huge turbines located nearby could blight their homes and have a significant impact on property values
- Concerns about the hum of turbines disturbing both people and wildlife
- Skylines in scenically beautiful areas might be spoiled forever
- Turbines can kill birds – migratory flocks tend to follow strong winds, but wind companies argue they steer clear of migratory routes
- Suitable areas for wind farms are often near the coast where land is expensive
- Turbines can affect TV reception nearby
- The opportunity cost of heavy investment in wind compared to the alternatives.

The recent rapid increase in demand for turbines has resulted in a shortage of supply. New projects now have to make orders for turbines in large blocks up to several years in advance to ensure firm delivery dates from manufacturers. Likewise, the investment from manufacturers is having to rise significantly to keep pace with such buoyant demand.

New developments in wind energy include:

- in 2008, a Dutch company installed the world's first floating wind turbine off the southern coast of Italy in water 110 metres deep – the technology is known as the Submerged Deepwater Platform System.
- the Swedish company Nordic has recently brought a two-bladed turbine onto the market.

Biofuels

Biofuels are fossil fuel substitutes that can be made from a range of agri-crop materials including oilseeds, wheat, corn and sugar. They can be blended with petrol and diesel.

In recent years, increasing amounts of cropland have been used to produce biofuels. Initially, environmental groups such as Friends of the Earth and Greenpeace were very much in favour of biofuels, but as the damaging environmental consequences became clear, such environmental organisations were the first to demand a rethink of this energy strategy.

The main methods of producing biofuels are:

- crops that are high in sugar (sugar cane, sugar beet, sweet sorghum) or starch (corn/maize) are grown and then yeast fermentation is used to produce ethanol
- plants containing high amounts of vegetable oil (such as oil palm, soybean and jatropha) are grown, and the oils derived from them are heated to reduce their viscosity; they can then be burned directly in a diesel engine, or chemically processed to produce fuels such as biodiesel

- wood can be converted into biofuels such as woodgas, methanol or ethanol fuel
- cellulosic ethanol can be produced from non-edible plant parts, but costs are not economical at present – this method is seen as the potential second generation of biofuels.

Ethanol is the most common biofuel globally, particularly in the USA and Brazil (Figure 12.15). It accounts for over 90 per cent of total biofuel production. Ethanol can be used in petrol engines when mixed with gasoline. Most existing petrol engines can run on blends of up to 15 per cent ethanol. Global production of ethanol has risen rapidly in recent decades, although since 2010 it has levelled off to a considerable extent. In the USA, about 40 per cent of the maize crop is used to produce ethanol. The USA and Brazil are by far the largest producers of ethanol in the world. Together, these two countries produce 87 per cent of the world total. However, production in the European Union and China is growing significantly.

In contrast to the USA, Brazil uses sugar cane to produce ethanol. More than half of Brazil's sugar cane crop is now used for this purpose. Sugar cane-based ethanol can be produced in Brazil at about half the cost of maize-based ethanol in the USA. This difference is due to:

- climatic factors
- land availability
- the greater efficiency of sugar in converting the Sun's energy into ethanol.

The USA has set a target of increasing the use of biofuels to 15 billion gallons by 2015. Subsidies are an important element in encouraging biofuel production.

Global biodiesel production and capacity have risen significantly in recent years. Biodiesel is the most common biofuel produced in Europe, with the continent accounting for over 60 per cent of global production. Germany and France are the leading producers within Europe. Biodiesel can be used in any diesel engine when mixed with mineral diesel, usually up to a limit of 15 per cent biodiesel. Rapeseed oil is the major source of Europe's biodiesel. After the EU, the USA is the second most important producer of biodiesel. In the latter, soybean oil is the main source for production.

Increasing investment is taking place in research and development of the so-called 'second generation' biodiesel projects including algae and cellulosic diesel. Other important trends in the industry are a transition to larger plants and consolidation among smaller producers.

Geothermal electricity

Geothermal energy is the natural heat found in the Earth's crust in the form of steam, hot water and hot rock. Rainwater may percolate several kilometres below the surface in permeable rocks, where it is heated due to the Earth's **geothermal gradient**. This is the rate at which temperature rises as depth below the surface increases. The average rise in temperature is about 30 °C per kilometre, but the gradient can reach 80 °C near plate boundaries.

This source of energy can be used to produce electricity, or its hot water can be used directly for industry, agriculture, bathing and cleansing (Figure 12.16). For example, in Iceland hot springs supply water at 86 °C to 95 per cent of the buildings in and around Reykjavik. At present, virtually all the geothermal power plants in the world operate on steam resources, and they have an extremely low environmental impact.

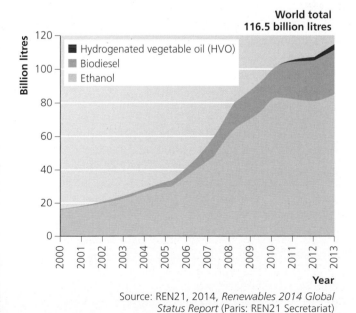

Source: REN21, 2014, *Renewables 2014 Global Status Report* (Paris: REN21 Secretariat)

Figure 12.15 Global biofuel production, 2000–13

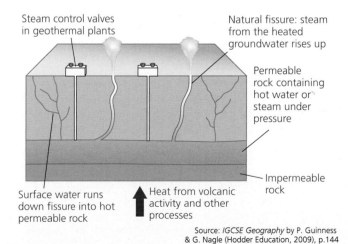

Source: *IGCSE Geography* by P. Guinness & G. Nagle (Hodder Education, 2009), p.144

Figure 12.16 Geothermal power

Figure 12.17 Geothermal power plant, Wairakei, New Zealand

First begun in Larderello, Italy, in 1904, total world installed geothermal capacity passed 12000 megawatts by the end of 2013. This is enough electricity to meet the needs of over 70 million people. About 700 geothermal projects were under development in 76 countries in 2013, amounting to about 30000 megawatt capacity. The USA is the world leader in geothermal electricity, with plants in Alaska, California, Hawaii, Nevada and Utah. Total production accounts for about 0.4 per cent of the electricity used in the USA. Other leading geothermal-electricity-using countries are the Philippines, Indonesia, Mexico, Italy, New Zealand (Figure 12.17), Iceland and Japan.

The advantages of geothermal power for those countries that have access to this form of energy are:

- extremely low environmental impact
- plants occupy relatively small land areas
- generation is not dependent on weather conditions (like wind and solar power)
- relatively low maintenance costs.

The limitations of this form of energy are:

- there are few locations worldwide where significant amounts of energy can be generated
- total global generation remains small
- some of the best locations are far from where the energy could be used
- installation costs of plant and piping are relatively high.

Solar power

From a relatively small base, the installed capacity of solar electricity is growing rapidly. Experts say that solar power has huge potential for technological improvement, which could make it a major source of global electricity in years to come (Figure 12.18). In 2000, global solar capacity was only 1275 megawatts. It grew to 5085 megawatts in 2005 and 40183 in 2010. Global solar power capacity passed the milestone of 100000 megawatts in 2012 and was close to 137000 by the end of 2013. This is a rapid rate of increase that is likely to continue. Current solar electricity generation amounts to about 0.5 per cent of all global electricity generation. Germany, China, Italy, Japan, USA and Spain currently lead the global market for solar power.

Figure 12.18 Solar electricity generated by photovoltaic panels in Spain

Solar electricity is currently produced in two ways:

- **Photovoltaic (PV) systems** – solar panels that convert sunlight directly into electricity.
- **Concentrating solar power (CSP) systems** – use mirrors or lenses and tracking systems to focus a large area of sunlight into a small beam. This concentrated light is then used as a heat source for a conventional thermal power plant. The most developed CSP systems are the solar trough, parabolic dish and solar power tower. Each method varies in the way it tracks the Sun and focuses light. In each system, a fluid is heated by the concentrated sunlight, and is then used for power generation or energy storage.

Another idea being considered is to build solar towers. Here, a large glassed-in area would be constructed with a very tall tower in the middle. The hot air in this 'greenhouse' would rise rapidly up the tower, driving turbines along the way.

Traditional solar panels comprise arrays of photovoltaic cells made from silicon. These cells absorb photons in light and transfer their energy to electrons, which form an electrical circuit. However, standard solar panels:

- are costly to install
- have to be tilted and carefully positioned so as not to shade neighbouring panels.

A number of companies are now developing a new technique to manufacture solar panels. This involves using different materials and building them in very thin layers or films, almost like printing on paper, to produce the photovoltaic effect. The cost of production is reduced because the layers or films use less material, and they can be deposited on bases such as plastic, glass or metal.

Tidal power

Although currently in its infancy, a study by the Electric Power Research Institute has estimated that as much as 10 per cent of US electricity could eventually be supplied by tidal energy. This potential could be equalled in the UK and surpassed in Canada.

Tidal power plants act like underwater windmills, transforming sea currents into electrical current. Tidal power is more predictable than solar or wind power, and the infrastructure is less obtrusive, but start-up costs are high. The 240 megawatt Rance facility in north-western France is the only utility-scale tidal power system in the world. However, the greatest potential is Canada's Bay of Fundy in Nova Scotia. A pilot plant was opened at Annapolis Royal in 1984, which at peak output can generate 20 megawatts. More ambitious projects at other sites along the Bay of Fundy are under consideration, but there are environmental concerns. The main concerns are potential effects on fish populations, levels of sedimentation building up behind facilities and the possible impact on tides along the coast.

Figure 12.19 Animal dung being dried for fuel in India

According to the World Energy Outlook, 1.3 billion people were still living without access to electricity in 2012. This is equal to 18 per cent of the world's population. Nearly 97 per cent of those without access to electricity live in Sub-Saharan Africa or in Asia. The largest populations without electricity are in India, Nigeria, Ethiopia, Bangladesh, Democratic Republic of Congo and Indonesia.

In LICs, the concept of the **energy ladder** is important. Here, a transition from fuelwood and animal dung to 'higher-level' sources of energy occurs as part of the process of economic development. Income, regional electrification and household size are the main factors affecting the demand for fuelwood. Forest depletion is therefore initially heavy near urban areas but slows down as cities become wealthier and change to other forms of energy. It is the more isolated rural areas that are most likely to lack connection to an electricity grid. It is in such areas that the reliance on fuelwood is greatest. Wood is likely to remain the main source of fuel for the global poor in the foreseeable future.

The collection of fuelwood does not cause deforestation on the same scale as the clearance of land for agriculture, but it can seriously deplete wooded areas. The use of fuelwood is the main cause of indoor air pollution in LICs. Indoor air pollution is responsible for 1.5 million deaths every year. More than half of these deaths are of children below the age of 5.

Section 12.1 Activities

1 Suggest reasons for the variations and trends in the consumption of hydro-electricity by world region.
2 Discuss recent changes in the installed capacity of wind energy.
3 a What are biofuels?
  b Why has biofuel production expanded so rapidly?
  c Examine the advantages and disadvantages of biofuels.
4 a What is geothermal energy?
  b Explain the geographical locations of the main producing countries.
5 Explain the difference between photovoltaic and concentrated solar power systems.

Fuelwood in LICs

In LICs, about 2.5 billion people rely on fuelwood, charcoal and animal dung for cooking (Figure 12.19). Fuelwood and charcoal are collectively called fuelwood, which accounts for just over half of global wood production. Fuelwood provides much of the energy needs for Sub-Saharan Africa. It is also the most important use of wood in Asia.

☐ Trends in high-, middle- and low-income countries

Deindustrialisation, increasing energy efficiency and relatively low population growth in HICs in general has resulted in a decrease in primary energy consumption in all four HICS considered in Table 12.8. The decrease in the USA is marginal, but more significant in Japan, Germany

Table 12.8 Primary energy consumption, 2002–12 (million tonnes oil equivalent)

	Country	2002	2012
HICs	USA	2295.5	2208.8
	Japan	513.3	478.2
	Germany	334.0	311.7
	UK	221.7	203.6
MICs	South Korea	203.0	271.1
	Malaysia	53.1	76.3
	China	1073.8	2735.2
	India	310.8	563.5
LICs	Bangladesh	14.8	26.3
	Pakistan	47.4	69.3
	Peru	12.1	22.3
	Algeria	28.6	44.6

and the UK. There are examples of HICs, such as Canada and Spain, where energy consumption increased during this ten-year time period, but such increases have been modest. In such countries, there is every prospect that they will follow the example of the four HICS considered in Table 12.8 in due course.

In contrast in the MICs, growth rates were considerable and consistent with high rates of economic growth. Consumption in China increased at an incredible rate. In 2002, energy consumption in China was less than half that of the USA. By 2012, it was almost 24 per cent higher! India, which consumed less primary energy than

Japan and Germany in 2002, now consumes considerably more than both of these countries. In the same time period, South Korea overtook the UK in primary energy consumption.

Although most LICs struggle to fund their energy requirements, all four countries in Table 12.8 show a significant increase in primary energy consumption. Energy is vital for economic growth and to satisfy the basic demands of growing populations. Bangladesh, which has a considerably higher population than both South Korea and Malaysia, consumed far less energy than both of its Asian neighbours in 2012, even though its own energy consumption rose significantly in the period under consideration.

There is a strong positive correlation between GNP per person and energy use. In poor countries, it is the high- and middle-income groups who generally have enough money to purchase sufficient energy, and they also tend to live in locations where electricity is available. It is the poor in such countries who lack access to the advantages that electricity brings.

□ The environmental impact of energy

Increasing energy insecurity has stimulated exploration of technically difficult and environmentally sensitive areas. Such exploration and development is economically feasible when energy prices are high, but becomes less so when prices fall. No energy production location has suffered more environmental damage than the Niger delta in West Africa.

Case Study: The Niger delta

The Niger delta (Figure 12.20) covers an area of 70 000 km², making up 7.5 per cent of Nigeria's land area. It contains over 75 per cent of Africa's remaining **mangroves**. A report published in 2006 estimated that up to 1.5 million tonnes of oil has been spilt in the delta over the last 50 years. The report, compiled by WWF, says that the delta is one of the five most polluted spots on Earth. Pollution is destroying the livelihoods of many of the 20 million people who live in the delta. The pollution damages crops and fishing grounds, and is a major contributor to the upsurge in violence in the region. People in the region are dissatisfied with bearing the considerable costs of the oil industry but seeing very little in terms of the benefits. The report accused the oil companies of not using the advanced technologies available to them to combat pollution. However, Shell claims that 95 per cent of oil discharges in the last five years have been caused by sabotage.

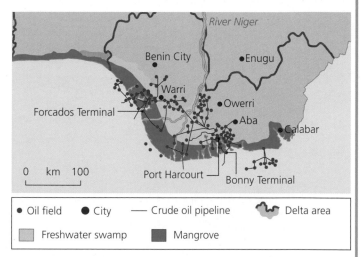

Figure 12.20 Map of oil fields in the Niger Delta

The flaring (burning) of unwanted natural gas found with the oil is a major regional and global environmental problem. The gas found here is not useful because there is no gas pipeline infrastructure to take it to consumer markets. It is estimated that 70 million m³ are flared off each day. This is equivalent to 40 per cent of Africa's natural gas consumption. Gas flaring in the Niger delta is the world's single largest source of greenhouse gas emissions.

One of the world's largest wetlands, and Africa's largest remaining mangrove forest, has suffered an environmental disaster:

- Oil spills, acid rain from gas flares and the stripping away of mangroves for pipeline routes have killed off fish.
- Between 1986 and 2003, more than 20 000 hectares of mangroves disappeared from the coast, mainly due to land clearing and canal dredging for oil and gas exploration.
- The oilfields contain large amounts of natural gas. This is generally burnt off as flares rather than being stored or reinjected into the ground. Hundreds of flares have burned continuously for decades. This causes acid rain and releases greenhouse gases.

- The government has recognised 6817 oil spills in the region since the beginning of oil production. Critics say the number is much higher.
- Construction and increased shipping have changed local wave patterns, causing shore erosion and the migration of fish into deeper water.
- Various types of construction have taken place without adequate environmental impact studies.

The federal environmental protection agency has only existed since 1988 and **environmental impact assessments** were not compulsory until 1992.

In early 2015, the major oil company Royal Dutch Shell agreed to an $84 million settlement with the Bodo community in the Niger delta for two oil spills that were among the biggest spills in decades in Nigeria. Thousands of hectares of mangrove were affected. The money will go to over 15 000 fishermen whose livelihoods were affected and to the community in general. According to Amnesty International, Royal Dutch Shell and the Italian company ENI have admitted to more than 550 oil spills in the Niger delta in 2014. By contrast, on average, there were only 10 spills a year across the whole of Europe between 1971 and 2011.

Case Study: Oil sands in Canada and Venezuela

Huge **oil-sand deposits** in Alberta, Canada, and Venezuela could be critical over the next 50 years as the world's production of conventional oil falls. The oil sands are a mixture of bitumen and sand. The bitumen will not flow unless heated or diluted with lighter hydrocarbons to make it transportable by pipelines and usable by refineries. Such synthetic oil, which can also be made from coal and natural gas, could provide a vital bridge to an era of new technologies. The government of Alberta estimates that recoverable oil reserves total about 200 billion barrels.

In 2012, the oil sands provided direct employment to over 22 000 workers, but indirect employment is many times higher. Current production is about 1.3 million barrels a day. This is expected to grow to 3 million barrels a day by 2020. In 2012, 56 per cent of Canada's total oil production came from oil sands. Alberta's oil sands are the third largest oil reserve in the world (Figure 12.21).

However, there are serious environmental concerns about the development of oil sands, which have a big carbon footprint:

- It takes 2 tonnes of mined sand to produce 1 barrel of synthetic crude, leaving lots of waste sand.
- It takes about three times as much energy to produce a barrel of Alberta oil-sands crude as it does a conventional barrel of oil. Thus, oil sands are large sources of greenhouse gas emissions.
- Oil sands require 2–4.5 barrels of water to produce a single barrel of oil.
- The development of oil sands has had a huge impact on the landscape, including the removal of a significant area of boreal forest.

The environmental organisation Greenpeace has called on the Canadian government and the oil companies to stop the development of oil sands on environmental, health and social grounds.

Venezuela's heavy oil production has not kept pace with that of Canada, but in 2013 it totalled over 1.25 million barrels a day.

Figure 12.21 Operations in Alberta's oil sands

Pathways crossing difficult environments

Energy pathways are supply routes between energy producers and consumers that may be pipelines, shipping routes or electricity cables. As energy companies have had to search further afield for new sources of oil, new energy pathways have had to be constructed. Some major oil and gas pipelines cross some of the world's most inhospitable terrain. The Trans-Alaskan Pipeline (TAP) crosses three mountain ranges and several large rivers. Much of the pipeline is above ground to avoid the permafrost problem. Here, the ground is permanently frozen down to about 300 metres, apart from the top metre, which melts during the summer. Building foundations and the uprights that hold the pipeline above ground have to extend well below the melting zone (called the 'active layer'). The oil takes about 6 days to make the 1270 kilometre journey. Engineers fly over the pipeline every day by helicopter to check for leaks and other problems. Problems such as subsidence have closed the pipeline for short periods.

Section 12.1 Activities

1 Why is fuelwood a vital source of energy in many poor countries?
2 With reference to Table 12.8, describe and explain contrasting trends in energy consumption between HICs, MICs and LICs.
3 The Niger delta has been described as an 'environmental disaster area'. Briefly discuss this assertion.
4 Outline the advantages and disadvantages of exploiting oil-sand deposits.
5 What are energy pathways and why are they so important?

12.2 The management of energy supply

Case Study: China

China's energy mix

China is the biggest consumer and producer of energy in the world (Table 12.9). It overtook the USA in total energy usage in 2009. The USA had held the top position in the energy usage league for more than a century. In 2012, China consumed 2.74 billion tonnes of oil equivalent, compared with 2.2 billion tonnes in the USA. The demand for energy in China continues to increase significantly as the country expands its industrial base. However, energy usage per person in the USA is much higher, with the average American using more than four times the Chinese average.

China's energy consumption rose by 45 per cent in the 7 years to 2013, according to data from the National Bureau of Statistics. However, in 2014 the Chinese government announced plans to cap the increasing rate at which it consumes energy to 28 per cent for the seven-year period to 2020.

Table 12.9 China's Key Energy Statistics

		World rank
Total primary energy production 2012	101.781 quadrillion British thermal units	1
Total primary energy consumption 2012	105.882 quadrillion British thermal units	1
Total primary coal production 2012	4 017 920 thousand short tons	1
Total petroleum consumption 2013	10 480 thousand barrels per day	2
Total electricity net generation 2012	4 768 billion kilowatt hours	1

Source: eia.gov/beta/international/country

In 2012, China's energy consumption breakdown by energy source was:

- **coal**: 68.4 per cent
- **oil**: 17.6 per cent
- **hydro-electricity**: 7.1 per cent
- **natural gas**: 4.7 per cent
- **nuclear energy**: 0.8 per cent
- **renewable energy** (other than hydro-electricity): 1.2 per cent.

An evolving energy policy

China's energy policy has evolved over time. As the economy expanded rapidly in the 1980s and 1990s, much emphasis was placed on China's main energy resource, coal, in terms of both increasing production and building more coal-fired power stations. However, this was at the expense of huge environmental impact and an alarmingly high casualty rate among coal miners. In 2012, China consumed just over half of the world's coal. Between 2003 and 2013, China accounted for 87 per cent of the growth in global coal consumption. Coal is the dirtiest of the fossil fuels and thus the environmental consequences of such a heavy reliance on coal were all too predictable. According to Greenpeace, 80 per cent of China's carbon dioxide emissions and 85 per cent of its sulphur dioxide pollution comes from burning coal. China is the world's leading energy-related CO_2 emitter. In November 2014, China unveiled an accord aimed at limiting carbon emissions.

China was also an exporter of oil until the early 1990s, although it is now a very significant importer (Figure 12.22). China is the world's second largest consumer of oil, and moved from second largest net importer of oil to the largest in 2014. This transformation has had a major impact on Chinese energy policy as the country has sought to secure overseas

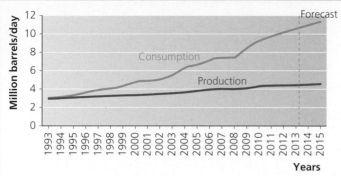

Source: EIA International Energy Statistics and Short-Term Energy Outlook, January 2014

Figure 12.22 Chinese oil consumption and production, 1993–2015

sources of supply. As a result, China has had an increasing influence on the global energy market. Long-term energy security is viewed as essential if the country is to maintain the pace of its industrial revolution.

In recent years, China has tried to take a more balanced approach to energy supply and at the same time reduce its environmental impact. The 11th Five-Year Plan (2006–10) focused on two major energy-related objectives: **a** to reduce energy use per unit GDP by 20 per cent and **b** to ensure a more secure supply of energy. Because of the dominant position of coal in China's energy mix, the development of clean coal technology is central to China's energy policy with regard to fossil fuels. China has emerged in the last two years as the world's leading builder of more efficient, less polluting coal power plants. China has begun constructing such clean coal plants at a rate of one a month. The government has begun to require that power companies retire an older, more polluting power plant for each new one they build.

The further development of nuclear and hydropower is another important strand of Chinese policy. The country also aims to stabilise and increase the production of oil while augmenting that of natural gas and improving the national oil and gas network. Nuclear power reached a capacity of 9.1 gigawatts by the end of 2008, with a target capacity of 40 gigawatts by 2020. By the end of 2009, China had 11 operational nuclear reactors with a further 17 under construction. The World Nuclear Association (WNA) says that China has a further 124 nuclear reactors on the drawing board. This will lead to a dramatic increase in China's demand for uranium, the raw material of nuclear reactors.

China's strategic petroleum reserve

As part of China's concerns about energy security and its increasing reliance on oil imports, the country is developing a strategic petroleum reserve (Figure 12.23). The plan is for China to build facilities that can hold 500 million barrels of crude oil by 2020 in three phases. This will be equivalent to about 90 days' supply. Phase 1, completed in 2009, consists of four sites with a total storage capacity of 103 million barrels. Phase 2, to be completed by the end of 2015, will add a further 170 million barrels of storage capacity.

Figure 12.23 China's strategic petroleum reserve (SPR)

China is following the USA and other countries in building up a petroleum reserve. This will protect China to a certain extent from fluctuations in the global oil price, which can arise for a variety of reasons.

Renewable energy policy

China aims to produce at least 15 per cent of overall energy output from renewable energy sources by 2020 as the government seeks to improve environmental conditions. Renewable energy currently contributes more than one-quarter of China's total installed energy capacity, with hydro-electricity by far the largest contributor. China produces more hydro-electricity than any other country in the world, with hydro-electricity accounting for more than 15 per cent of the country's total electricity generation. The world's largest hydro-electricity project, the Three Gorges Dam along the Yangtze River, was completed in 2012. It has 32 generators with a total maximum capacity of 22.5 gigawatts.

China is now the world leader in wind energy, currently accounting for 31 per cent of global installed wind-power capacity. The year 2008 saw the initial development of China's offshore wind farm policy. China's wind turbine manufacturing industry is now the largest in the world. Chinese policy is not just to gain the energy advantages of wind energy but also to develop it as a significant industrial sector. China is now also the largest manufacturer of solar PV. The solar hot-water market in China has also continued to boom, partly as a result of a new rural energy subsidy programme for home appliances, for which solar hot water qualifies. China aims to increase solar electricity capacity from 3 gigawatts in 2012 to 35 gigawatts by the end of 2015.

The Three Gorges Dam

The Three Gorges Dam across the Yangtze River in China is the world's largest electricity-generating plant of any kind (Figures 12.24 and 12.25). The dam is over 2 kilometres long and 100 metres high. The lake impounded behind it is over 600 kilometres long. All of the originally planned components

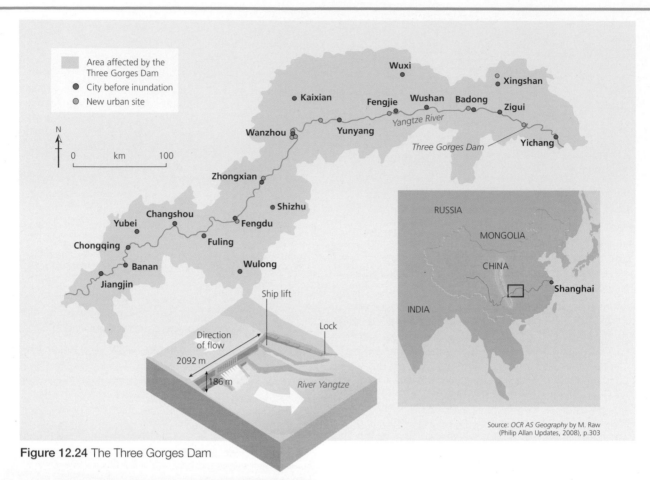

Figure 12.24 The Three Gorges Dam

Source: *OCR AS Geography* by M. Raw
(Philip Allan Updates, 2008), p.303

Figure 12.25 The Three Gorges Dam

were completed in late 2008. The Dam began running at full capacity in mid-2012 when the last of its 32 generators became operational. Total generating capacity is now 22.5 gigawatts.

One objective of such a large capacity is to reduce China's dependence on coal. The dam supplies Shanghai and Chongqing in particular with electricity. This is a multipurpose scheme that also increases the river's navigation capacity and reduces the potential for floods downstream. The dam has raised water levels by 90 metres upstream, transforming the rapids in the gorge into a lake, allowing shipping to function in this stretch of the river. The dam protects an estimated 10 million people from flooding.

However, there was considerable opposition to the dam because:

- over 1 million people had to be moved to make way for the dam and the lake
- much of the resettlement has been on land above 800 metres above sea level, which is colder and has less fertile soils
- the area is seismically active and landslides are frequent
- there are concerns that silting will quickly reduce the efficiency of the project
- significant archaeological treasures were drowned
- the dam interferes with aquatic life
- the total cost is estimated at $70 billion; many people argue that this money could have been better spent.

Section 12.2 Activities

1 Describe China's energy mix.
2 Comment on the trends in China's production and consumption of oil shown in Figure 12.22.
3 Why is China developing a strategic petroleum reserve?
4 Discuss China's renewable energy policy.
5 a Explain the objectives of the Three Gorges Dam.
  b For what reasons was the construction of the Three Gorges Dam opposed?

12.3 Environmental degradation

☐ Pollution: land, air and water

Pollution is the dominant factor in the **environmental degradation** of land, air and water and impacts significantly on human health. Figure 12.26 shows the considerable global variations in deaths from urban air pollution. Compare the relatively low incidence in southern and eastern Africa and western Europe with the very high level in China and a number of other Asian countries (Figure 12.27). A recent report by the World Health Organization estimates that about 8 million people died in 2012 as a result of air pollution. This figure comprised:

- **3.7 million deaths attributable to ambient air pollution** – 88 per cent of those premature deaths occurred in LICs and MICs
- **4.3 million deaths attributable to household air pollution** – almost all of these deaths occurred in LICs and MICs. About 3 billion people cook and heat their homes using solid fuels (that is, wood, crop wastes, charcoal, coal and dung) in open fires and leaky stoves.

These findings are higher than previous estimates and confirm that air pollution is the world's greatest single environmental health risk.

A study in China revealed that children exposed to highly polluted air while in the womb had more changes in their DNA, and a higher risk of developmental problems, than those whose mothers breathed cleaner air during pregnancy. Apart from the direct effects on health of pollution there are considerable indirect economic effects, which include:

- the cost of healthcare for pollution-related illnesses
- interruptions to the education of children, which may cause them to leave school with lower qualifications than expected
- lost labour productivity.

Pollution has a considerable negative impact on ecosystems all over the world. In some regions, the changes brought about by pollution have been little short of catastrophic.

Figure 12.27 Air pollution over Ulaanbaatar, Mongolia

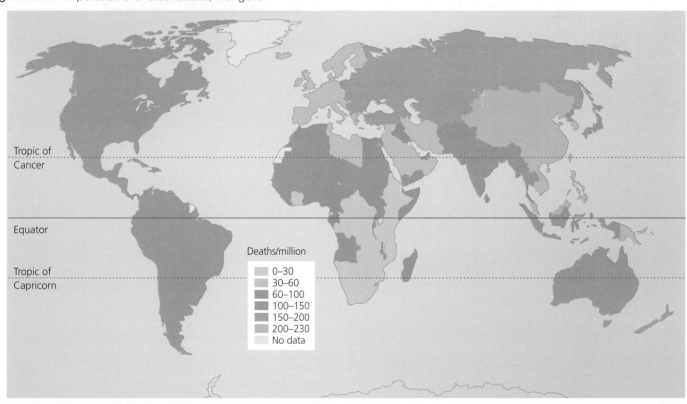

Figure 12.26 Global distribution of deaths from urban air pollution

Case Study: Environmental degradation in the Niger delta

The Niger delta covers an area of 70 000 km², making up 7.5 per cent of Nigeria's land area. It contains over 75 per cent of Africa's remaining mangrove. A report published in 2006 estimated that up to 1.5 million tonnes of oil has been spilt in the delta over the last 50 years. The report, compiled by WWF, says that the delta is one of the five most polluted spots on Earth. Pollution is destroying the livelihoods of many of the 20 million people who live there. The pollution damages crops and fishing grounds and is a major contributor to the upsurge in violence in the region. People in the region are dissatisfied with bearing the considerable costs of the oil industry but seeing very little in terms of the benefits. The report accused

the oil companies of not using the advanced technologies to combat pollution that are evident in other world regions. However, shell claims that 95 per cent of oil discharges in the last five years have been caused by sabotage.

The flaring (burning) of unwanted natural gas found with the oil is a major regional and global environmental problem. The gas found here is not useful because there is no gas pipeline infrastructure to take it to consumer markets. It is estimated that 70 million m³ are flared off each day. This is equivalent to 40 per cent of Africa's natural gas consumption. Gas flaring in the Niger delta is the world's single largest source of greenhouse-gas emissions.

Figure 12.28 Environmental problems in the Niger delta

Virtually every substance is **toxic** at a certain dosage. The most serious polluters are the large-scale processing industries, which tend to form agglomerations as they have similar locational requirements (Table 12.10). The impact of a large industrial agglomeration may spread well beyond the locality and region, to cross international borders. For example, prevailing winds in Europe generally carry pollution from west to east. Thus the problems caused by acid rain in Scandinavia have been due partly to industrial activity in the UK. Dry and wet deposition can be carried for considerable distances. For example, pollution found in Alaska in the 1970s was traced back to the Ruhr industrial area in Germany.

Pollution is the major **externality** of industrial and urban areas. It is at its most intense at the focus of pollution-causing activities, declining with distance from such concentrations. For some sources of pollution, it is possible to map the **externality gradient and field** (Figure 12.29). In general, health risk and environmental impact is greatest immediately around the source of pollution, and the risk decreases with distance from the source. However, atmospheric conditions and other factors can complicate this pattern. Exposure to pollution can result in health and environmental effects (Table 12.11) that range from fairly minor to severe.

Table 12.10 The most polluting industries

Industrial sector	Examples
Fuel and power	Power stations, oil refineries
Mineral industries	Cement, glass, ceramics
Waste disposal	Incineration, chemical recovery
Chemicals	Pesticides, pharmaceuticals, organic and inorganic compounds
Metal industries	Iron and steel, smelting, non-ferrous metals
Others	Paper manufacture, timber preparation, uranium processing

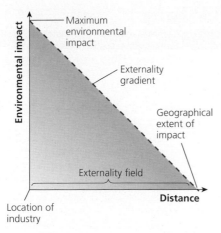

Figure 12.29 Externality gradient and field

Table 12.11 Major sources and health and environmental effects of air pollutants

	Major sources	Health effects	Environmental effects
Sulphur dioxide (SO_2)	Industry	Respiratory and cardiovascular illnesses	Precursor to acid rain, which damages lakes, rivers and trees; damage to cultural relics
Nitrous oxide (NO_x)	Vehicles, industry	Respiratory and cardiovascular illnesses	Nitrogen deposition leading to overfertilisation and eutrophication
Particulate matter	Vehicles, industry	Particles penetrate deep into lungs and can enter bloodstream	Visibility
Carbon monoxide (CO)	Vehicles	Headaches and fatigue, especially in people with weak cardiovascular health	
Lead (Pb)	Vehicles (burning leaded gasoline)	Accumulates in bloodstream over time; damages nervous system	Kills fish/animals
Ozone (O_3)	Formed from reaction of nitrous oxides and VOCs	Respiratory illnesses	Reduced crop production and forest growth; smog precursor
Volatile organic compounds (VOCs)	Vehicles, industrial processes	Eye and skin irritation; nausea, headaches; carcinogenic	Smog precursor

Strategies to tackle air pollution

Considering the intense use of energy and materials, levels of pollution have generally declined in HICs:

- In recent decades, increasingly strict environmental legislation has been passed in these countries. This is the beginning of a process to make polluters pay for the cost of their actions themselves, rather than expecting society as a whole to pay the costs.
- Industry has spent increasing amounts on research and development to reduce pollution – the so-called 'greening of industry'.
- The most polluting activities, such as commodity processing and heavy manufacturing, have been relocated to the emerging market economies.
- The expectation is that after a certain stage of economic development in a country, the level of pollution and the degradation it causes will decline (Figure 12.30).

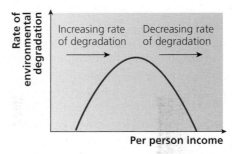

Figure 12.30 The environmental Kuznets curve

Types and amounts of pollution change with economic development. In low-income economies where primary industries frequently dominate, pollution related to agriculture and mining often predominates. As countries industrialise, manufacturing industries, energy production and transport become major polluters. The focal points of pollution will be the large urban-industrial complexes. The NICs of the world are in this stage. In contrast, the HICs have experienced deindustrialisation as many of their major polluting industries have filtered down to NICs. This has resulted in improved environmental conditions in many HICs in general, although pollution from transportation has often increased.

Figure 12.31 shows the measures advocated by the World Health Organization to reduce pollution and its multitude of adverse effects.

It is important to consider the different levels of impact between one-off pollution incidents (**incidental pollution**) and longer-term pollution (**sustained pollution**). The former is mainly linked to major accidents caused by technological failures and human error. Consequences of the latter include ozone depletion and global warming. Some of the worst examples of incidental pollution are shown in Table 12.12.

Major examples of incidental pollution, such as Chernobyl and Bhopal, can have extremely long-lasting consequences that are often difficult to determine in the earlier stages. The effects of both accidents are still being felt more than two decades after they occurred.

There are many examples of successful policies in transport, urban planning, power generation and industry that reduce air pollution:

- **For industry** – clean technologies that reduce industrial smokestack emissions; improved management of urban and agricultural waste, including capture of methane gas emitted from waste sites as an alternative to incineration (for use as biogas)
- **For transport** – shifting to clean modes of power generation; prioritising rapid urban transit, walking and cycling networks in cities as well as rail interurban freight and passenger travel; shifting to cleaner heavy-duty diesel vehicles and low-emissions vehicles and fuels, including fuels with reduced sulfur content
- **For urban planning** – improving the energy efficiency of buildings and making cities more compact, and thus energy efficient
- **For power generation** – increased use of low-emissions fuels and renewable combustion-free power sources (like solar, wind or hydropower); co-generation of heat and power; and distributed energy generation (for example mini-grids and rooftop solar-power generation)
- **For municipal and agricultural waste management** – strategies for waste reduction, waste separation, recycling and reuse or waste reprocessing, as well as improved methods of biological waste management such as anaerobic waste digestion to produce biogas, are feasible, low-cost alternatives to the open incineration of solid waste; where incineration is unavoidable, then combustion technologies with strict emission controls are critical.

Figure 12.31 The World Health Organization – reducing air pollution

It is usually the poorest people in a society who are exposed to the risks from both incidental and sustained pollution. In the USA, the geographic distribution of both minorities and the poor has been found to be highly correlated to the distribution of air pollution, municipal landfills and incinerators, abandoned toxic waste dumps and lead poisoning in children. The race correlation is even stronger than the class correlation. Unequal environmental protection undermines three basic types of equity:

- **procedural equity**, which refers to the extent that planning procedures, rules and regulations are applied in a non-discriminatory way
- **geographic equity**, which refers to the proximity of communities to environmental hazards and locally unwanted land uses such as smelters, refineries, sewage treatment plants and incinerators

Table 12.12 Major examples of incidental pollution

Location	Causes and consequences
Seveso, Italy	In July 1976, a reactor at a chemical factory near Seveso in northern Italy exploded, sending a toxic cloud into the atmosphere. An area of land 18 km^2 was contaminated with the dioxin TCDD. The immediate after-effects, seen in a small number of people with skin inflammation, were relatively mild. However, the long-term impact has been much worse. The population is suffering increased numbers of premature deaths from cancer, cardiovascular disease and diabetes.
Bhopal, India	A chemical factory owned by Union Carbide leaked deadly methyl isocyanate gas during the night of 3 December 1984. The plant was operated by a separate Indian subsidiary that worked to much lower safety standards than those required in the USA. It has been estimated that 8000 people died within two weeks, and a further 8000 have since died from gas-related diseases. The NGO Greenpeace puts the total fatality figure at over 20000. Bhopal is recognised as the world's worst industrial disaster.
Chernobyl, Ukraine	The world's worst nuclear power-plant accident occurred at Chernobyl, Ukraine, in April 1986. Reactor number four exploded, sending a plume of highly radioactive fallout into the atmosphere, which drifted over extensive parts of Europe and eastern North America. Two people died in the initial explosion and over 336000 people were evacuated and resettled. In total, 56 direct deaths and an estimated 4000 extra cancer deaths have been attributed to Chernobyl. The estimated cost of $200 billion makes Chernobyl the most expensive disaster in modern history.
Harbin, China	An explosion at a large petrochemical plant in the north-east Chinese city of Harbin released toxic pollutants into a major river. Benzene levels were 108 times above national safety levels. Benzene is a highly poisonous toxin that is also carcinogenic. Water supplies to the city were suspended. Five people were killed in the blast and more than 60 injured; 10000 residents were temporarily evacuated.

- **social equity**, which refers to the role of race and class in environmental decision-making.

Ironically, some government actions have created and exacerbated environmental inequity. More stringent environmental regulations have driven noxious facilities to follow the path of least resistance towards poor, overburdened communities where protesters lack the financial support and professional skills of more affluent areas, or where the prospect of bringing in much-needed jobs justifies the risks in the eyes of some residents.

Sustained pollution: ozone depletion and skin cancer

The **ozone layer** in the stratosphere prevents most harmful ultraviolet (UV) radiation from passing through the atmosphere. However, chlorofluorocarbons (CFCs) and other ozone-depleting substances have caused an estimated decline of about 4 per cent a decade in the ozone layer of the stratosphere since the late 1970s. Depletion of the ozone layer allows more UV radiation to reach the ground, leading to more cases of skin cancer, cataracts and other health and environmental problems. Widespread global concern resulted in the Montreal Protocol banning the production of CFCs and related ozone-depleting chemicals.

Skin cancer is the fastest-growing type of cancer in the USA. In the UK, it is the second most common cancer in young people aged 20–39. Overexposure to UV radiation is the major cause. Skin cancer generally has a 20- to 30-year latency period. There is a significant relationship between skin cancer and latitude in the USA and Canada. However, the use of tanning salons has also been criticised, with a number of studies linking the use of artificial tanning to cases of skin cancer. The World Health Organization's estimates on UV-related mortality and morbidity are that annually about 1.5 million DALYs (disability-adjusted life years) are lost through excessive UV exposure.

Sustained pollution, such as that caused by ozone-depleting substances, usually takes much longer to have a substantial impact on human populations than incidental pollution, but it is likely to affect many more people in the long term. Likewise, tackling the causes of sustained pollution will invariably be a much more difficult task, as the sources of incidental pollution are much more localised compared with the more ubiquitous nature of sustained pollution.

Case Study: China's 'cancer villages'

China's rapid economic growth has led to widespread environmental problems. Pollution problems are so severe in some areas that the term 'cancer village' has become commonplace. In the village of Xiditou, south-east of Beijing, the cancer rate is 30 times the national average. This has been blamed on water and air contaminated by chemical factories. Tests on tap water have found traces of highly carcinogenic benzene that were 50 per cent above national safe limits. In the rush for economic growth, local governments eagerly built factories, but they had very limited experience of environmental controls. Some facts support this:

- The Chinese government admits that 300 million people drink polluted water.
- This water comes from polluted rivers and groundwater.
- 30 000 children in China die of diarrhoea or other water-borne illnesses each year.
- The River Liao is the most polluted, followed by waterways around Tianjin and the River Huai.

In 2013, the Chinese government promised to tackle 'cancer villages', after a huge social media backlash from both ordinary Chinese people and global campaigners. This related mainly to Huangjiawa, a village in Shandong province, after it emerged that the area has one of the highest rates of stomach cancer in the world. It is thought the wells in the village have been contaminated with toxins from a nearby aluminium smelter, which also pumps pollution into the sky. News of this incident spread rapidly across Weibo, a Chinese social media site similar to Twitter. Greenpeace East Asia estimate that 320 million people are without access to clean drinking water in China and 190 million people are drinking water severely contaminated with hazardous chemicals.

Case Study: The BP oil spill in the Gulf of Mexico

On 20 April 2010, a large explosion occurred on the BP-licensed drilling rig Deepwater Horizon sited in the Gulf of Mexico (Figure 12.32). Eleven oil workers died and a blowout preventer, intended to prevent the release of crude oil, failed to activate. The rig sank in 1500 metres of water, with reports of an oil slick 8 kilometres long on the surface. Oil leaked from the well at sea-floor level for 87 days until 15 July when BP sealed it with a capping stack.

Figure 12.32 Deepwater Horizon exploding

Estimates of the amount of oil gushing from the well reached up to 40 000 barrels a day. The US government estimated that 4.9 million barrels of oil were spilled in total, making it the largest accidental oil spill in history. It was surpassed only by the 1991 Persian Gulf spill in which Iraq intentionally spilled twice this amount.

The environmental impact occurred in four ecosystems: the offshore waters, inshore coastal waters, seabed and shoreline wetlands and beaches. The spill also inflicted serious economic and psychological damage on communities along the Gulf Coast that depend on tourism, fishing and drilling. President Obama called the BP oil spill 'the worst environmental disaster America has ever faced'.

Most of the oil has now evaporated or dissolved, but an estimated 10 million gallons remain on the sea floor, and gobs of oil can still be found nestled into marshes along the coast. There is still debate about the impact on wildlife. BP has claimed that there is no evidence of any 'significant long-term population-level impact to any species'. However, other scientists have cautioned that the full scope of the impact will not be known for some time. There is evidence that bottlenose dolphins in the Gulf of Mexico have been dying at more than twice the normal rate over the last five years.

1 Define **a** *pollution* and **b** *environmental degradation*.
2 **a** With reference to Figure 12.26, describe the global distribution of deaths from urban air pollution.
  b Suggest reasons for the spatial variations you have identified.
3 Which industries are the largest polluters?
4 Write a brief explanation of Figure 12.29.
5 Explain the relationship illustrated by Figure 12.30.

☐ Water: demand, supply and quality

The global water crisis

> According to the WHO/UNICEF Joint Monitoring Programme for Water Supply and Sanitation, at least 1.8 billion people worldwide are estimated to drink water that is faecally contaminated. An even greater number drink water that is delivered through a system without adequate protection against sanitary hazards.
>
> *UN Water, 2014*

The longest a person can survive without water is about 10 days. All life and virtually every human activity needs water. It is the world's most essential resource and a pivotal element in poverty reduction. But for about 80 countries, with 40 per cent of the world's population, lack of water is a constant threat. And the situation is getting worse, with demand for water doubling every 20 years. In those parts of the world where there is enough water, it is being wasted, mismanaged and polluted (Figure 12.33) on a grand scale. In the poorest nations, it is not just a question of lack of water; the paltry supplies available are often polluted. The quality of drinking water is a major environmental determinent of health. **Water security** (Figure 12.34) has become a major issue in an increasing number of countries.

Figure 12.33 Polluted water – Christchurch, New Zealand

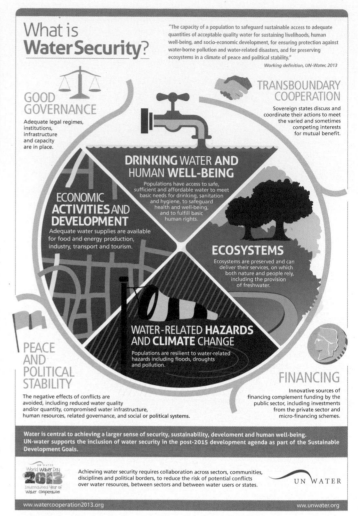

Source: www.unwater.org/fileadmin/user_upload/unwater_new/docs/water_security_poster_Oct2013.pdf

Figure 12.34 What is water security?

Securing access to clean water is a vital aspect of development.

- More than 840 000 people die each year from a water-related disease. While deaths associated with dirty water have been virtually eliminated from HICs, in LICs most deaths still result from water-borne disease.
- At any one time, half of the world's hospital beds are occupied by patients suffering from water-borne diseases.
- 750 million people around the world lack access to safe water – approximately one in nine people; 82 per cent of those who lack access to improved water live in rural areas, while just 18 per cent live in urban areas.
- In LICs, 70 per cent of industrial waste is dumped untreated into rivers and other water sources, which pollutes the usable water supply.
- Women and children spend 140 million hours a day collecting water.

Water scarcity has been presented as the 'sleeping tiger' of the world's environmental problems, threatening to put world food supplies in jeopardy, limit economic and social development and create serious conflicts between neighbouring drainage-basin countries. In the twentieth century, global water consumption grew six-fold, twice the rate of population growth. Much of this increased consumption was made possible by significant investment in water infrastructure, particularly dams and reservoirs affecting nearly 60 per cent of the world's major river basins.

The UN estimates that two-thirds of world population will be affected by 'severe water stress' by 2025. The situation will be particularly severe in Africa, the Middle East and South Asia. The UN notes that already a number of the world's great rivers such as the Colorado in the USA are running dry, and that **groundwater** is also being drained faster than it can be replenished. Many major **aquifers** have been seriously depleted, which will present serious consequences in the future. In an effort to add impetus to global water advancement, the UN proclaimed the period 2005–15 as the International Decade for Action, 'Water for Life'.

The Middle East and North Africa face the most serious problems. Since 1972, the Middle East has withdrawn more water from its rivers and aquifers each year than is being replenished. Yemen and Jordan are withdrawing 30 per cent more from groundwater resources annually than is being naturally replenished. Israel's annual demand exceeds its renewable supply by 15 per cent. In Africa, 206 million people live in water-stressed or water-scarce areas.

The Pilot Analysis of Global Ecosystems (PAGE), undertaken by the World Resources Institute, calculated water availability and demand by river basin. This analysis estimated that at present 2.3 billion people live in **water-stressed areas**, with 1.7 billion resident in **water-scarce areas**. The PAGE analysis forecasts that these figures will rise to 3.5 billion and 2.4 billion people respectively by 2025.

The link between poverty and water resources is very clear, with those living on less than $1.25 a day roughly equal to the number without access to safe drinking water. Improving access to safe water can be among the most cost-effective means of reducing illness and mortality (Figure 12.35). In LICs, it is common for water collectors, usually women and girls, to have to walk several kilometres every day to fetch water. Once filled, pots and jerry cans can weigh as much as 20 kilograms. In urban areas in LICs, water is still often distributed by donkey and cart (Figure 12.36).

Figure 12.35 The narrow irrigation zone along the banks of the River Nile, Egypt

Figure 12.36 Water collection/distribution in central Asia

Since 1930, global population has increased from 2 billion to over 7 billion, putting ever-increasing pressure on the world's water supplies. However, it is not just the increase in population that is influencing the demand for water, but also rising per person usage in many countries. As households become more affluent, they use more water in an increasing number of different ways.

Millennium Development Goal 7, target 10, stated: 'Halve, by 2015, the proportion of people without sustainable access to safe water and basic sanitation'. Although this goal has been achieved, much remains to be done to improve water security in many parts of the world.

Water utilisation at the regional scale

Every year, 110 000 km³ of precipitation falls onto the Earth's land surface. This would be more than adequate for the global population's needs, but much of it cannot be captured and the rest is very unevenly distributed. For example:

- Over 60 per cent of the world's population live in areas receiving only 25 per cent of global annual precipitation.
- The arid regions of the world cover 40 per cent of the world's land area, but receive only 2 per cent of global precipitation.
- The Congo River and its tributaries account for 30 per cent of Africa's annual runoff in an area containing 10 per cent of Africa's population.

Figure 12.37 shows what happens to the precipitation reaching land surfaces. **Green water** is that part of total precipitation that is absorbed by soil and plants, then released back into the air. As such, it is unavailable for human use. However, green water scarcity is the classic cause of famine. Green water accounts for 61.1 per cent of total precipitation. The remaining precipitation, known as **blue water**, collects in rivers, lakes, wetlands and groundwater. It is available for human use before it evaporates or reaches the ocean. As Figure 12.37 shows, only 1.5 per cent of total precipitation is directly used by people.

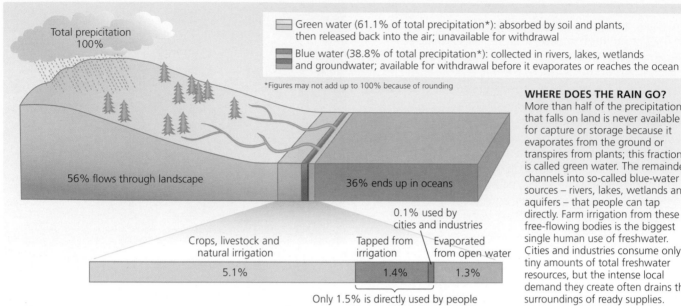

Green water (61.1% of total precipitation*): absorbed by soil and plants, then released back into the air; unavailable for withdrawal

Blue water (38.8% of total precipitation*): collected in rivers, lakes, wetlands and groundwater; available for withdrawal before it evaporates or reaches the ocean

Total prepicitation 100%

*Figures may not add up to 100% because of rounding

56% flows through landscape

36% ends up in oceans

0.1% used by cities and industries

Crops, livestock and natural irrigation
5.1%

Tapped from irrigation
1.4%

Evaporated from open water
1.3%

Only 1.5% is directly used by people

WHERE DOES THE RAIN GO?
More than half of the precipitation that falls on land is never available for capture or storage because it evaporates from the ground or transpires from plants; this fraction is called green water. The remainder channels into so-called blue-water sources – rivers, lakes, wetlands and aquifers – that people can tap directly. Farm irrigation from these free-flowing bodies is the biggest single human use of freshwater. Cities and industries consume only tiny amounts of total freshwater resources, but the intense local demand they create often drains the surroundings of ready supplies.

Figure 12.37 Where does the rain go?

Total world blue water withdrawals are estimated at 3390 km³, with 74 per cent for agriculture, mostly irrigation (Figure 12.38). About 20 per cent of this total comes from groundwater. Although agriculture is the dominant water user, industrial and domestic uses are growing at faster rates. Demand for industrial use has expanded particularly rapidly.

The amount of water used by a population depends not only on water availability but also on levels of **urbanisation** and economic development. As global urbanisation continues, the demand for **potable water** in cities and towns will rise rapidly. In many cases, demand will outstrip supply.

In terms of agriculture, more than 80 per cent of crop **evapotranspiration** comes directly from rainfall, with the remainder from irrigation water diverted from rivers and groundwater. However, this varies considerably by region. In the Middle East and North Africa, where rainfall is low and unreliable, more than 60 per cent of crop evapotranspiration originates from irrigation.

Figure 12.39 contrasts water use in HICs and LICs/MICs. In the latter, agriculture accounts for over 80 per cent of total water use, with industry using more of the remainder than domestic allocation. In the HICs, agriculture accounts for slightly more than 40 per cent of total water use. This is lower than the amount allocated to industry. As in LICs/MICs, domestic use is in third place.

As LICs industrialise and urban-industrial complexes expand, the demand for water grows rapidly in the industrial and domestic sectors. As a result, the competition with agriculture for water has intensified in many countries and regions. This is a scenario that has already played itself out in many HICs, where more and more difficult decisions are having to be made on how to allocate water.

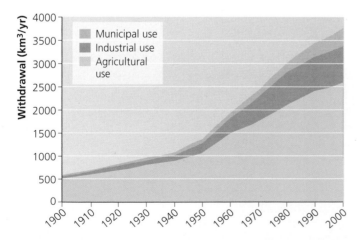

Figure 12.38 Global water use (agriculture, industry, domestic), 1900–2000

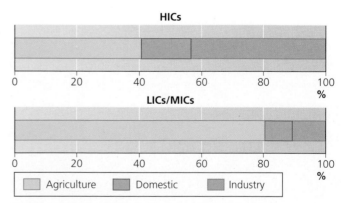

Figure 12.39 Water for agricultural, industrial and domestic uses in HICs and LICs/MICs

Large variations in water allocation can also exist within countries. For example, irrigation accounts for over 80 per cent of water demand in the west of the USA, but only about 6 per cent in the east.

The environmental and human factors affecting water scarcity

The world's population is increasing by about 80 million a year. This converts to an increased demand for freshwater of around 64 billion m³ per year, which equates to the total annual flow rate of the River Rhine.

A country is judged to experience water stress when water supply is below 1700 m³ per person per year. When water supply falls below 1000 m³ per person a year, a country faces water scarcity for all or part of the year. These concepts were developed by the Swedish hydrologist Malin Falkenmark.

Water scarcity is to do with the availability of potable water. **Physical water scarcity** is when physical access to water is limited. This is when demand outstrips a region's ability to provide the water needed by the population. It is the arid and semi-arid regions of the world that are most associated with physical water scarcity. Here, temperatures and evapotranspiration rates are very high and precipitation low. In the worst-affected areas, points of access to safe drinking water are few and far between

However, annual precipitation figures fail to tell the whole story. Much of the freshwater supply comes in the form of seasonal rainfall (Figure 12.40), as exemplified by the monsoon rains of Asia. India gets 90 per cent of its annual rainfall during the summer monsoon season from June to September. National figures can also mask

Figure 12.40 The dried-up bed of the Rio Oja, northern Spain

significant regional differences. Analysis of the supply and demand situation by river basin can reveal the true extent of such variations. For example, the USA has a relatively high average water-sufficiency figure of 8838 m³ per person a year. However, the Colorado river basin has a much lower figure of 2000, while the Rio Grande river basin is lower still at 621 m³ per person a year.

However, in increasing areas of the world, physical water scarcity is the result of human activity, largely overuse. Examples of physical water scarcity include:

- Egypt has to import more than half of its food because it does not have enough water to grow it domestically.
- The Murray–Darling basin in Australia has diverted large quantities of water to agriculture.
- The Colorado river basin in the USA once had an abundant supply of water but resources have been heavily overused, leading to very serious physical water scarcity downstream.

Figure 12.41 shows these regions and other parts of the world that suffer from physical water scarcity.

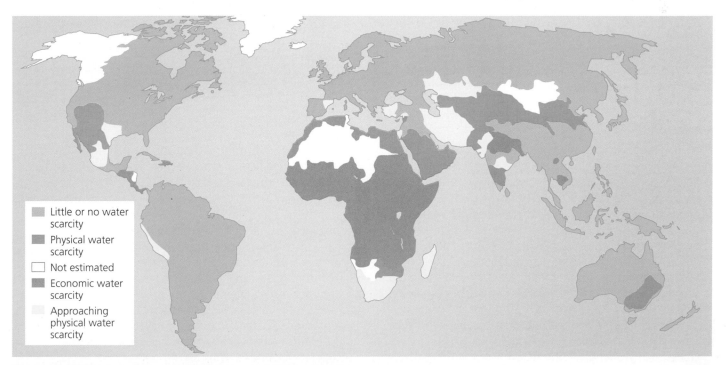

Little or no water scarcity

Physical water scarcity

Not estimated

Economic water scarcity

Approaching physical water scarcity

Figure 12.41 Physical water scarcity and economic water scarcity

Economic water scarcity exists when a population does not have the necessary monetary means to utilise an adequate supply of water. The unequal distribution of resources is central to economic water scarcity, where the crux of the problem is lack of investment. This occurs for a number of reasons, including political and ethnic conflict. Figure 12.41 shows that much of Sub-Saharan Africa is affected by this type of water scarcity.

Scientists expect water scarcity to become more severe, largely because:

- the world's population continues to increase significantly
- increasing affluence is inflating the per person demand for water
- there is an increasing demand for production of biofuels – biofuel crops are heavy users of water
- climate change is increasing aridity and reducing supply in many regions
- many water sources are threatened by various forms of pollution.

The Stockholm International Water Institute has estimated that each person on Earth needs a minimum of 1000 m^3 of water per year for drinking, hygiene and growing food for sustenance. Whether this water is available depends largely on where people live on the planet, as water supply is extremely inequitable. For example, major rivers such as the Yangtze, Ganges and Nile are severely overused and the levels of underground aquifers beneath major cities such as Beijing and New Delhi are falling.

In many parts of the world, the allocation of water depends largely on the ability to pay. A recent article in *Scientific American*, entitled 'Facing the freshwater crisis', quotes an old saying from the American West: 'Water usually runs downhill, but it always runs uphill to money' – meaning that poorer people and non-human consumers of water, the fauna and flora of nearby ecosystems, usually lose out when water is scarce.

Virtual water

The importance of the concept of **virtual water** is being increasingly recognised. Virtual water is the amount of water that is used to produce food or any other product and is thus essentially embedded in the item. One kilogram of wheat takes around 1000 litres of water to produce, so the import of this amount of wheat into a dry country saves that country this amount of water. According to *Scientific American* (August 2008, page 34), 'The virtual water concept and expanded trade have also led to the resolution of many international disputes caused by water scarcity. Imports of virtual water in products by Jordan have reduced the chance of water-based conflict with its neighbour Israel, for example.'

The size of global trade in virtual water is more than 800 billion m^3 of water a year. This is equivalent to the flow of ten Nile Rivers. Greater liberalisation of trade in agricultural products would further increase virtual water flows.

☐ The degradation of rural environments

Rural environments supply humankind with most of its food and gene pool and contain the vast majority of the world's forested land. However, rural areas all around the world have been degraded at a rapid rate over the last century. This has been due primarily to population growth and increasing pressures on the land, although urban activities through processes such as climate change can also have profound consequences for rural environments.

The UN Food and Agriculture Organization defines soil degradation as 'a change in the soil health status resulting in a diminished capacity of the ecosystem to provide goods and services for its beneficiaries'. Soil degradation involves both the physical loss (erosion) and the reduction in quality of topsoil associated with nutrient decline and contamination. It impacts significantly on agriculture and also has implications for the urban environment, pollution and flooding.

Globally, it is estimated that 2 billion hectares of soil resources have been degraded. This is equivalent to about 15 per cent of the Earth's land area. Such a scale of soil degradation has resulted in the loss of 15 per cent of world agricultural supply in the last 50 years. Some scientists consider this to be a 'slow-motion disaster'. In temperate areas, much soil degradation is a result of market forces and the attitudes adopted by commercial farmers and governments. In contrast, in the tropics much degradation results from high population pressure, land shortages and lack of awareness. The greater climate extremes and poorer soil structures in tropical areas give greater potential for degradation in such areas compared to temperate latitudes. This difference has been a significant factor in development or the lack of it.

The main cause of soil degradation is the removal of the natural vegetation cover, leaving the surface exposed to the elements. Figure 12.42 shows the human causes of degradation, with deforestation and overgrazing as the two main problems. The resulting loss of vegetation cover is a leading cause of wind and water erosion.

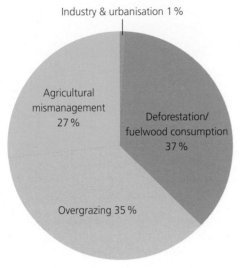

Figure 12.42 Causes of land degradation

Deforestation occurs for a number of reasons, including the clearing of land for agricultural use, for timber and for other activities such as mining. Such activities tend to happen quickly, whereas the loss of vegetation for fuelwood, a massive problem in many LICs, is generally a more gradual process. Deforestation means that rain is no longer intercepted by vegetation, with rainsplash loosening the topsoil and leaving it vulnerable to removal by overland flow.

Overgrazing is the grazing of natural pastures at stocking intensities above the livestock carrying capacity. Population pressure in many areas and poor agricultural practices have resulted in serious overgrazing. This is a major problem in many parts of the world, particularly in marginal ecosystems. The process occurs in this way:

1 Trampling by animals (and humans) damages plant leaves.
2 Some leaves die away, reducing the ability of plants to photosynthesise.
3 Now there are fewer leaves to intercept rainfall and the ground is more exposed.
4 Plant species sensitive to trampling quickly disappear.

5 Soil begins to erode when bare patches appear. Trampling will have compacted the soil and damaged its structure.
6 Loose surface-soil particles are the first to be carried away, either by wind or water.
7 The loss of soil structure means that less water can infiltrate to the lower soil horizons. The growth rate of plants is reduced and it is more difficult for damaged plants to recover.

Agricultural mismanagement is also a major problem due to a combination of lack of knowledge and the pursuit of short-term gain against consideration of longer-term damage. Such activities include shifting cultivation without adequate fallow periods, absence of soil conservation measures, cultivation of fragile or marginal lands, unbalanced fertiliser use and the use of poor irrigation techniques.

Soil degradation is more directly the result of:

- **erosion by wind and water** – these two agents of erosion account for approximately 80 per cent of the world's degraded landscapes
- **physical degradation** – loss of structure, surface sealing and compaction
- **chemical degradation** – through various forms of pollution; changes in pH, **acidification** and **salinisation** are examples of chemical degradation
- **biological degradation** – through loss of organic matter and biodiversity
- **climate and land-use change** – which may accelerate the factors above.

The environmental and socio-economic consequences of soil degradation are considerable. Such consequences can occur with little warning as damage to soil is often not perceived until it is far advanced.

The increasing world population and the rapidly changing diets of hundreds of millions of people as they become more affluent are placing more and more pressure on land resources. Some soil and agricultural experts say that a decline in long-term soil productivity is already seriously limiting food production in LICs.

The loss of the ability of degraded soils to store carbon is receiving significant attention. Over the last 50 years or so, global soils have lost about 100 billion tonnes of carbon, in the form of carbon dioxide, to the atmosphere, due to the depletion of soil structure.

The UN's Food and Agriculture Organization lists five root causes of unsustainable agricultural practices and degradation of the rural environment (Figure 12.43).

- **Policy failure**

  Leading among the causes of unsustainable agriculture are inadequate or inappropriate policies, which include pricing, subsidy and tax policies, which have encouraged the excessive and often uneconomic use of inputs, such as fertilisers and pesticides and the overexploitation of land. They may also include policies that favour farming systems that are inappropriate both to the circumstances of the farming community and to available resources.

- **Rural inequalities**

  Rural people often know best how to conserve their environment, but they may need to overexploit resources in order to survive. Meanwhile, commercial exploitation by large landowners and companies often causes environmental degradation in pursuit of higher profits.

- **Resource imbalances**

  Almost all of the future growth in the world's population will be in LICs, and the biggest increases will be in the poorest countries of all – those least equipped to meet their own needs or invest in the future.

- **Unsustainable technologies**

  New technologies have boosted agricultural production worldwide, but some have had harmful side-effects that must be contained and reversed, such as resistance of insects to pesticides, land degradation through wind or water erosion, nutrient depletion, poor irrigation management and the loss of biological diversity.

- **Trade relations**

  As the value of raw materials exported by LICs has fallen, their governments have sought to boost income by expansion of crop production and timber sales that have damaged the environment.

  Source: www. fao.org

Figure 12.43 Five root causes of unsustainable practices

The environmental impact of capital-intensive farming

There is a growing realisation that the modes of production, processing, distribution and consumption that prevail – because in the short- to medium-term they are the most profitable – are not necessarily the most healthy or the most environmentally sustainable (Figure 12.44). In many parts of the world, agro-industrialisation is having a devastating impact on the environment, causing:

- deforestation
- land degradation and desertification
- salinisation and contamination of water supplies
- air pollution
- increasing concerns about the health of long-term farm workers

Figure 12.44 Gobi desert – climate change has resulted in further land degradation

- landscape change
- declines in biodiversity.

About a third of the world's farmland is already affected by salinisation, erosion or other forms of degradation.

The global cattle population is currently around 1.5 billion. The pasture required amounts to about a third of all the world's agricultural land. A further third of this land is taken up by animal feedcrops. An estimated 1.3 billion people are employed in the livestock industry. The balance between livestock and grass is sustainable at present, but as the demand for meat increases, the pressures that cattle make on the land may well soon exceed supply. More cattle means more manure. Manure is often used to restore depleted soil, but can lead to pollution by heavy metals such as cadmium, nickel, chromium and copper.

In 2000, annual global meat consumption was 230 million tonnes. The forecast for 2050 is 465 million tonnes. There is a strong relationship between meat consumption and rising per person incomes (Figure 12.45), although anomalies do occur due to cultural traditions. It is no coincidence that many committed environmentalists are vegetarian. A study at the University of Chicago calculated that changing from the average American diet to a vegetarian one could cut annual emissions by almost 1.5 tonnes of carbon dioxide.

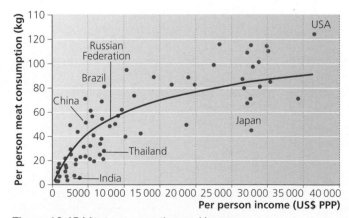

Figure 12.45 Meat consumption and income

Large-scale farming has been expanding geographically into a number of fragile environments, particularly into areas of rainforest. *The State of the World's Forests 2007*, published by the FAO, reported that between 1990 and 2005, the world's total forest area was reduced by 3 per cent. This is a rate of 7.3 million hectares per year.

Mainly because of the uniformity required by large food companies, important breeds of livestock are becoming extinct. The FAO's *State of the World's Animal Genetic Resources* report stated that at least one livestock breed a month had been lost over the previous seven years. Food scientists are concerned about this trend as genetic resources are the basis of food security.

Agro-industrialisation is characterised by large areas of monoculture that, among other things, leaves crops more vulnerable to disease due to the depletion of natural systems of pest control. Monoculture results in reliance on pesticides, which in turn causes a downward environmental cycle (Figure 12.46).

Figure 12.46 Land degradation in southern Italy

Poverty and rural degradation

The interactions shown in Figure 12.47 illustrate certain poverty-environment processes where poor households are 'compelled' to degrade environmental resources. However, this should not hide the fact that much environmental degradation is caused by large-scale commercial operations and government policy. There are also an increasing number of sustainable schemes being practised in poor rural areas.

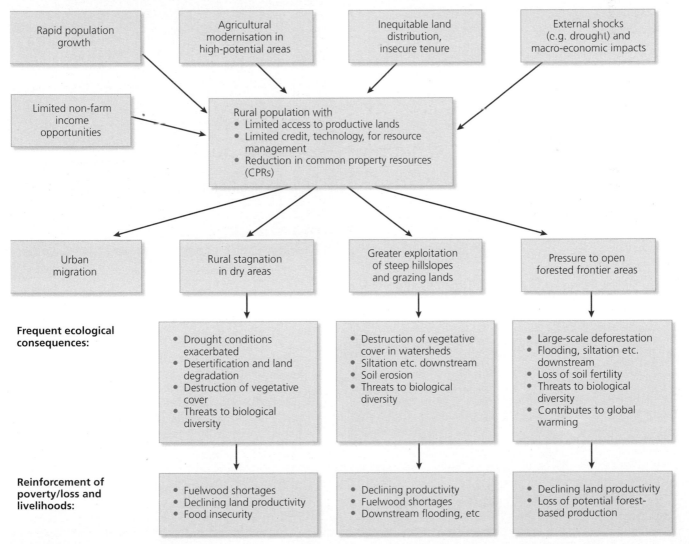

Figure 12.47 Poverty–environment links

Poor households can suffer significantly from the actions of large-scale rural operations. They may be pushed onto more marginal lands by logging, ranching or mining operations. Government policy can also have a significant negative effect on the poor, for example if land or tree tenure rights are insecure.

Urban/rural impact

Urban areas can affect the environmental degradation of their rural surroundings in a number of ways. For example, untreated wastewater is a major pollutant of rivers, which can contaminate estuaries and coastal fishing areas, and can pollute the drinking-water supplies of rural communities downstream. Urban use of groundwater can result in a depletion of the aquifer to the detriment of small farmers who rely on shallow wells. In arid areas, cities many kilometres inland can cause saltwater intrusion under coastal areas as a result of groundwater pumping. On a larger scale, the huge urban-industrial complexes of the world are the main cause of climate change, which is having an adverse impact on the entire planet.

Section 12.3 Activities

1 Discuss the five root causes of unsustainable practices set out in Figure 12.43.
2 Describe and explain the relationship between per person income and per person meat consumption illustrated in Figure 12.45. Suggest reasons for significant anomalies. What statistical technique could you use to assess the relationship between these two variables?
3 Produce a 100-word summary of the relationships illustrated in Figure 12.47.

↻ □ The degradation of urban environments

The environmental quality of urban areas has a huge impact on their populations (Figure 12.48). The degradation of urban environments occurs mainly through urbanisation, industrial development and inadequate infrastructure. Few urban residents can escape the effects of poor air and water quality, inadequate sanitation, a lack of proper solid waste management and the improper storage and emission of hazardous substances. Figure 12.49 shows how urban environmental problems can act at different scales, affecting households, communities and cities.

Figure 12.48 London's low-emission zone

Amenity loss	Traffic congestion	City	Loss of heritage and historical buildings	Reduced property and building values
Accidents and disasters	Polluted land	**Community**	Inappropriate and inadequate technology use	Inadequate tax/financial revenues
Flooding and surface drainage	Garbage dumping	**Household** Household health, garbage generation, air/water/noise pollution, spread of diseases	Lack of understanding of environmental problems	Lack of, and inappropriate, laws and legislation
Toxic and hazardous wastes/dumps	Flooding	Noise pollution	Natural disasters	High living densities
Loss of agricultural land and desertification	Air pollution	Water pollution	Inadequate supply and transmission loss of electricity	Misguided urban, government and management practices

Figure 12.49 The scales of urban environmental problems

The relationship between the urban poor and the environment is different from that between the rural poor and the environment. The urban poor are particularly affected by poor environmental services such as sub-standard housing, a lack of sanitation and other aspects of urban poverty.

Case Study: Urban degradation in Delhi

In 2014, the World Health Organization stated that Delhi was the most air-polluted city in the world. The WHO ranking was based on the concentration of $PM_{2.5}$ particles in a study of 1600 cities across 91 countries. These very fine particles, less than 2.5 micrometres in diameter, are linked to increased rates of heart disease, lung cancer and chronic bronchitis. According to the WHO, Delhi had an average $PM_{2.5}$ level of 153, about ten times that of London, UK. The WHO says that air pollution is the fifth biggest cause of death in India. However, Indian and Delhi officials have disputed the WHO findings, feeling that the analysis has overestimated the problem in Delhi compared to polluted cities in other countries such as Beijing.

The National Capital Region (NCR) of Delhi has grown rapidly in recent decades. Delhi now covers an area of approximately 900 km². The population of the NCR, which was estimated at 17.8 million in 2014, is expected to reach 22.5 million in 2025. Such a fast rate of expansion has increased the environmental impacts of transportation, industrial activity, power generation, construction, domestic activities and waste generation. Thus, Delhi's pollution problems have a wide variety of causes, as Table 12.13 shows.

- Transport is the main source of $PM_{2.5}$, NO_x and VOC emissions.
- Road dust is the main source of PM_{10} emissions.
- Power plants are responsible for over half of SO_2 emissions and almost 30 per cent of CO_2 emissions.

Economic growth has created an increasing number of people in the middle-class income bracket who have aspired to own their own car. However, it is not just private cars that have significantly increased in number, but also taxis, autorickshaws, buses and business vehicles of all kinds. There are now an estimated 7.2 million vehicles on Delhi's roads – about twice the number in 2000. Traffic speeds have fallen as the city has struggled to improve its road infrastructure to meet increasing demand. As traffic speeds have fallen idling time has increased, resulting in higher levels of air pollution.

Pollution levels are particularly high during winter due to:

- **night-time heating needs** – the poor in Delhi use open fires to keep warm in winter
- **seasonal weather conditions** – which trap pollutants very close to the ground.

Delhi also suffers from dust blown in from the deserts of the western state of Rajasthan.

Water pollution and a lack of solid-waste treatment facilities have caused serious damage to the Yamuna River. Roughly half of all the city's raw sewage goes straight into the river: 22 drains flow into the Yamuna from the city, containing industrial waste as well as domestic waste. Pickling, dyeing and electroplating factories are major sources of water pollution.

Surveys have shown that people in Delhi are also concerned about noise and light pollution. Residents perceive levels of both to be high. A study published in 2013 warned that very high levels of noise pollution were causing age-related hearing loss 15 years earlier than normal. Among other environmental issues, residents are also extremely critical of garbage disposal in the city – with very good reason (Figure 12.50).

Table 12.13 The sources of emissions in Delhi (tonnes/year and percentage), 2010

	PM$_{2.5}$	PM$_{10}$	SO$_2$	NO$_x$	CO	VOC
Transport	17750 (26%)	23800 (18%)	950 (2%)	329750 (67%)	421450 (28%)	208900 (63%)
Domestic	7300 (10%)	8800 (7%)	2050 (5%)	2350 (1%)	161200 (10%)	18300 (6%)
Diesel	3200 (5%)	4300 (3%)	1050 (3%)	81300 (16%)	85100 (6%)	31600 (9.49%)
Brick kilns	9250 (13%)	12400 (9%)	4000 (11%)	6750 (1%)	171850 (11%)	24200 (7%)
Industries	9000 (13%)	12650 (9%)	8500 (23%)	41500 (8%)	219600 (14%)	13250 (4%)
Construction	2450 (3%)	8050 (6%)	100 (1%)	2150 (1%)	2700 (1%)	50 (0.01%)
Waste burning	3850 (6%)	5450 (4%)	250 (1%)	1450 (1%)	20050 (1%)	1600 (0.5%)
Road dust	6300 (9%)	41750 (31%)	–	–	–	–
Power plant	10150 (15%)	16850 (13%)	20250 (54%)	27200 (5%)	442150 (29%)	34900 (10%)
Total	69050 (100%)	133900 (100%)	37000 (100%)	492250 (100%)	1524050 (100%)	332700 (100%)

Source: www.urbanemissions.info

DELHI MAY DROWN IN ITS OWN WASTE

Adapted (slightly) from Darpan Singh, *Hindustan Times*, 30 April 2013

Growing by heaps and mounds, Delhi's garbage crisis may soon reach its breaking point.

Three of the four stinking waste mountains (landfills) are long overdue for closure and there are no fresh landfills available to take in the current daily discard of 9000 tonnes. By 2020, the Capital needs an additional area of 28 sq km to dump 15 000 tonnes of garbage daily.

Since as much as 85 per cent of the city doesn't have a formal door-to-door trash pick-up system, the emerging scenario is both worrisome and scary.

The 2500-odd filthy community bins (*dhalaos*) that serve as secondary collection centres for the three municipal bodies in thousands of colonies will start overflowing, and garbage will spill on to the streets. Residents will have no option but to start throwing waste out, making Delhi drown in its own discard.

People in Delhi are bitterly opposed to new landfills coming up in their neighbourhood as they have seen the authorities did not maintain the past ones scientifically, turning them into massive, polluting heaps. The black thick liquid, leachate, created when rainwater filters down through the landfill, has made the soil highly toxic. Rainwater runoff goes into surface water drains while methane poisons the air.

Despite court intervention, the government and civic agencies have failed to find a way forward. The civic bodies have now told the Delhi high court that 'since there's no other option, we have been forced to put human life and property at risk'.

Delhi's non-dumping options to manage waste have also shrunk drastically. Burning waste no longer seems viable because of environmental concerns and poor segregation of waste. Compost plants are not doing well because manure doesn't sell, and again becomes garbage.

As much as 50 per cent of the waste is fit for composting. About 30 per cent of it can be recycled. Effective segregation at source, in transit and during disposal, will mean only 20 per cent of the refuse is needed to be sent to the landfill site. This will also mean a cleaner city with fewer dhalaos, garbage trucks and longer lifespan for landfills. But instead of proper segregation, only random picking continues.

Either private sweepers, who snap up the most sought-after refuse, or residents themselves take waste to *dhalaos*. There, rag-pickers slog through the muck to hunt for recyclable materials.

When waste is taken from dhalaos to landfills, another set of trash-pickers collect what their street counterparts miss, completing a cycle of 'illegal' segregation.

Civic bodies blame residents for not segregating waste but what's the point when everything will eventually be mixed-up? Segregation by residents will only work when the corporations have a complete door-to-door waste collection system and trash pickups have separate containers for dry and wet waste.

Delhi has miserably failed to manage its waste load. Only 15 per cent of R1350 crore that the three corporations spend on waste management and sanitation is spent on actual disposal. The rest goes into collection and transportation.

The authorities must ensure segregation and promote composting and recycling. They must quantify waste generation for setting effective reduction targets. But don't wait for the authorities to do everything. From segregation, recycling to composting — you can make a difference. And, yes, consume and waste less. Now is the time.

Figure 12.50 Delhi's domestic waste problem

Thousands of people live without adequate water supply and sanitation. Many parts of the city have no piped water connections and depend on tankers to deliver water. This affects almost 25 per cent of Delhi's population. About half of the water flowing through Delhi's water pipes is lost through leakage. A study published in 2012 stated that nearly 60 per cent of Delhi's slums did not have sewerage facilities.

Tackling air pollution

In 2014, the Indian environment ministry launched a National Air Quality Index that will rank 66 Indian cities. It will give real time information of air quality, to put pressure on local authorities to take concrete steps to reduce pollution. The index will provide associated health risks in a colour-coded manner that can be understood by everyone.

Investment in public transport has been slow to develop in Delhi. The Delhi Metro Rail (DMR) only opened in December 2002 with an 8.3 kilometre rail line, and has since been extended to 190 kilometres with the completion of Phase II in 2011. Construction of Phase III is now underway, to add a further 103 kilometres. Another significant development has been the Delhi bus rapid transit (BRT), which opened along a 5.6 kilometre initial corridor in 2008.

Shutting coal power plants, promoting motorless transport and imposing strict penalties for those violating pollution control norms are among the suggestions that the government is looking at to improve the air quality in the city over the next five years.

The impact of rural areas

Runoff from fertilisers and pesticides can contaminate downstream urban water supplies. Deforestation, watershed degradation and soil eroding-practices can exacerbate flood–drought cycles. Deteriorating conditions in rural areas can give added impetus to rural–urban migration, placing additional pressures on the urban environment. This is part of a process known as the 'urbanisation of poverty'.

> ### Section 12.3 Activities
>
> 1 Discuss the scales of urban environmental problems illustrated by Figure 12.49.
> 2 Outline one example of urban degradation in Delhi.
> 3 How can the activities of rural areas have an impact on urban environmental degradation?

☐ Constraints on improving degraded environments

There are numerous constraints on improving the quality of degraded environments:

- In many LICs, population growth continues at a high rate, putting increasing pressure on already fragile environments.
- High rates of rural–urban migration can lead to rapidly deteriorating environmental conditions in large urban areas, at least in the short term.
- Environmental hazards, often made worse by climate change, present an increasing challenge in some world regions. In many regions, natural hazards have increased in scale and unpredictability.
- Poor knowledge about the environmental impact of human actions is a significant factor in many locations where perhaps moderate adaptation of human behaviour could bring about beneficial changes.
- Poor management at both central and local government levels may result in problems that can at least be partially rectified not being addressed. The quality of governance has been recognised as a key factor in the general development process.
- Many degraded environments require substantial investment to bring in realistic solutions. Such finance is beyond the means of many poor countries. However, there may be a choice between low-cost and high-cost schemes, as Table 12.14 illustrates.
- Civil war has put back development by decades in some countries. Land mines that have yet to be cleared have put large areas off limits in some countries.

Corruption and crime can also reduce the effectiveness of schemes to reduce environmental degradation. An article in *The Guardian* newspaper in October 2009 stated that a revolutionary UN scheme to cut carbon emissions by paying poorer countries to preserve their forests was a recipe for corruption,

Table 12.14 Technical and institutional costs in resource management by and for poor people

	High institutional costs	Low institutional costs
Relatively high technical costs	• Large-scale irrigation • Arid or semi-arid land reforestation or pasture improvement • Sodic or saline land reclamation • Mangrove reforestation • Integrated river-basin management • Many transboundary resources, e.g. international rivers, air quality • Resettlement schemes • Water-pollution reduction programmes • Rural road maintenance • Ocean fisheries management	• Small-scale hill irrigation • Food crop systems on difficult soils • Localised water harvesting structures • Centralised provision of energy services • Solar energy for individual households • Pipe sewer systems • Emissions reduction devices • Improved public transport
Relatively low technical costs	• Aquifer management • Protection of critical areas • Coastal fisheries management • Coral-reef management • Pasture management • Land-reform programmes • Integrated pest management • Wild-game management	• Treadle pump irrigation • Humid tropics reforestation • Small water harvesting systems • Joint forest management regimes • Improved cooking stoves and cooking energy for poor families • Sloping agricultural land technology (SALT) • Small-scale quarrying • Household-based sanitation systems

Source: UNDP

and without strong safeguards could be hijacked by organised crime. Many countries and organisations have strongly backed the UN plans to expand the global carbon market to allow countries to trade the carbon stored in forests.

☐ The protection of environments at risk

Environments at risk can be protected in various ways. At the most extreme, human activity and access can be totally banned, such as in Wilderness Areas; or extremely limited, as is usually the case in National Parks. However, in many areas it is usually necessary to sustain significant populations and rates of economic activity, particularly in LICs. In these cases, various types of sustainable-development policies need to be implemented. Individual environments can be assessed in terms of:

- **needs** – what needs to be done to reduce environmental degradation as far as possible without destroying the livelihoods of the resident population?
- **measures** – what are the policies and practices that can be implemented to achieve these aims at various time scales?

- **outcomes** – how successful have these policies been at different stages of their implementation? Have policies been modified to cope with initially unforeseen circumstances?

An example of an endangered environment is Ecuador's Andean cloud forests. The Andean Corridor Project aims to protect the vital ecosystems of this diversity hotspot. Four reserves totalling 142 000 hectares have already been created and there are plans to add more reserves to protect the forests and wildlife from human activity.

Sometimes degraded environments cross international borders and effective action requires close cooperation between countries. Degradation in one country, for example the large-scale removal of forest cover in Nepal, can have severe implications for a country downstream of a river system, in this case India.

Section 12.3 Activities

1 Discuss the constraints on improving the quality of degraded environments.
2 Write a brief summary of the framework presented in Table 12.14.

12.4 The management of a degraded environment

Case Study: Namibia – community development

Namibia, in south-west Africa, is a very sparsely populated country with a generally dry climate. Almost 29 per cent of its 2.3 million people live below the international poverty line of $1.25 a day. Environmental degradation and sustainability are significant issues in its marginal landscapes, with the government attempting to tackle these issues and reduce poverty at the same time. The causes of degradation have been mainly uncontrolled exploitation by a low-income population forced to think only in the short term in order to survive, and lack of management at all levels of government in earlier years. As the degradation process intensified, the government, assisted by international agencies, identified the problems and put in place a significant strategy.

Namibia's Communal Conservancy Programme (Figure 12.51) is regarded as a successful model of community-based natural resource management, with an improving record for wildlife numbers and poverty reduction. The programme gives rural communities unprecedented management and use rights over wildlife, which have created new incentives for communities to protect this valuable resource and develop economic opportunities in tourism.

Figure 12.51 Wildlife in a community conservancy in Namibia

The Conservancy Programme began in 1996 and there are now 64 community conservancies covering over 17 per cent of the country and embracing one in four of rural Namibians. **Community conservancies** (Figure 12.52) are legally recognised common property resource management organisations in Namibia's communal lands. The use rights given to conservancies include the rights to hunt, capture, cull and sell 'huntable game'. However, the government determines the overall culling rate and establishes quotas for protecting game used for trophy hunting.

An obvious sign of success is the significant increase in the numbers of wildlife in the conservancies after decades of decline. In the north-west conservancies, elephant numbers more than doubled between 1982 and 2000, and populations of oryx, springbok and mountain zebra rose tenfold. This improvement results from a decline in illegal hunting and poaching due to the economic value that conservancy communities now place on healthy wildlife populations.

The conservancies benefit from a number of 'new' economic activities, including:

- contracts with tourism companies
- selling hunting concessions
- managing campsites
- selling wildlife to game ranchers
- selling crafts.

These activities are in addition to traditional farming practices, which were usually at the subsistence level. The diversification of economic activity made possible by the Conservancy Programme has increased employment opportunities where few existed beforehand, and also raised incomes.

The significant participation of conservancy populations has been central to the design of the Programme. Conservancies are built around the willingness of communities to work collectively. Often, they form when neighbouring villages and tribal authorities agree to trace a boundary around their shared borders and manage the wildlife within this area. The Conservancy Programme has inbuilt flexibility that allows communities to choose diverse strategies for wildlife management and distributing benefits.

Support from and cooperation between a number of different institutions has been important to the development of the programme (Figure 12.53). Such institutions bring substantial experience and skills in helping conservancies to develop. Running skills training programmes has been an important aspect of such support. For example, communal conservancies are able to call on the experience of various NGOs for help and advice. This enables good practice in one area to be applied in other areas. The Namibian Community-based Tourism Association has been instrumental in helping communities negotiate levies and income-sharing agreements with tourism companies.

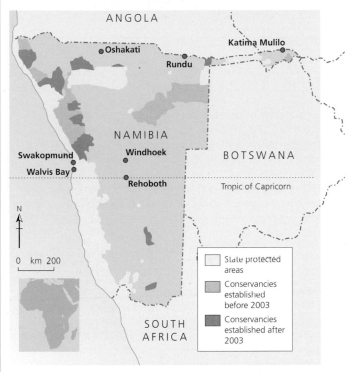

Figure 12.52 Namibia's community conservancies

Organisation	Support activities
Legal Assistance Centre	Supplies legal advice and advocacy on issues related to community-based natural resource management (CBNRM)
Namibia Community-Based Tourism Association	Serves as an umbrella organisation and support provider for community-based tourism initiatives
Namibia NGO Forum	Represents a broad range of NGOs and community-based organisations
Namibia Nature Foundation Rössing Foundation	Provides assistance through grants, financial administration, technical support, fundraising, and monitoring and evaluation
Multi-disciplinary Research Centre	Provides training and materials for CBNRM partners
Namibia Development Trust	Centre of the University of Namibia provides research-related support
Centre for Research Information	Provides assistance to established and emerging conservancies in southern Namibia
Action in Africa – Southern Africa Development and Consulting	Provides research, developmental assistance, and market linkages for natural plant products

!NARA	Conducts capacity training in participatory, democratic management for conservancy communities and institutions supporting communities
Desert Research Foundation of Namibia	Researches arid land management, conducts participatory learning projects with communities about sustainable management, and engages policymakers to improve regulatory framework for sustainable development
Rural People's Institute for Social Empowerment	Provides assistance to established and emerging conservancies in southern Kunene and Erongo regions
Integrated Rural Development and Nature Conservation	A field-based organisation working to support conservancy development in Kunene and Caprivi regions
Nyae Nyae Development Foundation	Supports San communities in the Otjozondjupa region in the Nyae Nyae Conservancy
Ministry of Environment and Tourism	MET is not a formal member, but attends meetings and participates in NACSO working groups. Provides a broad spectrum of support in terms of policy, wildlife monitoring and management, and publicity.

Sources: MET 2005; NEEN 2004 a, b, c; Weaver 2007; Jones 2008 in Update 'Scaling up Namibia's community conservancies', p.32

Figure 12.53 Namibian Association of CBNRM Support Organisations (NACSO)

Although rural poverty remains significant in Namibia, the Conservancy Programme has resulted in substantial progress, with income rising year on year. Table 12.15 shows the detailed breakdown of conservancy-related income in 2006. In 2006, conservancy income reached nearly N$19 million (US$1.4 million). Income from small businesses as.sociated with the conservancies but not directly owned by them brought in another N$8 million (US$580 000).

Table 12.15 Conservancy-related income, 2006

Source of income	Value in N$	% of total conservancy income
Miscellaneous	34 788	0.1
Premium hunting	43 600	0.2
Veld products	39 000	0.1
Thatching grass	2 450 481	9.1
Shoot and sell hunting	504 883	1.9
Interest earned	161 807	0.6
Craft sales	474 343	1.8
Campsites and community-based tourism enterprises	3 746 481	14.0
Trophy meat distribution	870 219	3.2
Game donation	860 950	3.2
Use of own game	739 629	2.8
Trophy hunting	6 113 923	22.8
Joint-venture tourism	10 794 668	40.2
Total	26 834 772	100.0

Source: WWF et al. 2007:113

An important aspect of development has been the involvement of women in the employment benefits. Such jobs have included being game guards and natural resource monitors, as well as serving tourists in campgrounds and lodges.

Rising income from conservancies has made possible increasing investment in social development projects. This has made conservancies an increasingly important element in rural development.

Scaling-up resource management

Following the perceived success of community conservancies, the Namibian government has extended the concept to **community forests** (Figure 12.54). Establishing a community forest is similar to the process of forming a conservancy. This is a good example of the **scaling-up process** from one natural resource system to another. Based on the Forest Act of 2001, the project helps local communities to establish their own community forests, to manage and utilise them in a sustainable manner. Because many rural Namibians are poor, it is important that they have a greater say in how forest resources are managed and share the benefits of properly managed forest resources.

Forest fires and uncontrolled cutting have been two of the main problems facing forest-protection efforts in Namibia for some time. About 4 million hectares of forest and veld are burnt annually, mostly as a result of fires started deliberately to improve grazing and to clear hunting grounds.

The advent of community forests has led to improved forest resource management. It has also improved the livelihoods of local people based on the empowerment of local communities with forest use rights. Villagers in community forests derive an income by marketing forestry products such as timber and firewood, poles, wild fruits, devil's claw, thatching grass, tourism, honey from beekeeping, wildlife, woven baskets and other crafts.

Based in large part on the success of CBNRM in the conservancies, the Namibian government enacted legislation in 2001 allowing the formation of community forests – areas within the country's communal lands for which a community has obtained management rights over forest resources such as timber, firewood, wild fruits, thatch grass, honey, and even some wildlife (MET 2003). The establishment of the community forest program shows how the scaling-up process can reach across natural resource systems, affecting natural resource policy at the broadest level. Although the community forest program and the conservancy program are now administered separately by different ministries, some groups have expressed interest in merging the programs to allow a more integrated approach to managing natural resources at the community level (Tjaronda 2008).

Establishing a community forest is similar to the process of forming a conservancy. Communities must:

- Submit a formal application to the government;
- Elect a forest management committee from the community;
- Develop a constitution;
- Select, map, and demark a community forest area;
- Submit a forest management plan describing how the community will harvest forest resources sustainably and manage other activities such as grazing and farming within the forest area;
- Specify use rights and bylaws necessary to act on their management plan;
- Craft a plan to ensure the equitable distribution of revenues to all community members; and
- Obtain permission from the area's traditional authority (MET 2003).

As of April 2008, a total of 45 community forests had been formed (although only 13 were officially gazetted), encompassing 2.2 million ha and benefiting some 150,000 Namibians. In the north-eastern region alone, 16 registered forests have generated more than N$300,000 (US$38,000) since 2005 (The Namibian 2008; Tjaronda 2008).

Figure 12.54 Extending the conservancy concept – community forests in Namibia

Country Pilot Partnership for Integrated Sustainable Land Management

The Government of Namibia identified land degradation as a serious problem that demanded remedial intervention. Five government departments, together with international agencies, established a Country Pilot Partnership for Integrated Sustainable Land Management. The activities were funded through the Global Environment Facility (GEF) in partnership with the United Nations Development Programme (UNDP) as implementing lead agency. Other organisations including the European Union were involved.

The pilot project began in early 2008 and ran until late 2011. The project objective was to develop and pilot a range of coping mechanisms for reducing the vulnerability of farmers and pastoralists to climate change, including variability. It took place both within and outside communal conservancies. Its organisation was at least partly based on what Namibia has learned from the establishment of communal and forest conservancies.

Section 12.4 Activities

1 Why was the Community Conservancy Programme introduced?
2 How have employment opportunities expanded under the conservancy programme?
3 Comment on the importance of the Namibian Association of CBNRM Organisations to the success of the Communal Conservancy Programme.
4 Comment on the distribution of conservancy-related income shown in Table 12.15.
5 Describe and explain the extension of the conservancy concept to forests in Namibia.

13 Global interdependence

13.1 Trade flows and trading patterns

☐ Visible and invisible trade (imports and exports)

Trade refers to the exchange of goods and services for money. The origin and continuing basis of **global interdependence** is trade. The global trading system developed at the time of European colonial expansion. Here, a 'colonial division of labour' emerged in which LICs exported primary products, agriculture and minerals, while Europe and North America exported manufactured goods. This remained the general pattern of world trade until the post-Second World War period, when a more complex pattern of international trade emerged. Trade is the most vital element in the growth of the global economy. World trade now accounts for over 30 per cent of GDP – about three times its share in 1960 (Figure 13.1).

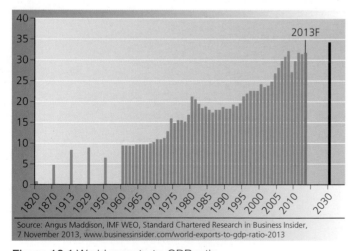

Source: Angus Maddison, IMF WEO, Standard Chartered Research in Business Insider, 7 November 2013, www.businessinsider.com/world-exports-to-gdp-ratio-2013

Figure 13.1 World exports-to-GDP ratio

Trade results from the uneven distribution of resources over the Earth's surface. Even countries with an abundance of resources and a wide industrial base cannot produce all of the goods and services that their populations desire. So they buy goods and services from other countries, providing they have the money to pay for them. Goods and services purchased from other countries are termed **imports**. In contrast, goods and services sold to other countries are called **exports**. Imports along with exports form the basis of international trade (Figure 13.2). The difference between the value of a country's imports and exports is known as the **balance of trade**.

A **trade deficit** occurs when the value of a country's imports exceeds the value of its exports. A country can make up this difference by using its savings or by borrowing, but clearly such a situation cannot continue indefinitely. In contrast, a positive or favourable balance of trade is known as a **trade surplus**. A trade surplus contributes to the GDP of a nation, but a trade deficit will reduce GDP.

Visible trade involves items that have a physical existence and can actually be seen. Thus raw materials (primary products) such as oil and food, and manufactured goods (secondary products) such as cars and furniture, are items of visible trade. **Invisible trade** is trade in services, which include travel and tourism, and business and financial services.

☐ Global patterns of and inequalities in trade flows

Figure 13.3 indicates the growth in the value of global **trade in goods (merchandise)** and services between 2003 and 2013. The value of the global trade in goods increased from less than $8 trillion in 2003 to more than $18.5 trillion in 2013. **Trade in services** also increased significantly from about $2 trillion in 2003 to about $4.7 trillion in 2013. For both the starting and end years of Figure 13.3, the value of world trade in services was roughly a quarter that of global trade in goods. The severe dip in the trade in goods in 2008–09 indicates the strong effect the global financial crisis had on world trade. The dip in the trade in services was of a much lower magnitude.

Figure 13.2 Vancouver, Canada – much of Canada's trade with Asia passes through the port of Vancouver

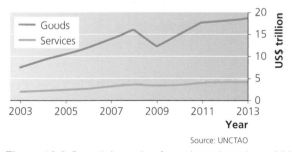

Source: UNCTAO

Figure 13.3 Growth in trade of goods and services, 2003–13

Figure 13.4 indicates the most important elements in the global trade in both goods (merchandise) and services. In terms of the former, manufactured goods dominate, followed by fuels and mining products, and then agriculture. Machinery and transport equipment is by far the most important element of manufactured goods. Figure 13.4 also shows the importance of **a** travel and **b** transport in terms of trade in commercial services.

Table 13.1 shows the spatial distribution of world trade in merchandise trade (visible trade) for the top 25 countries. China became the world's biggest trader in merchandise in 2013, with imports and exports totalling $4159 billion. China recorded a trade surplus of $259 billion, or 2.8 per cent of GDP. The USA was in second place overall, with imports and exports totalling $3909 billion. In contrast to China, the USA had a trade deficit of $750 billion (4.5 per cent of its GDP). Germany and Japan made up the other two places in the top four for both exports and imports. Overall, the top ten traders in merchandise accounted for 52 per cent of the world's total trade in 2013. NICs have increased their share of merchandise trade considerably in recent decades. The position of China as the world's largest trading country is the most obvious example of this trend, but other examples in Table 13.1 include South Korea, Mexico, India, Brazil, Thailand and Malaysia.

The trading positions of affluent countries with relatively small populations, such as the Netherlands, Belgium and Switzerland, is also worthy of note. The least developed countries' share of total global merchandise exports totalled only 1.1 per cent, compared to the 75.5 per cent of the G20 (group of 20 largest economies).

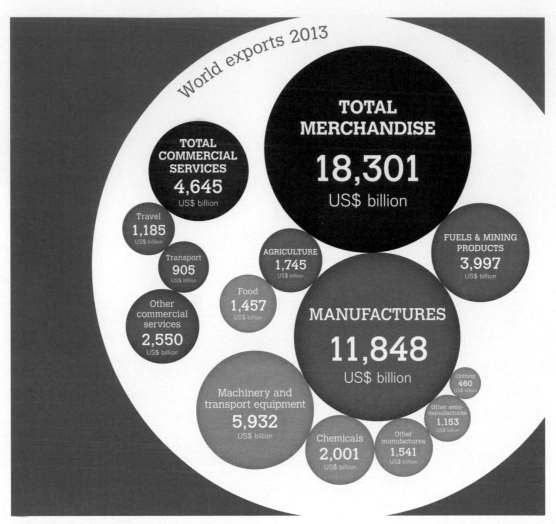

Figure 13.4 Major components of trade in merchandise and services

Figure 13.5 shows the economies of individual countries by the size of their merchandise trade in 2013. There are clear spatial patterns:

- Every single country in Africa is in the lowest class (under $250 billion). This class also includes most countries in the Middle East, Western and central Asia, Latin America and Eastern Europe.
- Brazil and Mexico are the only countries in Latin America outside of the lowest class.
- The countries of South East Asia are in the lower two classes, along with Australia and New Zealand.
- The major trading nations (the two higher classes) are in North America, Europe and East Asia. This group also includes the Russian Federation, India and Saudi Arabia.

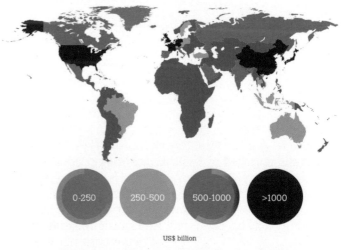

Source: www.wto.org/english/res_e/statis_e/its2014_e/its 2014_e.pdf page 10

Figure 13.5 World map showing economies by size of merchandise trade, 2013

Table 13.1 Leading exporters and importers in world merchandise trade, 2013

Rank	Exporters	Value ($billion)	Share (%)
1	China	2209	11.7
2	United States	1580	8.4
3	Germany	1453	7.7
4	Japan	715	3.8
5	Netherlands	672	3.6
6	France	580	3.1
7	Korea, Republic of	560	3.0
8	UK	542	2.9
9	Hong Kong, China	536	2.8
	– domestic exports	20	0.1
	– re-exports	516	2.7
10	Russian Federation	523	2.8
11	Italy	518	2.8
12	Belgium	469	2.5
13	Canada	458	2.4
14	Singapore	410	2.2
	– domestic exports	219	1.2
	– re-exports	191	1.0
15	Mexico	380	2.0
16	United Arab Emirates	379	2.0
17	Saudi Arabia, Kingdom of	376	2.0
18	Spain	317	1.7
19	India	313	1.7
20	Chinese Taipei	305	1.6
21	Australia	253	1.3
22	Brazil	242	1.3
23	Switzerland	229	1.2
24	Thailand	229	1.2
25	Malaysia	228	1.2

Rank	Importers	Value ($billion)	Share (%)
1	United States	2329	12.3
2	China	1950	10.3
3	Germany	1189	6.3
4	Japan	833	4.4
5	France	681	3.6
6	UK	655	3.5
7	Hong Kong, China	622	3.3
	– retained imports	141	0.7
8	Netherlands	590	3.1
9	Korea, Republic of	516	2.7
10	Italy	477	2.5
11	Canada	474	2.5
12	India	466	2.5
13	Belgium	451	2.4
14	Mexico	391	2.1
15	Singapore	373	2.0
	– retained imports	182	1.0
16	Russian Federation	343	1.8
17	Spain	339	1.8
18	Chinese Taipei	270	1.4
19	Turkey	252	1.3
20	United Arab Emirates	251	1.3
21	Thailand	251	1.3
22	Brazil	250	1.3
23	Australia	242	1.3
24	Malaysia	206	1.1
25	Poland	205	1.1

Source: World Trade Organization International Trade Statistics 2014

The USA remains firmly in the leading position in global trade in commercial services, particularly as an exporter. What might be surprising to some people is that:

- the UK is the second largest exporter
- Japan is not as significant in the trade in services as it is for merchandise
- India is in the top ten for both exports and imports.

The top ten countries in world trade in commercial services represented half of commercial services trade in 2013. At the other end of the scale, the least developed countries (LDCs) have shown pleasing progress. Between 2000 and 2013, trade in services from LDCs grew on average by 14 per cent per year. Examples have been:

- Cambodia as the leading LDC tourist destination
- Ethiopia's expansion of air transportation services.

However, such increases are from a very low base and the LDCs' share of world exports of commercial services totalled only 0.7 per cent in 2013.

The main trend in the global share of trade in commercial services has been the declining share of North America and Europe and the increasing share of Asia. Between 2005 and 2013, Asia's exports of commercial services rose from 21.7 per cent to 26.2 per cent. In contrast, Europe's share fell from 51.8 per cent to 47.2 per cent.

Section 13.1 Activities

1 Define **a** exports, **b** imports and **c** the balance of trade.
2 Explain how trade deficits and trade surpluses can arise.
3 What is the difference between visible and invisible trade?
4 Describe the trend shown in Figure 13.1.
5 Summarise the information on merchandise trade presented in Tables 13.1 and 13.2.

Table 13.2 Leading exporters and importers in world trade in commercial services, 2013

Rank	Exporters	Value ($ billion)	Share (%)
1	United States	662	14.3
2	UK	293	6.3
3	Germany	286	6.2
4	France	236	5.1
5	China	205	4.4
6	India	151	3.2
7	Netherlands	147	3.2
8	Japan	145	3.1
9	Spain	145	3.1
10	Hong Kong, China	133	2.9
11	Ireland	125	2.7
12	Singapore	122	2.6
13	Korea, Republic of	112	2.4
14	Italy	110	2.3
15	Belgium	106	2.0
16	Switzerland	93	1.7
17	Canada	78	1.7
18	Luxembourg	77	1.6
19	Sweden	75	1.5
20	Denmark	70	1.4
21	Russian Federation	65	1.4
22	Austria	65	1.3
23	Thailand	59	1.2
24	Macao, China	54	1.1
25	Australia	52	

Rank	Importers	Value ($ billion)	Share (%)
1	United States	432	9.8
2	China	329	7.5
3	Germany	317	7.2
4	France	189	4.3
5	UK	174	4.0
6	Japan	162	3.7
7	Singapore	128	2.9
8	Netherlands	127	2.9
9	India	125	2.8
10	Russian Federation	123	2.8
11	Ireland	118	2.7
12	Italy	107	2.4
13	Korea, Republic of	106	2.4
14	Canada	105	2.4
15	Belgium	98	2.2
16	Spain	92	2.1
17	Brazil	83	1.9
18	United Arab Emirates	70	1.6
19	Australia	62	1.4
20	Denmark	60	1.4
21	Hong Kong, China	60	1.4
22	Sweden	57	1.3
23	Thailand	55	1.3
24	Switzerland	53	1.2
25	Saudi Arabia, Kingdom of	52	1.2

Source: World Trade Organization International Trade Statistics 2014

☐ Factors affecting global trade

A range of factors influence the volume, nature and direction of global trade, including:

- resource endowment
- comparative advantage
- locational advantage
- investment
- historical factors
- **terms of trade**
- changes in the global market
- trade agreements.

Resource endowment

Resource endowment is a significant factor in world trade. For example, the Middle East countries dominate the export of oil. Along with a few other countries elsewhere in the world, such as Venezuela and Nigeria, they form **OPEC** (Figure 13.6), the Organization of Petroleum Exporting Countries.

Figure 13.6 The headquarters of OPEC in Vienna

OPEC is an intergovernmental organisation comprising 12 oil-producing nations. It was founded in 1960 after a US law imposed quotas on Venezuelan and Persian Gulf oil imports in favour of the Canadian and Mexican oil industries. OPEC's stated objective is 'to co-ordinate and unify the petroleum policies of member countries and ensure the stabilisation of oil markets in order to secure an efficient, economic and regular supply of petroleum to consumers, a steady income to producers and a fair return on capital to those investing in the petroleum industry'. The OPEC countries account for a major proportion of world crude oil reserves. OPEC has been heavily criticised at times for the allegedly political nature of some of its decisions. This has generally happened when the oil-rich Arab countries have wanted to put pressure on the USA and other Western countries with regard to the Israel–Palestine issue.

Countries endowed with other raw materials such as food products, timber, minerals and fish also figure prominently in world trade statistics (Figure 13.7). In HICs, the wealth of countries such as Canada and Australia has been built to a considerable extent on the export of raw materials in demand on the world market. MICs and LICs rich in raw materials, such as Brazil and South Africa, have been trying to follow a similar path. In both cases, wealth from raw materials has been used for economic diversification to produce a more broadly based economy.

Figure 13.7 Rubber production provides an important source of foreign currency to Vietnam

Comparative advantage

The concept of **comparative advantage** is an important part of classical theory on international trade. This states that different countries will specialise in producing those goods and services for which each is best endowed. Each country will then trade a proportion of these goods and services with other nations to obtain goods and services that it needs but for which it is not favourably endowed. The concept is very easy to understand with regard to raw materials, but it also applies to manufactured goods and services. It is saying that even in the complexity of the modern global economy, countries tend to concentrate on the goods and services they are best at producing. This results in specialisation in production and employment. The evidence of this is that some countries have a global reputation for particular products. Examples include German cars, Japanese high-tech products, Scotch whisky, Belgian chocolate and Swiss watches.

Locational advantage

The location of market demand influences trade patterns. It is advantageous for an exporting country to be close to the markets for its products as this reduces transport costs, along with other advantages gained from spatial proximity. For example, the tourist industry in France benefits from the large populations of neighbouring countries that can reach France relatively quickly and cheaply. Likewise, manufacturing industry in Canada benefits from the proximity of the huge American market.

Some countries and cities are strategically located along important trade routes, giving them significant advantages in international trade. For example, Singapore, at the southern tip of the Malay peninsula, is situated at a strategic location along the main trade route between the Indian and Pacific Oceans. Similarly, Rotterdam in the Netherlands is located near the mouth of the River Rhine. Many goods brought in by large ocean carriers are trans-shipped onto smaller river vessels and other modes of transport at Rotterdam, or refined or manufactured in various ways in the port's industrial area.

Investment

Investment in a country is the key to it increasing its trade. Some MICs such as Brazil and South Africa have increased their trade substantially. These countries have attracted the bulk of FDI. Such low-income 'globalisers' as China, Brazil, India and Mexico have increased their trade-to-GDP ratios significantly. On the other hand, hundreds of millions of people live in countries that have become less rather than more globalised (in an economic sense) as trade has fallen in relation to national income. However, in the poorest LICs businesses frequently operate in investment climates that undermine their incentive to invest and grow. Economic, social and political instability deters investment by making future benefits more uncertain or undermining the value of assets. Various studies show that the greater the level of instability, the lower the rate of private investment and growth. Crime and corruption represent a substantial risk to investment and increase the cost of doing business in countries where this is a substantial problem.

Historical factors

Historical relationships, often based on colonial ties, remain an important factor in global trade patterns. For example, the UK still maintains significant trading links with Commonwealth countries because of the trading relationships established at a time when these countries were colonies. Such links are weaker than they once were, but in many cases they remain significant (Figure 13.8). Other European countries such as France, Spain, the Netherlands, Portugal and Belgium also established colonial networks overseas and have maintained such ties to varying degrees in the post-colonial period.

Colonial expansion heralded a trading relationship dictated by the European countries mainly for their own benefit. The colonies played a subordinate role that brought them only very limited benefits at the expense of distortion of their economies. The historical legacy of this **trade dependency** is one of the reasons why, according to development economists, poorer tropical countries have such a limited share of world trade.

The terms of trade

The most vital element in the trade of any country is the terms on which it takes place. If countries rely on the export of commodities that are low in price, and need to import items that are relatively high in price, they need to export in large quantities to be able to afford a relatively low volume of imports. Many poor nations are **primary-product dependent**, which means they rely on one or a small number of primary products to obtain foreign currency through export. Although in the early years of this century some commodity prices rose significantly, over the long term the world market price of primary products has been generally low compared with that for manufactured goods and services. Also, the price of primary products is subject to considerable variation from year to year, making economic and social planning extremely difficult. In contrast, the manufacturing and service exports of HICs generally rise in price at a reasonably predictable rate, resulting in a more regular income and less uncertainty for the rich countries of the world. The terms of trade for many LICs are worse now than they were two decades ago. Thus it is not surprising that so many nations are struggling to get out of poverty. Because the terms of trade are generally disadvantageous to the poor countries of the world, many LICs have very high trade deficits.

However, it is not just poor countries that suffer because of the terms of trade. In December 2014, the Australian treasurer stated that Australia was in the grip of the biggest fall in the terms of trade since records began in 1959 because of a drastic fall in commodity prices.

Conventional neo-liberal economists generally welcome the large transfers of capital linked to high trade deficits. They say that trade deficits are strongly related to stages of economic development. The argument is that capital inflows swell the available pool of investment funds and thus generate future growth in the South. However, Marxist and populist writers argue that:

- if the expansion of trade volumes brings benefits to MICs and LICs, the accompanying expansion of trade deficits may bring considerable problems
- trade deficits have to be financed – one way is to borrow more money from abroad, but this will increase a country's debt; another is to divert investment away from important areas of the economy such as agriculture, industry, education and health
- in this way, high trade deficits in the South constrain growth and produce a high level of dependency.

Figure 13.8 Lloyds of London – insurance is a vital factor in the movement of world trade

Changes in the global market

The rapid growth of NICs (MICs) has brought about major changes in the economic strength of countries. An article in the *Sunday Telegraph* entitled 'Developing nations emerge from shadows as sun sets on the West' charted the financial problems that beset the West in the first decade of the new millennium, culminating with the impact of the global recession 2008–10. 'The West' is the term that many financial writers use to describe the mature economies of the USA, Canada, the EU, Japan, Australia and New Zealand.

The article highlighted the poor decisions made by Western policy-makers, contrasting this with the powerful economic growth figures of the BRIC nations (Brazil, Russia, India and China). These four countries, along with other high-growth nations outside of the established core group of nations, are known as **emerging markets**.

While the developed world (the core of the world economy) grew by an average of 2.1 per cent a year in the first decade of the twenty-first century, the emerging markets expanded by 4.2 per cent. Figure 13.9 shows the significant differences in economic growth of the advanced economies (the core) and the emerging-market economies in recent decades.

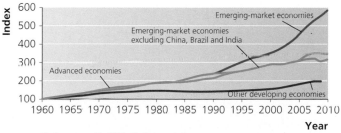

*Indexes are set to 100 in the base period

Figure 13.9 Line graph showing evolution of GDP, by group, 1960–2010

In 1990, the HICs controlled about 64 per cent of the global economy as measured by gross domestic product. This fell to 52 per cent by 2009 – one of the most rapid economic changes in history! Most of this global shift occurred in the last decade of that period. Such a huge global economic change has had major political consequences, with the emerging economies exerting much more power than they had previously in international negotiations.

Many major investors are turning their backs, at least partially, on Western nations and seeking out opportunities in the faster-growing emerging markets. There have been major changes in the distribution of the world's foreign exchange reserves. The G7 countries (USA, Canada, Japan, Germany, UK, France and Italy) held only 17 per cent of the global total between them in 2010. Japan is the only significant creditor nation in this group. In contrast, the BRICs held 42 per cent in 2010, with

China alone holding 30 per cent. It is not so long ago that the USA was the world's biggest creditor. Today, it is by far the world's biggest debtor. Much of the money borrowed by the USA and other Western economies has come from the reserves built up in emerging markets. The West no longer dominates the world's savings and as a result no longer dominates global investment and finance.

At the time of writing (mid-2015), the global economy seems to be at a crossroads. After a considerable rebound following the global financial crisis, global economic growth fell every year between 2010 and 2013 (from 5.5 per cent to 3.3 per cent). For example, all four BRIC countries saw growth rates fall sharply. However, it is likely that emerging markets and developing economies will continue to grow faster than the more mature economies.

In a recent article published by the BBC, Duncan Weldon, *Newsnight* economics correspondent, stated 'We may be living through a structural change in the global economy as big as any since World War II without fully realising it. The world economy may be becoming less integrated, with one of the important drivers of globalisation (trade) swinging into reverse.' The article noted that since the global financial crisis, world trade has been sluggish and outpaced by GDP. Thus, in very recent years, trade as a share of global GDP has been falling!

Trade agreements

A **trade bloc** is a group of countries that share trade agreements between each other. Since the Second World War, there have been many examples of groups of countries joining together to stimulate trade between themselves and to obtain other benefits from economic cooperation (Figure 13.10). The following forms of increasing economic integration between countries can be recognised:

- **Free trade areas** – members abolish tariffs and quotas on trade between themselves but maintain independent restrictions on imports from non-member countries. NAFTA is an example of a free trade area.
- **Customs unions** – a closer form of economic integration. Besides free trade between member nations, all members are obliged to operate a common external tariff on imports from non-member countries. Mercosur, established on 1 January 1995, is a customs union joining Brazil, Paraguay, Uruguay and Argentina in a single market of over 200 million people.
- **Common markets** – customs unions that, in addition to free trade in goods and services, also allow the free movement of labour and capital.
- **Economic unions** – organisations that have all the characteristics of a common market but also require members to adopt common economic policies on such matters as agriculture, transport, industry and regional policy. The EU is an example of an

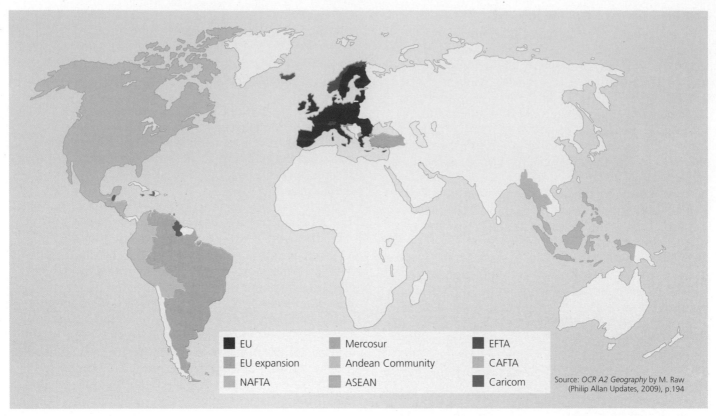

■ EU	■ Mercosur	■ EFTA
■ EU expansion	■ Andean Community	■ CAFTA
■ NAFTA	■ ASEAN	■ Caricom

Source: *OCR A2 Geography* by M. Raw
(Philip Allan Updates, 2009), p.194

Figure 13.10 World trade blocs

economic union, although it must be remembered that its present high level of economic integration was achieved in several stages. When Denmark, Ireland and the UK joined in 1973, the organisation could best be described as a common market. The increasing level of integration has been marked by changes in the name of the organisation. Initially known as the European Economic Community, it later became the European Community and finally, from November 1993, the European Union. Some nations in the EU have been more in favour of greater integration than others. Figure 13.11 shows that one of the most recent members, Romania, is very enthusiastic about a high level of integration.

Regional trade agreements have proliferated in the last two decades. In 1990, there were fewer than 25; by 1998, there were more than 90. The most notable of these are the European Union, NAFTA in North America, ASEAN in Asia and Mercosur in Latin America. The United Nations (UN) (1990) refer to such organisations as 'geographically discriminatory trading arrangements'. Nearly all of the World Trade Organization (WTO)'s members belong to at least one regional pact. All such arrangements have one unifying characteristic: the preferential terms that trade participants enjoy over non-participating countries. Although no regional group has as yet adopted rules contrary to those of the WTO, there are some concerns:

Figure 13.11 Street banner in Bucharest, Romania – 'The answer is more Europe'

- Regional agreements can divert trade, inducing a country to import from a member of its trading bloc rather than from a cheaper supplier elsewhere.
- Regional groups might raise barriers against each other, creating protectionist blocks.
- Regional trade rules may complicate the establishment of new global regulations.

There is a growing consensus that international regionalism is on the ascendency. The EU, NAFTA and ASEAN+ (associated agreements with other countries) triad of regional trading arrangements dominates the world economy, accounting for 67 per cent of all world trade. Whether the regional trade agreement trend causes the process of world trade liberalisation to falter in the future remains to be seen.

Apart from trade blocs, there are a number of looser trade groupings aiming to foster the mutual interests of member countries. These include:

- **the Asia–Pacific Economic Co-operation forum (APEC)** – its 21 members border the Pacific Ocean and include Canada, the USA, Peru, Chile, Japan, China and Australia; the member countries have pledged to facilitate free trade
- **the Cairns Group of agricultural exporting nations** – formed in 1986 to lobby for freer trade in agricultural

products; its members include Argentina, Brazil, Canada, New Zealand, Australia, the Philippines and South Africa.

Figure 13.12 details trade agreements between the EU and New Zealand, and the EU and the ACP.

Trade and development

There is a strong relationship between trade and economic development. In general, countries that have a high level of trade are richer than those with lower levels of trade. Countries that can produce goods and services in demand elsewhere in the world will benefit from strong inflows of foreign currency and from the employment their industries provide. Foreign currency allows a country to purchase from abroad goods and services it either does not produce itself or does not produce in large enough quantities.

An Oxfam report published in April 2002 stated that if Africa increased its share of world trade by just 1 per cent, it would earn an additional £49 billion a year – five times the amount it receives in aid. The World Bank has acknowledged that the benefits of globalisation are barely being passed on to Sub Saharan Africa and may actually have accentuated many of its problems.

EU and New Zealand bilateral trade agreement in agricultural products	
Background	**Agreement**
Before the UK joined the EU in 1973, New Zealand had a special trade relationship with the UK: 90 per cent of New Zealand's meat and dairy products were exported to the UK. Special trading arrangements were negotiated between the EU and New Zealand to secure the latter's main export market. Although New Zealand's major export markets have increasingly shifted to Asia and the Pacific rim, the EU remains New Zealand's second largest trading partner (for sheepmeat, dairy produce and wine). The UK remains New Zealand's most important trading partner within the EU.	Initially the EU imposed a common external tariff of 20 per cent on New Zealand imports. This was later reduced to 10 per cent and then to 0 per cent with a voluntary limit on the volume of New Zealand exports.
	The Uruguay round of WTO negotiations in the 1990s changed this arrangement and introduced Tariff Rate Quotas (TRQs). New Zealand's lamb exports to the EU were allocated a tariff-free quota of 227 000 tonnes/year. Any imports exceeding the quota attracted a 12.8 per cent tariff.
	New Zealand has, none the less, complained that trade in lamb is unfair. Whereas New Zealand sheep farmers receive no government subsidies, EU farmers can lower their prices because they get a ewe subsidy of €21 per head.
EU and bilateral trade agreement with ACP banana growers	
Background	**Agreement**
There is a long-running dispute between the EU, ACP banana growers, Latin American banana growers and the WTO.	The 2000 Cotonou trade agreement between the EU and ACP provided a 775 000 tonne tariff-free quota for ACP bananas. At the same time Latin American producers faced a €230/tonne tariff for their banana exports to the EU.
The UK and France have close political, historic and economic ties with many small countries in Africa, the Caribbean and the Pacific (ACP) which depend heavily on banana exports. The special trade agreements concluded between the UK, France and the ACP banana growers were adopted by the EU.	In 2007 WTO ruled that this agreement violated global trade rules, giving an unfair advantage to ACP growers. Although the tariff for Latin American bananas was reduced to €175/tonne the WTO insists that the revised trade arrangements remain unacceptable. By the end of 2008 the dispute was still unresolved.
Meanwhile, other banana exporters, especially in Latin America (e.g. Ecuador, Nicaragua, Mexico) complained that these arrangements were unfair. They argued that they should have the same access to the EU market as ACP growers. However, growing conditions in Latin America are more favourable, the scale of production is much greater (with large plantations owned by US TNCs) and therefore costs are low. Free trade would mean that ACP growers could not compete and that most would go out of business.	

Source: OCR A2 Geography by M. Raw (Philip Allan Updates, 2009), p.196

Figure 13.12 Examples of EU trade agreements

Section 13.1 Activities

1 Explain the importance of resource endowment as a factor in world trade.
2 What do you understand by the concept of *comparative advantage*?
3 How have **a** locational advantage and **b** historical factors influenced global trade patterns?
4 Define *terms of trade*.
5 Why have trade agreements increased significantly in recent decades?
6 Briefly examine the link between trade and development.

☐ The World Trade Organization

The World Trade Organization deals with the rules of world trade. Its primary function is to ensure that trade flows as freely as possible.

In 1947, a group of 23 nations agreed to reduce tariffs on each other's exports under the General Agreement on Tariffs and Trade (GATT). This was the first multilateral accord to lower trade barriers since Napoleonic times. Since the GATT was established, there have been nine 'rounds' of global trade talks, of which the most recent, the Doha (Qatar) round, began in 2001. A total of over 140 member countries have been represented at the talks in Doha. The Doha round was still in progress in 2015. Its work programme covers 20 areas of trade. This round of negotiation is also known as the Doha Development Agenda, as a major objective is to improve the trading prospects of LICs.

The most important recent development has been the creation of the World Trade Organization (WTO) in 1995. Unlike its predecessor, the loosely organised GATT, the WTO was set up as a permanent organisation with far greater powers to arbitrate trade disputes. Figure 13.13 shows the benefits of the global trading system according to the WTO.

1 The system helps promote peace.
2 Disputes are handled constructively.
3 Rules make life easier for all.
4 Freer trade cuts the cost of living.
5 It provides more choice of products and qualities.
6 Trade raises incomes.
7 Trade stimulates economic growth.
8 The basic principles make life more efficient.
9 Governments are shielded from lobbying.
10 The system encourages good government.

Source: WTO

Figure 13.13 The ten benefits of the WTO trading system

Although agreements have been difficult to broker at times, the overall success of GATT/WTO is undeniable: today, average tariffs are only a tenth of what they were when GATT came into force and world trade has been increasing at a much faster rate than GDP. However, in some areas **protectionism** is still an issue, particularly in the sectors of clothing, textiles and agriculture. In principle, every nation has an equal vote in the WTO. In practice, the rich world shuts out the poor world from key negotiations. In recent years, agreements have become more and more difficult to reach, with some economists forecasting the stagnation or even the break-up of the WTO.

Trade wars: Steel

Four months after the WTO launched a new round of global trade talks in Doha, the USA imposed tariffs of up to 30% on steel imports to protect its own fragile steel industry. More than 30 US steel producers went bankrupt between 1997 and 2002. Those that remained were considered to be inefficient and high cost compared with most of their foreign counterparts. Management consultants have largely put this down to the strength of the steel unions and their demands for high wages and health insurance. The crux of the problem is that world steelmaking capacity, estimated at between 900 million and 1000 million tonnes, is 20% higher than current demand. Although restructuring has already occurred, more is bound to happen both in the USA and in other parts of the world.

The reaction of America's trading partners was not difficult to predict. Trade unionists warned that the new trade barriers could result in 5000 job losses in the UK and 18 000 in the EU as a whole. The countries affected by the new tariffs argued that the USA was in breach of WTO rules. They also announced that they would demand compensation from the USA for the effect of the tariffs. However, as it could take up to two years for the WTO to reach a judgement, significant damage could be done in the intervening period to the steel industries of those nations affected. To its credit the EU stated that any retaliatory action would be within WTO rules. Overall this dispute was the last thing that the global steel industry, worth an estimated $500 billion, wanted.

Figure 13.14 Trade war in the WTO

Relations between the USA and the EU were soured in the early 2000s by the so-called 'banana war', and by disagreements over hormone-treated beef, GM foods and steel (Figure 13.14). Leading agricultural exporters such as the USA, Australia and Argentina want a considerable reduction in barriers to trade for agricultural products. Although the EU is committed in principle to reducing agricultural support, it wants to move slowly, arguing that farming merits special treatment because it is a 'multifunctional activity' that fulfils important social and environmental roles. Many MICs and LICs have criticised the WTO for being too heavily influenced by the interests of the USA and the EU.

The WTO exists to promote **free trade**. Most countries in the world are members and most of those countries that are not currently members want to join. The fundamental issue is: does free trade benefit all those concerned, or is it a subtle way in which the rich nations exploit their poorer counterparts? Most critics of free trade accept that it does generate wealth but they deny that all countries benefit from it. The non-governmental organisation (NGO) Oxfam is a major critic of the way the present trading system operates. Figure 13.15 shows the main goals of its 'Make Trade Fair' campaign.

Supporters of the WTO say that it is scarcely credible to argue that the poverty of poor countries is the result of globalisation since they are all outside the mainstream of free trade and economic globalisation. Critics of the WTO, on the other hand, say that the WTO and other international organisations should be paying more attention to the needs of these countries, making it easier for them to become more involved in, and gain tangible benefits from, the global economic system.

1. End the use of conditions attached to IMF–World Bank programmes which force poor countries to open their markets regardless of the impact.
2. Improve market access for poor countries and end the cycle of subsidised agricultural overproduction and export dumping by rich countries.
3. Change WTO rules so that developing countries can protect domestic food production.
4. Create a new international commodities institution to promote diversification and end oversupply in order to raise prices for producers and give them a reasonable standard of living.
5. Change corporate practices so that companies pay fair prices.
6. Establish new intellectual property rules to ensure that poor countries are able to afford new technologies and basic medicines.
7. Prohibit rules that force governments to liberalise or privatise basic services that are vital for poverty reduction.
8. Democratise the WTO to give poor countries a stronger voice.

Figure 13.15 Oxfam's 'Make Trade Fair' Campaign

Critics of the WTO also ask why it is that HICs have been given decades to adjust their economies to imports of textiles and agricultural products from LICs, when the latter are pressurised to open their borders immediately to banks, telecommunications companies and other components of the service sector in HICs. The removal of tariffs can have a significant impact on a nation's domestic industries. For example, India has been very concerned about the impact of opening its markets to foreign imports (Figure 13.16).

Since India was forced by a WTO ruling to accelerate the opening up of its markets, food imports have quadrupled. Large volumes of cheap, subsidised imports have flooded in from countries such as the USA, Malaysia and Thailand. The adverse impact has been considerable and includes the following:

- Prices and rural incomes have fallen sharply. The price paid for coconuts has dropped 80 per cent, for coffee 60 per cent, and pepper 45 per cent.
- Foreign imports, mainly subsidised soya from the USA and palm oil from Malaysia, have undercut local producers and have virtually wiped out the production of edible oil.

Figure 13.16 India – the impact of the removal of agricultural tariffs

The new emphasis on exports, in order for India to compete in the world market, is also threatening rural livelihoods. For example, in Andhra Pradesh, India, funding from the World Bank and the UK will encourage farm consolidation, mechanisation and modernisation. In this region, it is expected that the proportion of people living on the land will fall from 70 per cent to 40 per cent by 2020.

Farmers, trade unionists and many others are against these trends, or at least the speed at which they are taking place. They are calling for the reintroduction of import controls, thus challenging the basic principle of the globalisation process – the lowering of trade barriers.

Opposition to the WTO comes from a number of sources:

- many LICs and MICs, who feel that their concerns are largely ignored
- environmental groups concerned, for example, about a WTO ruling that failed to protect dolphins from tuna nets
- labour unions in some HICs, notably the USA, concerned about **a** the threat to their members' jobs as traditional manufacturing filters down to MICs and LICs and **b** violation of 'workers' rights' in MICs and LICs.

Tea, like coffee, bananas and other raw materials, exemplifies the relatively small proportion of the final price of the product that goes to producers. The great majority of the money generated by the tea industry goes to the post-raw-material stages (processing, distributing and retailing), usually benefiting companies in HICs rather than the producers in LICs.

A report by the Dutch Tea Institute in 2006 drew particular attention to:

- the problems of falling prices and rising input costs
- the consequent pressure to limit labour costs of tea production workers
- the urgent need for improvement of labour, social, ecological and economic conditions throughout the tea sector in the LICs.

The global tea market is dominated by a small number of companies including Unilever and Sara Lee. About half of all the tea produced is traded internationally. Annual export sales of tea in its raw material state are worth almost $3 billion. The retail value of the global tea business is of course much higher. The large tea companies wield immense power over the industry. As many countries now produce tea, they have to compete with each other in an increasingly competitive market. Global supply is rising at a faster rate than consumption, keeping prices low.

Tea producers complain that the global trading system prevents them from moving up the value chain (Figure 13.17)

by processing and packing the tea they grow. This is mainly because they would have to compete with very powerful brands, and they would find it very difficult to achieve the economies of scale of the global tea companies.

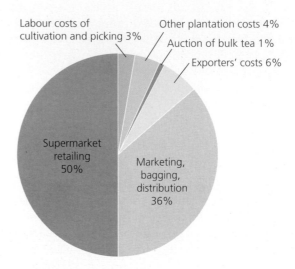

Figure 13.17 Tea value chain

☐ The nature and role of Fairtrade

Many supermarkets and other large stores in HICs now stock some 'fairly traded' products. Most are agricultural products such as bananas, orange juice, nuts, coffee and tea (Figure 13.18) but the market in non-food goods such as textiles and handicrafts is also increasing. The **Fairtrade** system operates as follows:

- Small-scale producers group together to form a cooperative or other democratically run association with high social and environmental standards.
- These cooperatives deal directly with companies (cutting out 'middlemen') such as large supermarkets in HICs.
- HIC companies (through their customers) pay significantly over the world market price for the products traded. The price difference can be as large as 100 per cent. This might mean, for example, supermarket customers paying a few pence more for a kilo of bananas.
- The higher price achieved by the LIC cooperatives provides both a better standard of living (often saving producers from bankruptcy and absolute poverty) and some money to reinvest in the farms of producers.

Advocates of the Fairtrade system argue that it is a model of how world trade can and should be organised to tackle global poverty. This system of trade began in the 1960s with Dutch consumers supporting Nicaraguan

Figure 13.18 Fairtrade tea

farmers. It is now a global market worth £315 million a year, involving over 400 HIC companies and an estimated 500 000 small farmers and their families in the world's poorest countries. Food sales are growing by more than 25 per cent a year, with Switzerland and the UK being the largest markets. Figure 13.19 compares the prices received by plantation workers under 'normal' trading conditions with those received by workers in a Fairtrade scheme.

In 1993 a group of farmers in Ghana formed a cooperative to sell their own cocoa. It was supported by SNV, a Dutch NGO, and the UK Department for International Development. The cooperative ensures farmers are paid for what they produce and are not cheated by middlemen. It includes:

- *Kuapa Kokoo Farmers' Union.* This is a national body made up of 45 000 cocoa farmers who elect representatives.
- *Kuapa Kokoo Farmers' Trust.* This is responsible for distributing money for community projects, generated from the Fairtrade Premium. Projects include providing clean water supplies and mobile health clinics, building schools and improving sanitation.

In 2008 Kuapa Kokoo sold 4250 tonnes of cocoa to the Fairtrade market. This means that the farmers receive a guaranteed price. For example, even if the world price of cocoa falls to US$1000 per tonne, the Fairtrade price remains at US$1600 per tonne. The minimum Fairtrade price is $1600 – if the world price goes higher farmers will receive the higher price, plus the social premium of $150 per tonne, and these prices have been reviewed and will be increased.

In 1998 Kuapa Kokoo came together with the NGO Twin, supported by The Body Shop, Christian Aid and Comic Relief, to found the Divine Chocolate company. As Kuapa Kokoo is part owner of Divine it not only gets a fair price for its cocoa but also has an influence on how the organisation is run and a share in the profits it has helped to create. Divine Fairtrade

Kuapa Kokoo farmers spread cocoa beans out to dry

chocolate is sold in the UK, the Netherlands, Scandinavia and the USA.

Like all food production, fair trade will only work as a solution if it is sustainable in the long term. Income, and therefore food security, depends on maintaining soil health and water supply through good agricultural management. In Africa, a continent riven by war, conflict and corruption, political stability is equally important. Fairtrade can contribute to this stability by reducing poverty.

Figure 13.19 Kuapa Kokoo Fairtrade Cooperative

Section 13.1 Activities

1 Examine the role of the World Trade Organization.
2 Why has the WTO been so heavily criticised?
3 Describe and explain the nature and role of Fairtrade.
4 Comment on the tea value chain presented in Figure 13.17.

13.2 International debt and international aid

☐ Debt: causes, nature and problems

Experts from a variety of disciplines blame the rules of the global economic system for excluding many countries from its potential benefits. Many single out **debt** as the major problem for the world's poorer nations. Here, debt is considered at the national scale rather than the personal level. The term 'debt' generally refers to **external debt (foreign debt)**, which is that part of the total debt in a country owed to creditors outside the country. Unpayable debt is a term used to describe external debt when the interest on the debt is beyond the means of a country, thus preventing the debt from ever being repaid. A country's external debt – both debt outstanding and debt service – affects a country's creditworthiness and thus its overall economic vulnerability.

Many poor countries are currently paying back large amounts in debt repayments to banks, lending agencies and governments in HICs while at the same time struggling to provide basic services for their populations. Sometimes an ever increasing proportion of new debt is used to service interest payments on old debts. The **debt service ratio** of many poor countries is at a very high level compared to their ability to pay. The debt service ratio is the proportion of a country's export earnings that it needs to use to meet its debt repayments. Some countries need to put aside 20–30 per cent of their export earnings to meet their debt repayments. A larger number of countries have a debt service ratio of between 10 and 20 per cent. These figures would be very significant for affluent countries, but can prove to be a crippling burden for nations with very low incomes. Figure 13.20 is a Christian Aid newspaper advertisement illustrating the plight of Haiti, one of the world's poorest countries, after the devastating earthquake of January 2010. Other organisations such as Oxfam, CAFOD and Islamic Relief mounted similar campaigns to cancel debt.

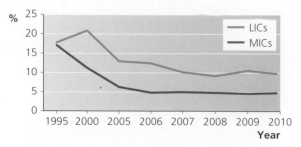

Figure 13.21 Debt service-to-export ratio, 1995–2010

THE ONLY THING STILL STANDING IS THE DEBT.

The suffering in Haiti is extraordinary. Tens of thousands dead. Many more injured and homeless. The British public have donated over £38m to the DEC to get emergency aid to the people suffering on the ground. It's been a fantastic response.

But when the emergency response teams have done their work, the Haitians will still face an uphill battle. Haiti has debts of over $800m. No wonder 80% of the population lives in poverty. As part of our ongoing fight against the systems that keep people poor, Christian Aid is lobbying Chancellor Alistair Darling (the UK's representative at the IMF) to lead the world in making sure that all of Haiti's debt is cancelled. You can help us.

Sign the petition online to drop the debt.
Visit www.christianaid.org.uk

POVERTY

christian aid

A009667

Figure 13.20 Christian Aid Haiti campaign

Table 13.3 shows the key debt indicators for developing countries by region, while Figure 13.21 illustrates the debt service-to-exports ratio for both LICs and MICs between 1995 and 2010.

Table 13.3 Debt indicators for LICs

Country group	Debt outstanding/ GNI, 2010	Debt outstanding/ exports, 2010
East Asia & Pacific	13.5	37.0
Europe & Central Asia	43.0	121.6
Latin America & Caribbean	21.7	102.1
Middle East & North Africa	14.1	42.5
South Asia	19.2	94.3
Sub-Saharan Africa	20.0	54.0
Top 10 borrowers	18.4	67.9
Other LICs	27.9	70.0

Source: World Bank Debtor Reporting System and International Monetary Fund, quoted in Global Development Finance 2012

Although external debt is still a major global problem, LICs and MICs improved the sustainability of their external debt between 2000 and 2010 (Figure 13.21). The average debt-service ratio for all developing countries in 2010 was 9.8 per cent. The higher figure for MICs such as China compared to LICs largely reflects the greater economic strength of the former group of countries.

The debt-service ratio of the LICs fell from 17.2 per cent in 1995 to 4.8 per cent in 2010 (Figure 13.21). Two significant reasons for such a considerable change have been:

- increased export earnings
- debt restructuring and outright debt relief from official and private creditors through the HIPC and MDRI (both discussed below).

While supporters of globalisation argue that economic growth through trade is the only answer, critics say that HICs should still do more to help the LICs through **debt relief** and by opening their markets to exports from LICs.

The USA owes more money to the rest of the world than any other country. Some other rich countries such as the UK and France also owe substantial amounts. However, these countries have huge assets against which they can borrow, so their debts are thought to be manageable, although the recent global financial crisis has called this into question. However, in general, debt repayment by rich countries is very different from the immense struggle that poor countries have in trying to pay their debts.

According to the World Bank, the total external debt stocks owed by LICs increased by $437 billion over 12 months, to reach $4 trillion at the end of 2010. When a LIC has to use a high proportion of its income to service debt, this takes money away from what could have been spent on education, health, housing, transport and other social and economic priorities. Multilateral debts are obligations to international financial institutions such as the World Bank, the International Monetary Fund and regional development banks. Multilateral debt service takes priority over private and bilateral debt service.

In 2012, the IMF highlighted countries it said were at risk of not being able to pay their debts: Afghanistan,

Burkina Faso, D.R. Congo, Djibouti, Gambia, Grenada, Kiribati, Laos, Maldives, São Tomé and Principe, Tajikistan, Tonga and Yemen.

How did the international **debt crisis** come about? Development economists have pointed to a sequence of events that began in the early 1970s as the main reason for the debt problems of many LICs. It began with the Arab–Israeli war of 1973–74, which resulted in a sharp increase in oil prices. Governments and individuals in the oil-producing countries invested so-called petrodollars (profits from oil sales) in the banks of affluent countries. Eager to profit from such a high level of investment, these banks offered relatively low-interest loans to poorer countries to fund their development. These countries were encouraged to exploit raw materials and grow cash crops so that they could pay back their loans with profits made from exports. However, periods of recession in the 1980s and 1990s led to rising inflation and interest rates in Western countries. At the same time, crop surpluses led to a fall in prices. As a result, the demand for exports from LICs fell and export earnings declined significantly. These factors, together with oil price increases, left many LICs unable to pay the interest on their debts.

Loans can help countries to expand their economic activities and set up an upward spiral of development if used wisely. However, many of the loans that burden the world's poorest countries were given under dubious circumstances and at very high rates of interest. Critics argue that banks frequently lent irresponsibly to governments that were known to be corrupt. The term **odious debt** has been used to describe debt incurred as HICs loaned to dictators or other corrupt leaders when it was known that the money would be wasted. For example, shortly after freedom from apartheid South Africa had to pay debts incurred by the apartheid regime. Often such loans led to little tangible improvement in the quality of life for the majority for the population, but instead saddled them with long-term debt. If such countries had been companies they would have been declared bankrupt. However, international law offers no 'fresh start' to countries in such a situation.

Many development economists also focus on the legacy of **colonialism**, arguing that the colonising powers left their former colonies with high and unfair levels of debt when they became independent. Such debts were often at very high interest rates. For example:

- In 1949, Indonesia, as a condition of independence, was required to assume the Dutch colonial government's debt, much of which had been acquired fighting pro-independence rebels in the previous four years.

- In order to receive independence from France, Haiti was required to pay France 150 million francs.

In recent years, much of the debt has been 'rescheduled' and new loans have been issued. However, new loans have frequently been granted only when LICs agreed to very strict conditions under 'structural adjustment programmes', which have included:

- agreeing to free-trade measures, which have opened up their markets to intense foreign competition
- severe cuts in spending on public services such as education and health
- the privatisation of public companies.

☐ Debt relief

Restructuring debt to LICs began in a limited way in the 1950s. The UN (Figure 13.22) and its related organisations have been fundamental in this process from the start. In 1956, Argentina was the first country to renegotiate the repayment of its debt with bilateral creditors within the framework of the Paris Club (set up for this purpose). Attempts were made by creditor nations to tackle the 'Third World debt crisis' through the 1980s and 1990s. However, these efforts were viewed as limited in nature and often self-serving. The overall debt of poorer countries continued to rise. The rescheduling of debt repayments often brought temporary relief, but with interest added over a longer time period the overall debt simply increased. It was not until the mid-1990s that a more comprehensive global plan to tackle the debt of the poorest countries was formulated (Figure 13.23).

Figure 13.22 The UN building, Manhattan, New York

Strengths	Weaknesses
• Allow a country's loans to be rescheduled in order to make them more manageable	• Often accompanied by a shift from domestic food cultivation to production of cash crops or commodities for export
• Make the country's economy more competitive	• Reduce government expenditure by cutting social programmes, e.g. health and education, and abolishing food and agricultural subsidies
• Improve foreign investment potential by removing trade and/or investment restrictions	• Privatisation of state enterprises to cut government expenditure results in assets being sold to TNCs
• Boost foreign exchange by promoting exports	• Increase pressure on countries to generate exports to pay off debt. This is likely to increase deforestation, land degradation and other environmental damage
• Reduce government deficits through cuts in spending	• Some MICs accused of protecting their own interests

Figure 13.23 Strengths and weaknesses of debt-reduction schemes

The Heavily Indebted Poor Countries (HIPC) Initiative

The Heavily Indebted Poor Countries (HIPC) Initiative was first established in 1996 by the International Monetary Fund (IMF) and the World Bank. Its aim was to provide a comprehensive approach to debt reduction for heavily indebted poor countries so that no poor country faced a debt burden it could not manage. To qualify for assistance, countries have to pursue IMF and World Bank supported adjustment and reform programmes. In 1999, a comprehensive review of the Initiative allowed the fund to provide faster, deeper and broader debt relief and strengthened the links between debt relief, poverty reduction and social policies. In 2006, the Multilateral Debt Relief Initiative (MDRI) was launched to provide additional support to HIPCs to reach the Millennium Development Goals. In 2007, the Inter-American Development Bank also decided to provide debt relief to the five HIPCs in the western hemisphere. According to a recent World Bank–IMF report, debt relief provided under both initiatives has substantially alleviated debt burdens in recipient countries.

To be considered eligible for HIPC Initiative assistance, a country should have a track record of macro-economic stability, have prepared an interim Poverty Reduction Strategy Paper (PRSP) and cleared any outstanding arrears. Completing these requirements means that the country can now receive full and irrevocable reduction in debt available under the HIPC Initiative and MDRI. To reach completion point, a country must maintain macro-economic stability under an IMF's Poverty Reduction and Growth Facility (PRGF)-supported programme, carry out key structural and social reforms as agreed upon at the decision point and implement a PRSP satisfactorily for one year.

Figure 13.24 shows the status of HIPC countries. As of April 2015, 36 nations were classed as 'post-completion-point countries' and three as 'pre-decision-point countries'. Of the former group of countries, thirty are in Africa, as are all three countries in the latter group! In terms of the former group, the HIPC initiative is providing $76 billion in debt-service relief over time. About 44 per cent of the funding comes from the IMF and other multilateral institutions, and the rest comes from bilateral creditors.

Post-completion-point countries		
Afghanistan	Ethiopia	Mauritania
Benin	The Gambia	Mozambique
Bolivia	Ghana	Nicaragua
Burkina Faso	Guinea	Niger
Burundi	Guinea-Bissau	Rwanda
Cameroon	Guyana	São Tomé & Principe
Central African Republic	Haiti	Senegal
Chad	Honduras	Sierra Leone
Comoros	Liberia	Tanzania
Republic of Congo	Madagascar	Togo
Democratic Republic of Congo	Malawi	Uganda
Côte d'Ivoire	Mali	Zambia
Pre-decision-point countries		
Eritrea	Somalia	Sudan

Figure 13.24 HIPC countries

Debt relief frees up resources for social spending

Debt relief is part of a much larger process, which includes international aid, designed to address the development needs of LICs. For debt reduction to have a meaningful impact on poverty, the additional funds made available need to be spent on programmes that are of real benefit to the poor.

Before the HIPC Initiative, eligible countries spent on average more on debt serving than on education and health combined. Now these countries have significantly increased their spending on education, health and other social services. On average, such spending is around five times the amount of debt-service payments.

Conclusion

There can be little doubt that the HIPC Initiative and MDRI have been more comprehensive debt-relief structures than anything that had gone before. However, the initiatives have drawn criticism, both in terms of the limited number of countries involved and the total extent of debt reduction. Even if all of the external debts of these countries were cancelled, most would still depend on significant levels of concessional external assistance, since their receipts of such assistance have been much larger than their debt-service payments for many years.

Since 1990, LICs have increased their buffer for external debt and its service. Total debt services have decreased significantly since 1999, due largely to debt-relief initiatives by multilateral and bilateral donors.

Section 13.2 Activities

1 Define **a** *debt* and **b** *debt-service ratio*.
2 What do you understand by the term *odious debt*?
3 How has the legacy of colonialism contributed to the debts of a considerable number of LICs?
4 Discuss the sequence of events that are generally accepted to have led to the debt crisis.
5 **a** What is debt relief?
  b Comment on the nature and effectiveness of the HIPC Initiative and the MDRI.

☐ International aid

The origins of foreign aid date back to the Marshall Plan of the late 1940s. This was when the USA set out to reconstruct the war-torn economies of Western Europe and Japan as a means of containing the international spread of communism. By the mid-1950s, the battle for influence between West and East in the developing world began to have a marked effect on the geography of aid. Even today, bilateral aid is strongly influenced by ties of colonialism and neo-colonialism and by strategic considerations. However, it would be wrong to deny that aid is also given for humanitarian and economic reasons (Figure 13.25).

Figure 13.25 A school in North Africa – education is often a focus of international aid

Aid is assistance in the form of grants or loans at below market rates. Most LICs have been keen to accept foreign aid because of the:

- **'foreign exchange gap'**, whereby many LICs lack the hard currency to pay for imports such as oil and machinery that are vital to development
- **'savings gap'**, where population pressures and other drains on expenditure prevent the accumulation of enough capital to invest in industry and infrastructure
- **'technical gap'** caused by a shortage of skills needed for development.

Many LICs rely on a very small range of exports for foreign currency. The prices of such products are often low compared with the goods and services they need to import, and the prices for such raw materials can also be very volatile.

But why do richer nations give aid? Is it down to altruism, or self-interest? Much of the evidence suggests the latter. Contrary to popular belief, most foreign aid is not in the form of a grant, nor is famine relief a major component. A significant proportion of foreign aid is 'tied' to the purchase of goods and services from the donor country and often given for use on jointly-agreed projects. However, the proportion of **tied aid** in relation to total international aid has been falling in recent decades.

Figure 13.26 shows how these factors combine to form the cycle of poverty.

The different types of international aid

Figure 13.27 shows the different types of **international aid**. The basic division is between official government aid and voluntary aid:

- Official government aid is where the amount of aid given and who it is given to is decided by the government of an individual country. The Department for International Development (DFID) runs the UK's international aid programme.
- Voluntary aid is run by NGOs or charities such as Oxfam, ActionAid and CAFOD. NGOs collect money from individuals and organisations. However, an increasing amount of government money goes to NGOs because of their special expertise in running aid programmes efficiently.

Official government aid can be divided into:

- **bilateral aid**, which is given directly from one country to another – a significant proportion of bilateral aid is 'tied'
- **multilateral aid**, which is provided by many countries and organised by an international body such as the UN.

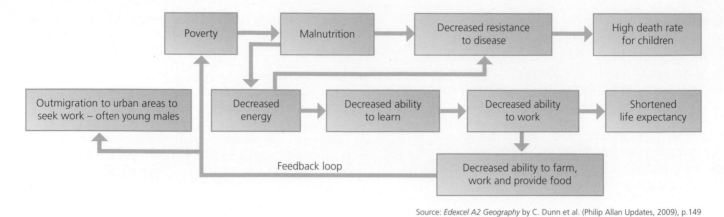

Source: *Edexcel A2 Geography* by C. Dunn et al. (Philip Allan Updates, 2009), p.149

Figure 13.26 The cycle of poverty

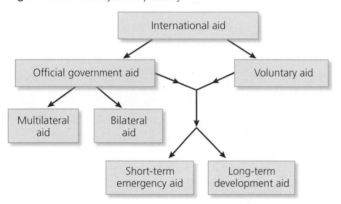

Figure 13.27 The different types of international aid

Aid supplied to poorer countries is of two types:

- Short-term emergency aid, often termed 'relief aid', is provided to help cope with unexpected disasters such as earthquakes, volcanic eruptions and tropical cyclones.
- Long-term **development aid** is directed towards the continuous improvement in the quality of life in a poorer country.

Figure 13.28 shows the types of bilateral aid provided by DFID. As Figure 13.28 shows, bilateral aid covers a number of different categories of assistance. Figure 13.29 shows how DFID can provide emergency aid to countries in need.

Financial Aid – Poverty Reduction Budget Support (PRBS) – Funds provided to developing countries for them to spend in support of a government policy and their expenditure programmes whose long-term objective is to reduce poverty; funds are spent using the overseas governments' own financial management, procurement and accountability systems to increase ownership and long-term sustainability. PRBS can take the form of a general contribution to the overall budget – **general budget support** – or support with a more restricted focus which is earmarked for a specific sector – **sector budget support**.

Other Financial Aid – Funding of projects and programmes such as Sector Wide Programmes not classified as PRBS. Financial aid in its broader sense covers all bilateral aid expenditure, other than technical cooperation and administrative costs, but in SID we separately categorise Humanitarian Assistance, DFID Debt Relief and 'other bilateral aid' as it is a rapidly declining flow.

Technical Cooperation – Activities designed to enhance the knowledge, intellectual skills, technical expertise or the productive capability of people in recipient countries. It also covers funding of services that contribute to the design or implementation of development projects or programmes.

This assistance is mainly delivered through research and development, the use of consultants, training (generally overseas partners visiting the UK or elsewhere for training programmes) and employment of 'other Personnel' (non-DFID experts on fixed-term contracts). This latter category is becoming less significant over time as existing contracted staff reach the end of their assignments.

Humanitarian Assistance – Provides food aid and other humanitarian assistance, including shelter, medical care and advice in emergency situations and their aftermath. Work of the conflict pools is also included.

DFID Debt Relief – Includes sums for debt relief on old DFID aid loans and cancellation of debt under the Commonwealth Debt Initiative (CDI). The non-CDI DIFD debt relief is reported on the basis of the 'benefit to the recipient country'. This means that figures shown represent the money available to the country in the year in question that would otherwise have been spent on debt servicing. The CDI debt cancellation is reported on a 'lump sum' basis where all outstanding amounts on a loan are shown at the time the agreement to cancel is made.

Other Bilateral Aid – Covers support to the development work of UK and international Civil Society Organisations (increasingly through partnership agreements with CSOs). It includes bilateral aid delivered through multilateral organisations including aid delivered through multi donor funds such as the Education Fast Track Initiative. 'Other bilateral aid' also includes any aid not elsewhere classified such as DFID's Development Awareness Fund.

Source: Statistics of International Development 2008, Department for International Development

Figure 13.28 UK Department for International Development – types of bilateral aid

4 December 2007

On 15 and 16 November, southern Bangladesh was hit by Cyclone Sidr. So far, over 6 million people have been affected and 2997 people have been confirmed dead. Many more have been injured, and the death toll could reach 10 000 (the death toll following the cyclone in 1991 was 140 000). Also, around 300 000 houses have been destroyed, as have many crops and large tracts of agricultural land.

Following an initial DFID contribution of £2.5 million, which is being channelled through the UN for immediate relief efforts, a further £2.5 million was pledged on 23 November. On 28 November, an additional £2 million was committed to help survivors to rebuild their homes and livelihoods. DFID has also provided 12 lightweight boats to reach inaccessible parts of Bangladesh, and despatched over 100 000 blankets for people made homeless.

Already DFID money is helping to rebuild more than 16 000 homes, provide food to 70 000 families and clean water to 260 000 families. The UK's disaster relief aid in Bangladesh now totals almost £12 million (US$24 million) for this year, with £4.7 million having been provided in response to the severe floods that occurred in August.

Secretary of State for International Development, Douglas Alexander, said yesterday:

'Unless emergency relief supplies get to victims it is all too likely that more people will die needlessly. That is why the UK continues to provide funds to get more food, clean water, basic shelter and other emergency supplies to tens of thousands of survivors. With half a million animals killed, nearly two million acres of crops and more than a million homes destroyed, the next challenge is to help people rebuild their homes and livelihoods. UK support is meeting immediate and longer term needs as well. I continue to be admiring of the resilience and determination of the Bangladeshi people as they face these challenges.'

Source: UK Department for International Development (DFID)

Figure 13.29 DFID provides emergency aid towards Bangladesh cyclone

There is no doubt that many countries have benefited from international aid. All the countries that have developed into MICs from LICs have received international aid. However, their development has been due to other reasons too. It is difficult to be precise about the contribution of international aid to the development of each country. According to some left-wing economists, aid does not do its intended job because:

- too often aid fails to reach the very poorest people and when it does the benefits are frequently short lived
- a significant proportion of foreign aid is 'tied' to the purchase of goods and services from the donor country and often given for use only on jointly-agreed projects
- the use of aid on large capital-intensive projects may actually worsen the conditions of the poorest people

- aid may delay the introduction of reforms, for example the substitution of food aid for land reform
- international aid can create a culture of dependency that can be difficult to break.

Arguments put forward by the political right-wing economists against aid are as follows:

- Aid encourages the growth of a larger than necessary public sector.
- The private sector is 'crowded out' by aid funds.
- Aid distorts the structure of prices and incentives.
- Aid is often wasted on grandiose projects of little or no benefit to the majority of the population.
- The West did not need aid to develop.

Many development economists argue there are two issues more important to development than aid:

- changing the terms of trade so that LICs get a fairer share of the benefits of world trade
- writing off the debts of the poorest countries.

Figure 13.30 shows the average official development assistance (ODA) received by the economic status of countries between 1970 and 2012. One-quarter of all ODA during this period was allocated to the LICs. Table 13.4 shows the top ten recipients of ODA in 2012–13, with the largest amounts of money destined for Afghanistan, Myanmar, Vietnam and India.

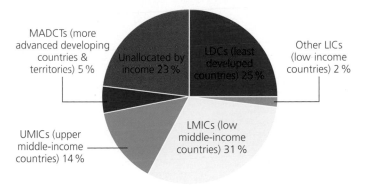

Source: OECD, September 2014

Figure 13.30 Average ODA by the economic status of countries, 1970–2012

Table 13.4 Top ten recipients of gross ODA (US$), 2012–13 average

Rank	Country
1	Afghanistan
2	Myanmar
3	Vietnam
4	India
5	Indonesia
6	Kenya
7	Tanzania
8	Côte d'Ivoire
9	Ethiopia
10	Pakistan

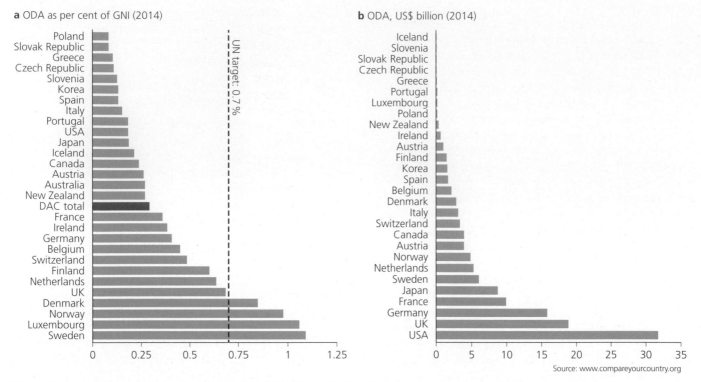

a ODA as per cent of GNI (2014)

b ODA, US$ billion (2014)

Source: www.compareyourcountry.org

Figure 13.31a ODA as a percentage of GNI, 2014, **b** ODA in $billion, 2014

The ODA provided by the more affluent countries of the world varies widely. Figure 3.31 shows the two standard ways in which ODA is measured. Figure 13.31a is the favoured method, which shows ODA as a percentage of gross national income (GNI) for 2014. In this year, only five countries met the UN target of 0.7 per cent of GNI. Figure 13.31b shows total ODA by country. Here, the USA is by far the largest donor, although it is only in 19th place as a percentage of GNI.

Over the last 30 years or so, some countries have changed from being recipients of ODA to donor countries. South Korea is a case in point. Between 1945 and the late 1990s, Korea received about $12.7 billion in aid, which assisted economic development and helped to alleviate poverty. This aid was mainly provided by the USA, Japan and Western Europe. The economic transformation of South Korea has been astounding.

As one of the four Asian 'tigers', it was a member of the first generation of NICs along with Singapore, Hong Kong and Taiwan. South Korea became a donor country in 1987. The amount of ODA provided by South Korea reached approximately $200 million in 1998, $800 million in 2008 and $1200 million in 2010.

The effectiveness of aid: top-down and bottom-up approaches

Over the years, most debate about aid has focused on the amount of aid made available. However, in recent years the focus has shifted more towards the effectiveness of aid. This has involved increasing criticism of the traditional top-down approach to aid.

The financing of the Pergau Dam in Malaysia with UK government aid is an example of a capital-intensive government-led aid programme, set up without consulting the local people. Work began in 1991 and around the same time Malaysia bought £1 billion-worth of arms from the UK, leading many people to believe that the £234 million in aid was 'tied' to the arms deal.

The Hunger Project is one of a number of organisations that have adopted a radically different approach (Figure 13.33). The Hunger Project has worked in partnership with grassroots organisations in Africa, Asia and Latin America to develop effective **bottom-up strategies**. The key strands in this approach have been:

- mobilising local people for self-reliant action
- intervening for gender equality
- strengthening local democracy.

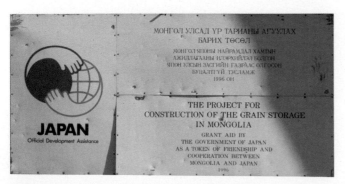

Figure 13.32 A Japanese aid project in Mongolia

	The conventional top-down, service-delivery model	The Hunger Project's bottom-up empowerment model
Who are hungry people?	**Beneficiaries** whose basic needs must be met.	**Principal authors and actors in development** – hardworking, creative individuals who lack opportunities.
What must be done?	**Provide services** through government or charities.	**Mobilise and empower** people's self-reliant action, and stand in solidarity with them for their success.
What's the primary resource for development?	**Money and the expertise** of consultants and programme managers.	**People:** their vision, mobilisation, entrepreneurial spirit and confidence.
Who is in charge?	**Donors,** who provide the money and hold implementers to account.	**Local people:** through elected local leaders whom they hold to account.
What are the main constraints?	**Bureaucracy:** the inefficiency of the delivery system.	**Social conditions:** resignation, discrimination (particularly gender), lack of local leadership, lack of rights.
What is the role of women?	**Vulnerable group** who must be especially targeted beneficiaries.	**Key producers** who must have a voice in decision-making.
What about social and cultural issues?	**Immutable conditions** that must be compensated for.	**Conditions** that people can transform.
How should we focus our work?	**Carefully target** beneficiaries on an objective-needs basis.	**Mobilise everyone** as broadly as possible, build spirit and momentum of accomplishment.
What is the role of central government?	**Operate** centrally managed service-delivery programmes.	**Decentralise** resources and decision making to local level; build local capacity; set standards; protect rights.
What is the role of local government?	**Implementing arm** of central programmes.	**Autonomous** leadership directly accountable to people.
What is the role of civil society?	**Implementing arm** of central programmes.	**Catalyst** to mobilise people; fight for their rights; empower people to keep government accountable.

Figure 13.33 Contrasting top-down and bottom-up aid models

Source: The Hunger Project

Non-governmental organisations: leading sustainable development

NGOs have often been much better at directing aid towards sustainable development than government agencies. The selective nature of such aid has targeted the poorest communities using **appropriate technology** and involving local people in decision-making.

Case Study: WaterAid in Mali

WaterAid was established in 1981. Its first project was in Zambia but its operations spread quickly to other countries. Mali is one of the countries in which WaterAid currently operates.

Mali, in West Africa, is one of the world's poorest nations (Figure 3.34). The natural environment is harsh, and is deteriorating. Rainfall levels, already low, are falling further and desertification is spreading. Currently, 65 per cent of the country is desert or semi-desert, and 11 million people still lack access to safe water. WaterAid has been active in the country since 2001.

Its main concern is that the fully privatised water industry frequently fails to provide services to the poorest urban and rural areas. WaterAid is running a pilot scheme in the slums surrounding Mali's capital, Bamako, providing clean water and sanitation services to the poorest people. Its objective is to demonstrate to both government and other donors that projects in slums can be successful, both socially and economically.

Figure 13.34 Mali

WaterAid has financed the construction of the area's water network. It is training local people to manage and maintain the system, and to raise the money needed to keep it operational. Encouraging the community to invest in its own infrastructure is an important part of the philosophy of the project. According to Idrissa Doucoure, WaterAid's West Africa Regional Manager, 'We are now putting our energy into education programmes and empowering the communities to continue their own development into the future. This will allow WaterAid to move on and help others.'

Already there have been significant improvements in the overall health of the community. The general view is that it takes a generation for health and sanitation to be properly embedded into people's day-to-day lives.

WaterAid is the UK's only major charity dedicated exclusively to the provision of safe domestic water, sanitation and hygiene education to the world's poorest people. These three crucial elements provide the building blocks for all other development. Without them, communities remain stuck in a cycle of disease and poverty. The combination of safe water, sanitation and hygiene education maximises health benefits and promotes development (Figures 13.35, 13.36 and 13.37). The combined benefits of safe water, sanitation and hygiene education can reduce incidences of childhood diarrhoea by up to 95 per cent. A child dies every 15 seconds from diseases associated with a lack of access to safe water and adequate sanitation.

Figure 13.35 Water supply is very basic in many poor countries

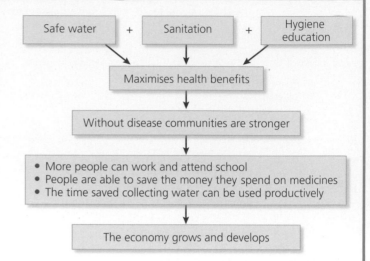

Figure 13.36 WaterAid's building blocks of development

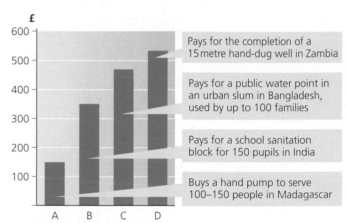

Figure 13.37 WaterAid cost examples

In the longer term, communities are able to plan and build infrastructure that enables them to cope better in times of hardship. In areas with WaterAid projects, life in times of drought is eased because:

- previously, in times of drought women in particular would spend hours in search of water, leaving little time to find food
- children would also miss out on education in the search for water
- cattle can be watered, rather than sold or left to die because of water shortage
- during famines, with sanitation, water and hygiene people are less sick and so are better able to fend off disease.

Microcredit and social business

The development of the Grameen Bank in Bangladesh illustrates the power of microcredit in the battle against poverty. The Grameen Foundation uses microfinance and innovative technology to fight global poverty and bring opportunities to the poorest people. The bank provides tiny loans and financial services to poor people to start their own businesses. Women are the beneficiaries of most of these loans. A typical loan might be used to buy a cow and sell milk to fellow villagers or to purchase a piece of machinery that can be hired out to other people in the community.

The concept has spread beyond Bangladesh to reach 3.6 million families in 25 countries. Muhammad Yunus highlights **social business** as the next phase in the battle against poverty in his book *Creating a World Without Poverty*. He presents a vision of a new business model that combines the operation of the free market with the quest for a more humane world.

Section 13.2 Activities

1 Define *international aid*.
2 Explain the difference between official government aid and voluntary aid.
3 Produce a flow diagram to show how aid can speed up development.
4 Discuss two possible disadvantages of international aid.
5 Look at Figure 13.36.
   a What do you understand by the following terms:
      i *safe water,* **ii** *sanitation,* **iii** *hygiene education*?
   b Why is it so important to combine these three factors to maximise the health benefits to a community?
6 Explain why healthier communities are more likely to be able to improve their living standards.
7 Explain **a** *microcredit* and **b** *social business*.

13.3 The development of international tourism

☐ Reasons for and trends in the growth of international tourism

Over the last 50 years, **tourism** has developed into a major global industry that is still expanding rapidly. Tourism is defined as travel away from the home environment **a** for leisure, recreation and holidays (Figure 13.38), **b** to visit friends and relatives and **c** for business and professional reasons.

Figure 13.38 A remote island in Indonesia – a destination for a small expedition cruise ship

Under one method of economic measurement, tourism is the world's major service industry. Tourism has an economic, social and environmental impact on virtually every country in the world. In some countries, it is a considerable political issue. Without doubt, it is one of the major elements in the process of globalisation.

International tourist arrivals reached a record of 1.135 billion in 2014 (Figure 13.39) This was the fifth consecutive year of above-average growth since the 2009 economic crisis. In the same year, **international tourist receipts** amounted to a record $1.245 billion. An additional $221 billion was generated from international passenger transport, bringing total exports from international tourism up to $1.5 trillion. Other figures for 2014 highlighted by the United Nations World Tourism Organization (UNWTO) include tourism accounting for:

- over 9 per cent of global GDP
- 1 in 11 of all jobs globally (over 275 million people)
- 6 per cent of global exports
- 30 per cent of global services exports.

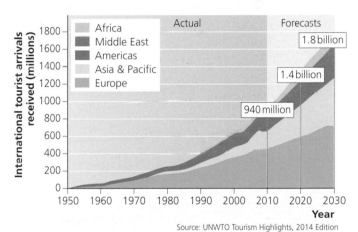

Source: UNWTO Tourism Highlights, 2014 Edition

Figure 13.39 Growth in world tourism and forecast to 2030

In regional terms, the highest rates of growth in the previous year were achieved by the Americas (up 7 per cent) and Asia and the Pacific (up 5 per cent). In contrast, Europe and the Middle East (up 4 per cent) and Africa (up 2 per cent) grew at a slightly more modest pace. Looking in more detail at sub-regions, North America (up 8 per cent) had the highest growth rate, followed by North-East Asia, South Asia, Southern and Mediterranean Europe, Northern Europe and the Caribbean, all increasing by 7 per cent.

Europe still attracts more than half of all international tourists (Figure 13.40), a total of 588 million in 2014. The number of tourists visiting the other regions of the world were, in descending order: Asia and the Pacific (263 million), the Americas (181 million), Africa (56 million) and the Middle East (50 million).

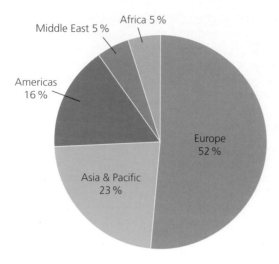

Figure 13.40 Regional share of world tourist arrivals, 2014

Table 13.5 Top ten countries – international tourism arrivals, 2013

Rank	Country	International tourist arrivals (millions)
1	France	83.0
2	USA	69.8
3	Spain	60.7
4	China	55.7
5	Italy	47.7
6	Turkey	37.8
7	Germany	31.5
8	UK	31.2
9	Russian Federation	28.4
10	Thailand	26.5

Table 13.5 shows international tourism arrivals by the main individual countries. Here, four countries recorded a total of over 50 million arrivals in 2013. In order of importance, these were France (Figure 13.41), Spain, the USA and China. Table 13.6 shows international tourism receipts by country for the same year. The trend is similar to that for tourism arrivals. However, it is not an exact relationship because:

■ the average number of days spent in some destinations is longer than in others
■ visitors spent more money in some destinations than others.

Table 13.6 Top ten countries – international tourism receipts, 2013

Rank	Country	International tourist arrivals ($ billion)
1	USA	139.6
2	Spain	60.4
3	France	56.1
4	China	51.7
5	Macao (China)	51.6
6	Italy	43.9
7	Thailand	42.1
8	Germany	41.2
9	UK	40.6
10	Hong Kong (China)	38.9

Source: UNWTO

The countries recording a total of over $50 billion in order of importance were: the USA, Spain, France and China.

Tourism is an increasingly important contributor to economic growth and employment in a significant number of countries. A range of factors have been responsible for the growth of global tourism. Figure 13.42 subdivides these factors into economic, social and political reasons, and also includes factors that can reduce levels of tourism, at least in the short term. Some of these factors have been active for a longer time period than others.

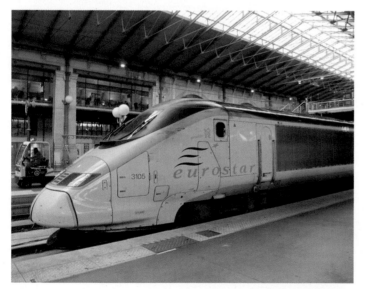

Figure 13.41 The fast Eurostar trains have been popular with tourists visiting France

Economic	Steadily rising real incomes – tourism grows on average 1.3 times faster than GDP
	The decreasing real costs (with inflation taken into account) of holidays
	The widening range of destinations within the middle-income range
	The heavy marketing of shorter foreign holidays aimed at those who have the time and disposable income to take an additional break
	The expansion of budget airlines
	'Air miles' and other retail reward schemes aimed at travel and tourism
	'Globalisation' has increased business travel considerably
	Periods of economic recession can reduce levels of tourism considerably
Social	An increase in the average number of days of paid leave
	An increasing desire to experience different cultures and landscapes
	Raised expectations of international travel with increasing media coverage of holidays, travel and nature
	High levels of international migration over the last decade or so means that more people have relatives and friends living abroad
	More people are avoiding certain destinations for ethical reasons
Political	Many governments have invested heavily to encourage tourism
	Government backing for major international events such as the Olympic Games and the World Cup
	The perceived greater likelihood of terrorist attacks in certain destinations
	Government restrictions on inbound/outbound tourism
	Calls by non-governmental organisations to boycott countries such as Burma

Figure 13.42 Factors affecting global tourism

The medical profession was largely responsible for the growth in people taking holidays away from home. During the seventeenth century, doctors increasingly began to recommend the benefits of mineral waters, and by the end of the eighteenth century there were hundreds of spas in existence in the UK. Bath (Figure 13.43) and Tunbridge Wells were among the most famous. The second stage in the development of holiday locations was the emergence of the seaside resort. Sea bathing is usually said to have begun at Scarborough in about 1730.

Figure 13.43 The historical mineral waters in the spa town of Bath

The annual holiday for the masses, away from work, was a product of the Industrial Revolution, which brought big social and economic changes. However, until the latter part of the nineteenth century only the very rich could afford to take a holiday away from home.

The first **package tours** were arranged by Thomas Cook in 1841. These took travellers from Leicester to Loughborough (UK), 19 kilometres away. At the time, it was the newly laid railway network that provided the transport infrastructure for Cook to expand his tour operations. Of equal importance was the emergence of a significant middle class who had the time and money to spare for extended recreation.

By far the greatest developments have occurred since the end of the Second World War, arising from the substantial growth in leisure time, affluence and mobility enjoyed in HICs. However, it took the jet plane to herald the era of international mass tourism. In 1970, when Pan Am flew the first Boeing 747 from New York to London, scheduled planes carried 307 million passengers. By 2013, the number had reached 3.1 billion according to the International Air Transport Association (IATA).

Travel motivators are the reasons that people travel. All the major tourism organisations recognise three major categories (Figure 13.44).

Prime reasons	Secondary subdivisions	Tertiary destination preferences	Externalities
Leisure	Holiday	Climate	Destination security
	Sport or cultural event		
	Educational trip		
	Pilgrimage	Attractions	
Business	Conference/ exhibition	Festivals and events	
	Individual meetings		Exchange rate
Visiting friends and relatives	Stay with family	Accommodation/ restaurants/ bars	
	Meet friends	Transport (to the destination and within it)	

Figure 13.44 Key travel motivators

Many LICs and MICs have become more open to FDI in tourism than they were two or three decades ago. In general, there are now fewer restrictions on foreign investment in tourism in LICs and MICs than for many other economic activities. In fact, many governments in LICs and MICs have very actively promoted a range of:

- **'soft' measures** such as tourism internet sites and support for trade fairs
- **'hard' measures** such as providing incentives for foreign investors.

Recent data from the World Tourism Organization (WTO) shows that tourism is one of the top five export categories for as many as 83 per cent of countries and is the main source of foreign exchange for at least 38 per cent of countries.

Section 13.3 Activities

1 Define *tourism*.
2 Explain the terms **a** *international tourism arrivals* and **b** *international tourism receipts*.
3 Describe the changes shown in Figure 13.39.
4 Suggest reasons for the global share of tourist arrivals shown in Figure 13.40.

Variations in the level of tourism over time and space

Unfortunately, more than many other industries, tourism is vulnerable to **external shocks**. Periods of economic recession characterised by high unemployment, modest wage rises and high interest rates affect the demand for tourism in most parts of the world. Because holidays are a high-cost purchase for most people, the tourist industry suffers when times are hard.

Tourism in individual countries and regions can be affected by considerable fluctuations caused by a variety of factors:

- **Natural disasters** – earthquakes, volcanic eruptions, floods and other natural events can have a major impact on tourism where they occur.
- **Natural processes** – coastal erosion and rising sea levels are threatening important tourist locations around the world.
- **Terrorism** – terrorist attacks, or the fear of them, can deter visitors from going to certain countries, in the short term at least.
- **Health scares** – for example the severe acute respiratory syndrome (SARS) epidemic in March 2003 had a considerable short-term impact on tourism in China and other countries in South East Asia.
- **Exchange-rate fluctuations** – for example if the value of the dollar falls against the euro and the pound, it makes it more expensive for Americans to holiday in Europe, but less expensive for Europeans to visit the USA.

- **Political uncertainties** – governments may advise their citizens not to travel to certain countries if the political situation is tense.
- **International image** – a 2006 US-made film called *Turistas* has caused major concern in Brazil; it depicts a group of US backpackers whose holiday in a Brazilian resort turns into a nightmare when they are drugged and kidnapped and then their organs are removed by organ traffickers.
- **Increasing competition** – as new, 'more exciting' destinations increase their market share, more traditional destinations may see visitor numbers fall considerably.

The World Travel and Tourism Council (WTTC), in its assessment of the global performance of tourism in 2014, made particular note of the Ukraine–Russia conflict, Ebola in West Africa, political instability in Thailand, the continuation of major instability in Syria and Libya and terror attacks in Nigeria and Kenya. Since then, the terror attacks in Tunisia and Sharm el-Sheikh, Egypt have also affected tourism in those regions.

☐ The impacts of tourism

Social and cultural impact

Many communities in LICs and MICs have suffered considerable adverse cultural changes, some of them through the imposition of the worst of Western values. The result has been, in varying degrees:

- the loss of locally owned land as tourism companies buy up large tracts in the most scenic and accessible locations
- the abandonment of traditional values and practices
- displacement of people to make way for tourist developments
- abuse of human rights by governments and companies in the quest to maximise profits
- alcoholism and drug abuse
- crime and prostitution, sometimes involving children
- visitor congestion at key locations, hindering the movement of local people
- denying local people access to beaches to provide 'exclusivity' for visitors
- the loss of housing for local people as more visitors buy second homes in popular tourist areas.

Figures 13.45 and 13.46 show how the attitudes to tourism of host countries and destination communities in particular can change over time. An industry that is usually seen as very beneficial initially can eventually become the source of considerable irritation, particularly where there is a big clash of cultures. Parents in particular are often fearful of the impact 'outside' cultures may have on their children.

Figure 13.45 Entrance to a national park in Andalucia, Spain – the graffiti refers to the number of foreigners buying up houses in the nearby village of Frigiliana

1 Euphoria
- Enthusiasm for tourist development
- Mutual feeling of satisfaction
- Opportunities for local participation
- Flows of money and interesting contacts

2 Apathy
- Industry expands
- Tourists taken for granted
- More interest in profit making
- Personal contact becomes more formal

3 Irritation
- Industry nearing saturation point
- Expansion of facilities required
- Encroachment of the ways of life

4 Antagonism
- Irritations become more overt
- The tourist is seen as the harbinger of all that is bad
- Mutual politeness gives way to antagonism

5 Final level
- Environment has changed irreversibly
- The resource base has changed and the type of tourist has also changed
- If the destination is large enough to cope with mass tourism it will continue to thrive

Figure 13.46 Doxey's index of irritation caused by tourism

The tourist industry and the various scales of government in host countries have become increasingly aware of these problems and are now using a range of management techniques in an attempt to mitigate such effects. Education is the most important element so that visitors are made aware of the most sensitive aspects of the **host culture**.

At its very worst, the impact of tourism amounts to gross abuse of human rights. For example, the actions of the military regime in Burma – forcing people from their homes to make way for tourism developments, and using forced labour to construct tourist facilities – have brought condemnation from all over the world. The tourist industry has a huge appetite for basic resources, which often impinge heavily on the needs of local people. A long-term protest against tourism in Goa highlighted how one five-star hotel consumes as much water as five local villages, with the average hotel resident using 28 times more electricity per day than a local person. In such situations, tourist numbers may exceed the **carrying capacity** of a destination by placing too much of a burden on local resources. The concept of carrying capacity has sometimes been taken beyond just the ability of the physical environment to accommodate tourists/visitors without resultant deterioration and degradation. One classification has identified four elements of the concept:

- **Physical** – the overall impact on the physical environment, for example footpath erosion
- **Ecological** – the number of tourists that can be accommodated without significant impact on the flora and fauna
- **Economic** – the number of tourists a destination can take without significant adverse economic implications
- **Perceptual** – the attitudes of the local people in terms of how they view increasing tourist numbers.

Changing community structure

Communities that were once very close socially and economically may be weakened considerably due to a major outside influence such as tourism. The traditional hierarchy of authority within the community can be altered as those whose incomes are enhanced by employment in tourism gain higher status in the community. The age and sex structure may change as young people in particular move away to be closer to work in **tourist enclaves**. Changing values and attitudes can bring conflict to previously settled communities. The close ties of the extended family often diminish as the economy of the area changes and material wealth becomes more important.

However, tourism can also have positive social and cultural impacts:

- Tourism development can increase the range of social facilities for local people.
- It can lead to greater understanding between people of different cultures.
- Family ties may be strengthened by visits to relatives living in other regions and countries.
- It can help develop foreign-language skills in host communities.
- It may encourage migration to major tourist-generating countries.
- A multitude of cultures congregating together for major international events such as the Olympic Games can have a very positive global impact.

Section 13.3 Activities

1 Discuss three negative social/cultural aspects of tourism.
2 Comment on Doxey's index (Figure 13.46).

Economic impact

It is east to underestimate the economic impact of tourism. What is commonly thought of as the tourist industry is only the tip of the iceberg. Figure 13.47 shows both the direct and indirect economic impacts of tourism.

Tourism undoubtedly brings valuable foreign currency to LICs and MICs, and a range of other obvious benefits, but critics argue that its value is often overrated because:

- **Economic leakages** (Figure 13.48) from LICs and MICs to HICs run at a rate of between 60 and 75 per cent. With cheap package holidays, by far the greater part of the money paid stays in the country where the holiday was purchased.
- Tourism is labour-intensive, providing a range of jobs especially for women and young people. However, most local jobs created are menial, low paid and seasonal. Overseas labour may be brought in to fill middle and senior management positions.
- Money borrowed to invest in the necessary infrastructure for tourism increases the national debt.
- At some destinations tourists spend most of their money in their hotels, with minimum benefit to the wider community.
- Tourism might not be the best use for local resources that could in the future create a larger multiplier effect (see below) if used by a different economic sector.
- Locations can become overdependent on tourism.
- International trade agreements, such as the General Agreement on Trade in Services (GATS), are a major impetus to globalisation and allow the global hotel giants to set up in most countries. Even if governments favour local investors, there is little they can do.

Figure 13.47 The economic impact of tourism

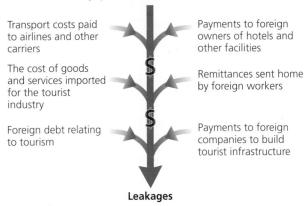

LIC tourist destinations

Total money spent on tourism to this destination

Transport costs paid to airlines and other carriers

Payments to foreign owners of hotels and other facilities

The cost of goods and services imported for the tourist industry

Remittances sent home by foreign workers

Foreign debt relating to tourism

Payments to foreign companies to build tourist infrastructure

Leakages

Figure 13.48 Economic leakages

However, supporters of the development potential of tourism argue that:

- tourism benefits other sectors of the economy, providing jobs and income through the supply chain; this is called the **multiplier effect** (Figure 13.49) because jobs and money multiply as a result of tourism development
- it is an important factor in the balance of payments of many nations
- it provides governments with considerable tax revenues
- by providing employment in rural areas, it can help to reduce rural–urban migration
- a major tourism development can act as a growth pole, stimulating the economy of the larger region
- it can create openings for small businesses in which start-up costs and barriers to entry are generally low
- it can support many jobs in the informal sector, where money goes directly to local people (Figure 13.50).

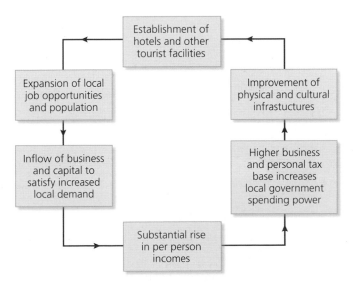

Establishment of hotels and other tourist facilities

Expansion of local job opportunities and population

Improvement of physical and cultural infrastructures

Inflow of business and capital to satisfy increased local demand

Higher business and personal tax base increases local government spending power

Substantial rise in per person incomes

Figure 13.49 The multiplier effect of tourism

Figure 13.50 Beach artist, Agadir, Morocco – an example of informal-sector employment

Environmental impact

The type of tourism that does not destroy what it sets out to explore has come to be known as 'sustainable' (Figure 13.51). The term comes from the 1987 UN Report on the Environment, which advocated the kind of development that meets present needs without compromising the prospects of future generations. Following the 1992 Earth Summit in Rio de Janeiro, the WTTC and the Earth Council drew up an environmental checklist for tourist development, which included waste minimisation, reuse and recycling, energy efficiency and water management. The WTTC has since established a more detailed programme called 'Green Globe', designed to act as an environmental blueprint for its members.

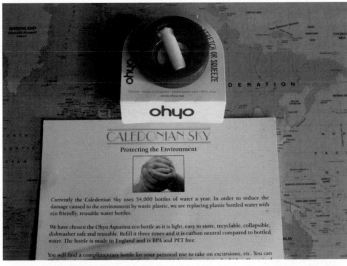

Figure 13.51 Eco-friendly reusable water bottle introduced by the expedition cruise ship Caledonian Sky

The pressure group Tourist Concern defines sustainable tourism as 'Tourism and associated infrastructures that: operate within capacities for the regeneration and future productivity of natural resources, recognise the contribution of local people and their cultures, accept that these people must have an equitable share in the economic benefits of tourism, and are guided by the wishes of local people and communities in the destination areas'. This definition emphasises the important issues of equity and local control, which are difficult to achieve for a number of reasons:

- Governments are reluctant to limit the number of tourist arrivals because of the often desperate need for foreign currency.
- Local people cannot compete with foreign multinationals on price and marketing.
- It is difficult to force developers to consult local people.

Negative environmental impacts

In so many LICs and MICs, newly laid golf courses have taken land away from local communities while consuming large amounts of scarce freshwater. It has been estimated that the water required by a new golf course can supply a village of 5000 people. In both Belize and Costa Rica, coral reefs have been blasted to allow for unhindered watersports. Like fishing and grazing rights, access to such common goods as beach front and scenically desirable locations does not naturally limit itself. As with overfishing and overgrazing, the solution to 'overtouristing' will often be to establish ownership and charge for use. The optimists argue that because environmental goods such as clean water and beautiful scenery are fundamental to the tourist experience, both tourists and the industry have a vested interest in their preservation. That 'ecotourism' is a rapidly growing sector of the industry supports this viewpoint – at least to a certain extent.

Education about the environment visited is clearly the key. Scuba divers in the Ras Mohammad National Park in the Red Sea, who were made to attend a lecture on the ecology of the local reefs, were found to be eight times less likely to bump into coral (the cause of two-thirds of all damage to the reef), let alone deliberately to pick a piece. However, there is huge concern about the future of many coral reefs; no more so than the Great Barrier Reef, which receives 2 million visitors a year (Figure 13.52).

Figure 13.52 The Great Barrier Reef

Positive environmental impacts

The environmental impact of tourism is not always negative. Landscaping and sensitive improvements to the built environment have significantly improved the overall quality of some areas. On a larger scale, tourist revenues can fund the designation and management of protected areas such as national parks and national forests.

☐ The life cycle model of tourism

Butler's model of the evolution of tourist areas (Figure 13.53) attempts to illustrate how tourism develops and changes over time. In the first stage, the location is explored independently by a small number of visitors. If visitor impressions are good and local people perceive that real benefits are to be gained, then the number of visitors will increase as the local community becomes actively involved in the promotion of tourism. In the development stage, holiday companies from HICs take control of organisation and management, with package holidays becoming the norm. Eventually, growth ceases as the location loses some of its former attraction. At this stage, local people have become all too aware of the problems created by tourism. Finally, decline sets in, but because of the perceived economic importance of the industry efforts will be made to re-package the location, which, if successful, may either stabilise the situation or result in renewed growth ('rejuvenation').

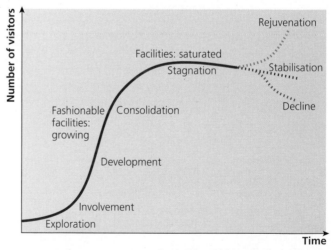

Source: *Advanced Geography: Concepts & Cases* by P. Guinness & G. Nagle (Hodder Education, 1999), p.217

Figure 13.53 Butler's model of the evolution of tourism in a region

The model provides a useful summary of the stages that a number of holiday resorts, particularly in the Mediterranean, have been through. For example, it has been applied to the Costa del Sol and the Costa Brava in Spain. However, research has shown that it does not apply well to all locations. Prosser (1995) summarised the criticisms of the model:

- Doubts on there being a single model of tourism development
- Limitations on the capacity issue
- Lack of empirical support for the concept
- Limited practical use of the model.

Also, it does not include the possible role of local and national governments in the destination country or the impact of, say, a low-cost airline choosing to add a destination to its network.

Figure 13.54 Hurgada on the Red Sea coast of Egypt

> ### Section 13.3 Activities
>
> 1 Discuss the indirect economic benefits of tourism.
> 2 Suggest how, by careful planning, you could minimise the economic leakage of a foreign holiday.
> 3 Find an example of the application of the Butler model to a particular destination. Write a brief summary of the example as a case study.

☐ Recent developments in international tourism

The growth of special interest (niche) tourism

In the last 20 years, more specialised types of tourism have become increasingly popular. An important factor seems to be a general re-assessment of the work–life balance. An increasing number of people are determined not to let work dominate their lives. One result of this has been the development of **niche tourism**. Niche market tour operators have increased in number to satisfy the rising demand for specialist holidays, which include:

- theme parks and holiday village enclaves
- cruising
- heritage and urban tourism

Figure 13.55 Tourists with a local guide visiting a First World War cemetery at Ypres, Belgium

- wilderness and ecotourism
- cultural/historical interest (Figure 13.55)
- medical and therapy travel
- conflict/dark tourism
- religious tourism
- working holidays
- sports tourism.

Some aspects of special-interest tourism have now reached a very significant size. For example, one in every twelve package holidays booked in early 2009 was for a cruise.

Ecotourism

As the level of global tourism increases rapidly, it is becoming more and more important for the industry to be responsibly planned, managed and monitored. Tourism operates in a world of finite resources where its impact is becoming of increasing concern to a growing number of people. At present, just over 5 per cent of the world's population have ever travelled by plane. However, this is undoubtedly going to increase substantially.

Leo Hickman, in his book *The Final Call*, claims: 'The net result of a widespread lack of government recognition is that tourism is currently one of the most unregulated industries in the world, largely controlled by a relatively small number of Western corporations such as hotel groups and tour operators. Are they really the best guardians of this evidently important but supremely fragile global industry?' Hickman argues that most countries only have a junior minister responsible for tourism rather than a secretary of state for tourism, which is what the size of the industry in most countries would justify.

Environmental groups are keen to make travellers aware of their **destination footprint**. They are urging people to:

- 'fly less and stay longer'
- carbon-offset their flights
- consider 'slow travel'.

Supporters of slow travel suggest that tourists should consider the impact of their activities both on individual holidays and in the longer term too. For example, they may decide that every second holiday will be in their own country (not using air transport). It could also involve using locally run guesthouses and small hotels as opposed to hotels run by international chains. This enables more money to remain in local communities.

Virtually every aspect of the industry now recognises that tourism must become more sustainable. **Ecotourism** is at the leading edge of **sustainable tourism**. An example of ecotourism in Ecuador is considered below.

A new form of ecotourism is developing in which volunteers help in cultural and environmental conservation and research. An example is the Earthwatch scientific research projects, which invite members of the general public to join the experts as fully fledged expedition members – on a paying basis, of course. Several Earthwatch projects in Australia have helped Aboriginal people to locate and document their prehistoric rock art and to preserve ancient rituals directly.

Case Study: Ecotourism in Ecuador

Ecuador's travel and tourism industry directly contributed 1.9 per cent to GDP in 2014, with the total travel and tourism economy (direct and indirect) contributing 5.3 per cent. The latter amounted to $4702 million. The 118 000 direct jobs in the industry accounted for 1.7 per cent of total employment, while the total number of jobs in the wider travel and tourism economy (338 000) made up 4.8 per cent of total employment.

International tourism is Ecuador's third largest source of foreign income, after the export of oil and bananas. The number of visitors has increased substantially in recent years, both to the mainland and to the Galapagos Islands where Darwin conducted research on evolution. The majority of tourists are drawn to Ecuador by its great diversity of flora and fauna. The country contains 10 per cent of the world's plant species. Much of the country is protected by national parks and nature reserves.

As visitor numbers began to rise, Ecuador was anxious not to suffer the negative externalities of mass tourism evident in many other countries. The country's tourism strategy has been to avoid becoming a mass-market destination and to market 'quality' and 'exclusivity' instead, in as eco-friendly a way as possible. Tourist industry leaders were all too aware that a very large influx of visitors could damage the country's most attractive ecosystems and harm its image as a 'green' destination for environmentally conscious visitors.

Ecotourism has helped to bring needed income to some of the poorest parts of the country. It has provided local people with a new alternative way of making a living. As such, it has reduced human pressure on ecologically sensitive areas.

The main geographical focus of ecotourism has been in the Amazon rainforest around Tena, which has become the main access point. The ecotourism schemes in the region are usually run by small groups of indigenous Quichua Indians (Figure 13.56). The indigenous movement in Ecuador is one of the strongest in South America.

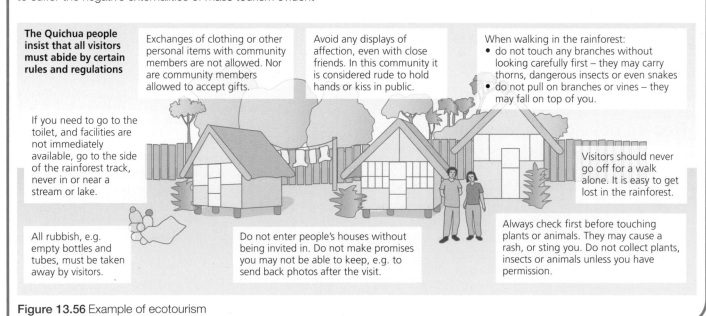

The Quichua people insist that all visitors must abide by certain rules and regulations

Exchanges of clothing or other personal items with community members are not allowed. Nor are community members allowed to accept gifts.

Avoid any displays of affection, even with close friends. In this community it is considered rude to hold hands or kiss in public.

When walking in the rainforest:
- do not touch any branches without looking carefully first – they may carry thorns, dangerous insects or even snakes
- do not pull on branches or vines – they may fall on top of you.

If you need to go to the toilet, and facilities are not immediately available, go to the side of the rainforest track, never in or near a stream or lake.

Visitors should never go off for a walk alone. It is easy to get lost in the rainforest.

All rubbish, e.g. empty bottles and tubes, must be taken away by visitors.

Do not enter people's houses without being invited in. Do not make promises you may not be able to keep, e.g. to send back photos after the visit.

Always check first before touching plants or animals. They may cause a rash, or sting you. Do not collect plants, insects or animals unless you have permission.

Figure 13.56 Example of ecotourism

Galapagos islands at risk

Illegal fishing, non-native species and the demands of more than 160 000 tourists each year threaten this irreplaceable ecosystem and the people who depend on it for their food and livelihoods.

Source: WWF

In early 2007, the government of Ecuador declared the Galapagos Islands at risk, warning that visitor permits and flights to the island could be suspended. The Galapagos Islands straddle the equator 1000 kilometres off the coast of Ecuador. All but 3 per cent of the islands are a national park. Five of the thirteen islands are inhabited. Visitor numbers are currently 160 000 a year and rising.

The volcanic islands can be visited all year round, but the period between November and June is the most popular. Boat trips generally cost from £700 to over £2000. An additional national park entrance fee of £65 is payable on arrival. Among the many attractions are giant tortoises, marine iguanas and blue-footed boobies.

In signing the emergency decree to protect the islands, the President of Ecuador stated: 'We are pushing for a series of actions to overcome the huge institutional, environmental and social crisis in the islands'.

The identified problems include the following:

- The population has been increasing rapidly, doubling every 11 years. The total population was estimated at 40 000 in 2014. This is putting increasing pressure on natural resources and the disposal of domestic waste.

- Illegal fishing of sharks and sea cucumbers is believed to be at an all-time high.
- The number of visiting cruise ships continues to rise.
- There are growing concerns over the increase in non-native species on the islands.
- There are internal arguments within the management structure of the national park.
- Controversially, a hotel opened on the islands in 2006.
- In mid-2007, a UN delegation visited the islands to determine whether they should be declared 'in danger'.

It would seem that the tourism carrying capacity of the Galapagos Islands has been reached or even exceeded. Yet many more people will want to visit this unique environment in the future. The management of tourists in the islands has evolved with the increasing pressure of numbers. However, some radical approaches are likely to be required in the future.

Section 13.3 Activities

1 Write a ten bullet-point list on ecotourism in Ecuador.
2 Why is there so much concern about the threat from tourism on the Galapagos Islands?

13.4 The management of a tourist destination

Case Study: Jamaica

Jamaica is the third largest of the Caribbean islands, and the largest English-speaking island in the Caribbean Sea. It is situated 145 kilometres south of Cuba and 965 kilometres south of Florida, USA. Tourism in Jamaica (Figure 13.57) originated in the latter part of the nineteenth century when a limited number of affluent people, many with medical conditions, came to Jamaica to avoid the cold winters in the UK and North America. Figure 13.58 illustrates the attractions of Jamaica's climate. The first tourist hotels were built in Montego Bay and Port Antonio. The industry expanded after the First World War with advances in transportation, although it has been estimated that only a few thousand foreign tourists visited Jamaica each year in the 1920s. By 1938, the figure had risen to 64 000, and by 1952 it had reached 104 000. Growth continued in the following decades, with 345 000 visitors in 1966 and over 600 000 in 1982. Since the 1987–88

season, the number of foreign tourists has exceeded 1 million a year, partly as a result of the significant increase in the arrivals of cruise-ship passengers.

In 2014, Jamaica welcomed 2.08 million stopover visitors, along with 1 423 797 cruise visitors. This gave a total of 3.5 million visitors for the year. Tourism's direct contribution to GDP in 2013 amounted to JM$109.3 billion, or 7.7 per cent of the national total (Table 13.7). Adding all the indirect economic benefits increased this figure to JM$364.8 billion. Direct employment in the industry amounted to 82 000 (7 per cent of total employment) but the overall figure, which includes indirect employment, is over three times as large (23.45 per cent of total employment). For example, farmers supply food to the hotels and carpenters make furniture for the industry. In the most popular tourist areas, the level of reliance on the industry is extremely high indeed. $\Rightarrow$

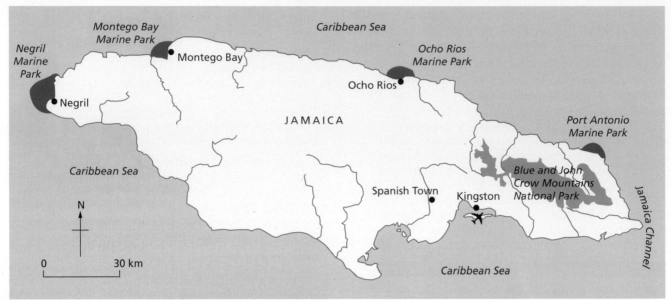

Source: *IGCSE Geography* 2nd edition, P. Guinness & G. Nagle (Hodder Education, 1999) p.205

Figure 13.57 Marine parks and national parks in Jamaica

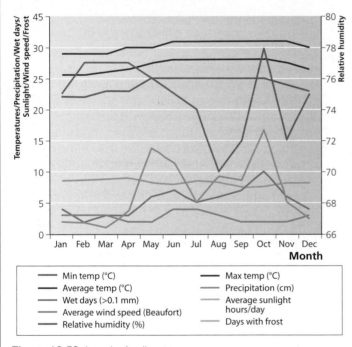

Figure 13.58 Jamaica's climate

Table 13.7 The importance of the travel and tourism industry to the Jamaican economy, 2013

	Travel and tourism industry	Travel and tourism economy
GDP % of total	7.7	25.6
Employment	82 000	275 000

Source: From WTTC data

Figure 13.59 illustrates the importance of foreign visitor exports as a percentage of total exports for the period 2004–14, along with the forecast for 2024. It was

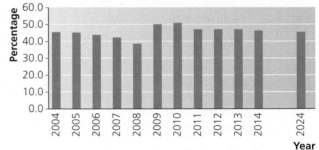

Source: WTTC Travel & Tourism Economic Impact 2014

Figure 13.59 Jamaica – foreign visitor exports as a percentage of total exports

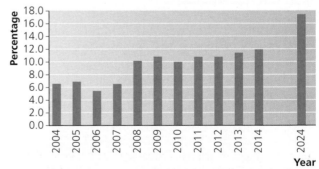

Source: WTTC Travel & Tourism Economic Impact 2014

Figure 13.60 Jamaica – tourism as a percentage of the whole economy GDP

46.4 per cent in 2013, but has reached 50 per cent in the last decade. Figure 13.60 shows the contribution of the industry to the whole economy (GDP) for the same time period. The tourism industry accounted for over 11 per cent of total investment in Jamaica in 2013.

Jamaica is an example of a tourist area where there has been clear evidence of growth and development. The term **growth** refers to the increase in numbers, while **development** refers to the expansion of tourism activities such as the development of adventure tourism and ecotourism. As the tourist industry has expanded, its linkages with other industries have developed as well. But as tourism develops it also has an impact on the environment, and the question of its sustainability comes to the fore. The careful management of tourism can do much to mitigate the impact. Management techniques that Jamaica has used include:

- trying to avoid the pitfalls of mass tourism, such as the construction of high-rise hotels
- creating national parks, marine parks and other protected areas
- promoting ecotourism and other environmentally friendly forms of tourism
- encouraging community tourism
- linking the profits of tourism to social development in the country.

Jamaica's north coast, with its pleasant weather and white-sand beaches, is the centre of the island's tourist industry (Figure 13.61). The main resorts are Montego Bay, Ocho Rios and Port Antonio, although many tourists also visit the capital city, Kingston. Accommodation varies between modern high-rise hotels, elegant old-world style buildings, villas, apartments and guesthouses. The number of rooms available in Jamaica is approximately 30 000. Jamaica has a relatively good road network, with Highway 2000 linking settlements in the south and the North Coast Highway serving the north of the island.

Figure 13.61 A beach fringed with palm trees in Runaway Bay

About two-thirds of visitors to the island arrive by air. Jamaica is served by two international airports: Norman Manley International in Kingston and Sangster International in Montego Bay. There is a private jet centre in Montego Bay and four aerodromes serving small carriers for inter-island travel. Cruise passenger terminals are located in Ocho Rios, Montego Bay, Port Antonio and Kingston.

While sun and sand are the main attractions of a holiday in Jamaica, the island also has other attributes, including dolphin parks, nature reserves, museums and galleries. There is a wide variety of flora and fauna, with 252 species of birds (27 are endemic), 200 native species of orchids, 500 species of true ferns and about 50 species of coral. There are excellent facilities for a range of sports including tennis, golf and equestrian activities. Jamaica's cuisine is an attraction for many visitors. There are many festivals and entertainment events during the year, often featuring Jamaica's native music, reggae.

Among the island's protected areas are the Cockpit Country, Hellshire Hills and Litchfield Forest Reserves. In 1992, Jamaica's first marine park, covering 15 km², was established in Montego Bay. The following year, the Blue and John Crow Mountains National Park was established on 780 km² of wilderness that supports thousands of tree and fern species, rare animals and insects, such as the Homerus swallowtail, the western hemisphere's largest butterfly. The Negril Marine Park was established in 1998. These decisions have been essential for the sustainability of the environment and the tourist industry itself. Jamaica wants to increase its tourist business, but more people will only visit if the physical environment retains its attractions and diverse attributes.

The industry has brought considerable opportunities to Jamaica's population, although it has also had its problems. During the 1970s, the Jamaican government introduced 'Jamaicanisation' policies designed to attract much-needed foreign investment in tourism. Policies included comparatively high wages and special industry taxes that went directly into social development, healthcare and education. These sectors are often referred to by economists as 'soft infrastructure'. However, tourism has spurred the development of vital 'hard infrastructure' too, such as roads, telecommunications, water supply and airports. Jamaica has been determined to learn from the 'mistakes' of other countries and ensure that the population will gain real benefits from the growth of tourism.

Tourism is the largest source of foreign exchange for Jamaica. The revenue from tourism plays a significant part in helping central and local government fund economic and social policies. Also, as attitudes within the industry itself are changing, larger hotels and other aspects of the industry have become more socially conscious. Classic examples are the funding of local social projects.

A paper on tourism by the People's National Party (PNP) stated that 'The momentum generated by the current round of investment in resort development has created an enormous pull factor in terms of investor confidence. This has set the stage for an even more powerful wave of investment in the next 10 years.'

The Jamaica Tourist Board (JTB) is responsible for marketing the country abroad. It has used Jamaica's status as one of the host countries for the 2007 Cricket World

Cup to good effect. The JTB also promotes the positive aspects of Jamaican culture, and the Bob Marley Museum in Kingston has become a popular attraction. Such attractions are an important aid in supporting Jamaica's objective of reducing seasonality.

The high or 'winter' season runs from mid-December to mid-April, when hotel prices are highest. The rainy season extends from May to November. It has been estimated that 25 per cent of hotel workers are laid off during the off-season.

Jamaica's government is working to reduce the environmental impact of tourism. Figure 13.57 shows the location of Jamaica's national and marine parks. A further six sites have been identified for future protection. The Jamaican government sees the designation of the parks as a positive environmental impact of tourism. Entry fees to the national parks pay for conservation. The desire of tourists to visit these areas and the need to conserve the environment to attract future tourism drives the designation and management process.

The marine parks are attempting to conserve the coral reef environments off the coast of Jamaica. They are at risk from damage from overfishing, industrial pollution and mass tourism. The Jamaica Conservation and Development Trust is responsible for the management of the national parks, while the National Environmental Planning Agency has overseen the government's sustainable development strategy since 2001.

Negril

Negril is a large beach resort town located on the west coast of Jamaica. The town's development as a resort location began during the late 1950s, although access to the area proved difficult as ferries were required to drop off passengers in Negril Bay, forcing them to wade to shore. When the road between Montego Bay and Negril was improved in the early 1970s, it helped to increase Negril's position as a new resort location. A small airport was built for North American winter tourists. Europeans also came to Negril, and several hotels were built to cater directly to those guests. Figure 3.62 shows some of the largest hotels and the transport infrastructure of Negril.

This stretch of coastline arguably has the island's best beaches. Negril's beach has been rated as one of the top ten beaches in the world by many travel magazines. To the east

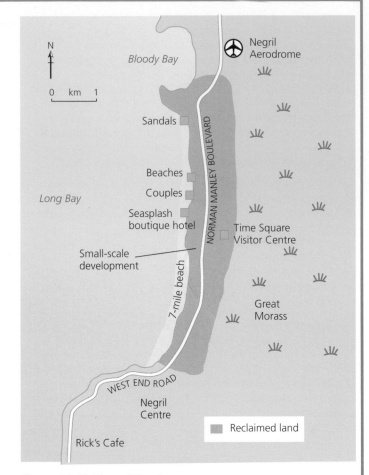

Figure 13.62 Map of Negril

of the shore lies a swamp called the Great Morass, amidst which is the Royal Palm Reserve, with wetlands that are protected. In 1990, the Negril Coral Reef Preservation Society (Figure 13.63) was formed as an NGO to address ongoing degradation of the coral-reef ecosystem. This was the precursor of the Negril Marine Park, which was established in 1998. Educating people about the fragility of coral reefs (Figure 13.64) and other endangered environments is a crucial aspect of sustainability.

Figure 13.63 Negril Coral Reef Preservation Society

Eric the eel says...

'Reef savers do it without touching'

1. Do not touch any part of the reef
2. Do not litter the sea or the beach
3. Do not take or buy any shells or shell jewellery
4. Do not touch or spear any marine life
5. Fishing on the reef is prohibited
6. Do not feed the fish
7. Do not take any sand from the beach
8. For your protection, please stay within the swim lanes

negril
CORAL REEF PRESERVATION SOCIETY

Figure 13.64 Coral reef protection society sign

Ecotourism and community tourism

Ecotourism is a developing sector of the industry with, for example, raft trips on the Rio Grande river increasing in popularity. Tourists are taken downstream in very small groups. The rafts, which rely solely on manpower, leave singly with a significant time gap between them to minimise any disturbance to the peace of the forest. Ecotourism is seen as the most sustainable form of tourist activity on the island.

Considerable efforts are being made to promote **community tourism** so that more money filters down to the local people and small communities. The Sustainable Communities Foundation through Tourism (SCF) programme has been particularly active in central and south-west Jamaica. Community tourism (Figure 13.65) is seen as an important aspect of **pro-poor tourism**.

The Astra Country Inn in Mandeville has been recognised as a pioneer hotel in community tourism. Its work with surrounding communities has included:

- promoting bed and breakfast accommodation in private homes
- training local guides
- developing community-based tourist attractions
- encouraging the development of local suppliers.

- Community tourism should involve local people in decision making and ownership.
- The local community should receive a fair share of the profits from tourism ventures.
- Tour companies should try to work with communities rather than individuals to avoid creating divisions.
- Tourism should be environmentally sustainable and not put excess pressure on natural resources.
- Tourism should support traditional cultures. It should encourage people to value and respect their cultural heritage.
- Where possible, tour operators should keep groups small to minimise cultural and environmental impacts.
- Tour guides should brief tourists on what to expect and on appropriate behaviour before arriving in a community.
- Local people should have the right to say no to tourism.

Source: *Geography Review*, January 2005, Philip Allan updates

Figure 13.65 Principles of community-based tourism

Challenges ahead

However, tourism has had its problems too. The behaviour of some tourists clashes with the island's traditional morals; people have a negative image of Jamaica because of its levels of violent crime and harassment; and despite the recent initiatives of the Jamaican government to protect the environment, much valuable biodiversity has already been lost. On a positive note, Jamaica is one of the few Caribbean tourist destinations that has done relatively well during the recent recession that has led to a decrease in visitor arrivals for the Caribbean region.

While Jamaica has undertaken several initiatives with regard to the sustainable development of tourism, the success of such initiatives has been mixed. A book entitled *Barriers to Sustainable Tourism Development in Jamaica* published in 2007 noted that initiatives often lacked adequate management and cohesion, and often had to work under significant financial constraints.

Section 13.4 Activities

1. With the aid of an atlas, describe the location of Jamaica.
2. Produce a bullet-point analysis of Figure 13.58.
3. Explain the location of the island's main resorts.
4. Describe the transport infrastructure of Jamaica. Why is this such an important factor in the development of tourism?
5. Discuss the importance of tourism to the economy of Jamaica.
6. What measures have been taken to advance the sustainability of tourism on the island?
7. Briefly discuss the development of ecotourism and community tourism in Jamaica.

Advanced : Human Geography Options

14 Economic transition

14.1 National development

☐ Employment structure and its role in economic development

In HICs and an increasing number of MICs and LICs, people do hundreds of different jobs. All of these jobs can be placed into four broad employment sectors:

- The **primary sector** produces raw materials from the land and the sea. Farming, fishing, forestry, mining, quarrying and fishing make up most of the jobs in this sector. Some primary products are sold directly to the consumer but most go to secondary industries for processing.
- The **secondary sector** manufactures primary materials into finished products (Figure 14.1). Activities in this sector include the production of processed food, furniture and motor vehicles. Secondary products are classed either as **consumer goods** (produced for sale to the public) or **capital goods** (produced for sale to other industries).
- The **tertiary sector** provides services to businesses and to people. Retail employees, drivers, architects, teachers and nurses are examples of jobs in this sector.
- The **quaternary sector** uses high technology to provide information and expertise. Research and development is an important part of this sector. Jobs in this sector include aerospace engineers, research scientists, computer scientists and biotechnology workers. Quaternary industries have only been recognised as a separate group since the late 1960s. Before then, jobs now classed as quaternary were placed in either the secondary or tertiary sectors depending on whether a tangible product was produced or not. However, even today much of the available information on employment does not consider the quaternary sector.

Figure 14.1 Large footwear factory in Vietnam – the secondary sector of employment

The **product chain**, which considers the full sequence of activities needed to turn raw materials into a finished product, can be used to illustrate the four sectors of employment. The food industry provides a good example (Figure 14.2). Some companies are involved in all four stages of the food product chain. Research and development (the quaternary sector) can improve the performance of all the other three sectors.

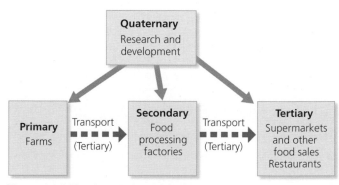

Figure 14.2 The food industry's product chain

How employment structures have changed over time

As an economy advances, the proportion of people employed in each sector changes (Figure 14.3). Countries such as the UK and the USA are 'post-industrial societies', where most people work in the tertiary sector. Yet in 1900, as much as 40 per cent of employment in the USA was in the primary sector. However, the mechanisation of farming, mining, forestry and fishing drastically reduced the demand for labour in these industries. As these jobs disappeared, people moved to urban areas where jobs in the secondary and tertiary sectors were expanding. Less than 4 per cent of employment in the USA is now in the primary sector.

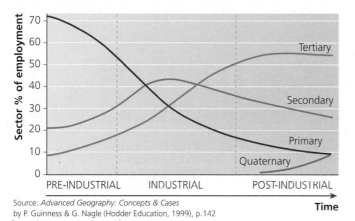

Source: *Advanced Geography: Concepts & Cases*
by P. Guinness & G. Nagle (Hodder Education, 1999), p.142

Figure 14.3 The sector model

Human labour has been replaced in manufacturing too. In more and more factories, robots and other advanced machinery handle assembly-line jobs that once employed large numbers of people. Also, many manufacturing jobs once done in America are now done in NICs. In 1950, the same number of Americans were employed in the secondary and tertiary sectors. By 1980, two-thirds were working in services. Today, 78 per cent of Americans work in the tertiary sector.

The tertiary sector is also changing. In banking, insurance and many other types of business, computer networks have reduced the number of people required. But elsewhere, service employment is rising, such as in health, education and tourism.

How employment structures vary between LICs, MICs and HICs

Low-income countries

People in the poorest countries of the world (LICs) are heavily dependent on the primary sector for employment. This is especially so in the least developed countries (LDCs). Most of these people work in agriculture and many are **subsistence farmers**. In some densely populated areas where the amount of land is very limited, there will not be enough work available for everyone to work a full week.

The work available is often shared and people are said to be **underemployed**.

In some regions of LICs, mining, quarrying, forestry or fishing may dominate the economy. Work in mining in LICs is often better paid than jobs elsewhere in the primary sector, but the working conditions are often very harsh. In poor countries, higher-paid jobs in the secondary, tertiary and quaternary sectors are usually very few in number. The tertiary jobs that are available are often in the public sector. Public-sector jobs such as teaching, nursing and refuse collection are paid by the government. However, wages in these jobs are usually low as the funds available to the governments of LICs are very limited. In some poor countries, salaries may not be paid on time due to a lack of government funds.

Many of the world's poorest countries are primary-product dependent, meaning that they rely on one or a small number of primary products for most of their export earnings. This makes them very vulnerable to changes in world markets. For example, if a country relies on coffee for most of its export earnings and the price of coffee falls substantially, that country will have far less money to pay for the imports it needs and less to invest in health, education and other important aspects affecting the quality of life of its people. Examples of primary-product dependency include:

- Primary products comprised 79 per cent of exports of the Pacific Island states in 2010.
- 80 per cent of Chad's total employment is in agriculture.
- Oil accounts for 97 per cent of the value of Angola's exports.

Middle-income countries

In MICs, employment in manufacturing has increased rapidly in recent decades. MICs have reached the stage of development where they attract **foreign direct investment (FDI)** from transnational corporations (TNCs) in both the manufacturing and service sectors. For most MICs, manufacturing industry has been the bedrock for exports and development, but there has been another way. In India, for example, the service sector has done much to lead economic development. The business environment in MICs is such that they also develop their own domestic companies. Such companies usually start in a small way, but some go on to reach a considerable size. Both processes create employment in manufacturing and services.

The increasing wealth of MICs allows for greater investment in agriculture. This includes mechanisation, which results in a falling demand for labour on the land. So, as employment in the secondary and tertiary sectors rises, employment in the primary sector falls. Eventually, MICs may become so advanced that the quaternary sector begins to develop. Examples of MICs where this has happened are South Korea, Singapore and Taiwan.

High-income countries

Developed countries (HICs) are often referred to as **post-industrial societies** because far fewer people are now employed in manufacturing industries than in the past. Most people work in the tertiary sector, with an increasing number in the quaternary sector. Jobs in manufacturing industries have fallen for two reasons:

- Many manufacturing industries have moved to take advantage of lower costs in MICs and LICs. Cheaper labour is often the main attraction, but many other costs are also lower.
- Investment in robotics and other advanced technology has replaced much human labour in many manufacturing industries that remain in HICs.

Table 14.1 compares the employment structure of a HIC, a MIC and a LIC.

A graphical method often used to compare the employment structure of a large number of countries is the triangular graph (Figure 14.4). One side (axis) of the triangle is used to show the data for each of the primary, secondary and tertiary sectors. Each axis is scaled from 0 to 100 per cent. The indicators on the graph show how the data for the UK can be read.

Table 14.1 Employment structure of a HIC, a MIC and a LIC

Country	% primary	% secondary	% tertiary
UK (HIC)	1	17	82
China (MIC)	43	25	32
Bangladesh (LIC)	63	11	26

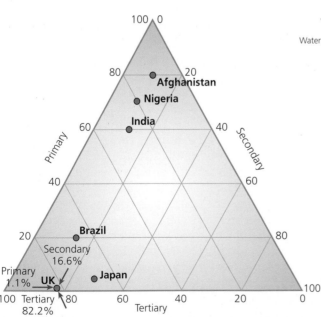

Figure 14.4 Triangular graph showing the employment structure of six countries

Figure 14.5 illustrates changes in the composition of employment in the UK between 1841 and 2011. Today, the service sector dominates total employment, with very few workers now employed in the primary sector. Most HICs have followed a broadly similar path. Figure 14.6 focuses on changes between 1978 and 2013. The decline in the primary and secondary sectors is clear to see. Employment in the service sector in general increased by about 60 per cent, although the rate of increase varies considerably within the service sector itself!

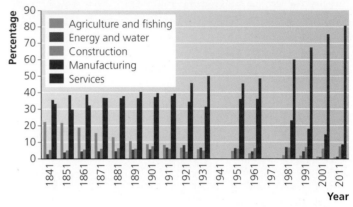

Source: Office for National Statistics

Figure 14.5 Changing employment structure in the UK, 1841–2011

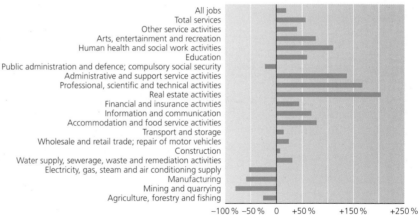

Source: Office for National Statistics

Figure 14.6 The UK – change in jobs by industry, 1978–2013

Outsourcing

A major change in employment has been the increase in **outsourcing**. Companies do this to save money. Work can be outsourced to companies in the same country or it can go abroad where labour and other costs are much lower. For example, many British and American companies have outsourced their call centres to India. It has been the revolution in information and communications technology that has enabled outsourcing to develop so rapidly into a major global industry. As higher-level ICT has spread down the global economic hierarchy from core to periphery, more and more countries have been competing for this valuable business.

Employment structure: the future

The nature of work in HICs has changed markedly over the last 50 years. It will continue to change in the future as the process of globalisation continues. The key questions are:

- Will even fewer people work in the primary sector and which tasks will be performed by those who remain?
- How much further will manufacturing employment fall and which products will HICs still produce?
- Which service sector jobs will decline and which will increase in importance?
- Which totally new services will begin to provide employment in the future?
- How many people will be unemployed at various stages in the future and what status and standard of living will they have?
- What changes will occur in **a** the working week, **b** paid holidays, **c** retirement age, **d** pensions, **e** the school leaving age, **f** working conditions and **g** the location of employment?
- What control will national governments have over these issues?

Employment is one of the most important factors in most people's lives. It is the income from employment that influences so many aspects of an individual's quality of life. The world of 2025 is likely to be very different from its present state.

With further advances in ICT, there will be a greater opportunity for more people to work from home. This is often referred to as **teleworking**. ICT will allow many people to perform the same tasks from home that they now do in their office. However, a decade ago it was thought that higher technology home working would be more important now than it has actually turned out to be. It seems that the physical clustering of people in organisations has proved more difficult to break down than many commentators thought.

It seems likely that international commuting and employment migration (geographical mobility) will increase as economic and psychological barriers to movement recede. The degree of occupational mobility should also increase as the pace of change quickens.

Section 14.1 Activities

1 Explain the product chain illustrated by Figure 14.2.
2 Why does the primary sector dominate employment in the poorest countries of the world?
3 On a copy of Figure 14.4, plot the positions of China and Bangladesh using the data in Table 14.1.
4 Explain the changes shown in Figure 14.3.
5 Why does outsourcing occur on such a large scale?
6 Discuss the changes that are likely to occur in employment in the future, in the country in which you live.

☐ Global inequalities in social and economic well-being

Development and its traditional income measures

Development, or improvement in the quality of life, is a wide-ranging concept. It includes wealth, but it includes other important aspects of our lives too. For example, many people would consider good health to be more important than wealth. People who live in countries that are not democracies, where freedom of speech cannot be taken for granted, often envy those who do live in democratic countries. Development occurs when there are improvements to individual factors making up the quality of life. Figure 14.7 shows one view of the factors that comprise the quality of life (Figure 14.8). For example, development occurs in a LIC when:

- the local food supply improves due to investment in machinery and fertilisers
- the electricity grid extends outwards from the main urban areas to rural areas
- a new road or railway improves the accessibility of a remote province
- levels of literacy improve throughout the country.

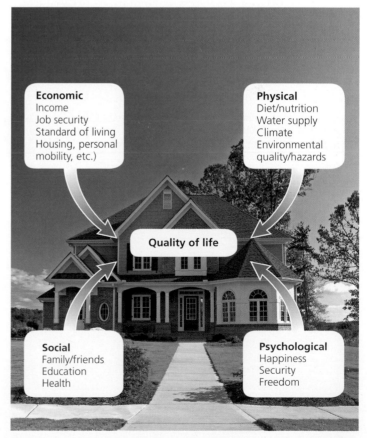

Figure 14.7 Factors comprising the quality of life

Figure 14.8 An open-pit toilet: sanitation is lacking in some LICs, reducing people's quality of life

The traditional indicator of a country's wealth has been the **gross domestic product (GDP)**. The GDP is the total value of goods and services produced by a country in a year. A more recent measure, gross national income, has to some extent taken over from GDP as a preferred measure of national wealth. **Gross national income (GNI)** comprises the total value of goods and services produced within a country; that is, its gross domestic product, together with its income received from other countries (notably interest and dividends), less similar payments made to other countries.

To take account of the different populations of countries, the **gross national income per person** is often used. Here, the total GNI of a country is divided by the total population. Per person figures allow for more valid comparisons between countries when their total populations are very different. However, 'raw' or 'nominal' GNI data does not take into account the way in which the cost of living can vary between countries. For example, a dollar buys much more in China than it does in the USA. To account for this, the GNI at 'purchasing power parity' (PPP) is calculated. Figure 14.9 shows how **GNI at purchasing power parity per person** varied globally in 2013. The lowest GNI figures are concentrated in Africa and parts of Asia. The highest figures are in North America, the EU, Japan, Australia and New Zealand.

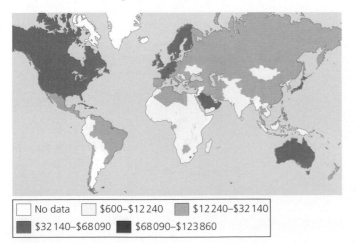

| No data | $600–$12 240 | $12 240–$32 140 |
| $32 140–$68 090 | $68 090–$123 860 | |

Figure 14.9 Worldwide GNI PPP per person, 2013

Table 14.2 shows the top 20 and bottom 20 countries in GNI PPP per person for 2013. The **development gap** between the world's wealthiest and poorest countries is huge. All bottom 20 countries are in Africa! However, a major limitation of GNI and other national data is that these are 'average' figures for a country, which tell us nothing about:

- the way in which wealth is distributed within a country – in some countries, the gap between rich and poor is much greater than in others
- how government invests the money at its disposal – for example, Cuba has a low GNI per person but high standards of health and education because these have been government priorities for a long time.

Table 14.2 Top 20 and bottom 20 countries in GNI (PPP) per person in 2013

Top 20 countries (US$)		Bottom 20 countries (US$)	
Qatar	123 860	Gambia	1620
Macao, SAR	112 180	Burkina Faso	1560
Kuwait	88 170	Comoros	1560
Singapore	76 850	Zimbabwe	1560
Brunei	68 090	Mali	1540
Norway	66 520	Rwanda	1430
Luxembourg	59 750	Uganda	1370
UAE	58 090	Madagascar	1350
Hong Kong, SAR	54 260	Ethiopia	1350
USA	53 960	Guinea-Bissau	1240
Switzerland	53 920	Togo	1180
Saudi Arabia	53 780	Eritrea	1180
Oman	52 170	Guinea	1160
Sweden	44 660	Mozambique	1040
Germany	44 540	Niger	910
Denmark	44 440	Burundi	820
Austria	43 810	Liberia	790
Netherlands	43 210	Malawi	760
Canada	42 590	Congo D.R.	680
Australia	42 540	Central African Republic	600

Source: 2014 World Population Data Sheet, Population Reference Bureau

Development not only varies between countries, it can also vary significantly within countries. Most of the measures that can be used to examine the contrasts between countries can also be used to look at regional variations within countries.

Broader measures of development: the Human Development Index

The way that the quality of life has been measured has changed over time. In the 1980s, the Physical Quality of Life Index (PQLI) was devised. The PQLI was the average of three development factors: literacy, life expectancy and infant mortality. However, in 1990 the **Human Development Index (HDI)** was devised by the UN as a better measure to show the disparities between countries. The HDI is a composite index that has changed slightly in character in recent years. The current index contains four indicators of development (Figure 14.10):

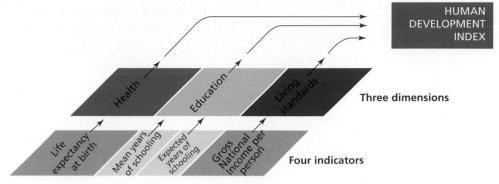

Source: *IGCSE Geography* 2nd edition, P. Guinness & G. Nagle (Hodder Education, 2014) p.164

Figure 14.10 The components of the HDI

- life expectancy at birth
- mean years of schooling for adults aged 25 years
- expected years of schooling for children of school entering age
- GNI per person (PPP).

The actual figures for each of these four measures are converted into an index (Figure 14.10), each of which has a maximum value of 1.0. The index values are then combined and averaged to give an overall HDI value. This also has a maximum value of 1.0. Every year, the UN publishes the *Human Development Report*, which uses the HDI to rank all the countries of the world in their level of development. The countries of the world are divided into four groups:

- Very high human development (Figure 14.11)
- High human development
- Medium human development
- Low human development.

Figure 14.12 shows the global distribution of these four groups in 2011.

Figure 14.11 The Waterfront, Vancouver – Canada has a very high level of human development

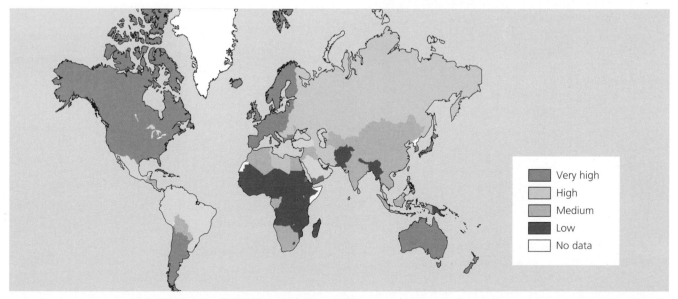

Source: *IGCSE Geography* 2nd edition, P. Guinness & G. Nagle (Hodder Education, 2014) p.165

Figure 14.12 Map of the HDI, 2011

Every measure of development has merits and limitations. No single measure can provide a complete picture of the differences in development between countries. This is why the UN combines different aspects of the quality of life to arrive at a figure of human development for each country. Although the development gap can be measured in a variety of ways, it is generally taken to be increasing (Figure 14.13). Many people are concerned about this situation, either because they see it as very unfair or because it can create political instability.

Overall, inequality has been getting worse, though trends in inequality are complicated. Basically:

If you compare *individuals* – average incomes per head of the world's richest and poorest people – the gap has narrowed, largely because China and India made immense reductions in poverty.

If you compare *countries* – the average income of one country and another – the gap has widened: more countries are lagging behind the rich nations than are catching up.

If you compare incomes *within countries* – between the richest people and the poorest – then again the gap is widening: from within China to the USA, the rich are pulling away from the poor.

One review of the literature looked at inequality from a variety of angles. It concluded that people round the centre of income distribution worldwide have been drawing together to some extent, yet the extremes have been flying apart.

The gap between the richest and the poorest has been widening, but income difference for those in the middle has slightly narrowed. 'There is no sign at all that either the extreme impoverishment at the bottom or the extreme enrichment at the top of the world distribution are coming to an end.'

Source: The Tomorrow Project

Figure 14.13 Globalisation – will the gap between the rich and poor narrow?

The Human Development Report

The HDI is published annually. It is a key part of the *Human Development Report* (HDR). According to a recent edition of the *HDR*, 'Human development is about putting people at the centre of development. It is about people realising their potential, increasing their choices and enjoying the freedom to lead lives they value. Since 1990, annual *Human Development Reports* have explored challenges including poverty, gender, democracy, human rights, cultural liberty, globalisation, water scarcity and climate change.'

In assessing the progress made in reducing global poverty, the *HDR* has noted that:

- in the last 60 years, poverty has fallen more than in the previous 500 years
- poverty has been reduced in some respects in almost all countries
- child death rates in developing countries have been cut by more than half since 1960
- malnutrition rates have declined by almost a third since 1960
- the proportion of children not in primary education has fallen from more than half to less than a quarter since 1960
- the share of rural families without access to safe water has been cut from nine-tenths to about a quarter since 1960.

These are just some of the achievements made during what the *HDR* calls the 'second Great Ascent from poverty', which started in the 1950s in the developing world, Eastern Europe and the former Soviet Union. The first Great Ascent from poverty began in Europe and North America in the late nineteenth century in the wake of the Industrial Revolution.

However, although the global poverty situation is improving, approximately one person in six worldwide struggles on a daily basis in terms of:

- adequate nutrition
- uncontaminated drinking water
- safe shelter
- adequate sanitation
- access to basic healthcare.

These people have to survive on $1.25 a day or less and are largely denied access to public services for health, education and infrastructure.

Figure 14.14 Rural Mongolia – a part of the world where many people still live in poverty

Individual measures of development

There are many individual measures of socio-economic development, some of which have already been mentioned. Indicators not mentioned above include:

- the number of people per doctor
- energy use per person
- number of motor vehicles per 1000 people
- per person food intake in calories
- televisions/refrigerators per 1000 population
- per person export and import volumes
- environmental protection expenditure as a percentage of GDP.

Some of the most important indicators of development are considered in more detail below.

Infant mortality rate

The **infant mortality rate** is regarded as one of the most sensitive indicators of socio-economic progress. It is an important measure of health equity both between and within countries.

There are huge differences in the infant mortality rate around the world, despite the wide availability of public health knowledge. Differences in material resources certainly provide a large part of the explanation for how international populations can share the same knowledge but achieve disparate mortality rates. Differences in the efficiency of social institutions and health systems can also enable countries with similar resource levels to register disparate mortality levels. Infant mortality generally aligns well with other indicators of development. However, many countries have significant intranational disparities in infant mortality, where populations share similar resource levels and health technology but achieve different health outcomes in various regions of the same country. Data on infant mortality rates can be found in Topic 4, Section 4.1.

Education

Education is undoubtedly the key to socio-economic development. It can be defined as the process of acquiring knowledge, understanding and skills. Education has always been regarded as a very important individual indicator of development and it has figured prominently in aggregate measures. Quality education generally, and female literacy in particular, are central to development (Figure 14.15). The World Bank has concluded that improving female literacy is one of the most fundamental achievements for a developing nation to attain, because so many aspects of development depend upon it. For example, there is a very strong relationship between the extent of female literacy and infant and child mortality rates. People who are literate are able to access medical and other information that will help them towards a higher quality of life compared with those who are illiterate.

Figure 14.15 A secondary school in Morocco

The UN sees education for **sustainable development** as being absolutely vital for the future of the planet. Sustainable development seeks to meet the needs of the present without compromising those of future generations. The year 2005 was the beginning of the United Nations Decade for Sustainable Development, which ran until 2014.

Nutrition

Undernourishment is concentrated in the least developed countries, particularly in Sub-Saharan Africa and South Asia. The remaining problem areas are found in former Soviet Union countries. However, transitory areas of undernourishment can be caused by natural or human-made disasters.

Hunger may be defined as a condition resulting from chronic under-consumption of food and/or nutritious food products. It can be a short-term or long-term condition. If long-term, it is usually described as chronic hunger. Malnutrition is the condition that develops when the body does not get the right amount of the vitamins, minerals and other nutrients it needs to maintain healthy tissues and organ function. Malnutrition occurs in people who are undernourished. Undernutrition is a consequence of consuming too few essential nutrients or using or excreting them more rapidly than they can be replaced. The leading cause of death in children in LICs is protein-energy malnutrition. This type of malnutrition is the result of inadequate intake of calories from proteins, vitamins and minerals. Malnutrition is not confined to LICs – it can also be a condition of the very poor in more affluent nations.

The global recession of 2008–09 increased malnutrition for many of the most vulnerable people in LICs. A paper published by the United Nations Standing Committee on Nutrition found that:

- in many countries, the hours of work needed to feed a household of five increased by 10–20 per cent during 2008
- the nutritional consequences of food price increases were likely to be considerable
- currently, some 50 million, or 40 per cent, of pregnant women in LICs are anaemic – this number is likely to rise because of the current economic situation; nutritional problems very early in pregnancy will influence later foetal and infant growth.

Increased and diversified agricultural production is one of the most reliable, sustainable interventions to improve nutrition and reduce infant and child malnutrition and mortality (Figure 14.16).

Figure 14.16 Food market in Agadir, Morocco

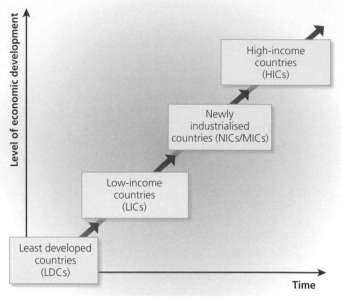

Figure 14.17 Stages of development

Section 14.1 Activities

1 Look at Figure 14.7. Select what you think are the four most important aspects of the quality of life. Justify your selections.
2 Define **a** *GNI per person*, **b** *GNI at purchasing power parity* and **c** the *development gap*.
3 Describe the global distribution of GNI PPP per person shown in Figure 14.9.
4 Why are organisations such as the UN increasingly using GNI data at purchasing power parity?
5 **a** Define the *infant mortality rate*.
   b Why is it judged to be a prime indicator of socio-economic development?
6 Why is the level of education considered to be such an important measure of a country's development?
7 Discuss the importance of nutrition as a measure of development.

☐ Different stages of development

Although the global development picture is complex, a general distinction can be made between the developed 'North' and the developing 'South'. These terms were first used in *North–South: A Programme for Survival*, published in 1980. This publication is generally known as the 'Brandt Report' after its chairperson Willy Brandt.

Other terms used to distinguish between the richer and poorer nations are:

- developed and developing
- more economically developed countries (MEDCs) and less economically developed countries (LEDCs)
- high-income countries (HICs), medium-income countries (MICs) and low-income countries (LICs).

Over the years, there have been a number of descriptions and explanations of how countries moved from one level of development to another. A reasonable division of the world in terms of stages of economic development is shown in Figure 14.17.

The concept of **least developed countries (LDCs)** was first identified in 1968 by the United Nations Conference on Trade and Development (UNCTAD). These are the poorest of the developing countries. They have major economic, institutional and human-resource problems. These are often made worse by geographical handicaps and natural and human-made disasters. *The Least Developed Countries Report 2009* identified 49 countries as LDCs. With 10.5 per cent of the world's population, these countries generate only one-tenth of 1 per cent of global income (0.1 per cent). The list of LDCs is reviewed every three years by the UN. When countries develop beyond a certain point, they are no longer considered to be LDCs.

Many of the LDCs are in Sub-Saharan Africa. Others are concentrated in the poverty belt of Asia (including Nepal and Afghanistan) or are small island nations in the South Pacific. As the gap between the richest and poorest countries in the world widens, LDCs are being increasingly **marginalised** in the world economy. Their share of world trade is declining and in many LDCs national debt now equals or exceeds GDP. Such a situation puts a stranglehold on all attempts to halt socio-economic decline.

LDCs are usually dependent on one or a small number of exports for their survival. Figure 14.18 shows a classification of LDCs according to their export specialisation.

Oil exporters	Agricultural exporters	Mineral exporters	Manufactures exporters	Services exporters	Mixed exporters
Angola	Afghanistan	Burundi	Bangladesh	Cape Verde	Lao People's Dem. Republic
Chad	Benin	Central African Republic	Bhutan	Comoros	Madagascar
Equatorial Guinea	Burkina Faso	Dem. Republic of the Congo	Cambodia	Djibouti	Myanmar
Sudan	Guinea-Bissau	Guinea	Haiti	Eritrea	Senegal
Timor-Leste	Kiribati	Mali	Lesotho	Ethiopia	Togo
Yemen	Liberia	Mauritania	Nepal	Gambia	
	Malawi	Mozambique		Maldives	
	Solomon Islands	Niger		Rwanda	
	Somalia	Sierra Leone		Samoa	
	Tuvalu	Zambia		São Tomé and Principe	
	Uganda			United Republic of Tanzania	
				Vanuatu	

Source: *The Least Developed Countries Report 2008*; UNCTAD

Figure 14.18 Classification of LDCs

The first countries to become **newly industrialised countries (NICs)** were South Korea, Singapore, Taiwan and Hong Kong. The media referred to them as the four 'Asian tigers'. A 'tiger economy' is one that grows very rapidly. This group is now often referred to as the first generation of NICs. The reasons for the success of these countries were:

- a good initial level of infrastructure
- a skilled but relatively low-cost workforce
- cultural traditions that revere education and achievement
- governments welcoming FDI from TNCs
- all four countries having distinct advantages in terms of geographical location
- the ready availability of bank loans, which were often extended at government behest and at attractive interest rates.

The success of these four countries provided a model for others to follow, such as Malaysia, Brazil, China and India. In the last 15 years, the growth of China has been particularly impressive. South Korea and Singapore have developed so much that many people now consider them to be developed countries.

Explaining the development gap

There has been much debate about the causes of the development gap. Detailed studies have shown that variations between countries are due to a variety of factors (Figure 14.19).

Physical geography

- Landlocked countries have generally developed more slowly than coastal ones.
- Small island countries face considerable disadvantages in development.
- Tropical countries have grown more slowly than those in temperate latitudes, reflecting the poor health and unproductive farming in the tropical regions. However, richer non-agricultural tropical countries such as Singapore do not suffer a geographical deficit of this kind.
- A generous allocation of natural resources has spurred economic growth in a number of countries.

Economic policies

- Open economies that welcomed and encouraged foreign investment have developed faster than closed economies.
- Fast-growing countries tend to have high rates of saving and low spending relative to GDP.
- Institutional quality in terms of good government, law and order and lack of corruption generally result in a high rate of growth.

Demography

- Progress through demographic transition is a significant factor, with the highest rates of growth experienced by those nations where the birth rate has fallen the most.

Figure 14.19 Factors affecting rates of development

Figure 14.20 shows how such factors have combined in MICs/LICs to produce higher and lower levels of development. In diagram **a**, the area where the three factors coincide gives the highest level of development. In contrast, in diagram **b** the area where the three factors combine gives the lowest level of development.

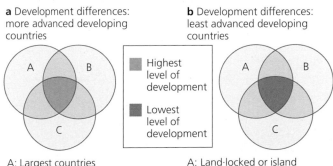

a Development differences: more advanced developing countries

b Development differences: least advanced developing countries

Highest level of development

Lowest level of development

A: Largest countries in region
B: Countries with abundant natural resources
C: Newly industrialised countries

A: Land-locked or island developing countries
B: Countries with few natural resources
C: Countries seriously affected by natural hazards

Figure 14.20 Fast and slow development MICs/LICs

Consequences of the development gap

The development gap has significant consequences for people in the most disadvantaged countries. The consequences of poverty can be economic, social, environmental and political (Figure 14.21). Development may not bring improvements in all four areas at first, but over time all four categories should witness advances.

Economic	Global integration is spatially selective: some countries benefit, others it seems do not. One in five of the world's people lives on less than $1 a day, almost half on less than $2 a day. Poor countries frequently lack the ability to pay for food, agricultural innovation, and investment in rural development.
Social	More than 850 million people in poor countries cannot read or write. Nearly a billion people do not have access to clean water and 2.4 billion do not have even basic sanitation. Eleven million children under 5 die from preventable diseases each year. People in these countries do not have the ability to combat the effects of HIV/AIDS.
Environmental	Poor countries have increased vulnerability to natural disasters. They lack the capacity to adapt to climate change or deal with consequent droughts. Poor farming practices lead to environmental degradation. Often, raw materials are exploited with very limited economic benefit to poor countries and little concern for the environment. Landscapes can be devastated by mining, vast areas of rainforest felled for logging and clearance for agriculture, and rivers and land polluted by oil exploitation.
Political	Poor countries that are low on the development scale often have a non-democratic government or they are democracies that function poorly. There is usually a reasonably strong link between development and improvement in the quality of government. In general, the poorer the country the worse the plight of minority groups.

Figure 14.21 Consequences of poverty

Section 14.1 Activities

1 Suggest three countries for each stage of development shown in Figure 14.17.
2 With reference to Figure 14.20, suggest why some poorer countries have been able to develop into MICs/NICs while many have not.
3 Review the physical, economic and demographic factors responsible for the development gap.

14.2 The globalisation of economic activity

☐ Global patterns of resources, production and markets

Globalisation is the increasing interconnectedness and interdependence of the world economically, culturally and politically. There are many aspects of globalisation, which are summarised in Figure 14.22. The word 'globalisation' did not come into common usage until about 1960. In

Dimension	Characteristics
Economic	Under the auspices of first GATT and latterly the WTO, world trade has expanded rapidly. Transnational corporations have been the major force in the process of increasing economic interdependence, and the emergence of different generations of NICs has been the main evidence of success in the global economy. However, the frequency of 'anti-capitalist' demonstrations in recent years shows that many people have grave concerns about the direction the global economy is taking. Many LEDCs and a significant number of regions within MEDCs feel excluded from the benefits of globalisation.
Urban	A hierarchy of global cities has emerged to act as the command centres of the global economy. New York, London and Tokyo are at the highest level of this hierarchy. Competition within and between the different levels of the global urban hierarchy is intensifying.
Social/cultural	Western culture has diffused to all parts of the world to a considerable degree through TV, cinema, the internet, newspapers and magazines. The international interest in brand-name clothes and shoes, fast food and branded soft drinks and beers, pop music and major sports stars has never been greater. However, cultural transmission is not a one-way process. The popularity of Islam has increased considerably in many Western countries, as has Asian, Latin American and African cuisine.
Linguistic	English has clearly emerged as the working language of the 'global village'. Of the 1.9 billion English speakers, some 1.5 billion people around the world speak English as a second language. In a number of countries there is great concern about the future of the native language.
Political	The power of nation states has been diminished in many parts of the world as more and more countries organise themselves into trade blocs. The European Union is the most advanced model for this process of integration, taking on many of the powers that were once the sole preserve of its member nation states. The United Nations has intervened militarily in an increasing number of countries in recent times, leading some writers to talk about the gradual movement to 'world government'. On the other side of the coin is the growth of global terrorism.
Demographic	The movement of people across international borders and the desire to move across such borders has increased considerably in recent decades. More and more communities are becoming multicultural in nature.
Environmental	Increasingly, economic activity in one country has had an impact on the environment in other nations. The long-range transportation of airborne pollutants is the most obvious evidence of this process. The global environmental conferences in Rio de Janeiro (1992) and Johannesburg (2002) is evidence that most countries see the scale of the problems as so large that only coordinated international action can bring realistic solutions.

Figure 14.22 The dimensions of globalisation

1961, Webster became the first major dictionary to give a definition of globalisation. However, the word was not recognised as academically significant until the early to mid-1980s. Since then, its use has increased dramatically.

Figure 14.23 shows Peter Dicken's view of the global economy. TNCs and nation states are the two major decision-makers. Nation states individually and collectively set the rules for the global economy but

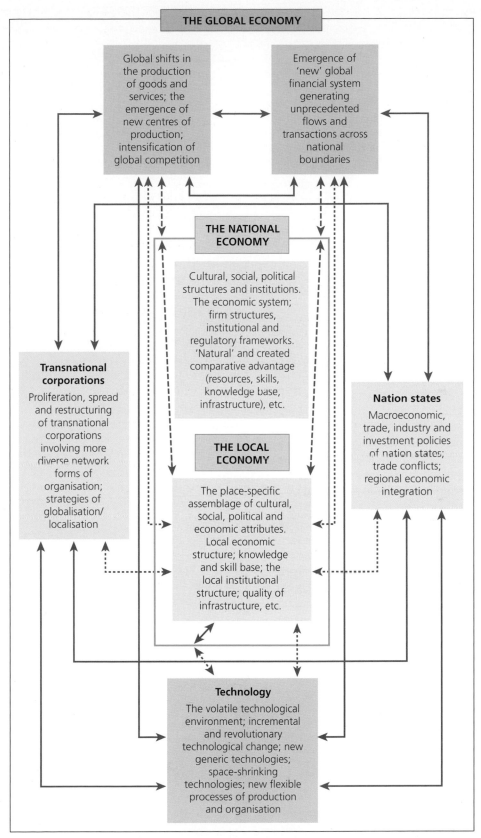

THE GLOBAL ECONOMY

Global shifts in the production of goods and services; the emergence of new centres of production; intensification of global competition

Emergence of 'new' global financial system generating unprecedented flows and transactions across national boundaries

THE NATIONAL ECONOMY

Cultural, social, political structures and institutions. The economic system; firm structures, institutional and regulatory frameworks. 'Natural' and created comparative advantage (resources, skills, knowledge base, infrastructure), etc.

Transnational corporations

Proliferation, spread and restructuring of transnational corporations involving more diverse network forms of organisation; strategies of globalisation/localisation

THE LOCAL ECONOMY

The place-specific assemblage of cultural, social, political and economic attributes. Local economic structure; knowledge and skill base; the local institutional structure; quality of infrastructure, etc.

Nation states

Macroeconomic, trade, industry and investment policies of nation states; trade conflicts; regional economic integration

Technology

The volatile technological environment; incremental and revolutionary technological change; new generic technologies; space-shrinking technologies; new flexible processes of production and organisation

Source: *Access to Geography: Globalisation* by P. Guinness (Hodder Education, 2003), p.4

Figure 14.23 The global economy

the bulk of investment is through TNCs, which are the main drivers of **global shift**. This is the movement of economic activity, particularly in manufacturing, from HICs to NICs and LICs. It is this process that has resulted in the emergence of an increasing number of NICs since the 1960s.

The development of globalisation

Globalisation is a relatively recent phenomenon (post-1960), which is very different from anything the world had previously experienced. It developed out of **internationalisation**. A key period in the process of internationalisation occurred between 1870 and 1914, when:

- transport and communications networks expanded rapidly
- world trade grew significantly, with a considerable increase in the level of interdependence between rich and poor nations
- there were very large flows of capital from European companies to other parts of the world.

International trade tripled between 1870 and 1913. At this time, the world trading system was dominated and organised by four nations: Britain, France, Germany and the USA. However, the global shocks of the First World War and the Great Depression (1929 to late 1930s) put a stop to this period of phenomenal economic growth. It was not until the 1950s that international interdependence was back on track. Since then, world trade has grown consistently faster than world GDP. However, even by 1990 the level was unremarkable compared with that of the late nineteenth and early twentieth centuries. It is not surprising, therefore, that some writers argue that the level of integration before 1914 was similar to that of today.

However, today's globalisation is very different from the global relationships of 50 or 100 years ago. Peter Dicken makes the distinction between the 'shallow integration' of the pre-1914 period and the 'deep integration' of the present period. The global economy is more extensive and complicated than it has ever been before.

Economic globalisation

Figure 14.24 shows the main influences on the globalisation of economic activity. This is not a definitive list and you may be able to think of others. The factors responsible for economic globalisation can be stated as follows:

- Until the post-1950 period, the production process itself was mainly organised within national economies. This has changed rapidly in the last 50 years or so with the emergence of a **new international division of labour (NIDL)** reflecting a change in the geographical pattern of specialisation, with the fragmentation of many production processes across national boundaries. The widespread use of terms such as 'outsourcing' and 'offshoring' signify the importance of this process. (For more information on outsourcing, refer back to page 452.)
- International trade flows have become increasingly complex as this process has developed.
- There have been major advances in trade liberalisation under the World Trade Organization.
- There was an emergence of fundamentalist free-market governments in the USA (Ronald Reagan) and the UK (Margaret Thatcher) around 1980. The economic policies developed by these governments influenced policy-making in many other countries.
- An increasing number of NICs emerged.
- The old Soviet Union and its Eastern European communist satellites merged into the capitalist system (Figure 14.25). Now, no significant group of countries stands outside the free-market global system.
- Other economies opened up, notably those of China and India.
- The world financial markets were deregulated.
- The 'transport and communications revolution' has made possible the management of the complicated networks of production and trade that exist today.

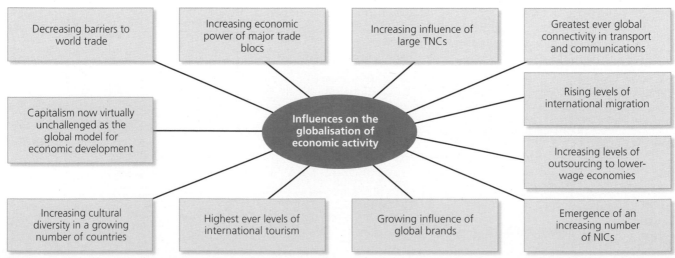

Figure 14.24 Globalisation trends

Figure 14.25 The fall of the Berlin Wall signalled the beginning of the integration of Eastern Europe into the free-market system

a A part of the wall that still exists
b A plaque showing where the wall used to be

The advantages for economic activity in working at the global scale

Large companies in particular recognise many advantages in working at the global scale as opposed to at national or continental scales:

- Sourcing of raw materials and components on a global basis reduces costs.
- TNCs can seek out the lowest-cost locations for labour and other factors.
- High-volume production at low cost in countries such as China helps to reduce the rate of inflation in other countries and helps living standards to rise.
- Collaborative arrangements with international partners can increase the efficiency of operations considerably.
- Selling goods and services to a global market allows TNCs to achieve very significant economies of scale.
- Global marketing helps to establish brands with huge appeal all around the world.

Which are the most globalised countries?

Figure 14.26 shows the most globalised countries in the world according to the 2007 A.T. Kearney/Foreign Affairs Index of Globalization. The 2007 index ranked 72 countries according to their degree of globalisation. These countries accounted for 97 per cent of the world's GDP and 88 per cent of the global population. The A.T. Kearney Index of Globalization comprised four key elements of global integration:

- economic integration
- personal contact
- technological connectivity
- political engagement.

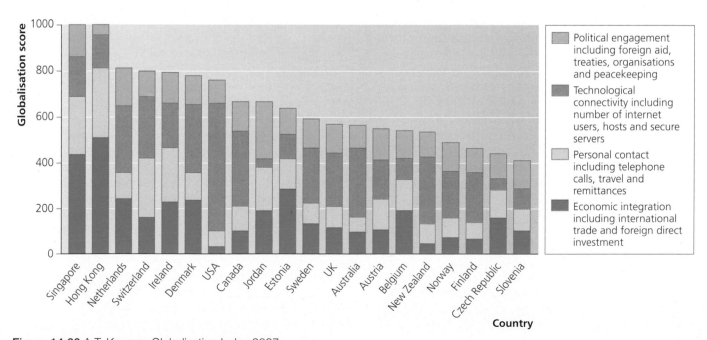

Figure 14.26 A.T. Kearney Globalization Index 2007

Legend:
- Political engagement including foreign aid, treaties, organisations and peacekeeping
- Technological connectivity including number of internet users, hosts and secure servers
- Personal contact including telephone calls, travel and remittances
- Economic integration including international trade and foreign direct investment

Figure 14.27 Post Office – Tiananmen Square, Beijing

Section 14.2 Activities

1 Define **a** *globalisation* and **b** *global shift*.
2 Discuss two of the dimensions of globalisation covered in Figure 14.22.
3 Explain the *new international division of labour*.
4 Briefly discuss the factors responsible for economic globalisation.
5 Explain the advantages for economic activity in working at the global scale.
6 Discuss the factors used by the A.T. Kearney Index to measure globalisation.

Economic integration brings together data on trade and FDI. The latter includes both inflows and outflows. International travel and tourism is also included in this category. Personal contact encompasses international telephone calls and cross-border remittances. As countries become more globalised, personal contacts increase accordingly. The average community today will have far more former members living abroad than was the case 30 years ago. International telephone calls are a good way of comparing countries at one point in time and also recording changes over time. Remittances have become an increasingly important source of money flowing between countries because of the growing number of migrant workers (Figure 14.27).

Technological connectivity concerns the number of internet users and internet hosts. The internet has arguably been the single most important advance in the globalisation process.

Political engagement considers a country's membership of a variety of international organisations. International contacts are vital to virtually all aspects of the development process.

Figure 14.26 shows that the most globalised countries in rank order are Singapore, Hong Kong, the Netherlands, Switzerland, Ireland, Denmark and the USA. However, the composition of their total globalisation scores is markedly different in some cases. Economic integration is the most important factor for Singapore and Hong Kong. It is much less important for the other five most globalised countries, and its contribution is particularly low for the USA. It can be argued that large economies like the USA are able to satisfy far more of their economic needs from within their own borders than small countries with limited land areas and resources in general.

For the USA, technological connectivity is by far the most dominant of the four elements of globalisation. This reflects the high level of affluence in the USA and the eagerness to embrace new technology at the business and individual levels.

☐ Transnational corporations and foreign direct investment

Major transnational corporations and foreign direct investment flows

Investment involves expenditure on a project in the expectation of financial (or social) returns. **Transnational corporations (TNCs)** are the main source of foreign direct investment (FDI). TNCs invest to make profits and are the driving force behind economic globalisation. They are capitalist enterprises that organise the production of goods and services in more than one country. As the rules regulating the movement of goods and investment have been relaxed in recent decades, TNCs have extended their global reach. As the growth of FDI has expanded, the sources and destinations of that investment have become more and more diverse. FDI is not dominated by flows from core to periphery in the same way that it was even 20 years ago. Investment flows from NICs such as South Korea, Taiwan, China, India and Brazil have increased markedly. The investment flow network is more complex today than it has ever been.

There are now few parts of the world where the direct or indirect influence of TNCs is not important. In some countries and regions, their influence on the economy is huge. Apart from their direct ownership of productive activities, many TNCs are involved in a web of collaborative relationships with other companies across the globe. Such relationships have become more and more important as competition has become increasingly global in its extent.

TNCs have a substantial influence on the global economy in general, and in the countries in which they choose to locate in particular. They play a major role in world trade in terms of what and where they buy and sell. A not inconsiderable proportion of world trade is intra-firm; that is, taking place within TNCs. The organisation of the car giants exemplifies intra-firm trade, with engines, gearboxes and other key components produced in one country and exported for assembly elsewhere. Table 14.3

shows the world's ten largest TNCs by revenue and profits, according to Global 500 published by *Fortune* magazine. The list is led by Royal Dutch Shell, Wal-Mart Stores and Exxon Mobil. The top five companies all recorded revenue in excess of $400 billion. Exxon Mobil recorded the largest profit of any company in the world at almost $45 billion. However, it should be noted that sometimes large corporations make a loss.

According to the *World Investment Report 2013*, there are over 100 000 TNCs worldwide. The 100 largest TNCs represent a significant proportion of total global production.

Global FDI inflows amounted to $1.45 trillion in 2013 (Figure 14.29). This comprised:

- $566 billion into developed countries (39 per cent)
- $778 billion into developing economies (54 per cent)
- $108 billion to transition economies (7 per cent).

Table 14.3 The world's 10 largest corporations, 2013

Rank	Company	Revenue ($ billion)	Profits ($ billion)	HQ country
1	Royal Dutch Shell	481.7	26.6	Netherlands
2	Wal-Mart Stores	469.2	17.0	USA
3	Exxon Mobil	449.9	44.9	USA
4	Sinopec Group	428.2	8.2	China
5	China National Petroleum	408.6	18.2	China
6	BP	388.3	11.6	UK
7	State Grid	298.4	12.3	China
8	Toyota Motor	265.7	11.6	Japan
9	Volkswagen	247.6	27.9	Germany
10	Total	234.3	13.7	France

Figure 14.28 The financial district, Chicago

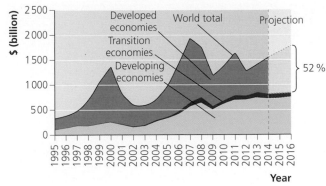

Source: *World Investment Report, 2013*

Figure 14.29 FDI inflows, global and by groups of economies, 1995–2016

The share of developing and 'transition economies' in global FDI inflows has risen significantly in recent decades. The transition economies are those in South East Europe and the Commonwealth of Independent States (CIS). They have been given the term 'transition economies' by the World Investment Report to recognise that they are still in the process of change from centrally planned communist economies to full members of the capitalist global economy. Developing Asia remains the world's largest recipient of FDI inflows.

Figure 14.30 shows the rapidly changing nature of **global FDI outflows** by group of economies. Between 1999 and 2013, the share of FDI outflows of the developed economies declined from 93 per cent to 61 per cent. In contrast, the share of the developing and transition economies increased from 7 per cent to 39 per cent.

Capital flow can help MICs and LICs with economic development by furnishing them with necessary capital and technology. However, capital flow from HICs to MICs and LICs has been skewed, with some countries far more favoured than others. Most Sub-Saharan African countries, which urgently needed foreign capital for economic betterment, have been largely excluded from globalised investment in the past, although there

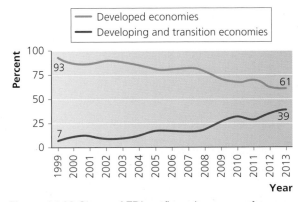

Figure 14.30 Share of FDI outflows by group of economies

Figure 14.31 Japanese investment in the UK – Suzuki in Crawley

is evidence that this is beginning to change. While controlling for other social and economic factors, democratised LICs appear to attract more foreign capital than undemocratic countries because their democratic institutions can provide secure and profitable environments for investment with protection of property rights and social spending on human capital.

The development of TNCs over time

Figure 14.32 shows the main stages in the historical evolution of TNCs. Although the first companies to produce outside their home nation did not emerge until the latter half of the nineteenth century, by 1914 US, British and mainland European firms were involved in substantial overseas manufacturing production. Prior to the First World War, the UK was the major source of overseas investment, the pattern of which was firmly based on its empire. Between the wars, TNC manufacturing investment, particularly American, increased substantially. By 1939, the USA had become the main source of foreign investment in manufacturing. The USA was to become even more powerful in the global economy after the Second World War, for it was the only

industrial power to emerge from the conflict stronger rather than weaker.

However, the USA does not dominate the global economy today in the way it did in the immediate post-war period. The reconstruction of the Japanese and German economies resulted in both countries playing a significant transnational role by the 1970s, which was to expand considerably in the following decades. In fact, the large Japanese TNCs were to become models for their international competitors as they revolutionised business organisation. Other HICs such as the UK, France, Italy, the Netherlands, Switzerland, Sweden and Canada also played significant roles in the geographical spread of FDI. More recently, NICs such as South Korea and Taiwan have expanded their corporate reach, not just to lower-wage economies but also into HICs. Figure 14.33 illustrates the sequential development of a TNC, which begins with operation in the domestic market only. Large companies often reach the stage when they want to produce outside of their home country and take the decision to become transnational. The benefits of such a move include:

- using cheaper labour, particularly in LICs
- exploiting new resource locations
- circumventing trade barriers
- tapping market potential in other world regions
- avoiding strict domestic environmental regulations
- maximising exchange-rate advantages.

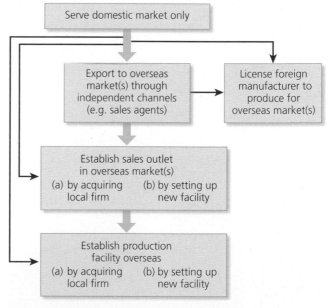

Figure 14.33 Sequential development of a TNC

Contrasting spatial and organisational structures

TNCs vary widely in overall size and international scope. Variations include:

- the number of countries
- the number of subsidiaries
- the share of production accounted for by foreign activities

Period	Type	Characteristics
1500–1800	Mercantile capitalism and colonialism	Government-backed chartered companies
1800–75	Entrepreneurial and financial capitalism	Early development of supplier and consumer markets Infrastructural investment by financial houses
1875–1945	International capitalism	Rapid growth of market-seeking and resource-based investments
1945–60	Transnational capitalism	FDI dominated by USA TNCs expand in size
1960–present	Globalisation capitalism	Expansion of European and Japanese FDI Growth of inter-firm alliances, joint ventures and outsourcing

Figure 14.32 Stages in the evolution of TNCs

- the degree to which ownership and management are internationalised
- the division of research activities and routine tasks by country
- the balance of advantages and disadvantages to the countries in which they operate.

Large TNCs often exhibit three organisational levels: headquarters; research and development; and branch plants. The headquarters of a TNC will generally be in the HIC city where the company was established. Research and development is most likely to be located here too, or in other areas within this country. It is the branch plants that are the first to be located overseas. However, some of the largest and most successful TNCs have divided their industrial empires into world regions, each with research and development facilities and a high level of independent decision-making. Figure 14.34 shows the locational changes that tend to occur as TNCs develop over time.

Figure 14.35 shows four simplified models that illustrate major ways of organising the geography of TNC production units. Toyota and the other global car manufacturers are closest to model **c**, a system often referred to as a 'horizontal organisational structure'. In contrast, Nike is a good example of model **d**, illustrating a vertical organisational structure. However, Nike is not integrated in the traditional sense – it does not own the various stages of production because it subcontracts the manufacturing stages of its product range.

a Globally concentrated production

All production at a single location. Products are exported to world markets.

b Host-market production

Each production unit produces a range of products and serves the national market in which it is located. No sales across national boundaries. Individual plant size limited by the size of the national market.

c Product specialisation for a global or regional market

Each production unit produces only one product for sale throughout a regional market of several countries. Individual plant size very large because of scale economies offered by the large regional market.

d Transnational vertical integration

Each production unit performs a separate part of a production sequence. Units are linked across national boundaries in a 'chain-like' sequence – the output of one plant is the input of the next plant

Each production unit performs a separate operation in a production process and ships its output to a final assembly plant in another country.

Figure 14.35 Ways of organising TNC production units

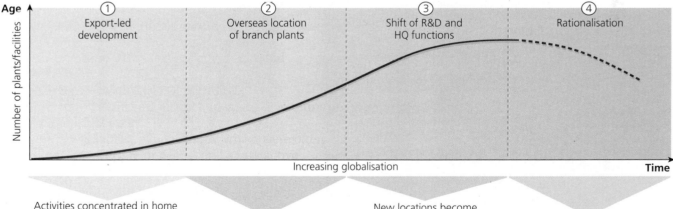

Figure 14.34 The development of TNCs – locational changes

Activities concentrated in home country where labour and sourcing are established. However, exports may be subject to tariffs and other restrictions

Incentives include cheaper labour, access to markets, and financial assistance from host governments

New locations become semi-autonomous as products are more carefully tailored to new markets

Increasing competition or recession necessitates concentrating activities in the best locations

Source: *Advanced Geography: Concepts & Cases* by P. Guinness & G. Nagle (Hodder Education, 1999), p.167

Nike is the world's leading supplier of sports footwear, apparel and equipment, and one of the best-known global brands. It was founded in 1972 and the company went public in 1980. The company is an example of a vertical organisational structure across international boundaries, characterised by a high level of subcontracting activity. Nike does not make any shoes or clothes itself, but contracts out production to South Korean and Taiwanese companies. Nike employs 650 000 contract workers in 700 factories worldwide. The company list includes 124 plants in China, 73 in Thailand, 35 in South Korea and 34 in Vietnam. More than 75 per cent of the workforce is based in Asia. The majority of workers are women under the age of 25.

The subcontracted companies operate not only in their home countries but also in lower-wage Asian economies such as Vietnam, the Philippines and Indonesia – 150 Asian factories employing 350 000 workers manufacture products for Nike. The company has a reputation for searching out cheap pools of labour. Nike's expertise is in design and marketing. Figure 14.36 shows Nike's 'commodity circuit'. It is a clear example of the New International Division of Labour (NIDL).

Nike illustrates both 'Fordist' and 'Flexible' characteristics. An example of its Fordist nature is the Air Max Penney basketball shoes, which consist of 52 component parts from five different countries. The shoes pass through 120 people during production, on a clearly demarcated global production chain.

However, Nike also exhibits Flexible characteristics. The company aims to produce new shoes on a regular basis to cater for niche markets. To achieve this objective, it utilises a just-in-time innovation structure, buying in necessary expertise at short notice. This involves short-term subcontracts, often allocated to firms based close to Nike's research and development headquarters near Beaverton in the state of Oregon, USA.

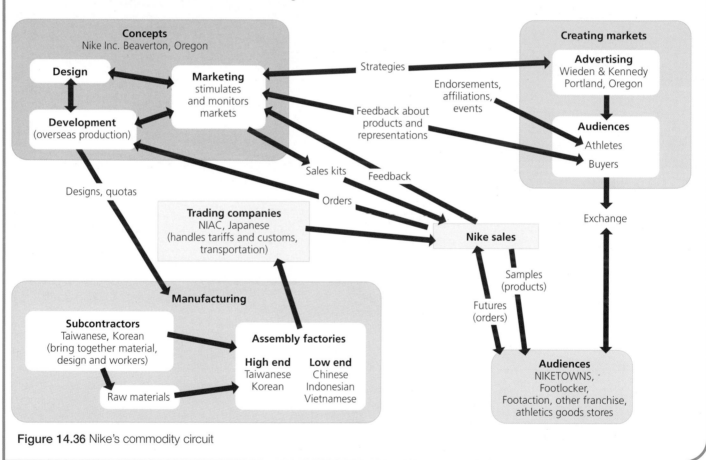

Figure 14.36 Nike's commodity circuit

Section 14.2 Activities

1 Define **a** *TNCs* and **b** *FDI*.
2 Describe the FDI inflow trends shown in Figure 14.29.
3 Comment on the trends shown in Figure 14.30.
4 Describe and explain the development of TNCs over time.
5 Comment on the different forms of TNC organisation shown in Figure 14.35.
6 Produce a brief bullet-point summary of the Nike case study.

The emergence and growth of newly-industrialised countries

In Asia, four **generations of NIC** can be recognised in terms of the timing of industrial development and their current economic characteristics. Within this region, only Japan is at a higher economic level than the NICs (Table 14.4) but there are a number of countries at much lower levels of economic development. The latter group are the poorest countries in the region.

Table 14.4 Levels of economic development in Asia

Level	Countries	GNI PPP per person, 2013 ($)
1	Japan – a HIC	37 630
2	First generation NICs, e.g. South Korea	33 440
3	Second generation NICs, e.g. Malaysia	22 460
4	Third generation NICs, e.g. China	11 850
5	Fourth generation NICs, e.g. Vietnam	5 030
6	Poorest countries, e.g. Cambodia	2 890

Nowhere else in the world is the filter-down concept of industrial location better illustrated. When Japanese companies first decided to locate abroad in the quest for cheap labour, they looked to the most developed of their neighbouring countries, particularly South Korea and Taiwan. Most other countries in the region lacked the physical infrastructure and skills levels required by Japanese companies. Companies from elsewhere in HICs, especially the USA, also recognised the advantages of locating branch plants in such countries. As the economies of the first generation NICs developed the level of wages increased, resulting in:

- Japanese and Western TNCs seeking locations in second-generation NICs where improvements in physical and human infrastructures now satisfied their demands but where wages were still low
- indigenous companies from the first-generation NICs also moving routine tasks to their cheaper-labour neighbours such as Malaysia and Thailand.

With time, the process also included the third-generation NICs, a significant factor in the recent very high growth rates in China and India. The least developed countries in the region, nearly all hindered by conflict of one sort or another at some time in recent decades, are now beginning to be drawn into the system. The recent high level of FDI into Vietnam makes it reasonable to think of the country as an example of a fourth-generation Asian NIC.

First-generation NICs

What were the reasons for the phenomenal rates of economic growth recorded in South Korea, Taiwan, Hong Kong and Singapore from the 1960s? What was it that set this group of 'Asian tigers' apart from so many others?

From the vast literature that has appeared on the subject, the following factors are usually given prominence:

- A good initial level of hard and soft infrastructure provided the preconditions for structural economic change.
- As in Japan previously, the land-poor NICs emphasised people as their greatest resource, particularly through the expansion of primary and secondary education but also through specialised programmes to develop scientific, engineering and technical skills.
- These countries have cultural traditions that revere education and achievement.
- The Asian NICs became globally integrated at a 'moment of opportunity' in the structure of the world system, distinguished by the geostrategic and economic interests of core capitalist countries (especially the USA and Japan) in extending their influence in East and South East Asia.
- All four countries had distinct advantages in terms of geographical location. Singapore is strategically situated to funnel trade flows between the Indian and Pacific Oceans, and its central location in the region has facilitated its development as a major financial, commercial and administrative/managerial centre. Hong Kong has benefited from its position astride the trade routes between North East and South East Asia, as well as acting as the main link to the outside world for south-east China. South Korea and Taiwan were ideally located to expand trade and other ties with Japan.
- The ready availability of bank loans, often extended at government behest and at attractive interest rates, allowed South Korea's *chaebol* in particular to pursue market share and to expand into new fields.

As their industrialisation processes have matured, the NICs have occupied a more intermediate position in the regional division of labour between Japan and other less developed Asian countries.

Deindustrialisation

In the USA and the UK, the proportion of workers employed in manufacturing has fallen from around 40 per cent at the beginning of the twentieth century to less than half that now. Even in Japan and Germany, where so much industry was rebuilt after 1945, manufacturing's share of total employment has dropped below 25 per cent. All HICs have followed this trend, known as deindustrialisation. The causal factors of deindustrialisation are:

- technological change enabling manufacturing to become more capital-intensive and more mobile
- the filter-down of manufacturing industry from HICs to lower wage economies, such as those of South East Asia
- the increasing importance of the service sector in the HICs.

There can be little surprise in the decline of manufacturing employment, for it has mirrored the previous decline in employment in agriculture in HICs. So if the decline of manufacturing in HICs is part of an expected cycle, the consequence of technological improvement and rising affluence, why is so much concern expressed about this trend? The main reasons would appear to be:

- The traditional industries of the Industrial Revolution were highly concentrated, so the impact of manufacturing decline has had severe implications in terms of unemployment and other social pathologies in a number of regions.
- The rapid pace of contraction of manufacturing has often made adjustment difficult.
- There are defence concerns if the production of some industries falls below a certain level.
- Some economists argue that over-reliance on services makes an economy unnecessarily vulnerable.
- Rather than being a smooth transition, manufacturing decline tends to concentrate during periods of economic recession.

The filter-down process of industrial relocation

The filter-down process, detailed by W.R. Thompson and others, operates at both global and national scales. Economic core regions have long been vulnerable to the migration of labour-intensive manufacturing to lower-wage areas of the periphery, as exemplified in the USA by the historical drift of the textile and shoe industries from New England, and apparel manufacture from New York, to North and South Carolina. The filter-down process is based on the notion that corporate organisations respond to changing critical input requirements by altering the geographical location of production to minimise costs and thereby ensure competitiveness in a tightening market.

The economic core (at national and global levels) has monopolised invention and innovation, and has thus continually benefited from the rapid growth rates characteristic of the early stages of an industry's life cycle (the product life cycle), one of exploitation of a new market. Production is likely to occur where the firm's main plants and corporate headquarters are located. Figure 14.37, illustrating the **product life cycle**, indicates that in the early phase scientific-engineering skills at a high level and external economies are the prime location factors.

In the growth phase, methods of mass production are gradually introduced and the number of firms involved in production generally expands as product information spreads. In this stage, management skills are the critical human inputs. Production technology tends to stabilise in the mature phase. Capital investment remains high and the availability of unskilled and semiskilled labour becomes a major locating factor. As the industry matures into a replacement market, the production process

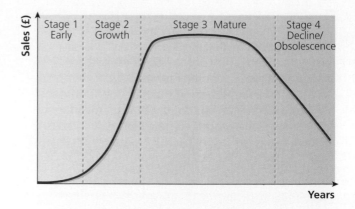

Requirements	Product–Cycle Phase		
	Early	Growth	Mature
Management	2	1	3
Scientific-engineering know-how	1	2	3
Unskilled and semi-skilled labour	3	2	1
External economies	1	2	3
Capital	3	1a	1a

Source: Based on Oakey 1984 and Erickson & Leinbach 1979

Figure 14.37 The product life cycle

becomes rationalised and often routine. The high wages of the innovating area, quite consistent with the high-level skills required in the formative stages of the learning process, become excessive when the skill requirements decline and the industry, or a section of it, 'filters-down' to smaller, less industrially sophisticated areas where cheaper labour is available, but which can now handle the lower skills required in the manufacture of the product.

On a global scale, large TNCs have increasingly operated in this way by moving routine operations to LICs since the 1950s. However, the role of indigenous companies in developing countries should not be ignored. Important examples are the *chaebols* of South Korea, such as Samsung and Hyundai, and Taiwanese firms such as Acer. Here, the process of filter-down has come about by direct competition from LICs rather than from the corporate strategy of huge North American, European and Japanese transnationals.

It has been the revolution in transport and communications that made such substantial filter-down of manufacturing to LICs possible. Containerisation and the general increase in the scale of shipping have cut the cost of the overseas distribution of goods substantially, while advances in telecommunications have made global management a reality. In some cases, whole industries have virtually migrated, as did shipbuilding from Europe to Asia in the 1970s. In others, the most specialised work is done in HICs by skilled workers, and the simpler tasks elsewhere in the global supply chain.

Although the theory of the product life cycle was developed in the discipline of business studies to explain how the sales of individual products evolve, it can usefully be applied at higher scales. A firm with a range of ten products, half in stage 3 and half in stage 4, would have no long-term future. A healthy multi-product firm will have a strong research and development department ensuring a steady movement of successful products onto the market to give a positive distribution across the four stages of the model. Likewise, the industry mix of a region or a country can be plotted on the product life cycle diagram. Regions with significant socio-economic problems are invariably over-represented in stages 3 and 4. In contrast, regions with dynamic economies will have a more even spread across the model, with particularly good representation in the first two stages.

Figure 14.38 summarises the positive and negative effects of globalisation on **a** HICs and **b** NICs and LICs, including the impact of deindustrialisation on more affluent countries. Economists have recognised two types of deindustrialisation: positive and negative:

- Positive deindustrialisation occurs when the share of employment in manufacturing falls because of rapid productivity growth but where displaced labour is absorbed into the non-manufacturing sector. In such a situation, the economy is at or near full employment and GDP per person is rising steadily.
- Negative deindustrialisation results from a decline in the share of manufacturing in total employment, owing to a slow growth or decline in demand for manufacturing output, and where displaced labour results in unemployment.

Unfortunately in many HICs, the deindustrialisation experienced has been predominantly of the negative kind. Regional development policies have tried to address these problems with varying degrees of success.

Section 14.2 Activities

1 Describe and explain the information shown in Table 14.4.
2 Discuss the reasons for the development of the first generation of TNCs.
3 With reference to Figure 14.37, explain the product life cycle.
4 Examine the connection between industrial growth in NICs and some LICs, and deindustrialisation in HICs.
5 With reference to Figure 14.38, discuss the positive effects of deindustrialisation in HICs.

	Positive	Negative
In HICs	• Cheaper imports of all relatively labour-intensive products can keep cost of living down and lead to a buoyant retailing sector. • Greater efficiency apparent in surviving outlets. This can release labour for higher productivity sectors (this assumes low unemployment). • Growth in NICs and LICs may lead to a demand for exports from HICs. • Promotion of labour market flexibility and efficiency, greater worker mobility to area with relative scarcities of labour should be good for the country. • Greater industrial efficiency should lead to development of new technologies, promotion of entrepreneurship and should attract foreign investment. • Loss of industries can lead to improved environmental quality (e.g. Consett).	• Rising job exports leads to inevitable job losses. Competition-driven changes in technology add to this. • Job losses are often of unskilled workers. • Big gaps develop between skilled and unskilled workers who may experience extreme redeployment differences. • Employment gains from new efficiencies will only occur if industrialised countries can keep their wage demands down. • Job losses are invariably concentrated in certain areas and certain industries. This can lead to deindustrialisation and structural unemployment in certain regions. • Branch plants are particularly vulnerable as in times of economic recession they are the first to close, often with large numbers of job losses.
In NICs and LICs	• Higher export-generated income promotes export-led growth – thus promotes investment in productive capacity. Potentially leads to a multiplier effect on national economy. • Can trickle down to local areas with many new highly paid jobs. • Can reduce negative trade balances. • Can lead to exposure to new technology, improvement of skills and labour productivity. • Employment growth in relatively labour-intensive manufacturing spreads wealth, and does redress global injustice (development gap).	• Unlikely to decrease inequality – as jobs tend to be concentrated in core region of urban areas. May promote in-migration. • Disruptive social impacts, e.g. role of TNCs potentially exploitative and may lead to sweatshops. Also branch plants may move on in LICs too, leading to instability (e.g. in Philippines). • Can lead to overdependence on a narrow economic base. • Can destabilise food supplies, as people give up agriculture. • Environmental issues associated with over-rapid industrialisation. • Health and safety issues because of tax legislation.

Source: Sue Wam, 'The Global Shift', *Geo Factsheet*, Curriculum Press

Figure 14.38 The positive and negative impacts of global shift

14.3 Regional development

☐ The extent of income disparities within countries

The scale of disparities within countries is often as much an issue as the considerable variations between countries (Figure 14.39). The **Gini coefficient** is a technique frequently used to show the extent of income inequality. It allows:

■ analysis of changes in income inequality over time in individual countries

■ comparison between countries.

Figure 14.40 shows global variations in the Gini coefficient for 2007–08. It is defined as a ratio with values between 0 and 1.0; it can also be expressed as a percentage. A low value indicates a more equal income distribution while a high value shows more unequal income distribution. A Gini coefficient of 0 would mean that everyone in a country had exactly the same income (perfect equality). At the other extreme, a Gini coefficient of 1 (or 100%) would mean that one person had all the income in a country (perfect inequality). Figure 14.40 shows that in general more affluent countries have a lower income gap than lower-income countries. In 2007–08, the global gap ranged from 0.232 (23.2%) in Denmark to 0.707 (70.7%) in Namibia.

Table 14.5 shows the ten most unequal countries in the World Bank's data set, which includes 112 countries for which data was available for at least one year between 2008 and 2013. In contrast, Table 14.6 shows data for the ten least unequal countries. The Gini coefficient average among all countries in the entire data set was 38.8% (0.388).

Figure 14.39 Graffiti on a fence in a lower-income part of Ulaanbaatar, Mongolia

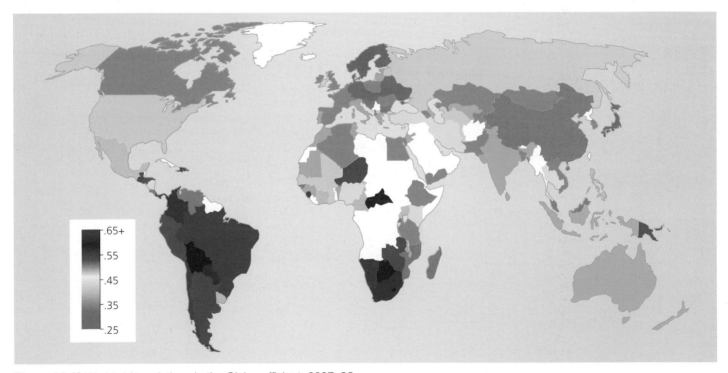

Figure 14.40 Worldwide variations in the Gini coefficient, 2007–08

Table 14.5 The ten most unequal countries

Country	Gini index	Bottom 10% share (%)	Top 10% share (%)
South Africa	65.0	1.1	53.8
Namibia	61.3	1.5	51.8
Botswana	60.5	1.1	49.6
Zambia	57.5	1.5	47.4
Honduras	57.4	0.8	45.7
Central African Republic	56.3	1.2	46.1
Lesotho	54.2	1.0	41.0
Colombia	53.5	1.1	42.0
Brazil	52.7	1.0	41.7
Guatemala	52.4	1.3	41.8

Table 14.6 The ten least unequal countries

Country	Gini index	Bottom 10% share (%)	Top 10% share (%)
Ukraine	24.8	4.4	21.0
Slovenia	24.9	3.9	20.8
Iceland	26.3	3.9	22.2
Czech Republic	26.4	3.7	22.2
Belarus	26.5	3.9	21.5
Slovakia	26.6	3.2	21.0
Norway	26.8	3.3	21.9
Denmark	26.9	3.3	22.1
Romania	27.3	3.7	21.5
Finland	27.8	3.7	22.6

The Lorenz curve is a graphical technique that shows the degree of inequality between two variables. It is often used to show the extent of income inequality in a population. The diagonal line represents perfect equality in income distribution. The further the curve away from the diagonal line, the greater the degree of income inequality. Thus in Figure 14.41, income inequality in Brazil was less in 2005 than in 1996. However, the significant gap between the 2005 curve and the diagonal line indicates that income inequality in Brazil remains very substantial indeed.

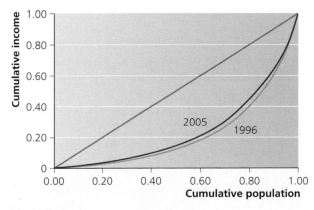

Figure 14.41 Lorenz curve for Brazil

A report published in October 2008 entitled 'Growing Unequal? Income Distribution and Poverty in OECD (Organization for Economic Cooperation and Development) Countries' found that:

- the gap between rich and poor has grown in more than three-quarters of OECD countries over the last two decades
- the economic growth of recent decades has benefited the rich more than the poor; in some countries, such as Canada, Finland, Germany, Italy, Norway and the USA, the gap also increased between the rich and the middle class
- countries with a wide distribution of income tend to have more widespread income poverty
- social mobility is lower in countries with high inequality, such as Italy, the UK and the USA, and higher in the Nordic countries where income is distributed more evenly.

Disparities within countries are rarely uniform throughout countries and thus a significant regional component usually exists. In China, the income gap between urban residents and the huge farm population reached its widest ever level in 2008 as rural unemployment in particular rose steeply. The ratio between more affluent urban dwellers and their rural counterparts reached 3.36 to 1, up from 3.33 to 1 in 2007. This substantial income gap is a very sensitive issue in China as more and more rural people feel they have been left behind in China's economic boom. The size of the income gap is not just a political problem; it is also causing considerable national economic concern. Falling purchasing power in rural areas is hindering efforts to boost domestic consumer spending. The government wants to do this to help compensate for declining exports caused by the global recession.

Section 14.3 Activities

1 What is the *Gini coefficient*?
2 Describe the global variation in the Gini coefficient shown in Figure 14.40.
3 What does the Lorenz curve in Figure 14.41 show?
4 Comment on the findings of the 2008 OECD report.

☐ Theory of regional disparities

The Swedish economist Gunnar Myrdal produced his **cumulative causation** theory in 1957. Figure 14.42 is a simplified version of the model Myrdal produced. Cumulative causation theory was set in the context of LICs but the theory can also be applied reasonably to more advanced nations. According to Myrdal, a three-stage sequence can be recognised:

- the pre-industrial stage, when regional differences are minimal
- a period of rapid economic growth, characterised by increasing **regional economic divergence**

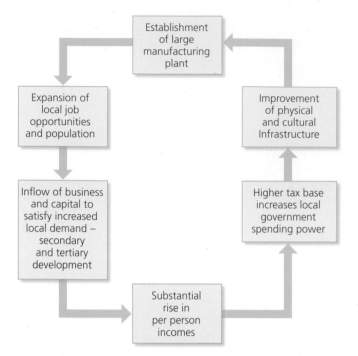

Figure 14.42 Simplified model of cumulative causation

- a stage of **regional economic convergence**, when the significant wealth generated in the most affluent region(s) spreads to other parts of the country.

Figure 14.43 shows how the regional economic divergence of the earlier stages of economic development can eventually change to regional economic convergence.

In Myrdal's model, economic growth begins with the location of new manufacturing industry in a region with a combination of advantages greater than elsewhere in the country. Once growth has been initiated in a dominant region, spatial flows of labour, capital and raw materials develop to support it and the growth region undergoes further expansion by the cumulative causation process. A detrimental 'backwash effect' is

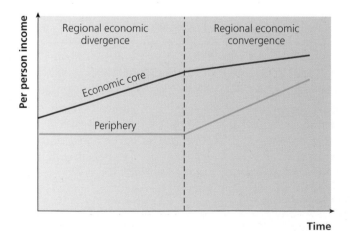

Figure 14.43 Regional economic divergence and convergence

Figure 14.44 A village in eastern Siberia – the standard of living in most parts of Asiatic Russia (the periphery) is lower than in European Russia

transmitted to the less developed regions as skilled labour and locally generated capital is attracted away. Manufactured goods and services produced and operating under the scale economies of the economic 'heartland' flood the market of the relatively underdeveloped 'hinterland', undercutting smaller-scale enterprises in such areas (Figure 14.44).

However, increasing demand for raw materials from resource-rich parts of the hinterland may stimulate growth in other sectors of the economy of such regions. If the impact is strong enough to overcome local **backwash effects** a process of cumulative causation may begin, leading to the development of new centres of self-sustained economic growth. Such **spread effects** are spatially selective and will only benefit those parts of the hinterland with valuable raw materials or other significant advantages.

The American economist Hirschman (1958) produced similar conclusions to Myrdal, although he adopted a different terminology. Hirschman labelled the growth of the **economic core region** (heartland) as 'polarisation', which benefited from 'virtuous circles' or upward spirals of development, whereas the **periphery** (hinterland) was impeded by 'vicious circles' or downward spirals. The term 'trickle-down' was used to describe the spread of growth from core to periphery. The major difference between Myrdal and Hirschman was that the latter stressed to a far greater extent the effect of counterbalancing forces overcoming polarisation (backwash), eventually leading to economic equilibrium being established. The subsequent literature has favoured the terms 'core' and 'periphery' rather than Myrdal's alternatives.

1 Suggest reasons why income disparities are narrowing in some countries but getting wider in others.
2 Define the terms **a** *economic core region* and **b** *periphery*.
3 Explain in your own words the process shown in Figure 14.42.
4 Describe and explain the trends shown in Figure 14.43.
5 What is the evidence in Figure 14.44 that this region is part of the economic periphery of Russia?

☐ Factors affecting internal disparities

Residence

Where people are born and where they live can have a significant impact on their quality of life (Figure 14.45). The focus of such study has been mainly on:

- regional differences within countries
- urban/rural disparities
- intra-urban contrasts.

Figure 14.45 Manholes in Ulaanbaatar, Mongolia – people in poverty sometimes live down these manholes because they provide access to the underground hot-water pipes that can provide warmth in the harsh winters

Case Study: Regional contrasts in Brazil

South-east Brazil (Figure 14.46) is the economic core region of Brazil. Over time, the south-east region has benefited from spatial flows of raw materials, capital and labour (Figure 14.47a). Capital and labour have come from abroad as well as from internal sources. The region grew rapidly through the process of cumulative causation. This process not only resulted in significant economic growth in the core, but also had a considerable negative impact on the periphery. The overall result was widening regional disparity. However, more recently some parts of the periphery, with a combination of advantages above the level of the periphery as a whole, have benefited from spread effects (trickle-down) emanating from the core (Figure 14.47b). Such spread effects are spatially selective and may be the result of either market forces or regional economic policy or, as is often the case, a combination of the two. The south region has been the most important recipient of spread effects from the south east, but the other regions have also benefited to an extent. This process has caused the regional gap to narrow at times, but often not for very long. However, in Brazil income inequality still remains very wide, although the gap has narrowed somewhat in recent years.

The south east's primary, secondary, tertiary and quaternary industries generate large amounts of money for Brazil. The natural environment of the south east provided the region with a number of advantages for the development of primary industries:

- The warm temperature, adequate rainfall and rich terra rossa soils (weathered from lava) have provided many opportunities for farming. The region is important for coffee, beef, rice, cacao, sugar cane and fruit.
- Large deposits of iron ore, manganese and bauxite have made mining a significant industry. Gold is still mined.
- The region is energy rich, with large deposits of oil and gas offshore. Hydro-electric power is generated from large rivers flowing over steep slopes.

Figure 14.46 South-east Brazil

- The temperate rainforest provides the raw material for forestry.
- Fishing is important for many of the coastal settlements.

The south east is the centre of both foreign and domestic investment in manufacturing industry. In the 1950s and 1960s, the government wanted Brazil to become a NIC. Because the south east had the best potential of all Brazil's five regions, investment was concentrated here. The region is the focus of the country's road and rail networks. It has the country's main airports and seaports. It also has a significant pipeline network for oil and gas. More TNCs are located in the south east than in the rest of Brazil. With the highest population density in Brazil, the labour supply is plentiful. The region also has the highest educational and skill levels in the country (Table 14.7).

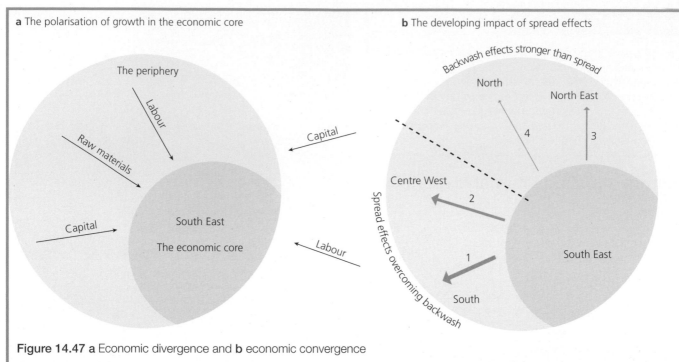

a The polarisation of growth in the economic core

The periphery

Labour

Raw materials

Capital

Capital

Labour

South East

The economic core

b The developing impact of spread effects

Backwash effects stronger than spread

North

North East

4

3

Centre West

2

Spread effects overcoming backwash

1

South

South East

Figure 14.47 **a** Economic divergence and **b** economic convergence

Table 14.7 The population of Brazil's five regions, Census 2010 (millions)

South east	84.4
North east	53.0
South	27.7
North	16.0
Centre west	14.4

The car industry is a major activity in the region. Most of the world's large car makers are here, including Ford, GM, Toyota, VW and Fiat. Other manufacturing industries include food processing, textiles, furniture, clothing, printing, brewing and shoemaking. The raw materials located in the region and the large market have provided favourable conditions for many of these industries. However, cheaper imports of shoes, clothes and textiles from Asia have led to a number of companies in the region closing.

São Paulo is by far the largest financial centre in South America. The headquarters of most Brazilian banks are in São Paulo. Most major foreign banks are also located there. This is not surprising as Brazil dominates the economy of South America, and São Paulo is the largest city in South America. The south east is the centre of research and development in both the public and private sectors; 80 kilometres from São Paulo is São José dos Campos, where the Aerospace Technical Centre is located. It conducts teaching and research and development in aviation and outer-space studies. Many people would be surprised to know that aircraft and aircraft parts make up Brazil's largest export category.

The success of the first large wave of investment by foreign TNCs in the south east encouraged other TNCs to follow suit. For the last 50 years, the south east has experienced an upward cycle of growth (cumulative causation).

Figure 14.48 shows variations in GDP per person by state in 2014. The range is considerable, with a very clear concentration of low GDP per person in the north-east region, with the highest figures in the south east and south.

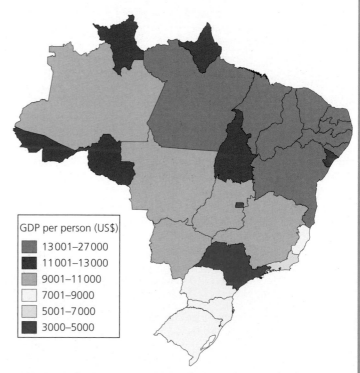

GDP per person (US$)

- 13 001–27 000
- 11 001–13 000
- 9001–11 000
- 7001–9000
- 5001–7000
- 3000–5000

Figure 14.48 Brazilian states by GDP per person, 2014

Intra-urban variations: the growth of slums and urban poverty

Residence as a factor in inequality within countries can also be examined at a more detailed scale (Figures 14.49 and 14.50). The focus of such analysis has been on intra-urban variations and the large number of people living in slum housing: 32 per cent of the world's urban population – almost 1 billion people – are housed in slums, with the great majority living in LICs. A **slum** is a heavily populated urban area characterised by substandard housing and squalor. However, virtually all large cities in HICs also have slum districts. The UN recognises that the focus of global poverty is moving from rural to urban areas, a process known as the **urbanisation of poverty**. Without significant global action, the number of slum dwellers will double over the next 30 years. The urban poor live in inner-city slums, peripheral shanty towns and in almost every other conceivable space, such as on pavements, traffic roundabouts, under bridges and in sewers.

Figure 14.49 *Favela* in São Paulo

Figure 14.50 The middle-income Jardins district of São Paulo

The numbers of people living in urban poverty are increased by a combination of economic problems, growing inequality and population growth, particularly growth due to in-migration (Figure 14.51). As 'The Challenge of Slums' (UN HABITAT, 2003) states: 'Slums result from a combination of poverty or low incomes with inadequacies in the housing provision system, so that poorer people are forced to seek affordable accommodation and land that become increasingly inadequate.' The report identifies women, children, widows and female-headed households as the most vulnerable among the poor. In urban African slums, women head over 30 per cent of all households.

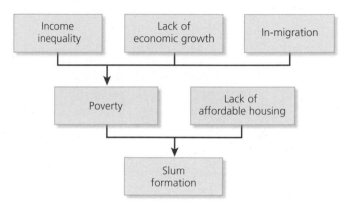

Figure 14.51 Slum formation

'The Challenge of Slums' groups the dimensions of urban poverty as follows:

- **Low income** – consisting of those who are unable to participate in labour markets and lack other means of support, and those whose wage income is so low that they are below a nominal poverty line.
- **Low human capital** – low education and poor health; health 'shock' in particular can lead to chronic poverty.
- **Low social capital** – this involves a shortage of networks to protect households from shocks, weak patronage on the labour market, labelling and exclusion; this particularly applies to minority groups.
- **Low financial capital** – lack of productive assets that might be used to generate income or avoid paying major costs.

Figure 14.52 sums up the constituents of urban poverty. The complexities of urban poverty indicate how difficult it is for individuals to improve their socio-economic situation. In many countries, social mobility has become more difficult rather than easier in recent times.

- Inadequate income (and thus inadequate consumption of necessities including food and, often, safe and sufficient water; often problems of indebtedness, with debt payments significantly reducing income available for necessities).

- Inadequate, unstable or risky asset base (non-material and material including educational attainment and housing) for individuals, households or communities.

- Inadequate shelter (typically poor quality, overcrowded and insecure).

- Inadequate provision of 'public' infrastructure (e.g. piped water, sanitation, drainage, roads, footpaths) which increases the health burden and often the work burden.

- Inadequate provision for basic services such as daycare/ schools/ vocational training, healthcare, emergency services, public transport, communications, law enforcement.

- Limited or no safety net to ensure basic consumption can be maintained when income falls; also to ensure access to shelter and healthcare when these can no longer be paid for.

- Inadequate protection of poorer groups' rights through the operation of the law, including laws and regulations regarding civil and political rights, occupational health and safety, pollution control, environmental health, protection from violence and other crimes, protection from discrimination and exploitation.

- Voicelessness and powerlessness within political systems and bureaucratic structures, leading to little or no possibility of receiving entitlements.

Source: Paul Guinness, 'Slum Housing Global Patterns Case Studies'
Geo Factsheet No 180, Curriculum Press

Figure 14.52 The constituents of urban poverty

Section 14.3 Activities

1 Explain the processes illustrated in Figure 14.47.
2 Produce a graph to illustrate the regional breakdown of Brazil's population (Table 14.7).
3 Examine the factors that lead to the formation of slums in LICs (Figure 14.51).
4 Compare Figures 14.49 and 14.50, which show different residential districts in São Paulo.
5 Discuss the constituents of urban poverty shown in Figure 14.52.

Ethnicity and employment

The development gap often has an ethnic and/or religious dimension whereby some ethnic groups in a population have income levels significantly below the dominant group(s) in the same population. This is often the case with **indigenous populations**. It is invariably the result of discrimination, which limits the economic, social and political opportunities available to the disadvantaged groups. Examples include South Africa, Indonesia and Bolivia. Because of such obvious differences in status, tensions can arise between majority and minority groups, resulting in:

- social unrest
- migration
- new political movements.

In South Africa, the wide gap in income originated in the apartheid era, but since then it has proved extremely difficult to close for a variety of reasons. Political change often occurs well in advance of significant economic and social change. Figure 14.53, using data from the 2011 Census, shows that average annual household income for black people was R60613 ($4400). This was about one-sixth that of white households, and a quarter of that of Asian households.

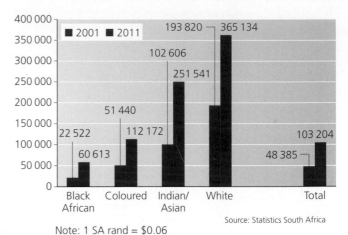

Note: 1 SA rand = $0.06

Source: Statistics South Africa

Figure 14.53 Average annual household income (SA rand) by population group of household head in South Africa

Inequality of wealth distribution is higher in Latin America than in any other part of the world. Indian and black people make up a third of the population, but have very limited parliamentary representation. Figure 14.54 shows the situation in five Latin American countries in 2005, prior to political transformation in Bolivia. The changes that have occurred in Bolivia have given hope to indigenous peoples elsewhere in Latin America. Such ethnic differences often have a strong regional component, as ethnic groups tend to concentrate in certain rural and urban areas.

Indians and black people – poorly represented in parliament

In Ecuador, Guatemala and Peru, indigenous people make up 34–60 per cent of the population but have had few seats in parliament. Even in Bolivia the majority Indian population only has 26 per cent of seats – though its power to change government policy through mass protest has been growing, an alarming development for governments fearful of 'mob rule'. Part of the popular enthusiasm for Hugo Chávez – and the fear and loathing he inspires in traditional elites – arises from the fact that he is part Indian and part black, thus representing two of the most disadvantaged groups in Latin American history.

Country (ethnic group)	% of population	% representation in lower house
Bolivia (indigenous)	61	26
Ecuador (indigenous)	34	3
Guatemala	60	12
Peru (indigenous)	43	1
Brazil (African descent)	44	3

Figure 14.54 Indians and black people in South America – poorly represented in parliament

Education

Education is a key factor in explaining disparities within countries. Those with higher levels of education invariably gain better-paid employment. In LICs, there is a clear link between education levels and family size, with those with the least education having the largest families. Maintaining a large family usually means that saving is impossible and varying levels of debt likely. In contrast, people with higher educational attainment have smaller families and are thus able to save and invest more for the future. Such differences serve to widen rather than narrow disparities. Educational provision can vary significantly not just by social class, but also by region.

Brazil has a greater disparity in income levels than most other countries. An important research study in the late 1990s concluded that the main cause was the huge variation in access to education. One of the authors of the study, Ricardo Paes de Barros, stated: 'There are not two Brazils. The poor and the rich live together in the same cities. They often work in the same multinational companies. The problem is that their educational background is absurdly unequal, and this results from the very poor quality of the public basic education system.' The report concluded that educational attainment explains 35–50 per cent of income inequality.

Figure 14.55 shows the population of secondary education or higher by region. The highest percentage is in the south east and the lowest in the north east. In the SAEB (Portuguese and mathematics) scores for students

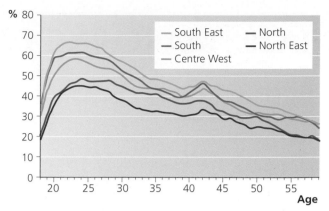

Figure 14.55 Percentage population of secondary education or higher in Brazil, by region

completing secondary education, the scores for 2005 had a regional ranking of:

1 South   **3** Centre West   **5** North
2 South East   **4** North East

Land ownership (tenure)

The distribution of land ownership has had a major impact on disparities in many countries. It can have a significant regional component. The greatest disparities tend to occur alongside the largest inequities in land ownership. The ownership of even a very small plot of land provides a certain level of security that those in the countryside without land cannot possibly aspire to.

Case Study: Brazil – inequities in land ownership

The distribution of land in Brazil in terms of ownership has been a divisive issue since the colonial era. Then, the monarchy rewarded those in special favour with huge tracts of land, leaving a legacy of highly concentrated ownership. For example, 44 per cent of all arable land in Brazil is owned by just 1 per cent of the nation's farmers (Figure 14.56), while 15 million peasants own little or no land. Many of these landless people are impoverished roving migrants who have lost their jobs as agricultural labourers due to the spread of mechanisation in virtually all types of agriculture.

At least a partial solution to the problem is **land reform** (Figure 14.57). This involves breaking up large estates and redistributing land to the rural landless. Although successive governments have vowed to tackle the problem, progress has been limited due to the economic and political power of the big fazenda or farm owners, who have not been slow to use aggressive tactics (legal or otherwise) to evict squatters and delay expropriation.

In the mid-1990s, land reform clearly emerged as Brazil's leading social problem, highlighted by a number of widely publicised squatter invasions. Such land occupations have occurred in both remote regions and established, prosperous farmlands in the south and south east. Each year in April, the Landless Rural Workers' Movement (known as the 'MST')

Figure 14.56 Crop production in Brazil

organises a series of land invasions, takeovers of buildings and other protests. The purpose is twofold:

■ to keep the issue high on the national political agenda
■ to commemorate the killing in 1996 of 19 landless protesters by police in the state of Para.

$\Rightarrow$

BRAZIL: MARCHING FOR REAL LAND REFORM

By Fabiana Frayssinet

RESENDE, Brazil, Aug 12 (IPS) – After 10 years of waiting for secure title to the land they occupy and farm, 35 families in Resende, in the southeastern Brazilian state of Rio de Janeiro, have joined a huge march organized by the Landless Workers Movement (MST) in Brasilia to demand effective agrarian reform.

Mario Laurindo knows all about protest demonstrations. Some 14 years ago, he and others in the MST set up a roadside camp and were evicted. For the past 10 years he and his family have lived in the "Terra Libre" (Free Land) settlement, 176 kilometres from the city of Rio de Janeiro, the state capital.

"We may grow old in the attempt, but we will continue the struggle," Laurino told IPS. A long time ago, he left the 'favela' (shantytown) where he lived, because he had no job, food or health care, and wanted to escape the high levels of urban violence.

Now, at least, he has plenty of food. With his wife and two children – they had two more, but they died – the family produces enough to subsist on, from honey to bananas. They also keep chickens and a few dairy cows.

Like other families in the settlement, Laurindo sells his surplus produce at a nearby town where he goes every day, crossing a river on boats built by another neighbour. Barter with other settlers complements the family diet.

"I'll never work for someone else again. Now I'm my own boss," says Laurindo, who has taken up the way of life of a small farmer and ekes out the family income with odd jobs such as bricklaying, but always on a self-employed basis, he stresses.

Like Osvaldo Cutis, a teacher and the spokesman for Terra Libre, Laurindo shares the goals of the settlements and of the MST, which is mobilizing 3,000 of its activists in Brasilia from Aug. 10–19.

The demonstration in the capital, which included marches, debates, cultural events and other activities, is an effort by the MST to put pressure on the government of leftwing President Luiz Inàcio Lula da Silva to distribute land within the next six months to at least 90,000 families who have been squatting in different parts of the country since 2003, many of them camping by the roadside.

The landless movement has carried out land occupations for the past 25 years "calling for fulfilment of law," Cutis told IPS. It also seeks better living conditions for another 45,000 families "who have been resettled on paper only," and are "suffering hardship" because they are still waiting for resources for housing, infrastructure and production, he said.

People in the Terra Libre settlement are all too familiar with this situation, The state Institute for Agrarian Reform (INCRA) has not legalized their ownership of the land where they have lived "on a temporary basis" for over a decade because of red tape and endless battles over inheritance and compensation for expropriation in the courts.

Terra Libre occupies 460 hectares of an old estate, which was deemed unproductive according to offi cial criteria set out in the law on agrarian reform, and which owed its workers the equivalent of one million dollars before it was taken over by the MST.

The problem is that until they have legal title to the land, the settlers do not have access to credits and tools from INCRA. But according to MST, many families to whom the government has already granted title deeds have not yet received this assistance.

"It's hard to convince some farmers to put effort and work into a plot of land that they might be forced to leave tomorrow," Curtis said.

Figure 14.57 Marching for real land reform

1 Summarise the data illustrated in Figure 14.53.
2 Describe the data presented in Figure 14.54.
3 With reference to Figure 14.55, explain how variations in educational attainment affect regional development.

4 Write a 100-word summary of the article on land reform in Brazil (Figure 14.57).
5 Suggest how land tenure can have an impact on regional disparities.

14.4 The management of regional development

Case Study: Canada

Canada is the second largest country in the world in land area after Russia. It is comprised of ten provinces and three territories (Figure 14.58) – that it crosses six time zones is a good indication of its size. Its greatest east–west expanse is 5200 kilometres; the greatest north–south distance is 4600 kilometres. For a country with such a large land area, Canada has a small population of only 35.5 million and huge variations in population distribution and density. The Canadian northlands make up that part of Canada lying north of 55 °N. Virtually the whole of this area has a population density of less than one person per km².

Figure 14.58 Map of Canada's 10 provinces and 3 territories

Canada is a country of major environmental contrasts, both east to west and north to south. As European settlers imposed themselves on the indigenous population, some regions offered far greater opportunities for settlement and economic activity than others, thus it is not surprising that regional differences in human well-being soon became apparent. Such differences have persisted to the present time, although over time they have lessened significantly in nature.

Much has been written on the causes and consequences of core–periphery contrasts in Canada. The urban–industrial core of the country is in the southern regions of Ontario and Quebec, where climatic conditions for agriculture and other activities are at their best. Canada's two main urban areas are Toronto and Montreal. Other important urban areas in this core region are Ottowa (the federal capital), Quebec, London and Windsor. All of these urban areas are located on or relatively close to the Great Lakes/Saint Lawrence Seaway – the system of locks, canals and channels that allow ocean-going ships to travel from Lake Superior to the Atlantic Ocean (Figure 14.59). The economic importance of this major routeway, both in terms of Canada's economic history and its economy today, cannot be underestimated. Canada's two other major routeways – the Trans-Canada Highway and Transcontinental railway system (Figure 14.60) are also of major economic importance. Table 14.8 shows all the metropolitan areas in Canada with a population of over 1 million.

Figure 14.59 Toronto, located beside Lake Ontario

Figure 14.60 Transcontinental railway travelling from Toronto to Vancouver at a stop in Manitoba

Table 14.8 The largest urban areas in Canada, 2011

Metropolitan area	Population (millions)
Toronto (Ontario)	5.58
Montreal (Quebec)	3.82
Vancouver (British Columbia)	2.31
Ottawa (Ontario)/Gatineau (Quebec)	1.24
Calgary (Alberta)	1.21
Edmonton (Alberta)	1.16

Source: Statistics Canada

A major urban sub-core is the Vancouver city region. Vancouver is Canada's third largest metropolitan area and its main port on the Pacific Ocean. The port is Canada's largest and most diversified. Vancouver ranks consistently as one of the highest quality of life urban areas in the world. It hosted the 2010 Winter Olympics and Paralympics.

The incidence of above-average incomes elsewhere in Canada, notably in Alberta, is due primarily to natural resource endowment. Also, with such a large proportion of the country located in the cold, inhospitable northlands, it is not surprising that Canada's core regions have a southern spatial bias.

In the literature on regional disparities in Canada, the main causal factors identified are:

- the huge size of the country
- the great contrasts in physical environment
- the uneven access to resources
- major contrasts in population density and distribution
- significant differences in secondary and tertiary employment opportunities
- large variations in the provision of transport infrastructure.

Table 14.9 presents a brief regional economic profile of Canada's provinces and territories, while Table 14.10 shows the current extent of reginal disparities in Canada.

Table 14.9 Regional profiles

Region	Characteristics
Ontario	The total GDP of Ontario is almost double that of Quebec, the next province in the economic rankings. Ontario is the industrial powerhouse of Canada. However, in recent decades Ontario has suffered from the negative impact of unstable US growth on the province's manufacturing sector, the global shift of some of its traditional industries to MICs, high fiscal deficits and rising debt. Such factors have restricted investment and innovation.
Quebec	Quebec's economy has been affected by problems similar to those in Ontario. Investment in Quebec has also been adversely affected by the political debate within the province regarding independence from the rest of Canada.
The Atlantic Provinces	The four Atlantic provinces have long been perceived as Canada's problem region. Three of these provinces – Nova Scotia, New Brunswick and Prince Edward Island – have the lowest per person incomes in the country. They have had weaker growth in GDP than the rest of the country for decades. Until the early 2000s, the remaining Atlantic province – Newfoundland and Labrador – was in a similar situation. However, the recent development of oil reserves has boosted per person incomes.
The Prairie Provinces	The Prairie provinces of Manitoba, Saskatchewan and Alberta form Canada's rich agricultural heartland. Three-quarters of Canada's farmland lies in these provinces. The hot, dry summers favour wheat, but other farming activities include dairying, cattle ranching, feed grains and oilseed crops. Alberta also benefits from considerable energy resources that have resulted in a large increase in investment, employment and income growth. Between 1981 and 2013, Alberta's population increased by 76 per cent, compared to the Canadian average of 42 per cent. Alberta's total GDP is over twice that of the combined GDP of Saskatchewan and Manitoba.
British Columbia	British Columbia has the fourth largest economy in Canada. Its economy is largely resource-based. Vancouver, Canada's third largest urban area, is a major Pacific port and the terminus of the transcontinental railway. Vancouver is the headquarters of many western-based natural resource companies. Although only 5 per cent of the province is arable land, farm productivity is high. The climate and landscape encourage tourism (Figure 14.61).
The territories	The three territories are very sparsely populated. Their decision-making powers are those delegated to them by the federal government. Their combined GDP is only about 0.5 of the national total. Per person incomes are high, particularly in Northwest territories and the Yukon, because their economies are largely resource based. The economy of Nunavut is not as developed as the other two territories.

Figure 14.61 Tourists at Lake Louise, British Columbia

Table 14.10 Regional disparities

Region	GDP 2014 (US $ million)	GDP per person (US $), 2014	Unem-ployment rate, 2013 (%)	Human Development Index, 2011
Canada	1 420 320	39 960	7.1	0.908
Newfoundland & Labrador	24 130	45 600	11.4	0.894
PEI	4 320	29 560	11.5	0.877
Nova Scotia	28 130	29 850	9.0	0.886
New Brunswick	23 080	30 580	10.4	0.882
Quebec	266 400	32 430	7.6	0.903
Ontario	519 720	38 000	7.5	0.913
Manitoba	46 130	36 030	5.4	0.885
Saskatchewan	59 590	53 100	4.0	0.898
Alberta	270 490	65 640	4.6	0.917
British Columbia	170 740	36 810	6.6	0.91
Northwest territories	3 410	77 400	6.8	0.911
Yukon	1 870	50 640	7.1	0.889
Nunavut	1 790	49 590	13.4	0.820

Regional development policies

For over half a century, Canada has had an explicit regional development policy whereby the federal government has been committed to reducing as far as possible the differences in living standards between its provinces and territories.

Prior to the 1950s, no explicit federal regional policy had been pursued in Canada, although certain programmes such as the Prairie Farm Rehabilitation Act had firm regional implications. The first direct effort to compensate for regional disparities was the equalization program established in 1957. A number of phases in the development of regional development policy have been identified:

■ Phase 1: the equalization program

Equalisation is based on the concept that the federal system should enable every province to provide services of average Canadian standards to its population without having to impose heavier than average tax burdens. Although equalization remains an integral part of the federal system, critics argue that it is not a regional development programme in the true sense, in that payments are not conditional on development use of the funds. Since 1957, the annual equalisation transfers to low-income provinces have risen from US$100 million to $12 billion in 2013–14.

In 2013–14, six provinces received equalization payments from the federal government as follows: Quebec $5.6 billion, Ontario $2.3 billion, Manitoba $1.3 billion, New Brunswick $1.1 billion, Prince Edward Island $0.3 billion. The following provinces did not qualify for equalization payments: Alberta; Saskatchewan; Newfoundland and Labrador; and British Columbia.

■ Phase 2: the introduction of regional incentives

This second phase of policy was initiated by a New Products Program for surplus manpower areas, which began in 1960 to help areas of high unemployment and slow economic growth. The scheme allowed firms to obtain double the normal rate of capital cost allowances on most of the assets acquired to manufacture products that were new to designated areas. This use of tax incentives mirrored existing schemes in a number of European countries. The Agriculture and Rural Development Act (ARDA) of 1961 was designed to alleviate the high incidence of low incomes in rural areas through federal–provincial programmes to increase small farmers' output and productivity.

A totally area-specific scheme was established by the Atlantic Development Board (ADB) in 1962 to improve the economic structure of the Atlantic provinces, the poorest part of Canada. A similar agency was established in Quebec in the form of the Eastern Quebec Development Board.

■ Phase 3: the Department of Regional Economic Expansion

DREE was established in 1969 to assist the various regions to realise their economic and social potential and to provide the national coordination frequently lacking from earlier schemes. The emphasis was placed on areas that had the potential for significant economic growth. In 1972, a new level of federal–provincial cooperation emerged with the introduction of General Development Agreements (GDAs). These ten-year programmes covered a wide range of development projects.

■ Phase 4: the Department of Regional Industrial Expansion

During the 1970s, DREE's regional development approach was increasingly considered to be too restricted in scope. In 1982, a new strategic approach was announced. The Department of Regional Industrial Expansion (DRIE) was set up, merging the regional programmes of the existing DREE with the industry, small firms and tourist components of the Department of Industry, Trade and Commerce. The revised federal–provincial agreements became known as Economic and Regional Development Agreements.

■ **Phase 5: new policy directions**

There was growing recognition that, despite a variety of efforts over the previous 25 years, unacceptable levels of regional disparity continued to exist. These concerns led to a fundamental restructuring of regional development policy announced in 1986. Significantly, there was a decentralisation away from Ottawa, the federal capital, to give regional agencies the primary responsibility for development within their local area. In 1987, the new policy resulted in the creation of three major regional development agencies:
- Atlantic Canada Opportunities Agency (ACOA)
- Western Economic Diversification (WD)
- Federal Economic Development Initiative in Northern Ontario (FedNor).

In 2005, the Canadian Economic Development Agency for the Regions of Quebec (CED-Q) was established to carry on development activities in Quebec from previous agencies. In 2009, two new regional agencies were established to bring the total to six. These were:

- The Federal Economic Development Agency for Southern Ontario
- The Canadian Northern Economic Development Agency

Figure 14.62 shows the level of funding for four of the major development agencies. For example, Western Economic Diversification Canada recently launched the Western Innovation Initiative. This is a $70 million, five-year programme that offers repayable contributions for small and medium-sized enterprises, for developing new and innovative technologies from the later stages of research and development to the marketplace. There has been a renewed focus on rural development, particularly support for the development of rural businesses.

During this latter phase of regional development policy, there has been a general desire to move from what has generally been termed a 'top-down' strategy to one that is more 'bottom-up' in nature.

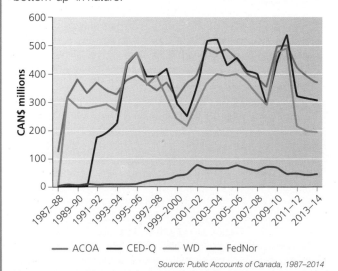

Source: Public Accounts of Canada, 1987–2014

Figure 14.62 Funding for the four original major development agencies – note that CAN$1 is about US$0.70

A general trend towards convergence

Economic data clearly shows that over the last 50 years or so there has been a broad tendency towards regional economic convergence. However, there have been periods when this general trend has been interrupted and reversed; notably:

- in the early 1980s
- from the early 2000s to the present time.

The main causal factors for these reversals have been identified as periods of rapidly rising commodity prices and periods of global economic uncertainty, particularly the global economic downturn that began in 2008. For example, high oil prices benefit the oil-producing provinces – namely Alberta, Saskatchewan and Newfoundland and Labrador – but they push up costs for the non-oil-producing provinces.

Disparities: interprovincial or urban/rural?

Urbanisation has often been put forward as an important factor to explain variations in economic development due to the high concentration of investment and innovation in large urban areas. A recent analysis of regional disparities in Canada concluded that regional disparities are as much, if not more, urban/rural than interprovincial in nature. The three largest metropolitan areas of Toronto (Ontario), Montreal (Quebec) and Vancouver (British Columbia) are by far the largest urban–industrial structures in the country. This analysis argued that:

- the absence of large metropolitan areas in a number of provinces put them at a disadvantage in both fiscal and development terms
- future regional policies should give more recognition to the importance of urbanisation in explaining disparities
- further research was needed to identify other sources of disparity, such as industrial structure
- regional differences play an important part in economic, cultural and political life
- the need for regional assistance in the form of equalization and other measures remains.

The Atlantic Provinces: Canada's main problem region

The four provinces of New Brunswick, Nova Scotia, Prince Edward Island and Newfoundland and Labrador make up the region of Atlantic Canada. Long regarded as the major problem region of the country, the provinces are characterised by slow economic growth, heavy reliance on primary industries, low per person incomes and persistently high unemployment rates. In an attempt to rectify such relative deprivation, the region has figured prominently in Canadian regional development programmes. Although this injection of federal funding has resulted in significant improvements covering many aspects of the regional economy, the Atlantic provinces still lag behind the rest of the nation according to most socio-economic indicators. It has been the recent development of oil resources that has changed the situation in Newfoundland and Labrador over the last decade or so.

Development in the Atlantic provinces has been hindered by a number of factors, particularly the paucity of natural resources, the low level of manufacturing industry and capital investment and the scattered nature of rural settlement. Although the provinces are the most rural of Canada's regions, the generally infertile soils and cool summers have restricted agricultural improvement. A consequence of such a lack of agricultural potential is the highest percentage of rural non-farm population in the country.

The principal urban nuclei of the region are the seaports of Halifax in Nova Scotia, St John's in Newfoundland and St John in New Brunswick. However, these urban areas had populations of only 390 000, 197 000 and 128 000 respectively in 2011. The region's major urban nuclei account for the lowest percentage of population for any region of the country. Such a small and dispersed population provides a very limited attraction for industries attempting to achieve economies of scale. Low capital intensity in the private sector and poor public services have often been cited as disincentives to new industry.

ACOA initiatives include investment to build industrial clusters that can anchor maritime innovation, including ocean technologies, aquaculture, bio-technology and environmental technology.

Indigenous peoples

While not a regional issue in the conventional sense, the relatively low standard of living of indigenous peoples affects some regions far more than others. The indigenous or Aboriginal peoples of Canada comprise the First Nations, Inuit and Metis. According to the 2011 census, they totalled 1.4 million; that is, 4.3 per cent of the national population. In terms of regional concentrations, the highest percentages of province or territory populations are in Nunavut (86 per cent), Northwest territories (52 per cent), Yukon (23 per cent), Manitoba (17 per cent) and Saskatchewan (16 per cent). The largest numbers of indigenous people are in Ontario (301 000), British Columbia (232 000), Alberta (220 000) and Manitoba (200 000).

The living conditions of indigenous peoples fall well below the Canadian norm. The pressure group Amnesty International has identified:

- widespread impoverishment
- inadequate housing
- food insecurity
- ill health
- unsafe drinking water.

There is considerable pressure on the federal and provincial governments to do more to raise the living standards of indigenous peoples to the Canadian norm.

Conclusion

Canada displays large regional variations that have persisted over time despite considerable government attempts to reduce these differences. However, advocates of regional development funding argue that the development gap would be much larger than it is in the absence of such funding. The general trend is towards regional economic convergence, but there have been significant interruptions to this trend.

There will always be differences of opinion about how best to spend government money. There is continued debate about whether resources should be focused on regions where growth can be created most successfully or whether it should be concentrated on the least favoured regions. Continuing research into factors such as urban/rural disparities and industrial structure should help to target funding more successfully in the future.

Section 14.4 Activities

1 **a** Where is Canada's main economic core region?
  b What are the reasons for its geographical location?
2 Suggest reasons for the development of an economic sub-core around Vancouver.
3 Comment on the regional disparities shown in Table 14.10.
4 Describe the main changes in regional policy since the 1950s.
5 Why are the Atlantic provinces considered to be Canada's number one problem region?

Appendix

Resources for the Geographical Skills Workbook

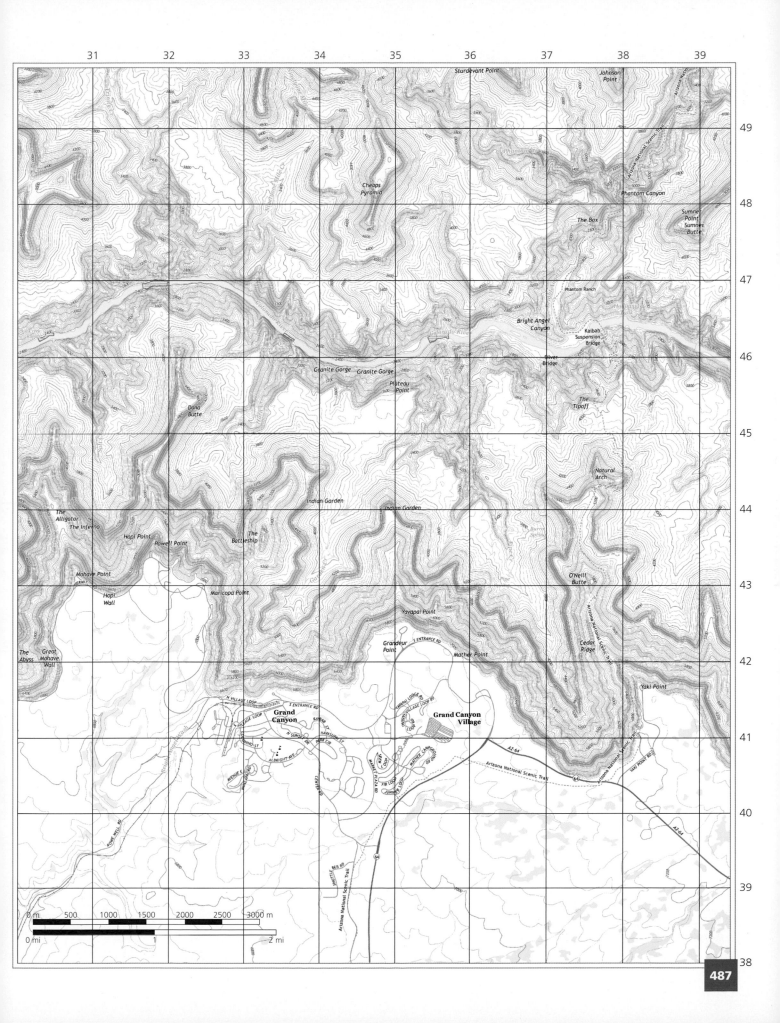

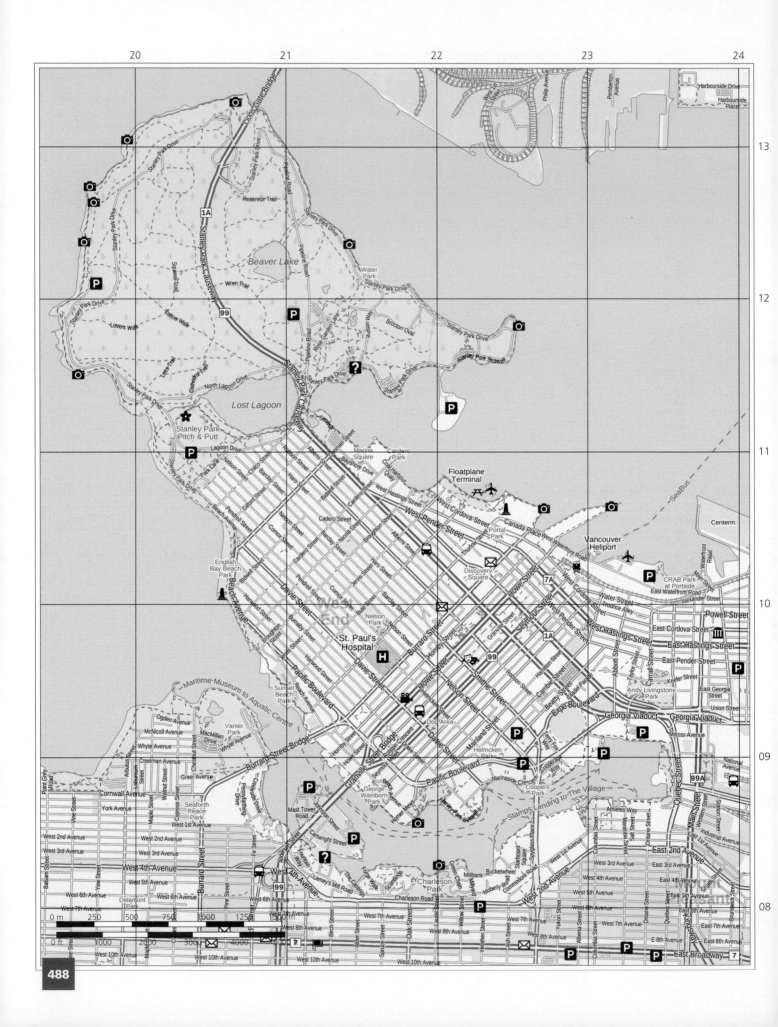

☐ Satellite images of the Aral Sea

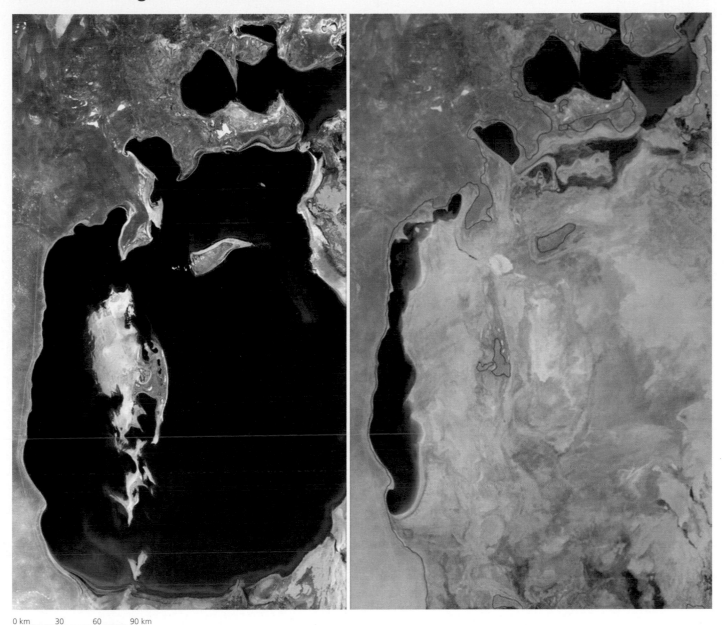

0 km   30   60   90 km

1989 (left) and 2014

Index

Acknowledgements

The publishers would like to thank the following for permission to reproduce copyright material:

Photo credits

Garrett Nagle: **p.11**, **p.12**, **p.13**, **p.17**, **p.20**, **p.22** *both*, **p.24**, **p.30**, **p.43**, **p.44** *both*, **p.46**, **p.48** *all*, **p.49**, **p.52**, **p.58** *left*, **p.62**, **p.66**, **p.67**, **p.69**, **p.72** *both*, **p.80**, **p.82** *all*, **p.200**, **p.201**, **p.202**, **p.204** *all*, **p.208**, **p.210**, **p.211**, **p.215** *both*, **p.216** *both*, **p.219** *all*, **p.220** *all*, **p.226**, **p.235**, **p.238**, **p.241**, **p.242**, **p.245**, **p.246** *both*, **p.247**, **p.248**, **p.251** *both*, **p.254**, **p.255**, **p.256** *all*, **p.257**, **p.273** *both*, **p.275**, **p.281**, **p.282** *both*, **p.283** *all*, **p.288**, **p.292**, **p.294**, **p.302**, **p.303** *both*, **p.304** *both*, **p.305**, **p.313**, **p.319**, **p.327**, **p.333** *all*, **p.334**, **p.335**

Paul Guinness: **p.85**, **p.88** *bottom*, **p.89**, **p.90**, **p.93** *both*, **p.94**, **p.98**, **p.102**, **p.103**, **p.104**, **p.105**, **p.107**, **p.112**, **p.113**, **p.116**, **p.118**, **p.119**, **p.120**, **p.123**, **p.125**, **p.126**, **p.128**, **p.131**, **p.132**, **p.134**, **p.137**, **p.141**, **p.143**, **p.144**, **p.145**, **p.151**, **p.153**, **p.156**, **p.157**, **p.158** *right*, **p.159**, **p.161**, **p.163** *both*, **p.167** *both*, **p.168** *both*, **p.170**, **p.171**, **p.172**, **p.174**, **p.175**, **p.177** *both*, **p.179**, **p.182**, **p.185**, **p.338** *left and top right*, **p.339**, **p.340**, **p.341**, **p.343**, **p.348**, **p.356**, **p.357**, **p.358** *both*, **p.359**, **p.362**, **p.363**, **p.372**, **p.373**, **p.376**, **p.378**, **p.380**, **p.381**, **p.384** *both*, **p.385**, **p.396**, **p.397** *both*, **p.399**, **p.403**, **p.404**, **p.413**, **p.416**, **p.417**, **p.418**, **p.420**, **p.424**, **p.427**, **p.429**, **p.432**, **p.434**, **p.435**, **p.436**, **p.437**, **p.439**, **p.441** *both*, **p.442**, **p.443** *both*, **p.450**, **p.454**, **p.455**, **p.457**, **p.458**, **p.463** *both*, **p.464**, **p.465**, **p.466**, **p.472**, **p.474**, **p.475**, **p.477** *both*, **p.479**, **p.481**, **p.482**, **p.483**

Chris Guinness: **p.84**, **p.91**, **p.95**, **p.110**, **p.114**, **p.122**, **p.129**, **p.158** *left*, **p.338** *bottom right*, **p.366**, **p.391**, **p.402**, **p.456**

p.25 © Peter Essick – Getty Images; **p.26** © Planet Observer – Getty Images; **p.58** *left* © CHOI JAE-KU/AFP/Getty Images; **p.88** *top* © The World FactBook (CIA), **p.100** © H. Mark Weidman Photography/Alamy Stock Photo; **p.146** © Dabrowski/Keystone/Rex Features, **p.183** © Soeren Stache/dpa/Corbis; **p.193** *both* © NASA, **p.205** © Schafer & Hill/Getty Images; **p.225** © Paula French/REX Shutterstock; **p.260** © RSMcLeod – Thinkstock/Getty Images; **p.287** © United States Geological Survey Publications; **p.344** © Paul Nevin/The Image Bank/Getty Images; **p.346** © Frédéric Soltan/Sygma/Corbis; **p.350** © Eye Ubiquitous/Alamy; **p.351** © Ariana Cubillos/AP/Press Association Images; **pp.354 and 355** © C Goodwin/Courtesy of Wikipedia; **p.368** © Bloomberg via Getty Images; **p.387** © Ashley Cooper/Alamy Stock Photo; **p.390** © DigitalGlobe via Getty Images; **p.392** © Esiri/Reuters/Corbis; **p.395** © US Coast Guard Photo/Alamy Stock Photo; **p.408** © Ikiwaner via Wikipedia Commons (www.gnu.org/licenses/old-licenses/fdl-1.2.html); **p.425** © Kim Naylor/Divine Chocolate; **p.426** © Christian Aid (www.christianaid.org.uk); **p.441** *bottom right* © Caledonian Sky/Ohyo; **p.447** © International Photobank/Alamy; **p.448** © Negril Coral Reef Preservation Society; **p.453** © aberenyi/Fotolia.com; **p.489** © NASA. Collage on Wikipedia by Producercunningham

Acknowledgements

Artwork: **p.13** *left* A. Goudie and R. Gardner, Waterfall formation, from *Discovering Landscape in England and Wales* **p.15** *bottom* A. Goudie, River delta shapes related to river, wave and tidal processes, from *The Nature of the Environment* (Blackwell, 1993), reproduced by permission of the publisher; pp. 18–19 'High Force: a remarkable cataract' © A.S. Goudie and R.A.M. Gardner 1992, reproduced with permission of Springer; **p.19** *bottom right* Phillippe Rekacewicz, The economic impacts of the shrinking sea, from An Assassinated Sea, in *Histoire-Géographie, initiation économique*, page 333, Classe de Troisième, Hatier, Paris, 1993 (data updated in 2002); L'état du Monde, 1992 and 2001 editions, La Découverte, Paris; **p.27** *top left* Hofer, T. and Messerli, B., The Ganges Drainage Basin, from *Floods in Bangladesh: History, Dynamics and Rethinking the Role of the Himalayas* (United Nations University Press/FAO 2006) and *Philip's Interactive Modern School Atlas* (Hodder Education, 2006) © Philip's; **p.27** *bottom right* Hofer, T. and Messerli, B., Major impacts of the 1998 floods, from *Floods in Bangladesh: History, Dynamics and Rethinking the Role of the Himalayas* (United Nations University Press/FAO 2006); **p.33** University of Oxford, 1989, Entrance examination for Geography; **p.34** *right* G. Nagle, Latitudinal contrasts in insolation, from *GCSE Geography Through Diagrams* (ORG) (Oxford University Press, 1998), © Garrett Nagle, reproduced by permission of the publisher; **p.35** R. Barry & R. Chorley, Contrasts in insolation by season and latitude, from *Atmosphere, Weather and Climate* (Routledge, 1998), reproduced by permission of the publisher; **p.36** D. Briggs et al., Seasonal temperature patterns, from *Fundamentals of the Physical Environment* (Routledge, 1997), reproduced by permission of the publisher; **p.37** *left* R. Barry & R. Chorley, Variations in pressure, from *Atmosphere, Weather and Climate* (Routledge, 1998), reproduced by permission of the publisher; **p.37** *right* E. Linacre & B. Geerts, Surface winds, from *Climates and Weather Explained* (Routledge, 1997), reproduced by permission of the publisher; **p.39** *right* G. Nagle, The effects of the North Atlantic Drift/Gulf Stream, from *GCSE Geography Through Diagrams* (ORG) (Oxford University Press, 1998), © Garrett Nagle, reproduced by permission of the publisher; **p.40** R. Barry & R. Chorley, The ocean conveyor belt, from *Atmosphere, Weather and Climate* (Routledge, 1998), reproduced by permission of the publisher; **p.41** *right* E. Linacre & B. Geerts, The Coriolis force, from *Climates and Weather Explained* (Routledge, 1997), reproduced by permission of the publisher; **p.44** *right* D. Briggs et al., Maximum vapour pressure, from *Fundamentals of the Physical Environment* (Routledge, 1997), reproduced by permission of the publisher; **p.46** G. Nagle, Types of precipitation, from *GCSE Geography Through Diagrams (ORG)* (Oxford University Press, 1998), © Garrett Nagle, reproduced by permission of the publisher; **p.55** *top* D. Briggs et al., The urban heat island (Chester, UK), from *Fundamentals of the Physical Environment* (Routledge, 1997), reproduced by permission of the publisher; **p.55** *bottom* D. Briggs et al., The effect of terrain roughness on wind speed, from *Fundamentals of the Physical Environment* (Routledge, 1997), reproduced by permission of the publisher; **pp.56–57** Met Office, London's heat-island effect and Variations in rainfall around London, from *National Meteorological Library and Archive Fact sheet 14 — Microclimates*. Contains public sector information licensed under the Open Government Licence v2.0. (http://www.nationalarchives.gov.uk/doc/open-government-licence/version/2/); **p.81** Hong Kong Special Administrative Region 1:100 000 Scale Orthophoto Map © Survey and Mapping Office, Lands Department/The Government of the Hong Kong Special Administrative Region; **p.135** *bottom* City of Toronto Planning & Development Department, Toronto – changing social structure in a growing city, from *Toronto in Transition* (April 1980); **p.152** P. J. Cloke, An index of rurality for England and Wales, from *Regional Studies*, Volume 11 (1977) © Regional Studies Association, reprinted by permission of Taylor & Francis Ltd, www.tandfonline.com on behalf of Regional Studies Association; **p.164** United Nations, World urban population, *2014 World Urbanization Prospects: The 2014 Revision*; **p.186** Map of the Cairo metro, from http://www.urbanrail.net/af/cairo/cairo-map.gif 2014 © UrbanRail.net; **p.188** D. Briggs et al., Air masses, from *Fundamentals of the Physical Environment* (Routledge, 1997), reproduced by permision of the publisher; **p.189** *right* E. Linacre & B. Geerts, ITCZ and surface winds, from *Climates and Weather Explained* (Routledge, 1997), reproduced by permission of the publisher; **p.195** R. Barry & R. Chorley, Köppen's climate classification, from *Atmosphere, Weather and Climate* (Routledge, 1998), reproduced by permission of the publisher; **p.198** E. Linacre & B. Geerts, El Niño events, from *Climates and Weather Explained* (Routledge, 1997), reproduced by permission of the publisher; **p.206** *bottom* D. Briggs et al., A model of succession, from *Fundamentals of the Physical Environment* (Routledge, 1997), reproduced by permission of the publisher; **p.207** *top left* D. Briggs et al., Community changes through succession, from *Fundamentals of the Physical Environment* (Routledge, 1997), reproduced by permission of the publisher; **p.209** *top* D. Briggs et al., Tropical rainforest types, from *Fundamentals of the Physical Environment* (Routledge, 1997), reproduced by permission of the publisher; **p.210** D. Briggs et al., Vegetation structure in a rainforest, from *Fundamentals of the Physical Environment* (Routledge, 1997), reproduced by permission of the publisher; **p.213** D. Briggs et al., Types of savanna, from *Fundamentals of the Physical Environment*

(Routledge, 1997), reproduced by permission of the publisher; **p.214** D. Briggs et al., Savanna soils, from *Fundamentals of the Physical Environment* (Routledge, 1997), reproduced by permission of the publisher; **p.227**; *bottom left* P. French, Wave terminology, from *Coastal and Estuarine Management* (Routledge, 1997), reproduced by permission of the publisher; **p.234** Human activity and longshore drift along the coast of West Africa, from *New Scientist* © Reed Business Information – UK. All rights reserved. Distributed by Tribune Content Agency; **p.235** *bottom right* A. Goudie, Evolution of wave-cut platforms, from *The Nature of the Environment* (Blackwell, 1993), reproduced by permission of the publisher; **p.239** *top* A. Goudie, Beach deposits, from *The Nature of the Environment* (Blackwell, 1993), reproduced by permission of the publisher; **p.245** *bottom* A. Goudie, Saltmarsh formation at Scolt Head Island, from *Landforms of England and Wales* (John Wiley & Sons, 1993), reproduced by permission of the publisher; **p.257** Relative sea-level change in the USA, from *New Scientist* © Reed Business Information – UK. All rights reserved. Distributed by Tribune Content Agency; **p.290** Survey map of the Alps – area affected by 1999 avalanches © IGN France; **p.325** A.S. Goudie, A model of desertification, from *The Human Impact on the Natural Environment* (Blackwell, 1993), reproduced by permission of the publisher; **p.357** *right* C. Whynne-Hammond, Transport costs and distance, from *Elements of Human Geography* (Allen & Unwin, 1979); **p.375** *bottom right* Oil consumption by world region, 1987–2012, from *BP Statistical Review of World Energy 2013*, reproduced by permission of BP p.l.c.; **p.383** *left* REN21, Global biofuel production, 2000–13, from *Renewables 2014 Global Status Report* (http://www.ren21.net/Portals/0/documents/Resources/GSR/2014/GSR2014_full%20report_low%20res.pdf); **p.396** *right* The diagram 'What is water security?' taken from: http://www.unwater.org/fileadmin/user_upload/unwater_new/docs/water_security_poster_Oct2013.pdf © UN-Water; **p.414** *top* World Trade Organization, Major components of trade in merchandise and services, from *International Trade Statistics 2014* (https://www.wto.org/english/res_e/statis_e/its2014_e/its2014_e.pdf); **p.414** *bottom* World Trade Organization, World map showing economies by size of merchandise trade, 2013, from *International Trade Statistics 2014* (https://www.wto.org/english/res_e/statis_e/its2014_e/its2014_e.pdf); **p.440** WTTC, The economic impact of tourism, from *Economic Impact of Travel and Tourism 2015, Annual Update*. Used by permission of the World Travel & Tourism Council. The original version was sourced from www.wttc.org; **p.484** *bottom left* PWGSC, Funding for the four original major development agencies, from *Public Accounts of Canada, 1987–2014*; **p.487** Map of the Grand Canyon provided by MapSherpa.com, with portions copyright USGS, © 1987–2014 HERE, © 2006–2014 TomTom, and US Forest Service Lands: FSTopo Data, 2009–2014; **p.488** Map of Vancouver provided by MapSherpa.com, with portions © OpenStreetMap contributors.

Tables: **p.27** *bottom left* Hofer, T. and Messerli, B., Watershed characteristics of the Ganga and the Brahmaputra/Meghna (Br/M) rivers and a comparison with the Nile, the Amazon and the Mississippi, from *Floods in Bangladesh: History, Dynamics and Rethinking the Role of the Himalayas* (United Nations University Press/FAO 2006); **p.28** Hofer, T. and Messerli, B., Major impacts of the 1998 floods, from *Floods in Bangladesh: History, Dynamics and Rethinking the Role of the Himalayas* (United Nations University Press/FAO 2006); **p.59** Field Studies Council, Characteristics of the London Plane tree, adapted from the Field Studies Council's Urban Ecosystems website (www.field-studies-council.org/urbaneco) (FSC Publications, 2009), reproduced by permission of the publisher; **p.71** *bottom left* D. Brunsden, Average porosity and permeability for common rock types, from 'Weathering processes' in C. Embleton and J. Thornes (eds) *Processes in Geomorphology* (Edward Arnold 1979); **p.79** *top right* Goudie, Examples of methods of controlling mass movement, from *The Nature of the Environment* (Wiley-Blackwell, 1993); **p.86** *bottom left* Carl Haub and Toshiko Kaneda,World population clock, 2014, from *2014 World Population Data Sheet*, Population Reference Bureau (Washington, DC: Population Reference Bureau, 2014); **p.87** *left* Carl Haub and Toshiko Kaneda, Birth rate, death rate and rate of natural change by world region, 2014, from *2014 World Population Data Sheet*, Population Reference Bureau (Washington, DC: Population Reference Bureau, 2014); **p.87** *right* Carl Haub and Toshiko Kaneda, Variations in total fertility rate and the percentage of women using contraception by world region, 2014, from *2014 World Population Data Sheet*, Population Reference Bureau (Washington, DC: Population Reference Bureau, 2014); **p.89** *right* Carl Haub and Toshiko Kaneda, Death rate, infant mortality rate and life expectancy at birth by world region, 2014, from *2014 World Population Data Sheet*, Population Reference Bureau (Washington, DC: Population Reference Bureau, 2014); **p.94** *top left* Carl Haub and Toshiko Kaneda, Dependency ratio calculations, from *2014 World Population Data Sheet*, Population Reference Bureau (Washington, DC: Population Reference Bureau, 2014); **p.97** *bottom right* Carl Haub and Toshiko Kaneda, The percentage of total population aged 65 years and over, 2014, from *2014 World Population Data Sheet*, Population Reference Bureau (Washington, DC: Population Reference Bureau, 2014); **p.98** Warren Sanderson and Sergei Scherbov, Remaining life expectancy among French women, 1952 and 2005, from 'Rethinking Age and Aging' *Population Bulletin* 63, no. 4 (Washington, DC: Population Reference Bureau, 2008); **p.99** *right* Carl Haub and Toshiko Kaneda, Selected data from *2014 World Population Data Sheet*, Population Reference Bureau; (Washington, DC: Population Reference Bureau, 2014); **p.102** *top* United Nations Development Programme, Top 25 countries in the Human Development Report 2014, from *Human Development Report 2014*, CC BY-SA 3.0 IGO (http://creativecommons.org/licenses/by/3.0/igo/); **p.103** *left* United Nations Development Programme, Human Development Index values Human Development Report 2014, CC BY 3.0 IGO (http://creativecommons.org/licenses/by/3.0/igo/); **p.116** *bottom* China's administrative regions by population taken from https://en.wikipedia.org/wiki/List_of_Chinese_administrative_divisions_by_population, CC BY-SA 3.0 (https://creativecommons.org/licenses/by-sa/3.0/); **p.130** *top* Reasons for migration from rural areas in Peru and Thailand, from J. Laite 'The migrant response in central Peru', in J. Gugler (ed.) *The Urbanization of the Third World* (Oxford University Press, 1988), reproduced by permission of the publisher; **p.130** *bottom* M. Parnwell, Reasons for migration from rural areas in Peru and Thailand, from *Population Movements and the Third World* (Routledge, 1993), reproduced by permission of the publisher; **p.146** Adapted from Philip Martin and Gottfried Zürcher, Factors encouraging migration from Mexico, by type of migrant, from 'Managing Migration: The Global Challenge', *Population Bulletin* 63, no. 1 (Washington, DC: Population Reference Bureau, 2008); **p.165** *bottom* The world's 50 largest cities in 2012, from worldatlas.com, U.S. Census Bureau and Times Atlas of the World; **p.169** Research Network, Loughborough University, Alpha global cities, reproduced by permission of Globalisation and World Cities (GaWC); **p.170** Research Network, Loughborough University, Beta and gamma global cities, reproduced by permission of Globalisation and World Cities (GaWC); **p.217** S. Nortcliff, Distribution of the main types of soil in the humid tropics, in 'The clearance of the tropical rainforest', from *Teaching Geography*, April 1987, Vol 12, No 3, pp.110–113, reproduced by permission of the Geographical Association (www.geography.org.uk); **p.218** *top left* S. Nortcliff, Characteristics of an oxisol under tropical rainforest in Amazonas, Brazil, in 'The clearance of the tropical rainforest', from *Teaching Geography*, April 1987, Vol 12, No 3, pp.110–113, reproduced by permission of the Geographical Association (www.geography.org.uk); **p.221** WWF, *Business solutions: delivering the Heart of Borneo declaration* (https://www.pwc.co.uk/assets/pdf/hob-business-solutions.pdf); **p.281** *left* C. Park, Monitoring for earthquake prediction, from *The Environment* (Routledge, 1997), reproduced by permission of the publisher; **p.330** *top* Hugget et al, Factors relating to the universal soil loss equation, from *Physical Geography – a Human Perspective* (Arnold, 2004), reproduced by permission of the publisher; **p.338** *top left* Richard Buckley, Main types of irrigation, from 'The Water Crisis: A Matter of Life and Death', in *Understanding Global Issues*, (European Schoolbooks Limited, 2001); **p.339** *right* Timeline of agricultural technology, from: https://en.wikipedia.org/wiki/Timeline_of_agriculture_and_food_technology CC BY-SA 3.0 (https://creativecommons.org/licenses/by-sa/3.0/); **p.351** *right* Area in farming in Jamaica, 1996 and 2007, from www.statinja.gov.jm, reproduced by permission of the Statistical Institute of Jamaica; **p.352** Exports of traditional and non-traditional commodities, January–December 2009, reproduced by permission of the Statistical Institute of

Jamaica; **p.365** ILO, Export processing zones – types of zones, from www.ilo.org, reproduced by permission of the International Labour Organization; **p.376** Oil reserves to production ratio at end of 2012, from *BP Statistical Review of World Energy 2013*, reproduced by permission of BP p.l.c.; **p.377** Natural gas production by world region, 2002–12, from *BP Statistical Review of World Energy 2013*, reproduced by permission of BP p.l.c.; **p.380** *left* REN21, Capacity of renewable energy sources 2004 and 2013, from *Renewables 2014 Global Status Report* (http://www.ren21.net/Portals/0/documents/Resources/GSR/2014/GSR2014_full%20report_low%20res.pdf); **p.405** The sources of emissions in Delhi (tonnes/year and percentage), 2010, from www.urbanemissions.info, reproduced by permission of urbanemissions.info; **pp.409–10** NACSO, Update 'Scaling up Namibia's community conservancies', p.32, MET 2005; NEEN 2004a, b, c; Weaver 2007; Jones 2008; **p.410** *bottom left* WWF et al. Conservancy-related income, 2006, from *World Resources 2008: Roots of Resilience. Growing the Wealth of the Poor*, p.35 (World Resources Institute, 2008 www.wri.org) CC BY 4.0 (http://creativecommons.org/licenses/by/4.0/legalcode); **p.415** World Trade Organization, Leading exporters and importers in world merchandise trade, 2013, from *International Trade Statistics 2014* (https://www.wto.org/english/res_e/statis_e/its2014_e/its2014_e.pdf); **p.416** World Trade Organization, Leading exporters and importers in world trade in commercial services, 2013, from *International Trade Statistics 2014* (https://www.wto.org/english/res_e/statis_e/its2014_e/its2014_e.pdf); **p.426** *bottom left* The World Bank, Debt indicators for LICs from World Bank Debtor Reporting System and International Monetary Fund, quoted in *Global Development Finance 2012* (http://data.worldbank.org/sites/default/files/gdf_2012.pdf); **p.433** *top* The Hunger Project, Contrasting top-down and bottom-up aid models (www.thp.org); **p.446** *bottom left, middle right and bottom right* WTTC data, from *Travel & Tourism Economic Impact 2014*, used by permission of the World Travel & Tourism Council. The original version was sourced from www.wttc.org; **p.454** *right* Carl Haub and Toshiko Kaneda, Top 20 and bottom 20 countries in GNI (PPP) per capita in 2013, from *2014 World Population Data Sheet*, Population Reference Bureau (Washington, DC:

Population Reference Bureau, 2014); **p.459** *top left* UNCTAD, Classification of LDCs, from *The Least Developed Countries Report 2008* (http://unctad.org/en/docs/ldc2008_en.pdf); **p.471** Sue Warn, The positive and negative impacts of global shift, in *The Global Shift*, from Geo Factsheet, reproduced by permission of Curriculum Press; **p.482** *left* Statistics Canada, The largest urban areas in Canada, 2011. Reproduced and distributed on an "as is" basis with the permission of Statistics Canada; **p.483** *left* Statistics Canada, Regional disparities (adapted). This does not constitute an endorsement by Statistics Canada of this product.

Text extracts: **p.151** *right* C. Bull, P. Daniel & M. Hopkinson, Principal characteristics of traditional rural society, from *The Geography of Rural Resources* (Oliver & Boyd, 1984); **pp.219–223** WWF, *Business solutions: delivering the Heart of Borneo declaration* (https://www.pwc.co.uk/assets/pdf/hob-business-solutions.pdf); **p.340** George Mwangi, China Agrees To Help Improve African Food Production – Kenya, contributing to Dow Jones Newswires; +254 735 781853, gmwangio@gmail.com, copyright © Automated Trader Ltd. 2010; **p.394** *left* The World Health Organization, Reducing air pollution, from *Ambient (outdoor) air quality and health* (www.who.int/mediacentre/factsheets/fs313) March 2014; **p.402** *left* FAO, Five root causes of unsustainable practices, from *The challenge of sustainability*; **p.406** Darpan Singh, Delhi's domestic waste problem, adapted from the article 'Delhi may drown in its own waste', from *Hindustan Times*, 30 April 2013; **p.430** *bottom* UK Department for International Development, Types of bilateral aid, *Statistics on International Development 2008*; **p.431** *left* UK Department for International Development, DFID provides emergency aid towards Bangladesh, 4 December 2007; **p.443** Leo Hickman, *The Final Call* (Eden Project Books, 2007); **p.478** *left* Paul Guinness, extract from The constituents of urban poverty, in *Slum Housing Global Patterns Case Studies*, from Geo FactSheet, No. 180, reproduced by permission of Curriculum Press; **p.480** Fabiana Frayssinet, Brazil: Marching for Real Land Reform, 2009, reproduced by permission of IPS Inter Press Service News Agency.